Global Biodiversity

Volume 4

Selected Countries in the
Americas and Australia

Global Biodiversity

Volume 4

Selected Countries in the Americas and Australia

Edited By
T. Pullaiah, PhD

Apple Academic Press Inc.
3333 Mistwell Crescent
Oakville, ON L6L 0A2
Canada

Apple Academic Press Inc.
9 Spinnaker Way
Waretown, NJ 08758
USA

First issued in paperback 2021

Exclusive worldwide distribution by CRC Press, a member of Taylor & Francis Group

No claim to original U.S. Government works

Global Biodiversity, Volume 4: Selected Countries in the Americas and Australia

ISBN 13: 978-1-77463-133-1 (pbk)
ISBN 13: 978-1-77188-750-2 (hbk)

Global Biodiversity, 4-volume set

ISBN 13: 978-1-77188-751-9 (hbk)

Library and Archives Canada Cataloguing in Publication

Global biodiversity (Oakville, Ont.) Global biodiversity / edited by T. Pullaiah, PhD.
Includes bibliographical references and indexes.
Contents: Volume 4. Selected countries in the Americas and Australia.
Issued in print and electronic formats.
ISBN 978-1-77188-750-2 (v. 4 : hardcover).--ISBN 978-0-429-43363-4 (v. 4 : PDF)

1. Biodiversity--Asia. 2. Biodiversity--Europe. 3. Biodiversity--Africa. 4. Biodiversity--America.
5. Biodiversity--Australia. I. Pullaiah, T., editor II. Title.

QH541.15.B56G66 2018   578.7   C2018-905091-8   C2018-905092-6

CIP data on file with US Library of Congress

Apple Academic Press also publishes its books in a variety of electronic formats. Some content that appears in print may not be available in electronic format. For information about Apple Academic Press products, visit our website at **www.appleacademicpress.com** and the CRC Press website at **www.crcpress.com**

Contents

Contents

About the Editor

T. Pullaiah, PhD
Former Professor, Department of Botany,
Sri Krishnadevaraya University, Anantapur, Andhra Pradesh, India,
E-mail: pullaiah.thammineni@gmail.com

T. Pullaiah, PhD, is a former Professor at the Department of Botany at Sri Krishnadevaraya University in Andhra Pradesh, India, where he has taught for more than 35 years. He has held several positions at the university, including Dean, Faculty of Biosciences, Head of the Department of Botany, Head of the Department of Biotechnology, and Member of Academic Senate. He was President of Indian Botanical Society (2014), President of the Indian Association for Angiosperm Taxonomy (2013) and Fellow of Andhra Pradesh Akademi of Sciences. He was awarded the Panchanan Maheshwari Gold Medal, the Dr. G. Panigrahi Memorial Lecture award of the Indian Botanical Society and Prof. Y.D. Tyagi Gold Medal of the Indian Association for Angiosperm Taxonomy, and a Best Teacher Award from Government of Andhra Pradesh. Under his guidance 54 students obtained their doctoral degrees. He has authored 46 books, edited 17 books, and published over 330 research papers, including reviews and book chapters. His books include *Ethnobotany of India* (5 volumes published by Apple Academic Press), *Flora of Andhra Pradesh* (5 volumes), *Flora of Eastern Ghats* (4 volumes), *Flora of Telangana* (3 volumes), *Encyclopaedia of World Medicinal Plants* (5 volumes, 2nd edition), and *Encyclopaedia of Herbal Antioxidants* (3 volumes). He was also a member of Species Survival Commission of the International Union for Conservation of Nature (IUCN). Professor Pullaiah received his PhD from Andhra University, India, attended Moscow State University, Russia, and worked as a Postdoctoral Fellow during 1976–78.

Contributors

Francisca Acevedo
National Commission for the Knowledge and Use of Biodiversity (CONABIO),
Liga Periférico Insurgentes Sur 4903, Col. Parques del Pedregal, Tlalpan 14010, Mexico City, Mexico

Eliel de Jesus Amaral
University of Brasília Institute of Biological Sciences - Block E, 1st Floor University of Brasília
Darcy Ribeiro University Campus, Asa Norte, Brasília-Distrito Federal, 70910-900, Brasília

Gerardo Avalos
Full Professor of Tropical Ecology, School of Biology, University of Costa Rica, and Director,
Center for Sustainable Development Studies, The School for Field Studies, 100 Cummings Center,
Suite 534-G Beverly, MA 01915, USA, E-mail: gavalos@fieldstudies.org

Frederick Bauer
Museo Nacional de Historia Natural del Paraguay, km 10.5, Ruta Mcal. José F. Estigarribia, 2160,
San Lorenzo-Paraguay/Facultad de Ciencias Exactas y Naturales (FACEN), Departamento de Biología,
Área Zoología, Campus UNA, San Lorenzo, Paraguay

Pier Cacciali
Instituto de Investigación Biológica del Paraguay, Del Escudo 1607, 1425, Asunción, Paraguay

Michelle Campi
Laboratorio de Análisis de Recursos Vegetales Área Micología FaCEN-UNA

Bernardo Cañiza
Laboratorio de Análisis de Recursos Vegetales FaCEN-UNA, 1039, San Lorenzo-Paraguay

Pablo Lozano Carpio
Universidad Estatal Amazónica, Paso Lateral km 2 ½ via Napo, Pastaza-Ecuador,
E-mail: plozano@uea.edu.ec, pablo_lozanoc@yahoo.com

Franklin E. Castañeda
Panthera, Tegucigalpa, Honduras

Angélica Cervantes
National Commission for the Knowledge and Use of Biodiversity (CONABIO),
Liga Periférico Insurgentes Sur 4903, Col. Parques del Pedregal, Tlalpan 14010, Mexico City, Mexico

Daryl D. Cruz-Flores
Researcher at Ecology and Systematic Institute, Cuba

Dennis Denis
Professor at the University of Havana, Cuba

C. R. Dickman
Professor in Ecology, School of Life and Environmental Sciences, The University of Sydney,
NSW 2006, Australia, E-mail: chris.dickman@sydney.edu.au

Sigrid Drechsel
Asociación Etnobotánica Paraguaya (AEPY), Ecuador 450 c/Bruno Guggiari. 2420,
Lambaré-Paraguay

Christian Dujak
Currently Centre for Research in Agricultural Genomics (CRAG) CSIC-IRTA-UAB-UB,
Campus UAB, Bellaterra, Barcelona, Spain

Marcelo Dujak
Categorized researcher CONACYT-PRONII, Asunción, Paraguay. E-mail: marcelodujak@gmail.com

Nereyda Estrada
School of Biology, Faculty of Sciences, National Autonomous University of Honduras, Tegucigalpa, Honduras

Lilian Ferrufino
Herbario Cyril Hardy Nelson Sutherland, School of Biology, Faculty of Sciences,
National Autonomous University of Honduras, Tegucigalpa, Honduras

Sarita C. Frontana-Uribe
National Commission for the Knowledge and Use of Biodiversity (CONABIO),
Liga Periférico Insurgentes Sur 4903, Col. Parques del Pedregal, Tlalpan 14010, Mexico City, Mexico

David A. Galbraith
Science Department, Royal Botanical Gardens, 680 Plains Road West, Burlington, Ontario L7T4H4,
Canada, E-mail: dgalbraith@rbg.ca

Kanchi N. Gandhi
Senior Nomenclatural Registrar, Harvard University Herbaria, Harvard University, Cambridge,
MA 02138, USA; Nomenclature Editor & Member of Board of Directors,
Flora of North America Association

Daniel Germer
Barrio el Bosque, Calle Principal, Casa 2022, Tegucigalpa, Honduras

José Grande
Professor of Botany and Ecology, Herbario MERF, Facultad de Farmacia y Bioanálisis,
Universidad de Los Andes, Núcleo Campo de Oro, Postal Code 5101, Mérida, Venezuela,
Email: jose.r.grande@gmail.com

A. Margarita Hermoso-Salazar
National Commission for the Knowledge and Use of Biodiversity (CONABIO),
Liga Periférico Insurgentes Sur 4903, Col. Parques del Pedregal, Tlalpan 14010, Mexico City, Mexico

Diana R. Hernández-Robles
National Commission for the Knowledge and Use of Biodiversity (CONABIO), Liga Periférico
Insurgentes Sur 4903, Col. Parques del Pedregal, Tlalpan 14010, Mexico City, Mexico

Pamela Marchi
Asociación Etnobotánica Paraguaya (AEPY), Ecuador 450 c/Bruno Guggiari. 2420, Lambaré-Paraguay,
E-mail: pamepy@gmail.com

Alicia Mastretta-Yanes
National Commission for the Knowledge and Use of Biodiversity (CONABIO), Liga Periférico
Insurgentes Sur 4903, Col. Parques del Pedregal, Tlalpan 14010, Mexico City, Mexico

Wilfredo Matamoros
Universidadde Ciencias y Artes de Chiapas, Instituto de Ciencias Biológicas, Museo de Zoología,
Colección de Ictiología, Tuxtla Gutierrez, Chiapas, México

James R. McCranie
10770 SW 164 Street, Miami, FL, Panthera, Tegucigalpa, Honduras, 33157, USA,
E-mail: jmccrani@bellsouth.net

Susana Ocegueda-Cruz
National Commission for the Knowledge and Use of Biodiversity (CONABIO),
Liga Periférico Insurgentes Sur 4903, Col. Parques del Pedregal, Tlalpan 14010, Mexico City, Mexico

Adriana Luiza Ribeiro de Oliveira

University of Brasília Institute of Biological Sciences - Block E, 1st Floor University of Brasília Darcy Ribeiro University Campus, Asa Norte, Brasília-Distrito Federal, 70910-900, Brasília, E-mail: ribeirosenna1@gmail.com

Dulce Parra-Toríz

National Commission for the Knowledge and Use of Biodiversity (CONABIO), Liga Periférico Insurgentes Sur 4903, Col. Parques del Pedregal, Tlalpan 14010, Mexico City, Mexico

Esther Quintero

National Commission for the Knowledge and Use of Biodiversity (CONABIO), Liga Periférico Insurgentes Sur 4903, Col. Parques del Pedregal, Tlalpan 14010, Mexico City, Mexico

Mónica Moraes R.

Research Scientist and Professor, Institute of Ecology, Biology Department, Universidad Mayor de San Andrés, Casilla 10077, Correo Central, La Paz, Bolivia, E-mail: monicamoraes45@gmail.com

Rafael Ramírez

National Commission for the Knowledge and Use of Biodiversity (CONABIO), Liga Periférico Insurgentes Sur 4903, Col. Parques del Pedregal, Tlalpan 14010, Mexico City, Mexico

M. Alicia Reséndiz-López

National Commission for the Knowledge and Use of Biodiversity (CONABIO), Liga Periférico Insurgentes Sur 4903, Col. Parques del Pedregal, Tlalpan 14010, Mexico City, Mexico

Alberto Romo-Galicia

National Commission for the Knowledge and Use of Biodiversity (CONABIO), Liga Periférico Insurgentes Sur 4903, Col. Parques del Pedregal, Tlalpan 14010, Mexico City, Mexico

Kevin O. Sagastume-Espinoza

School of Biology, Faculty of Sciences, National Autonomous University of Honduras, Tegucigalpa, Honduras

Daniel Silva Santiago

University of Brasília Institute of Biological Sciences - Block E, 1st Floor University of Brasília Darcy Ribeiro University Campus, Asa Norte, Brasília-Distrito Federal, 70910-900, Brasília

Jaime Sarmiento

Research Scientist, Ichthyology Department, National Museum of Natural History, Casilla 8706, La Paz, E-mail: jsarmientotavel@gmail.com

Ernesto Testé

Researcher at National Botanical Garden of Cuba, E-mail: dda@fbio.uh.cu

Adriana Valera

National Commission for the Knowledge and Use of Biodiversity (CONABIO), Liga Periférico Insurgentes Sur 4903, Col. Parques del Pedregal, Tlalpan 14010, Mexico City, Mexico

Alberto Yanosky

Director of Guyra Paraguay/Researcher CONACYT-PRONII, Biocentro Guyra Paraguay, Asunción-Paraguay

Abbreviations

ACG	Guanacaste Conservation Area (Área de Conservación Guanacaste)
APG IV	Angiosperm Phylogeny Group
AVHRR	Advanced Very High Resolution Radiometer Images
CBD	Convention on Biological Diversity
CCEA	Canadian Council on Ecological Areas
CITES	International Convention on the Traffic in Endangered Species
CR	critically endangered species
CSLME	Caribbean Sea Large Marine Ecosystem
DGEEC	General Direction of Statistics Surveys and Censuses Dirección General de Estadística, Encuestas y Censos
EN	endangered species
FACEN	Faculty of Exact and Natural Sciences (Facultad de Ciencias Exactas y Naturales)
GABI	Great American Biotic Interchange
GBIF	Global Biodiversity Information Facility
IABIN	Inter American Biodiversity Information Network
IBGE	Instituto Brasileiro de Geografia e Estatística
INBIO	National Institute of Biodiversity
INECC	National Institute of Ecology and Climate Change
ITCZ	Intertropical Convergence Zone
IUCN	International Union for the Conservation of Nature
LAF	lowland arid forest formation
LDF	lowland dry forest formation
LMF	lowland moist forest formation
LMWF	lower montane wet forest formation
MIT	Mexican insular territory
NTFP	non-timber forest product
ORDCP	Orinoco river delta coastal plain
PDF	premontane dry forest formation
PMF	premontane moist forest formation
PWF	premontane wet forest
SCP	Systematic Conservation Planning
SINAC	System of Conservation Areas
UNCED	United Nations Conference on Environment and Development
VU	vulnerable species

Abbreviations

ACC	Community Conservation Area (Área de Conservación Comunitaria)
APG	Angiosperm Phylogeny Group
AVHRR	Advanced Very High Resolution Radiometer Images
CBD	Convention on Biological Diversity
CCEA	Canadian Council on Ecological Areas
CITES	International Trade Convention on the Traffic in Endangered Species
CR	critically endangered species
CSLME	Caribbean Sea Large Marine Ecosystem
DGEEC	General Direction of Statistics, Surveys and Census and Dirección General de Estadística, Encuestas y Censos
EN	endangered species
FACEN	Faculty of Exact and Natural Sciences (Facultad de Ciencias Exactas y Naturales)
GAHI	Great American House Info Image
GBIF	Global Biodiversity Information Facility
IABIN	Inter American Biodiversity Information Network
IBGE	Instituto Brasileiro de Geografia e Estatística
INBIO	National Institute of Biodiversity
INECC	National Institute of Ecology and Climate Change
IUCZ	Intertropical Convergence Zone
IUCN	International Union for the Conservation of Nature
LCF	lowland and tidal formation
LDF	lowland dry forest formation
LMF	lowland moist forest formation
LWF	lowland rainforest wet forest formation
MIT	Mexican insular territory
NTFP	non-timber forest product
OMDCR	Dirección...
SDTF	seasonal dry forest formation
PMF	premontane moist forest formation
PWF	premontane wet forest formation
SCP	Systematic Conservation Planning
SINAC	System of Conservation Areas
UNCED	United Nations Conference on Environment and Development
VU	vulnerable species

Preface

The term 'biodiversity' came into common usage in the conservation community after the 1986 National Forum on BioDiversity, held in Washington, DC, and publication of selected papers from that event, titled *Biodiversity*, edited by Wilson (1988). Wilson credits Walter G. Rosen for coining the term. Biodiversity and conservation came into prominence after the Earth Summit, held at Rio de Janeiro in 1992. Most of the nations passed biodiversity and conservation acts in their countries. Biodiversity is now the buzzword of everyone from parliamentarians to laymen, professors, and scientists to amateurs. There is a need to take stock on biodiversity of each nation. The present attempt is in this direction.

The main aim of the book is to provide data on biodiversity of each nation. It summarizes all the available data on plants, animals, cultivated plants, domesticated animals, their wild relatives, and microbes of different nations. Another aim of the book series is to educate people about the wealth of biodiversity of different countries. It also aims to project the gaps in knowledge and conservation. The ultimate aim of the book is for the conservation of biodiversity and its sustainable utilization.

The present series of the four edited volumes is a humble attempt to summarize the biodiversity of different nations. Volume 1 covers *Biodiversity of Selected Countries in Asia*, Volume 2 presents *Biodiversity of Selected Countries in Europe,* Volume 3 looks at *Biodiversity of Selected Countries in Africa*, and Volume 4 contains *Biodiversity of Selected Countriess in the Americas and Australia*. In these four volumes, each chapter discusses the biodiversity of one country. Competent authors have been selected to summarize information on the various aspects of biodiversity. This includes brief details of the country, ecosystem diversity/vegetation/biomes, and species diversity, which include plants, animals and microbes. The chapters give statistical data on plants, animals, and microbes of that country, and supported by relevant tables and figures. They also give accounts on genetic diversity with emphasis on crop plants or cultivated plants, domesticated animals, and their wild relatives. Also mentioned are the endangered plants and animals and their protected areas. The book is profusely illustrated. We hope it will be a desktop reference book for years to come.

Biodiversity of some countries could not be presented in this book. This needs explanation. I tried to contact as many specialists as possible from these countries but was unable to convince these experts to write chapter on biodiversity of their country.

The book will be useful to professors, biology teachers, researchers, scientists, students of biology, foresters, agricultural scientists, wild life managers, botanical gardens, zoos, and aquaria. Outside the scientific field it will be useful for lawmakers (parliamentarians), local administrators, nature lovers, trekkers, economists, and even sociologists.

Since it is a voluminous subject, we might have not covered the entire gamut; however, we tried to put together as much information as possible. Readers are requested to give their suggestions for improvement for future editions.

I would like to express my grateful thanks to all the authors who contributed on the biodiversity of their countries. I thank them for their cooperation and erudition.

I wish to express my appreciation and help rendered by Ms. Sandra Sickels, Rakesh Kumar, and the staff of Apple Academic Press. Their patience and perseverance has made this book a reality and is greatly appreciated.

—*T. Pullaiah, PhD*

CHAPTER 1

Biodiversity in Bolivia

MÓNICA MORAES R.[1] and JAIME SARMIENTO[2]

[1]Research Scientist and Professor, Institute of Ecology, Herbario Nacional de Bolivia, Universidad Mayor de San Andrés, Casilla 10077 – Correo Central, La Paz, Bolivia, E-mail: monicamoraes45@gmail.com

[2]Research Scientist, Ichthyology Department, National Museum of Natural History, Casilla 8706, La Paz, Bolivia, E-mail: jsarmientotavel@gmail.com

1.1 Introduction

1.1.1 Geography, Geology, and Land Characteristics

Bolivia is located in the Mediterranean center of South America with an area of 1,098,581 km^2 and occupies 6% at the South American level. To the northwest, it borders with Peru, north to northeast with Brazil, southeast with Paraguay, south with Argentina, and southwest with Chile. This country represents the geographical and geological synthesis of the South American continent since its geological rocks belong to all the geological ages represented (Montes de Oca, 2005). The Bolivian territory has 8 physiographic provinces, 27 major landscapes, and 162 landscapes determined by relief and climate; which in turn acts as a modeler of the relief from the parent rock and which together determine the presence of plant cover (MDSP, 2001). The dominant relief features of Bolivia are the complex body of the Andes with the Altiplano (high plateau), the Sub-Andean hills and the eastern Beni and Chaco plains in the lowlands with the Brazilian Shield and the Chiquitanian hills (Ahlfield, 1970). The Andes, including the adjacent lowlands in the West and East, belongs to a very ancient geosyncline structure, which already developed in the early Paleozoic as an intracratonic pit between the Brazilian Shield in the East and the Arequipa massif in the West.

The exposure of the oldest rock formations is mainly found in the east to the northeastern region of the country in the Precambrian Brazilian shield with 4,500 to 600 million years (Argollo, 1996). Whereas the western region corresponds to the Andes zone and is formed by two important mountain ranges that surround a high plateau: the eastern Cordillera corresponds to the so-called Paleozoic block where rocks corresponding to all the systems of this era and sedimentary and igneous rocks of the Mesozoic and Cenozoic ages appear (Figures 1.1 and 1.2); while the western Cordillera is characterized by a volcanic origin and is formed by numerous cones and lava flows of

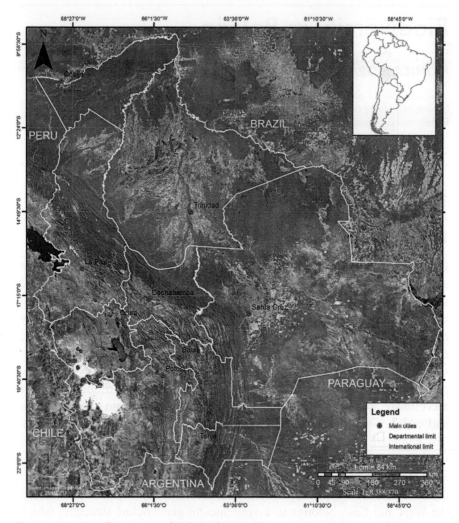

Figure 1.1   Satellite image of Bolivia showing the nine departments and the distribution of forest areas vs. open areas.

Tertiary and Quaternary ages. Between both mountain ranges lies the Altiplano, which is an extensive plateau (shared with southern Peru and northern Argentina and Chile) between 14–28°S and constitutes the second largest formation in the world, at an average altitude of 3,700 m in an area of 300 km wide and 2,000 km long (Argollo, 1996). This plateau is the result of the last orogenic episode (Noblet et al., 1996), dated at 0–27 million years based on structural and chronostratigraphic data (Rochat, 2000).

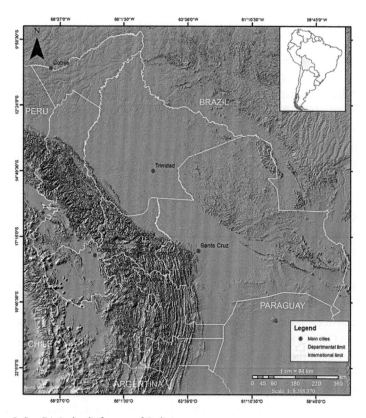

Figure 1.2 Digital relief map of Bolivia.

Between both mountain ranges is the endorheic or closed basin of the high plateau (so-called "Altiplano" with 3,700 to 4,200 m elevation) in the Andean geologic and physiographic formation with large areas of flat terrain, product of the accumulation of Quaternary sediments in an old lacustrine environment (Argollo, 1996). This plain is interrupted by many hills where rocks of different ages arise but mainly Tertiary; it is also related to known and exploited mineralogical belts. The eastern foothills of the Andes, which border the eastern plains of the Amazon, form the sub-Andean, which in its southern and central part represent the main oil region of Bolivia. The unit of greater geological uniformity corresponds to the eastern plains of Chaco and Beni plains, also known as the lowlands or the Bolivian Amazon and which represents the rest of an ancient Amazonian sea, characterized by the accumulation of sand – clayey sediments of Quaternary age. In the Devonian, the sediments also developed very uniformly in large parts of the eastern Cordillera, these consist of sandstones, calcareous sandstones, quartzite, and shales. In the Escoma-Carabuco region, Devonian sediments reach a depth of 1,748 m (Rivas, 1968) and around Lake Titicaca 3,000 m (Martínez, 1980). The highest mountains of the Andes of Bolivia are located mainly in the Western Cordillera (of volcanic origin) and are: Sajama (6,542 m), Parinacota (6,348 m), and Pomerape (6,282 m), whereas Illimani (6,438 m), Ancohuma (6,427 m) and Huayna Potosí (6,088 m) belong to the Eastern Andean mountains.

The alluvial plains in the lowlands of Bolivia are located on the eastern side of the sub-Andean region; its altitude ranges between 200 and 600 m (Montes de Oca, 2005). They have been formed by the accumulation of hundreds of meters of fine sediments, especially Quaternary. There is a difference between a pre-Andean or lowland strip that descends from the sub-Andean region, where the flood depressions in the Beni and Pantanal are distinguished, the alluvial terraces little dissected, the alluvial plains especially in the Department of Pando (northernmost of Bolivia), and the Chaco plain (southern to SE) which is different from the northern plains, mainly due to the arid climate (infrequent rivers, infiltrating sandy areas or marshes) (Montes de Oca, 2005).

The region of the Brazilian Shield is characterized by plains that partially present very shallow soils and outcroppings of very old rocks, such as granites and basalts (Montes de Oca, 2005). It includes the Chiquitanian highlands a dozen more or less low chains, representing the highest elevations in the eastern lowlands of Bolivia and adjoining areas, housing important habitat islands for special elements of flora and fauna (from 700 to 1,400 m in the Sunsas hills) (Ibisch et al., 2002). The substrate layer is constituted by Precambrian rocks, the oldest in the Bolivian territory and have a NE-SW orientation, generally presenting a wavy relief with outcrops of more or less solitary hills.

From the Plio-Pleistocene, the mesorelief was generated. The consolidated moraines of the Middle Pleistocene form a gently rolling landscape, while the moraines of the recent embankments are loose products with flanks towards the steeper valleys and with sharp ridges (Lauer and Rafiqpoor, 1986). During the Tertiary the mountains were raised in different phases and, in the most recent phase of their development, they became an erosion region (Martínez, 1980). With the beginning of the glacial period, approximately 2.0–1.8 million years ago, an important climatic change began after the Pliocene; four hot oscillations that were accompanied by erosion and soil generation; while the hot phases are divided by five glacial

Table 1.1 Number of Ecoregions of Bolivia With a Short Description of Vegetation Types and Landscapes, Altitudinal Ranges/Surface, and Protected Areas (Modified From Ibisch et al., 2003)

Ecoregions	Altitudinal ranges/Surface	Protected areas
1. Amazonian flood Forests in Beni, Cochabamba, La Paz, Pando and Santa Cruz: Amazonian forest plain, basins of the Precambrian Shield. In strips and watersheds of very variable size along the rivers.	100–500 m/ 63,588 km²	Manuripi Amazon National Wildlife Reserve, Noel Kempff Mercado National Park.
2. Sub-Andean Amazon Forests of Santa Cruz: Sub-Andean zones north of the Andes elbow in Bolivia.	500–1,000 m/23,529 km²	Madidi National Park and Integrated Management Natural Area, Pilón Lajas Indigenous Territory/Biosphere Reserve, Isiboro – Sécure and Carrasco National Parks, Amboró National Park and Integrated Management Natural Area.
3. Pre-Andean Amazon Forests in Beni and Pando: 100 km from the last Andean foothills.	150–500 m/58,308 km²	Madidi National Park and Integrated Management Natural Area, Pilón Lajas Indigenous Territory and Biosphere Reserve, Isiboro – Sécure National Park and National Park, Carrasco National Park, Amboró National Park and Integrated Management Natural Area.
4. Amazonian forests of Pando, Beni and La Paz: Amazon plain: in the west slightly waved, towards the east plane with outcrops of the Precambrian shield.	100–300 m/71,217 km²	Manuripi National Wildlife Reserve
5. Amazonian forests of Beni and Santa Cruz: Plains, Precambrian landscapes.	150–400 m/59,905 km²	National Park Noel Kempff Mercado.
6. Cerrado from La Paz: Plains of varying heights and shallows, of acid soils, affected by rainfall and floods, above all, by overflowing rivers of clear water.	180–500 to 1,000–2,000 m/ 9,837 km²	None.

Table 1.1 (Continued)

Ecoregions	Altitudinal ranges/Surface	Protected areas
7. Cerrado from Beni and Pando: Flat and undulating savannas with differences of level to more than 20 m in the north, with termiters floods by rainfall; Strongly weathered, nutrient poor soils, lateritic layers with pisolites.	100–200 m/27,171 km^2	Espiritu Wildlife Refuge
8 Chiquitano Cerrado of Santa Cruz: Plains, landscapes of hills and serranias, slabs (inselbergs).	120–1,000 m/23,491 km^2	Noel Kempff Mercado National Park, San Matías Integrated Natural Area.
9. Chacoan Cerrado of Santa Cruz: Plain with few hills and small hills.	170–1,100 m/24,468 km^2	National Parks and Natural Areas of Integrated Management Kaa-Iya, and Otuquis.
10. Flood savannahs of the Llanos de Moxos in Beni, Cochabamba and Santa Cruz: Grasses dominated by grasses and Cyperaceae; Aquatic and marsh plants ("yomomo, curichi"); Different types of forest islands, open forests ("tajibales"), palm forests and thorny low ("tusecales"). Gallery forests along the rivers.	100–200 m/94,660 km^2	Biosphere Reserve Beni Biological Station. Small reserves that are not formally part of the SNAP: Wildlife Espiritu Wildlife, Pedro Ignacio Muiba Regional Park, Kenneth Lee Reservation.
11. Wetlands of the Pantanal in Santa Cruz: Especially plains with extensive areas of flood and large lagoons by the Paraguay river. Palm forests. Alluvial soils.	100–800 m/33,328 km^2	Natural Area of Integrated Management San Matías, National Park-Natural Area of Integrated Management Otuquis.
12. Chiquitan dry Forest in Santa Cruz: Plains, low mountain ranges, slabs (inselbergs – Precambrian Shield).	100–1,400 m/101,769 km^2	Natural Area of Integrated Management San Matías, Municipal Reserve Valle de Tucavaca.
13. Gran Chaco in Santa Cruz, Tarija and Chuquisaca: Plain with few hills and small hills.	200–600 m/105,006 km^2	National Park and Natural Area of Integrated Management Kaa-Iya of the Great Chaco.

Ecoregions	Altitudinal ranges/Surface	Protected areas
14. Yungas in Santa Cruz, La Paz and Cochabamba: A region of almost perennial Andean forests on the northeastern slope of the Andes. Partially very steep northeastern wetlands of the Bolivian (and Peruvian) Andes. Dissected valleys.	1,000–4,200 m/55,556 km^2	Natural Area of Integrated Management Apolobamba, National Park and Natural Area of Integrated Management Madidi, Natural Area of Integrated Management (Biosphere Reserve and Indigenous Territory) Pilon Lajas, National Park and Natural Area of Integrated Management Cotapata, National Park and Isiboro Indigenous Territory -Secure, Carrasco National Park, National Park and Natural Area of Integrated Management Amboró.
15. Tucuman-Bolivian Forest: in Santa Cruz, Tarija and Chuquisaca. Due to thermal and water seasonality (and lower minimum temperatures), they are clearly distinguished from the moist montane forests north of the Elbow of the Andes, which in this work are considered as the Yungas (Bolivian-Peruvian). In Bolivia, this ecoregion of southern humid forests such as the Tucuman-Bolivian Forest has been known for some time.	800–3,900 m/29,386 km^2	National Reserve of Flora and Fauna Tariquía, National Park and Natural Area of Integrated Management Amboró; Natural Area of Integrated Management El Palmar and Serranía Iñao.
16. Montane Chaco in Santa Cruz: Tarija and Chuquisaca: Low mountain ranges of the last foothills of the Eastern Cordillera of the Andes, low valleys, foothills.	700–2,000 m/23,176 km^2	National Reserve of Flora and Fauna Tariquia
17. Dry Inter-Andean Forests in Cochabamba, Tarija, La Paz, Potosi and Chuquisaca: Large variation of deciduous plant formations ranging from dry forests in the Yungas region to the extensive valleys in the central and southern parts of the country. Valleys more or less dissected, small plains.	500–3,300 m/44,805 km^2	National Parks Amboró, Carrasco, Tunari and Toro Toro. Natural Areas of Integral Management El Palmar and Serranía Iñao.

Table 1.1 (Continued)

Ecoregions	Altitudinal ranges/Surface	Protected areas
18. Prepuna in Tarija, Potosi and Chuquisaca: Semi-desert of valleys more or less wide to dissected, small plains.	2,300–3,400 m/8,516 km²	None.
19. Wet Puna in La Paz: Natural potential vegetation is evergreen forest (dominated by *Polylepis* species) and is now found in less populated areas. Phytogeographically it is a region that shows affinities with the high Andean vegetation of the north of the Andes. Plain with hills around and to the south of Titicaca Lake, standing on the slopes of the Cordillera Real.	3,800–4,100 m/8,869 km²	Apolobamba Integrated Management Natural Area, Madidi National Park and Integrated Management Natural Area, Cotapata National Park and Integrated Management Natural Area.
20. Semihumid Puna in Cochabamba, Tarija, La Paz, Potosi, Oruro and Chuquisaca: Low mountains, high plateaus, valleys. Andean forests almost completely destroyed.	3,200–4,200 m/67,600 km²	Tunari National Park, Sama Mountain Range Biological Reserve.
21. High Andean Vegetation of the Eastern Cordillera with nival and subnival floors in La Paz and Cochabamba: Glacial valleys with lagoons, slopes, peaks, rocky peaks.	4,000–5,100 m/8,137 km²	Apolobamba National Integrated Management Area, Madidi Integrated National Park and Natural Area, National Park and Integrated Management Natural Area Cotapata, Tunari National Park.
22. Dry Puna in La Paz, Oruro and Cochabamba: High aridity, which may inhibit the development of extensive forest vegetation on its lower floors (there are only groves or chaparral areas in small areas with *Polylepis tarapacana* and *P. tomentella*). Low mountains, high altiplanic plateaus, wide valleys of the Desaguadero river.	3,500–4,100 m/35,973 km²	None.

Ecoregions	Altitudinal ranges/Surface	Protected areas
23. Desert Puna with nival and subnival floors of the Western Cordillera in La Paz, Oruro, and Potosi: It borders the Atacama Desert. Poor vegetation cover due to low rainfall and low temperatures is characteristic; There are only biotic elements present in one floor (*Nototriche turritella* in the Western Cordillera). Mountains/ volcanoes, extensive plains highlands, valleys with little vegetation, dunes, salt flats.	3,800–7,000 m/100,204 km^2	National Park and Natural Area of Integrated Management Sajama, National Reserve of Andean Fauna Eduardo Avaroa.

phases (Servant, 1977). It is thought that there was a wet glacial period about 27,000 years ago, which was followed by a dry glacial period approximately 20,000–13,000 years ago leaving four low moraines and two more phases, which occurred despite a general climate that gradually warmed (12,500 ago and 10,000 years) generating very marked moraines (Lauer and Rafiqpoor, 1986). About 10,000 years ago, the climate warmed up very quickly and the glaciers retreated a lot; during this postglacial phase, the glaciers advanced again or stopped their retreat. According to Graf (1981), 9,560 ± 90 years ago in the high Andes, the abundance of Asteraceae, Poaceae, Malvaceae, Amaranthaceae, Caryophyllaceae, and Ephedra increased, indicating a hotter and more humid phase. The pollen records of Mountain Pine (*Podocarpus*) and spores of arboreal ferns (*Cyathea*) in the northern Altiplano indicate that the cloud forests had advanced altitudinally, remaining at higher altitudes than today, coinciding with the intensive formation of wetlands or peat formations (so-called "bofedales"). The hottest time was recorded between 7,000 and 3,500 years ago. The last interruption of the peat formation is recorded between 3720 ± 65 and 3080 ± 65 years before today (Lauer and Rafiqpoor, 1986).

In the Bolivian territory there are 20 units of soils and there are also zonal and intrazonal units (e.g., Várzea soils such as fluvisols, the extensive shallows of vertisols, lateritic soils with plintisols) (Gerold, 2003). The wide differentiation of the horizons of the soil is explained by the geological and tectonic development of the physiographic units of Bolivia: the Western Cordillera (volcanic) with recent lava coverings (Andesite, Traquita, Ignimbrita) and pyroclastics; the coverings of the Altiplano with the depositions of spongy sediments of the Tertiary and Quaternary; the Eastern Cordillera with the Paleozoic metamorphic rocks (clay schists, quartziferous sandstones, phyllite) and granodiorite intrusions; the areas of the Sub-Andean mountain ranges with Mesozoic sedimentary rocks (sandstones, calcareous sandstones, limestone and conglomerates); the lowlands of Beni and Chaco, with strong accumulations of clays, silts, sands and gravels; the Chiquitanian mountains with sedimentary rocks (sandstones, shales and limestone rocks) and the Eastern Brazilian Shield with Precambrian rocks (gneiss and granite).

Bolivia is influenced by three hydrographic basins: Amazon, Paraná-Paraguay and endorheic basin of the Altiplano in addition to the Pacific slope (that is part of the Silala spring), much smaller but of great economic importance, which in total are made up of 10 subbasins, 270 main rivers, 184 lakes and lagoons, approximately 260 wetlands and 6 salt flats (Montes de Oca,

2005). The total length of the main rivers in the 10 subbasins is estimated at 57,000 km, the surface of lakes and lagoons of 11,193 km², the snow-capped mountains of 2,184 km² and the salares 13,091 km² (SNHN, 1998). The Bolivian lowlands are infiltrated by numerous rivers that are mostly born in the mountain range. Two fluvial systems are of importance: the rivers belonging to the Amazon basin and the Pilcomayo River that drains a large part of the southern section of the Eastern Cordillera, heading towards the Paraná-Paraguay fluvial system. The Amazonian rivers that come directly from the northeastern foothills of the Eastern Cordillera – due to the high rainfall, have the least periodic flows in the country. During the rainy season, often cause extensive flooding. In the lower basins, due to a small elevational gradient in the alluvial plain, the rivers begin to wander due to the high flow with abundant material of the Andes and form meanders creating vast floodplains under an intense process of plant succession and trophic chains, next to the seasonal dynamics and the distribution of rains with influence of the winds of the south (Pouilly et al., 2004).

According to the classification of Ibisch and Mérida (2003), Bolivia is represented by 23 ecological regions (Table 1.1). Five major zones are recognized: 1 Southwest of the Amazon, 2 the Cerrado, 3 Flooded savannas, 4 Andean eastern slope and inter-Andean valleys and 5 high mountain ranges and Altiplano included in three altitudinal levels: the lowlands with five ecoregions, the eastern Andean slopes (Andean plateau, including inter-Andean valleys) with five ecoregions and the high Cordillera and the Altiplano with two ecoregions (Figure 1.3). Evergreen forests of the Amazon basin contain biogeographical characteristic elements (e.g., Rubber, *Hevea brasiliensis* and Amazon Chestnut, *Bertholletia excelsa*). The southwest ecoregion of the Amazon is – together with the Yungas – one of the most complex and richest species of plants and animals in the country.

The Cerrado includes open vegetation types, such as savannas in plains, undulating hills and mountainous areas with nutrient-poor soils of different paleo-historical origins. The best-known enclosures extend over the Precambrian Shield, but similar physiognomic areas exist in the Andean region and in the Amazon and Chaco plain. Villarroel et al. (2016, http://www.museonoelkempff.org/sitio/Informacion/KEMPFFIANA/kempffiana12(1)/47–80%20Villarroel%20et%20al.%202016.pdf) recognized in the east of Bolivia eight physiognomic types, which are differentiated by woody cover and abundance, as well as by their edaphic characteristics and the level of drainage; among the representative species of the Cerrado (s.s.) are *Qualea* spp. (Vochysiaceae), Donkey egg (*Caryocar brasiliensis*,

Table 1.2 Number of Known and Expected Species (In Parenthesis) Within Main Taxonomic Groups in Bolivia

Taxon	Known and expected number of species	References
Algae	14,718 (19,500)	Cadima et al., 2005; Maldonado, 2005
Lichen	2,000 (4,800)	http://bio.botany.pl/lichens-bolivia
Mosses	1,399 (1,428)	Ibisch, 1998; Churchill et al., 2009
Vascular plants	12,245 (20,000)	Jorgensen et al., 2014
Fungi	799 (956)	https://www.inaturalist.org/ projects/fungi-of-bolivia
Mammals	389 (432)	Tarifa and Aguirre, 2009; Aguirre et al., 2009
Birds	1,435 (1,500–1,600)	Herzog et al., 2017
Fish (freshwater)	908 (1,200–1,400)	Sarmiento et al., 2014
Reptiles	313 (391)	Uetz et al., 2017
Amphibians	251 (334)	Amphibia Web, 2017
Insects	13,719 (28,000)	http://insectoid.info/checklist/ insecta/bolivia
Invertebrates (Crustacea)	46 (92)	http://insectoid.info/checklist/ malacostraca/bolivia
Mollusca	103 (206)	http://insectoid.info/checklist/ branchiopoda/bolivia; Galli, 2017

Caryocaraceae), Sucupira (*Bowdichia virgilioides*, Fabaceae), Chiquitanian almond (*Dypteryx alata*, Fabaceae), among many others. While in the dry forest or dry bush of the Cerradao, timber species are as follows: the cuchi (*Astronium urundeuva*, Anacardiaceae), the oak (*Amburana cearensis*, Fabaceae) and *Machaerium eriocarpum* (Fabaceae). Between Santa Cruz and Beni departments, it covers an area of 30,179 km² (Villarroel et al., 2016). The Chiquitanian dry forest covers an area of 20 million hectares and is the largest tropical dry forest in the world and is valuable for its unique ecology and its historical and cultural importance. The Dry Forest Chiquitano is an ecoregion characterized by its transitional location between the humid climate of the Amazon and the semi-dry of the Gran Chaco. Geologically it is marked by an undulating topography, the presence of small mountain ranges and the Precambrian Shield, a granite rock formation, which underlies the soils of the region, with spectacular occasional outcrops. The

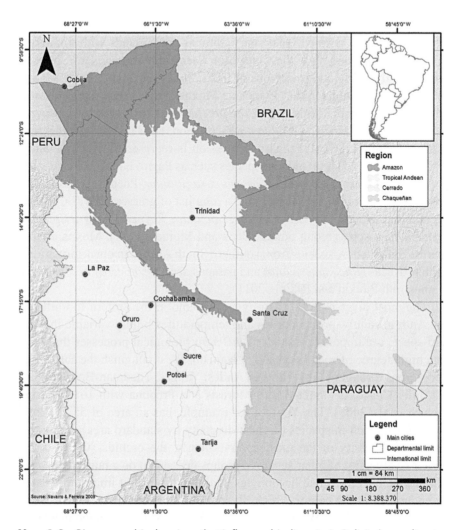

Map 1.3 Biogeographical regions that influence biodiversity in Bolivia (according to Navarro and Ferreira, 2009).

vegetation of the area is semi-deciduous, until deciduous. Exclusive species of this ecoregion are the Morado (*Machaerium nycticans*), the Momoqui (*Caesalpinia pluviosa*), the Yellow Tarara (*Centrolobium microchaete*), and the South American oak (*Amburana cearensis*), among others. More than 75% of the timber species in the forests of the Chiquitaniía Region currently have commercial value.

The flooded savannas are confirmed by plains of recent, relatively fertile alluvial soils, flooded mainly by overflowing rivers and comprises 128,000 km². They are seasonal savannas and wetlands with forest islands and gallery forests on poorly drained Quaternary alluvial bottomlands with a marked dry season and flooding in the wet season (Beck and Moraes, 1997, https://www.worldwildlife.org/ecoregions/nt0702). It also includes palm savannas and palm swamps in the 'Llanos de Moxos' plain, primarily east of the Mamoré River, pure stands of *Euterpe precatoria*, *Copernicia alba*, *Mauritia flexuosa* or *Mauritiella aculeata* are common, each in unique ecological settings (Moraes, 2016). Primary succession is represented predominantly by herbaceous and shrub pioneer species such as Pajaro bobo (*Tessaria integrifolia*, Asteraceae), Chuchio (*Gynerium sagittatum*, Poaceae), and Sauce (*Salix humboldtiana*, Salicaceae), and a long list of grasses such as *Cyperus giganteus* (Cyperaceae), *Eleocharis geniculata*, *Rhynchospora trispicata*, *Hymenachne* spp., among others (Beck and Moraes, 1997). Moxos savannas are inhabited by species from southern South American open-formations such as Rheas (*Rhea americana*) and pampas deer (*Ozotoceros bezoarticus*) (Langstroth Plotkin and Riding, 2011).

The most diverse areas of Bolivia are found in the eastern foothills of the Andes. Mainly, precipitations and temperatures highly variable in time and space, are important factors of different biological processes that lead to a high degree of speciation and endemism. We distinguish the following ecoregions throughout the Bolivian Andes: Yungas, Tucuman-Bolivian forest, hilly Chaco, dry inter-Andean forests and Prepuna with 161,438 km² (Ibisch et al., 2003). The Yungas, for example, has an area of 55,556 km² and are very rich in species (highest diversity by standard area), being the center of diversity of the most diverse family, the orchids, and also the center of diversity of other sensitive groups dependent on a humid seasonal climate such as ferns and bryophytes, making it the most important center of endemism in the country; high diversity of endemic species, especially at medium elevations. While the dry inter-Andean valleys include in a fragmented form and as patches interrupted between mountains a large variation of deciduous plant formations that go from the dry forests in the region of the Yungas, to the extensive valleys in the center and south of the country.

Finally, the high Andean region corresponds to several types of ecoregions characterized by a tropical vegetation of High Mountain, physically resembling the steppes of temperate zones. It extends from northern Peru to central Argentina, along the Andean Cordillera. In Bolivia with 220,783 km², three to five types of puna are distinguished according to climatic and edaphic conditions (Ibisch et al., 2003). Lately, Navarro (2002) presented a biogeographic separation in Provinces of Peruvian Puna, Bolivian-Tucuman and the Altiplano. Ibisch and Mérida (2003) proposed two ecoregions that are distinguished by ecological, geological-geomorphological and biogeo-graphic criteria: northern Puna and southern Puna (Table 1.2).

1.1.2 Light and Climate Conditions

Bolivia belongs to the tropical region of the world and this implies that the climate regime influences on the basis of two winter and summer seasons with respective transitional stages. The angle of the sun's radiation reaching the earth is never below 43° and for Bolivia which is close to the Capricorn Tropic line, the annual variation of the duration of the day is 3 hours; The tropical region is characterized by a positive annual balance in terms of radiation (Rafiqpoor et al., 2003). In winter (May–July), the lowest amount of solar radiation is recorded, while in summer (October–December) the highest amount of solar radiation is recorded. From the Eastern Cordillera to the NE of Bolivia, there are the lowest annual average global radiation values, between 3.9–5.1 kW-h/m²-day because it is found in the lowlands, due to the tropical climate and high humidity generating a greater dispersion of solar radiation, whereas in the SW part of the country, solar radiation increases in the highlands, with dry climate and values between 5.1–7.2 kW-h/m²-day (Andrade et al., 1998).

The location of Bolivia between 9°40'–22°50' south of the Andean mountains generates the effect of 1,000 hPa (approximately at sea level) by three supra-regional pressure systems (Emck et al., 2006): (i) Unstable anticyclone of the South Pacific Ocean that remains fixed by the mountains; it provides a constant current N-NNW and produces weather conditions in relation to the rest of the continent. (ii) Another relatively stable anticyclone in the South Atlantic Ocean that provides wind masses from the E to the eastern flank of the Cordillera de los Andes. (iii) Both anticyclones are separated by a low-pressure center from the Amazon basin to the south; which can cause a pattern of NE-SW currents. The circulation of the Southern Hemisphere shows strong dynamics due to the temperature contrast between the Antarctic and the equatorial regions, resulting in a pressure gradient; therefore, the South Pacific anticyclone is strong throughout the year and especially in the winter (Emck et al., 2006). During the winter, when the winds of the E are replaced by western winds, in the slope of the Eastern Cordillera of Bolivia they happen periods of more or less prolonged drought, reaching annual water deficits of −100 cm/year in southern Bolivia; except in the "island of humidity" that concentrates in the Chapare (center of the country) with 7,000 mm of annual precipitation, caused by the clash of divergent masses in front of the Andean wall in central Peru (Emck et al., 2006).

There is a gradient of vertical precipitation in tropical mountains with maximum rainfall, as in the Zongo Valley (W Bolivia) the highest precipitation is 3,000 mm/year and decreases significantly above 3,000 of altitude; although the general tendency in the central Andes is to increase W-E preprompts (Emck et al., 2006). In relation to the tree line, there are two types that depend on the availability of flora: Predominance of *Polylepis* (Rosaceae) forests in semihumid to arid regions above 4,000 m elevation, which then disappear in per-humid situations to the 3,400 m (Richter and Moreira-Muñoz, 2005).

The wind speed changes abruptly in the parallel of 20°S (in the elbow of the Andes: Amboró, in the W of the department of Santa Cruz, Bolivia): in January air masses enter from the north, while in July the winds are weaker and higher; in the winter the anticyclone of the South Pacific moves towards the Atlantic and allows the entry of Antarctic masses to the north, which is responsible for the cold winds or "surazos," as they are called in Bolivia. During the dry season, these "surazos" cause the intense descent of the temperature by 10–12°C in the alluvial region from the SE to the NE of Bolivia (Hanagarth, 1993).

In the lowlands and throughout the year, temperatures between 20–27°C are recorded, while in the mountains there are temperature gradients that depend, among others, on humidity, so that they can descend several degrees below 0. On the northeastern slope of the Yungas, a humid-diabatic gradient of 0.56°C/100 m is recorded, the upper limit of the hot ground is at an altitude of 1,200 m (average annual temperature 21°C); then between 1,200 and 2,700 m there are lower temperatures (Rafiqpoor et al., 2003). The frost limit is located on the altitudinal floor (from glacial zones to the valleys) with the annual average temperature of 13°C; the cold land is characterized by frost and has two floors: the upper one has temperatures of 10–6°C and corresponds to the misty forests with frosts of up to 20 days a year; finally, the lower limit of lands with snow coincides in the Eastern Cordillera at about 4,800 m.

Nevertheless, climate change is producing an accelerated retreat of tropical glaciers, compromising water for consumption, food security and potentially hydroelectric generation. Glaciers in theTuni Condoriri mountain basin in the eastern Cordillera have lost between 35 to 45% of their surface in the last 50 years (PNCC, 2007).

1.2 Main Ecosystem Characteristics

The Bolivian territory is continentally privileged due to the biogeographic influence of Amazonian, Andean, Chaco and Cerrado elements, which is why it is a mosaic of the biogeographic meeting point in contact zones and ecotones with unique characteristics (Moraes and Beck, 1992). The description of the main ecosystems in natural landscapes and modified by anthropic activities has been based on diverse contributions and publications expressed in different scales, considering climatic, orographic, geomorphological, edaphic, floristic and faunal components, among the main ones (Moraes and Beck, 1992; Beck et al., 1993; Ribera et al., 1996). Another group of resources also added bioclimatic and biocenotic factors, considering the structure of the vegetation and its classification through extensive phytosociological surveys and their relationship with local climatic patterns, such as the works by Navarro (2002), Josse et al. (2003), Navarro and Ferreira (2009), and Navarro (2011).

Bolivia is commonly described as having highlands (mostly in the western side, from 100 to 500 m) and lowlands in the eastern as major landscapes units, although there are Precambrian hills of 1,200 m elevation in the easternmost side of the country. Forests, savannahs, wetlands and marshes are found in the lowlands. Prairies, peats, valleys with fertile soils for important croplands drained by headwaters of Amazonian watersheds and a mountainous range are found in the high lands that reach up to 6,462 m elevation. Topographically, the west side presents the "Altiplano" which is surrounded by two mountain chains, clearly differentiated from a humid north and a dry south plateau. A similar pattern is found in the lowlands where a humid Amazonian wide region covers to the center, then turning to a drier vegetation cover with a lower forest in the east and a xeric one in the SW.

Among azonal formations and those that mix from the sub-Andean to the lowlands are the palm groves of the Palma Real (*Mauritia flexuosa*) whose distribution is related to wet-water wetlands and the high concentration of organic matter (Hanagarth, 1993; Moraes, 2016). While in the Andean areas are the highland peats mainly dominated by several species of Cyperaceae and associated with drainage systems of Andean glaciers, as forage for camelid and bird species.

1.2.1 Freshwater

Bolivia has unequally distributed water resources, but it is one of the countries with the highest freshwater supply per inhabitant in Latin America (approximately 50,000 m^3/inhabitant/year). The supply of water on a national scale is estimated at more than 500,000 million cubic meters per year and the current demand is more than 2,000 million cubic meters per year, that is, less than 0.5% of the total supply, which represents a great comparative advantage for the country (Paz et al., 2010). However, the spatial and temporal variability of precipitation within the territory is high and extreme hydrometeorological events (droughts, heavy rains, hail, snow) are frequent.

Despite this high amount of water, its availability and quality have been a constant critical issue in the country, in highlands, dry valleys and in the Chaco, water is the limiting factor of growth, a situation that has worsened in recent years, climate change has caused drought, loss of glaciers and ice reserves (Paz et al., 2010). The available water has also been threatened by the contamination of bodies of water, which have decreased not only their quality, but their availability to the population.

Most rivers of the Bolivian Amazonia are extremely muddy during most of the year. The Andean tributaries: Madre de Dios, Beni and Mamoré rivers, deliver most of the sediment loads; while the Iténez River and tributaries that arise in the Brazilian Shield have clear or black waters. Clearwater tributaries are also found in the Andean eastern slopes above approximately 700 m (Goulding et al., 2003).

The diversity of fish species in the Bolivian Amazon is represented by 56 families, 344 genera and 852 species (74 endemics) and is dominated by groups of ostariophysans and cichlids. Several groups are emblematic of the basin: freshwater stingrays of family Potamotrygonidae, the unique South American lungfish *Lepidosiren paradoxa*, and the electric eel *Electrophorus electricus* that, together with piranhas (*Pygocentrus* and *Serrasalmus*), are among the most feared fishes in the basin (Sarmiento et al., 2014). Also, long-range migratory catfishes of the genera *Brachyplatystoma*, some migrating 3,000–4,000 km between the lower Amazon and the foothills of the Andes in Bolivia for spawning, are noteworthy fishes; nearly to 10% of large and medium-sized fishes have migratory habits (Carvajal et al., 2014; Sarmiento et al., 2014, Reis et al., 2016). Also, four species of crocodilians (locally named as Caiman), the Bolivian Anaconda (*Eunectes beniensis*) and 11 species of aquatic turtles, including the giant South American river turtle (*Podocnemis expansa*) are found there. Also, as much as 72 species of aquatic birds are currently known from the basin (Sarmiento et al., 2016; Herzog et al., 2017), and at least 5 mammals, including the Bolivian dolphin (*Inia boliviensis*), the marsh deer (*Blastocerus dichotomus*), giant otter (*Pteronura brasiliensis*), as also the semi-aquatic giant rodent capibara (*Hydrochoerus hydrochaeris*), are related to aquatic habitats of the Bolivian Amazon. Similarly, the great diversity of aquatic habitats and morphological complexity of the Bolivian Amazon aquatic systems, is related to a very high diversity of aquatic insects and other taxa of invertebrates, like mollusks, crabs and shrimps. There is also a high diversity of both zoo and phytoplankton associated with meandering rivers and lakes (Maldonado, 2005).

1.2.2 Wetlands

The wetlands in Bolivia are areas of high biological diversity, an analysis of the specific richness and the degree of endemism in the different basins, shows that the highest rates of endemism were concentrated in the endorheic basins of the Altiplano (71–100%), in the Paraguay river basin (21–29%) and in the Amazon basin (5–21%) (Table 1.3). Of great importance are considered in the Llanos de Moxos numerous and parallel rectangular lagoons with a clear direction SW–NE between the Beni and Mamoré rivers, influenced by a geological fault, as well as by ancient human activity to generate zones of hydraulic management and a free of flood mesorelief for crops and human settlements (Hanagarth, 1993).

The Titicaca Lake, one of the most important centers of endemism in the continent, makes the difference with 80% of its exclusive ichthyofauna from the lake basin that is shared between Bolivia and Peru (Van Damme et al., 2009). Moreover and according to Dangles et al. (2017), over the last three decades round wetlands of *Distichia* spp. (Cyperaceae) and associated Poaceae located in highly glacierized catchments were less prone to drying, whereas small wetlands with irregularly shaped contours suffered the highest rates of drying; therefore, high Andean wetlands can, therefore, be considered as ecosystem sentinels for climate change, as they seem sensitive to glacier melting.

The Titicaca Lake basin in the high elevation plains of Bolivia is the smallest and the least diverse basin of Bolivia, it encompasses 154,176 km^2 or 13.8% of the territory (Montes de Oca, 2005). There are only two families of native fish represented in the basin, with nearly 29 fish species, most of them endemic of the basin and the Titicaca Lake itself. Remarkable species of the basin are the caraches, bogas, and ispis belonging to the genus *Orestias* (Carvajal et al., 2014; Sarmiento et al., 2014; Reis et al., 2016). Most species of *Orestias* are now vulnerable or critically endangered, and the umanto (*Orestias cuvieri*) is extinct (Van Damme et al., 2009). The fish fauna of the basin includes two introduced fishes: Rainbow Trout (*Oncorhynchus mykiss*) and the Pejerrey (*Odontesthes bonariensis*). Vertebrate aquatic biota includes the critically endangered Titicaca Frog (*Telmatobius culeus*), and some other species of *Telmatobius*; at least 42 bird species, mostly ducks and herons are known for the basin, notably the

Table 1.3 A Sample of Bolivian Small Mammals, From a Total of 132 Bats

Bats	Scientific name
Lesser bulldog bat	*Noctilio albiventris*
Greater bulldog bat	*Noctilio leporinus*
Silver-tipped myotis	*Myotis albescens*
Hairy-legged myotis	*Myotis keaysi*
Black myotis	*Myotis nigricans*
Brazilian brown bat	*Eptesicus brasiliensis*
Argentine brown bat	*Eptesicus furinalis*
Big-eared brown bat	*Histiotus macrotus*
Small big-eared brown bat	*Histiotus montanus*
Desert red bat	*Lasiurus blossevillii*
Hoary bat	*Lasiurus cinereus*
Cinnamon dog-faced bat	*Cynomops abrasus*
Southern dog-faced bat	*Cynomops planirostris*
Dwarf bonneted bat	*Eumops bonariensis*
Wagner's bonneted bat	*Eumops glaucinus*
Sanborn's bonneted bat	*Eumops hansae*
Western mastiff bat	*Eumops perotis*
Andersen's fruit-eating bat	*Artibeus anderseni*
Silver fruit-eating bat	*Artibeus glaucus*
Jamaican fruit bat	*Artibeus jamaicensis*
Great fruit-eating bat	*Artibeus lituratus*
Dark fruit-eating bat	*Artibeus obscurus*
Salvin's big-eyed bat	*Chiroderma salvini*
Little big-eyed bat	*Chiroderma trinitatum*
Hairy big-eyed bat	*Chiroderma villosum*
Velvety fruit-eating bat	*Enchisthenes hartii*
MacConnell's bat	*Mesophylla macconnelli*
Ipanema bat	*Pygoderma bilabiatum*
Visored bat	*Sphaeronycteris toxophyllum*
Bogota yellow-shouldered bat	*Sturnira bogotensis*
Hairy yellow-shouldered bat	*Sturnira erythromos*
Little yellow-shouldered bat	*Sturnira lilium*

Table 1.3 (Continued)

Bats	Scientific name
Greater yellow-shouldered bat	*Sturnira magna*
Tilda's yellow-shouldered bat	*Sturnira tildae*
Tent-making bat	*Uroderma bilobatum*
Brown tent-making bat	*Uroderma magnirostrum*
Bidentate yellow-eared bat	*Vampyressa bidens*
Southern little yellow-eared bat	*Vampyressa pusilla*
Great stripe-faced bat	*Vampyrodes caraccioli*
Short-headed broad-nosed bat	*Platyrrhinus brachycephalus*
Thomas's broad-nosed bat	*Platyrrhinus dorsalis*
Heller's broad-nosed bat	*Platyrrhinus helleri*
Buffy broad-nosed bat	*Platyrrhinus infuscus*
White-lined broad-nosed bat	*Platyrrhinus lineatus*
Common vampire bat	*Desmodus rotundus*
White-winged vampire bat	*Diaemus youngi*
Hairy-Legged vampire bat	*Diphylla ecaudata*
Muroids	
Highland grass mouse	*Akodon aerosus*
White-bellied grass mouse	*Akodon albiventer*
Azara's grass mouse	*Akodon albiventer*
Bolivian grass mouse	*Akodon boliviensis*
Cochabamba grass mouse	*Akodon siberiae*
White-throated grass mouse	*Akodon simulator*
Puno grass mouse	*Akodon subfuscus*
Chaco grass mouse	*Akodon toba*
Bolivian big-eared mouse	*Auliscomys boliviensis*
Painted big-eared mouse	*Auliscomys pictus*
Andean big-eared mouse	*Auliscomys sublimis*
Pleasant bolo mouse	*Bolomys amoenus*
Rufous-bellied bolo mouse	*Bolomys lactens*
Hairy-tailed bolo mouse	*Bolomys lasiurus*
Bolivian vesper mouse	*Calomys boliviae*
Large vesper mouse	*Calomys callosus*

Table 1.3 (Continued)

Bats	Scientific name
Small vesper mouse	*Calomys laucha*
Andean vesper mouse	*Calomys lepidus*
Bicolored arboreal rice rat	*Oecomys bicolor*
Unicolored arboreal rice rat	*Oecomys concolor*
Mamore arboreal rice rat	*Oecomys mamorae*
Robert's arboreal rice rat	*Oecomys roberti*
Tarija rice rat	*Oryzomys legatus*
Light-footed rice rat	*Oryzomys levipes*
Capricorn leaf-eared mouse	*Phyllotis caprinus*
Darwin's leaf-eared mouse	*Phyllotis darwini*
Bunchgrass leaf-eared mouse	*Phyllotis osilae*
Wolffsohn's leaf-eared mouse	*Phyllotis wolffsohni*

(Tarifa and Aguirre, 2009; https://en.wikipedia.org/wiki/List_of_mammals_of_Bolivia) and 64 Muroids (Tarifa and Aguirre, 2009; https://en.wikipedia.org/wiki/List_of_mammals_of_Bolivia)

Titicaca Grebe (*Rollandia microptera*), and two species of Andean flamingos (*Phoenicoparrus jamesi* and *P. andinus*) (Herzog et al., 2017). Aquatic invertebrates are mainly known from this lake itself, such as 449 aquatic invertebrates (Chironomidae, Coleoptera, Trichoptera and Ephemeroptera) that leave related to the Totora Rush (*Schoenoplectus tatora*), *Elodea potamogeton*, *Lemna* sp. and Characeae stands. Despite the relatively small number of endemic species in Lake Titicaca, there are possible species flocks of amphipods (genus *Hyalella*), and microgastropods (*Heleobia*) (Kroll et al., 2012).

Bolivia joined in November 1990 the Convention on Wetlands of International Importance, especially as Waterfowl Habitat, known as the RAMSAR Convention. To present there are 11 Bolivian wetlands included in the RAMSAR list, covering 14,842,405 hectares, positioning the country in fifth place in the list of countries with the largest amount of wetlands in the world (Table 1.2).

1.2.3 Forests

Bolivia ranks sixth in the world's tropical forest and 15th in the forest with an area of 439,200 km^2 (ca. 40% of the total). Native tropical forests cover a large part of the territory, mainly in the Yungas, Chapare, Northern Amazon, Llanos de Moxos in Beni and Santa Cruz, Chiquitanía and Chaco. Around 40% of the country's surface is covered by forests and its majority are used, but without changing in the marked form its structure and composition (Ibisch and Merida, 2003). Some Bolivian forests have a high value as centers of biological diversity, endemism and because they are threatened, which is why they are a priority at the global level (Moraes and Beck, 1992). They represent a high priority area for conservation, either through protected areas or sustainable use activities (Ibisch and Mérida, 2003). Degraded and fragmented forests with small and medium scale agriculture and artisanal use of forest resources (more than 18%) are an area that is relevant to ensure greater sustainability of agricultural production and where agroforestry systems should be established (Figure 1.4). More than 15% of the territory is suitable for reforestation activities to restore natural potential or generate ecosystemic functioning.

The forestry sector is the second most important sector in terms of export value (3% GDP), after soybean and its derivatives, having generated approximately US$ 200 million in exports of 1,000,000 m³/year. In 2002, according to the Forestry Superintendence, Bolivia had 1 million hectares of certified forest use, for 2007 it was estimated 2.2 million hectares, and in November 2010 there were 1,186,604 hectares of certified forest area registered (Quevedo and Urioste, 2010). In comparison with other studies carried out in Bolivia, the biomass of the Amazonian terra firme forest is 440.90 ± 29.30 Mg ha-1 and of the flooded forest: 396.02 ± 31.37 Mg ha-1; the necromass in relation to the gross aerial biomass or ≥ 10 cm of DAP is $14.48 \pm 3.92\%$ in the forest of the mainland and $8.23 \pm 1.07\%$ in the flooded forest (Araujo-Murakami et al., 2016).

1.2.4 Mountain Ecosystems

The mountainous ecosystems of Bolivia present approximately one-third of the total surface along with two types of systems: Andean (west side) and Precambrian Shield (east). In the Southern Puna, local quinoa crops (*Chenopodium quinoa*) and cañahua (*Ch. pallidicaule*), which are very important and appreciated as a food supplement. In Inter Andean ecosystems an important development potential is found for irrigated agriculture due to the availability of suitable soils, water sources, markets, road infrastructure and a high degree of organization, such as the Yungas mountain forests. A 31.9% of highlands and inter Andean valleys was used for cereal cultivation (390,668 ha with maize, and the remainder with grain sorghum, paddy

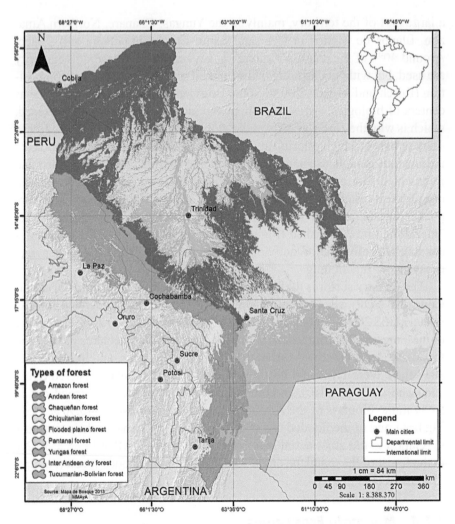

Map 1.4 Map of forests of Bolivia (Forests of Bolivia (MMAyA, 2013; https://geo. gob.bo/geonetwork/srv/esp/catalog.search#/metadata/e7a1ab97– 82f9–4d73–84b9-b6979b973a42).

rice, quinoa and wheat); tubers and roots were planted on 7.5% of the area (170,447 ha with potato), vegetables on 3.9%, fruit trees on 5.8%, fodder on 6.1% and stimulants on 1.4% (Tejada et al., 2017).

The Yungas (from 300–600 m until 3,000 to 3,800 m elevation) are climatologically and biogeographically very particular as they are confirmed by mountain, cloud, rainy and tropical forests on the humid eastern slopes and less affected by seasonal regimes because they have a northeastern exposure (NW–SE orientation) benefiting from the humidity brought by the trade winds (Ibisch and Mérida, 2003). Several altitudinal floors can be distinguished, floristically very different, that possibly deserve a subdivision: low and high with the mountain brow that would include the mixed mist forests surrounded higher by the forests of *Polylepis pepei*, today almost completely replaced by humid pastures representing the Yungueñan Paramo. The "Yungueñan paramo" are similar physiognomically and floristically to the true paramo formations of the northern Andes of South America, under intratropical climatic conditions. They are located in strips and spots on top of the mountain's eyebrow and form pajonales and almost always humid bushes. In the Yungas, the intraecoregional rate of replacement of taxa (beta diversity) is higher than in any other ecoregion (Ibisch and Mérida, 2003). The most diverse families are Araliaceae, Bromeliaceae, Euphorbiaceae, Lauraceae, Melastomataceae, Myrtaceae, Orchidaceae, Piperaceae, Podocarpaceae, Rubiaceae, and many more. Also, several endemisms and restricted geographically plant species belong to this ecosystem (Moraes and Beck, 1992; Killeen et al., 1993), as for example, among palm species with four endemic species: Janchicoco Palm (*Parajubaea torallyi*, which grows in dry inter Andean valleys), the Sunkha Palm (*P. sunkha*, in fertile valleys of W Santa Cruz), the Coquito Palm (*Syagrus cardenasii*, in drier inter Andean forests and rocky vegetation) and *Syagrus yungasensis* (a critically endangered CR species in La Paz department) (Moraes, 1999).

In the Ceja del Monte: Asteraceae, Cunoniaceae, Ericaceae, Solanaceae etc. (many families of ferns such as Aspleniaceae, Polypodiaceae, etc.). Several plantations of Cocoa trees (*Theobroma cacao*), Coca Leaf (*Erythroxylum coca*), and Papaya (*Carica papaya*), among other introduced species like citrics (*Citrus* spp.), Mango (*Mangifera indica*), Banana (*Musa paradisiaca*) and Coffee (*Coffea arabica*), as well as other native species have been established by local communities for economic incomes.

In relation to the typical fauna, there are bird species, such as *Lipaugus uropygialis* (1,800–2,600 m in Yungas forests) while *Cinclodes aricomae* and *Anairetes alpinus* (3,800–4,300 m) in the Yungueñan Paramo and *Polylepis* forests. Among the mammals of the humid forests of Yungas there are *Gracilinanus aceramarcae* (2,600–3,300 m), the Spectacled Bear (*Tremarctos ornatus*) and a deer (*Mazama chunyi*, 1,000–3,600 m).

1.2.5 Open Lowland

Extensive to semi-intensive livestock farming occupies more than 40% of Bolivia, being potential, in areas already deforested or naturally open (savannas) a use compatible with the conservation of the current ecosystems. In 10% of the territory, livestock that exceeds the natural potentials of ecosystems contributes significantly to the degradation of natural resources (dry valleys, Puna and Prepuna). According to Tejada et al. (2017), the main crop groups planted during the summer season, 43.4% of the area in the lowlands was cultivated with oilseeds and industrial crops (999,369 ha with soybean and the rest with sunflower, sugar cane and peanuts).

The Gran Chaco region has an area in SE Bolivia of 105,006 km². The Gran Chaco or dry Chaco (derived from Quechua language: *chaku*, "hunting land") is a sparsely populated, hot and semi-arid lowland natural region of the Río de la Plata basin, divided among Southeastern Bolivia, western Paraguay, northern Argentina and a portion of the SW Brazilian, where it is connected with the Pantanal region. This land is sometimes called the Chaco plain and is also one of the distinct physiographic provinces of the Parana-Paraguay plain division. The dominant vegetative structure is xerophytic deciduous forests with multiple layers including a canopy (trees), sub-canopy, shrub layer and herbaceous layer. There are ecosystems such as riverine forests, wetlands, savannas, and cactus stands as well. At higher elevations of the eastern zone of the humid/subhumid Chaco there are transitional mature forests from the wet forests of southern Brazil. These woodlands are dominated by canopy trees such as *Handroanthus impetiginosus* and characterized by frequent lianas and epiphytes.

1.3 Organisms and Taxonomic Groups

Bolivia is considered a megadiverse country (Ibisch, 1998) with numerous natural landscapes (ecoregions, ecosystems and vegetation types) and shows an estimated total of 48,325 known species (Tables 1.1–1.2, respectively). Non-vascular plants and invertebrate species are the less known groups with incipient scientific collections and few researches completed. Below are some peculiarities and characteristics of the groups of organisms in Bolivia.

1.3.1 Vegetation

Derived from the biogeographical influences of the Amazon, Andes, Cerrado and Chaco, the Bolivian vegetation formations are influenced by altitudinal gradients, latitudinal distribution, geomorphology, climate and dynamics associated with hydrography and seasonality. Approximately 15 major vegetation types are recognized in the country (Beck et al., 1993) according to their distribution, presence of representative species and coverage. The Puna and high Andean vegetation have mostly an extensive cover of Poaceae and herbaceous species as well as shrubs and very few trees: for example, species of the genera *Stipa, Festuca, Baccharis, Buddleja, Polylepis, Escallonia, Duranta, Senna*; the precipitation does not exceed 500 mm /year and is between 2,500–4,800 m elevation. The bushes of dry inter-Andean valleys are distributed in a dispersed and fragmented pattern between 500–3,300 m elevation with an annual rainfall of 500–600 mm and a wide thermal range of 2–28°C; the most characteristic groups are *Jacaranda mimosifolia, Erythrina falcata, Schinus molle, Kageneckia lanceolata, Prosopis* spp., *Astronium urundeuva*, many species of Cactaceae and epiphytes (mosses, lichens, orchids) and among shrubs; *Dodonaea viscosa, Baccharis* spp., *Nicotiana glauca* and *Tecoma stans*.

The Tucuman-Bolivian forest – influenced by the dry winds of the continental south and dry forests – corresponds to a semihumid formation between 800–3,000 m with precipitation of 1,000–1,700 mm/year; there are several species of Myrtaceae, *Juglans australis, Cedrela lilloi, Zanthoxylum coco, Alnus acuminata, Prunus tucumanensis, Podocarpus parlatorei, Escallonia hypoglauca* and *Myrica pubescens*. Humid mountainous forests (Yungas forests) cover a range from 500–2,800 m altitude and present a precipitation of 1,000–2,000 mm/year with a temperature of 17–24°C; typically are found mostly evergreen species such as *Cinchona* spp., *Inga* spp., *Clusia* spp., *Dendropanax* spp., *Poulsenia armata, Tetragastris altissima, Clarisia racemosa, Iriartea deltoidea, Dictyocaryum lamarckianum, Oenocarpus bataua, Persea americana*, among many others.

The dry forest of the Chaco or Chaco serrano is confirmed by few strata and rather abundance of thorn scrub, as well as succulent species in a range of 30–800 m of altitude and an annual precipitation of up to 300–500 mm; the characteristic species are *Prosopis alba, Ziziphus mistol, Geoffroea decorticans, Ruprechtia triflora, Schinopsis* spp., *Chorisia insignis, Celtis spinosa, Pereskia* spp. along with several Cactaceae, *Lithraea ternifolia, Trithrinax schizophylla*, among others. The wet and flooded savannas of Bolivia that include open landscapes with gallery forests, continuous forests, wetlands and pastures are between 150–400 m elevation and an annual rainfall of 1,000–2,500 mm per year; the largest cover in the savannahs is made up of species of Poaceae, Juncaceae and Cyperaceae, while in the marshes there are species of *Eichhornia* spp., *Thalia geniculata, Pontederia* spp. and *Victoria amazonica*; in very dense scrub formations there are species of *Bromelia serra, Machaerium* spp., *Curatella americana, Luehea paniculata* and *Pseudobombax* spp., palm formations of *Copernicia alba, Acrocomia totai*,

Mauritia flexuosa and *Mauritiella armata*, as well as forests with *Sterculia apetala, Triplaris* spp., *Genipa americana, Guazuma ulmifolia, Cupania cinerea, Apeiba tibourbou, Handroanthus* spp. and others.

Savannas ("campos") of the Cerrado on the Precambrian Shield between 500–850 m elevation and annual precipitation of 800–1,500 mm; savannas and semi-deciduous moist forests are developed with species adapted to fire dynamics, such as *Aspidosperma* spp., *Caryocar brasiliense, Qualea* spp., *Dipteryx alata, Amburana cearensis* and palms, such as *Allagoptera leucocalyx, Syagrus petraea, Attalea phalerata, A. speciosa*, among others. The Chiquitanian semi-deciduous forest located between 300–1,200 m elevation and with a precipitation of 1,000–1,500 mm per year is found in plains and hills; *Astronium urundeuva, Aspidosperma cylindrocarpon Chorisia speciosa, Machaerium* spp., *Chrysophyllum gonocarpum, Anadenanthera colubrina, Terminalia argentea* and others are represented. While the humid forest of the Cerrado (700–1,000 m altitude) presents species such as *Terminalia oblonga, Swietenia macrophylla, Ampelocera ruizii, Ocotea guianensis, Jacaratia spinosa*, among others.

Finally, there are two formations of the Amazon: forest and plain at 150–300 m elevation. The Amazon forest has an annual rainfall of 1,800–2,500 mm and a temperature between 25–27°C; presents species such as *Bertholletia excelsa, Hevea brasiliensis, Couratari guianensis, Dialum guianense, Mezilaurus itauba, Ceiba pentandra, Calycophyllum brasiliense, Phenakospermum guianensis, Dipteryx micrantha, Ficus* spp., as well as bamboo (*Guadua* spp.) and palms: *Chelyocarpus chuco, Astrocaryum ulei, Attalea* spp. and *Phytelephas macrocarpa*. The plain forest with a rainfall variation between 1,200–1,800 mm per year presents similar species, as well as *Hura crepitans, Swietenia macrophylla, Terminalia oblonga, Sloanea guianensis, Pourouma guianensis, Symphonia globulifera, Xylopia ligustrifolia*, and many others.

1.3.1.1 Vascular Plant Species

Bolivia occupies the tenth place in the world and the sixth in South America in plant richness. The number of native vascular plant species recorded in Bolivia is 12,165 (Jorgensen et al., 2014) from an estimated total of 20,000 and it is represented mostly by angiosperms (75%) followed by ferns and allies (11%). The Bolivian tree species is a palm – *Parajubaea torallyi* – which is an endemic species and is mainly distributed in dry inter Andean valleys of the center of the country.

Some genera of endemic plants are: *Tacoanihus* (Acanthaceae), *Corollonema, Dactyloslema, Fontelleae, Sleleoslemna* (Asclepiadaceae), *Polyclila* and *Rusbya* (Ericaceae), *Vasqueziella* (Orquidaceae), *Boelckea* (Brassicaceae), *Gerritea* (Poaceae), *Izozogia* (Zigophyllaceae) and *Sulcorebulia* (Cactaceae). A noticeable growing botanical research has contributed to the knowledge of tree species of Bolivia which was formerly inventoried to be confirmed by 120 plant families, 685 genera and 2,709 species (Killeen et al., 1993) and recently under the APGIV system the list includes 128 families, 604 genera and 2934 species of arboreal habit (meaning 24.12% of the Bolivian flora), of which 249 (8.49%) are endemic (Moraes et al., 2017). The Fabaceae have 424 species (representing 14.45% of the total) and with Mimosoideae with the highest number of species: 194 (6.61%), Rubiaceae with 154 (5.25%), Melastomataceae with 148 (5.04%) and Lauraceae with 131 (4.46%). There is a use of many species by local human communities based on the harvest for various purposes and categories of use: medicinal, food, construction, among others; for example, timber species are about 305 species (although they are the 35 that have greater commercial pressure) and medicinal over 3,000.

Bolivia is part of the center of origin of the potato (*Solanum*, Solanaceae) and its more than 4,300 varieties; Peanuts (*Arachis hypogaea*, Fabaceae) and other domesticated plants include 16 tubers, grains, fruits and vegetables. Among wild relatives include potatoes and peanuts, but also Sweet Potatoes (*Ipomoea*, Convolvulaceae), Beans (*Phaseolus*, Fabaceae), Cassava (*Manihot*, Euphorbiaceae), Pineapples (*Ananas*, Bromeliaceae), Peppers (*Capsicum*, Solanaceae), Papaya (*Carica*, Caricaceae), Passion fruit (*Passiflora*, Passifloraceae), Tobacco (*Nicotiana*, Solanaceae), Squash (*Cucurbita*, Cucurbitaceae), Cocoa (*Theobroma*, Sterculiaceae) and Vanilla (*Vanilla*, Orchidaceae) (Moraes et al., 2009). The 2013 Agricultural Census registered 34,970,168 hectares (ha), equivalent to 32.4% of the total area (109,858,100 ha), which contribute to the country's food security and sovereignty; arable land comprises 7,837,864 ha and of which 2,763,239 ha are planted in summer (Tejada et al., 2017).

1.3.1.2 Mosses

The diversity of bryophytes catalogs Bolivia among the 15–20 countries with the greatest diversity in the world (Ibisch, 1998). The mosses are represented with 918 species, *Anthoceros* with 4 (although they may rise up to 8–10) and liverworts with 477 (which may be more than 500) according to Churchill et al. (2009, see Table 1.2). According to these authors, in the

Table 1.4   Number of Bolivian Bird Species Within the 76 Families Recorded, a Total of 1,435 Species, Including One Extinct, Seven Globally Threatened, and One Introduced

Bird Family	Number of Species
1. Tinamidae	25
2. Rheidae	2
3. Cracidae	15
4. Odontophoridae	4
5. Anhimidae	2
6. Anatidae	23
7. Podicipedidae	5
8. Phoenicopteridae	3
9. Ciconiidae	3
10. Threskiornithidae	8
11. Ardeidae	16
12. Phalacrocoracidae	1
13. Anhingidae	1
14. Cathartidae	6
15. Falconidae	17
16. Accipitridae	46
17. Cariamidae	2
18. Eurypygidae	1
19. Rallidae	25
20. Heliornithidae	1
21. Psophiidae	3
22. Aramidae	1
23. Recurvirostridae	2
24. Charadriidae	11
25. Jacanidae	1
26. Thinocoridae	3
27. Scolopacidae	22
28. Laridae	9
29. Columbidae	27
30. Psittacidae	50
31. Opisthocomidae	1
32. Cuculidae	16
33. Tytonidae	1

Table 1.4 (Continued)

Bird Family	Number of Species
34. Strigidae	25
35. Steathornitidae	1
36. Nyctibiidae	5
37. Caprimulgidae	21
38. Apodidae	15
39. Trochilidae	83
40. Trogonidae	10
41. Alcedinidae	5
42. Momotidae	4
43. Ramphastidae	21
44. Picidae	36
45. Galbulidae	11
46. Bucconidae	19
47. Pipridae	15
48. Cotingidae	36
49. Pipritidae	1
50. Tyrannidae	201
51. Thamnophilidae	79
52. Conopophagidae	3
53. Rinocryptidae	13
54. Formicariidae	17
55. Furnariidae	101
56. Dendrocolaptidae	25
57. Vireonidae	12
58. Corvidae	6
59. Hirundinidae	19
60. Troglodytidae	14
61. Donacobiidae	1
62. Poliptilidae	4
63. Mimidae	3
64. Turdidae	21
65. Cinclidae	2
66. Passeridae	1 (introduced)
67. Motacillidae	5

Table 1.4 (Continued)

Bird Family	Number of Species
68. Fringillidae	15
69. Parulidae	21
70. Icteridae	28
71. Coerebidae	1
72. Emberizidae	76
73. Thraupidae	85
74. Insertae sedis 4	1
75. Passerelidae	2
76. Cardinalidae	18
Total	**1,435**

(Based on Herzog et al., 2017 and https://es.wikipedia.org/wiki/Anexo:Aves_de_Bolivia)

ecoregion represented by mountain formations of the Yungas (which represent 5% of Bolivia) were registered 1,059 bryophytes (with 77% of the total) and 40% are between 2,000–2,500 m elevation.

1.3.1.3 Lichen

Despite Bolivia's interesting location in the tropical zone and the biological uniqueness of its plant life, lichens of Bolivia remain almost completely unexplored; only 150 species were known from the country by 1998 (Flakus and Rodriguez, 2017). Some further 2,000 species were recorded and described in the following years and today the biota of lichens and lichenicolous fungi in Bolivia should consist in total of 4,000 lichenized and 800 lichenicolous species (http://bio.botany.pl/lichens-bolivia/, http://bio.botany.pl/lichens-bolivia/en,strona,catalog,5.html).

1.3.1.4 Fungi

The fungi of Bolivia are largely unknown, though incredibly diverse. Newman et al. published an online checklist of a total of 717 taxa (including subspecies and varieties) (http://mycoportal.org/portal/checklists/checklist.php?cl=49&pid=7) and could reach to 956; whereas the last accounting of fungal records for the country arrived at 799 species (https://www.inaturalist.org/projects/fungi-of-bolivia). For example, in the National Park Madidi there is a checklist of 121 species (http://mushroomobserver.org/species_list/show_species_list/121).

1.3.1.5 Algae

Freshwater algae species are represented in Bolivia. With emphasis on phytoplankton species, 1,216 genera and 14,718 species were inventoried (Cadima et al., 2005) and could reach to 19,500 (Maldonado, 2005); in the Chromophyta division, the genera *Lepocinclis, Phacus, Trachelomonas* and *Strombomonas* are mostly represented in the lowland alluvial plains and in the Precambrian Shield, while the greater number of species of the genera *Phymatodocis, Euastrum, Micrasterias* and *Xanthidium* (from the Chlorophyta) is found in the Amazonian lowlands, finally among the Cyanophyta, the *Microcystis* species are better adapted in areas with higher precipitation and *Nodularia* spp. in xeric sites.

1.3.2 Animals

1.3.2.1 Mammals

Anderson (1997) published the first inventory of mammals in Bolivia with 320 species, then increased to 356 (Salazar-Bravo and Emmons, 2003) and to present is 422 with 23 endemics and from the total with 116 medium-sized to large mammals (Aguirre et al., 2009). The richness of native mammals is strongly correlated to the geographical variation of Bolivia and the variation of habitats, from alluvial plains to the Andean highlands (Salazar-Bravo and Emmons, 2003). From this total, only one species is considered a native aquatic mammal of freshwater in the Amazon: the Pink Dolphin (*Inia boliviensis*, Cetacea order, Iniidae family), which is endemic to Bolivia. It has a diurnal and nocturnal habit, mostly solitary individuals and also groups of up to 19 individuals that can be comprised by several male individuals grouped to hunt, or females with their offspring.

Among Xenarthra superorder, which includes the armadillos (order Cingulata) with a single family (Dasypodidae) and 11 species among which are the Quirquincho (*Chaetophractus nationi*) in the Altiplano and the Cart Tatu (*Priodontes maximus*) considered vulnerable species. The other order (Pilosa) includes the sloths and anteater with 4 families and 5 species, the most emblematic species is the Giant Anteater (*Myrmecophaga tridactyla*). The Primates order (monkeys) include three families (Cebidae, Pitheciidae and Atelidae), with 25 species; the most peculiar species are the Spider Mon-

Table 1.5 Freshwater Fish Species in Bolivia

Order	Family	Species	Status	Fishbase Name
Perciformes	Cichlidae	*Acaronia nassa*	Native	Serepapa, acará
Characiformes	Acestrorhynchidae	*Acestrorhynchus pantaneiro*	Native	Cachorro, perro
Perciformes	Cichlidae	*Aequidens tetramerus*	Native	Mocotoro, serepapa, acará
Siluriformes	Auchenipetridae	*Ageneiosus inermis*	Native	Seferino, manduvé, boca de sapo, bocón
Siluriformes	Loricariidae	*Ancistrus hoplogenys*	Native	Carancho, vieja
Characiformes	Hemiodontidae	*Anodus elongatus*	Native	Saona, pichi, maduro, salmón
Characiformes	Characidae	*Aphyocharax alburnus*	Native	Picú
Osteoglossiformes	Arapaimidae	*Arapaima cf. gigas*	Introduced, invasive	Paiche
Perciformes	Cichlidae	*Astronotus crassipinnis*	Native	Palometa real, serepapa real, oscar
Siluriformes	Auchenipetridae	*Auchenipterichthys thoracatus*	Native	Torito
Siluriformes	Auchenipetridae	*Auchenipterus nuchalis*	Native	Chupa, bagre barbudo
Siluriformes	Pimelodidae	*Brachyplatystoma filamentosum*	Native	Pirahiba, azulejo, bacalao, blanquillo
Characiformes	Bryconidae	*Brycon cephalus*	Native	Yatorana, yaturana, matrinchán
Perciformes	Cichlidae	*Bujurquina vittata*	Native	Serepapa, mocotoro, acará
Siluriformes	Aspredinidae	*Bunocephalus coracoideus*	Native	Guitarrita, guitarrillo, riquiriqui, banjo catfish (Ingl)
Characiformes	Chilodontidae	*Caenotropus labyrinthicus*	Native	Sardina

Order	Family	Species	Status	Fishbase Name
Siluriformes	Pimelodidae	*Calophysus macropterus*	Native	Blanquillo
Characiformes	Gasteropelecidae	*Carnegiella strigata*	Native	Hacha
Characiformes	Serrasalmidae	*Catoprion mento*	Native	Piraña
Siluriformes	Auchenipetridae	*Centromochlus heckelii*	Native	Avioncito
Perciformes	Cichlidae	*Chaetobranchus flavescens*	Native	Serepapa
Characiformes	Crenuchidae	*Characidium bolivianum*	Native	Mojarra
Characiformes	Characidae	*Charax gibbosus*	Native	Cachorro
Characiformes	Chilodontidae	*Chilodus punctatus*	Native	Sardina
Perciformes	Cichlidae	*Cichla pleiozona*	Native	Tucunaré, yacundá
Characiformes	Serrasalmidae	*Colossoma macropomum*	Native	Pacú, pacú negro, tambaquí
Siluriformes	Callichthyidae	*Corydoras aeneus*	Native	Caracha
Perciformes	Cichlidae	*Crenicichla lugubris*	Native	Pez jabón
Characiformes	Curimatidae	*Curimata vittata*	Native	Llorona, branquiña
Characiformes	Curimatidae	*Curimatella meyeri*	Native	Boguita, llorona, sabalina
Characiformes	Cynodontidae	*Cynodon gibbus*	Native	Cachorro, dientón
Gymnotiformes	Sternopygidae	*Distocyclus conirostris*	Native	Anguila
Gymnotiformes	Sternopygidae	*Eigenmannia virescens*	Native	Anguila, cuchillo
Gymnotiformes	Gymnotidae	*Electrophorus electricus*	Native	Anguila eléctrica
Characiformes	Gasteropelecidae	*Gasteropelecus sternicla*	Native	Hacha
Gymnotiformes	Rhamphichthyidae	*Gymnorhamphichthys rondoni*	Native	Anguila, cuchillo
Gymnotiformes	Gymnotidae	*Gymnotus carapo*	Native	Anguila, cuchillo
Characiformes	Characidae	*Hemigrammus unilineatus*	Native	Lambari

Table 1.5 (Continued)

Order	Family	Species	Status	Fishbase Name
Characiformes	Hemiodontidae	*Hemiodus microlepis*	Native	Salmón
Siluriformes	Pimelodidae	*Hemisorubim platyrhynchos*	Native	Seferino, brazo de moza, estevan, conservero
Perciformes	Cichlidae	*Heros efasciatus*	Native	Serepapa
Characiformes	Erythrinidae	*Hoplerythrinus unitaeniatus*	Native	Yayú, yeyú
Characiformes	Erythrinidae	*Hoplias malabaricus*	Native	Bentón, comunario
Siluriformes	Callichthyidae	*Hoplosternum littorale*	Native	Buchere, simbao, simbado
Characiformes	Cynodontidae	*Hydrolycus scomberoides*	Native	Cachorro, dientón
Characiformes	Characidae	*Hyphessobrycon eques*	Native	Mojarra
Siluriformes	Pimelodidae	*Hypophthalmus marginatus*	Native	Suchi, mapara, vela
Siluriformes	Loricariidae	*Hypoptopoma gulare*	Native	Zapato, carancho, guaiguigué
Characiformes	Iguanodectidae	*Iguanodectes spilurus*	Native	Sardina, sardinita
Characiformes	Anostomidae	*Laemolyta proxima*	Native	Lisa
Siluriformes	Pimelodidae	*Leiarius marmoratus*	Native	Tujuno, bagre pintado, pira
Lepidosireniformes	Lepidosirenidae	*Lepidosiren paradoxa*	Native	Caparuch, pez pulmonado
Characiformes	Anostomidae	*Leporinus friderici*	Native	Lisa, boga, piau
Siluriformes	Doradidae	*Leptodoras acipenserinus*	Native	Tachacá, itagivá
Siluriformes	Callichthyidae	*Megalechis thoracata*	Native	Simabao, simbado
Perciformes	Cichlidae	*Mesonauta festivus*	Native	Serepapa

Order	Family	Species	Status	Fishbase Name
Characiformes	Serrasalmidae	*Metynnis maculatus*	Native	Pacupeba, pacupebita
Characiformes	Characidae	*Moenkhausia intermedia*	Native	Sardina, piky
Characiformes	Serrasalmidae	*Mylossoma duriventre*	Native	Pacupeba, plato, jatara
Characiformes	Lebiasinidae	*Nannostomus trifasciatus*	Native	Pez lápiz
Atheriniformes	Atherinopsidae	*Odontesthes bonariensis*	Introduced, invasive	Pejerrey
Salmon iformes	Salmonidae	*Oncorhynchus mykiss*	Introduced, invasive	Trucha, trucha arcoiris
Cyprinodontiformes	Cyprinodontidae	*Orestias agassii*	Native	Carachi, carachi negro
Siluriformes	Doradidae	*Oxydoras niger*	Native	Tachacá, cahuara, boni, turushucu
Myliobatiformes	Potamotrygonidae	*Paratrygon aiereba*	Native	Raya
Clupeiformes	Pristigasteridae	*Pellona castelnaeana*	Native	Sardinón, apapá
Siluriformes	Pimelodidae	*Phractocephalus hemioliopterus*	Native	General, coronel
Characiformes	Serrasalmidae	*Piaractus brachypomus*	Native	Tambaquí, pacú blanco
Siluriformes	Heptapteridae	*Pimelodella gracilis*	Native	Bagrecito
Siluriformes	Pimelodidae	*Pimelodus blochii*	Native	Chupa, griso, bagre chupa
Siluriformes	Pimelodidae	*Pinirampus pirinampu*	Native	Balnquillo, blanquillo blanco, barba chata
Perciformes	Sciaenidae	*Plagioscion squamosissimus*	Native	Corvina, curvina, curuvina

Table 1.5 (Continued)

Order	Family	Species	Status	Fishbase Name
Siluriformes	Doradidae	*Platydoras armatus*	Native	Itaguá
Characiformes	Characidae	*Poptella compressa*	Native	Sardina, panete
Beloniformes	Belonidae	*Potamorrhaphis eigenmanni*	Native	Pez aguja
Characiformes	Curimatidae	*Potamorhina altamazonica*	Native	Llorona, branquiña, sabalina
Myliobatiformes	Potamotrygonidae	*Potamotrygon motoro*	Native	Raya
Characiformes	Prochilodontidae	*Prochilodus nigricans*	Native	Sábalo
Characiformes	Curimatidae	*Psectrogaster rutiloides*	Native	Llorona, sabalina
Characiformes	Anostomidae	*Pseudanos gracilis*	Native	
Siluriformes	Pimelodidae	*Pseudoplatystoma tigrinum*	Native	Surubí, chuncuina, semicuyo, caparari
Siluriformes	Doradidae	*Pterodoras granulosus*	Native	Tachacá, itagivá
Characiformes	Serrasalmidae	*Pygocentrus nattereri*	Native	Pira roja, piraña colorada, ñata, palometa
Siluriformes	Heptapteridae	*Rhamdia quelen*	Native	Bagre, ñurundiá, suchi
Gymnotiformes	Rhamphichthyidae	*Rhamphichthys marmoratus*	Native	Anguila, cuchillo
Characiformes	Cynodontidae	*Rhaphiodon vulpinus*	Native	Cachorro
Characiformes	Anostomidae	*Rhytiodus microlepis*	Native	Boga
Characiformes	Characidae	*Roeboides affinis*	Native	Cachorro
Characiformes	Cynodontidae	*Roestes molossus*	Native	Cachorro

Order	Family	Species	Status	Fishbase Name
Characiformes	Bryconidae	*Salminus brasiliensis*	Native	Dorado, dorado de escama, salmón
Perciformes	Cichlidae	*Satanoperca jurupari*	Native	Serepapa, mocotoro
Characiformes	Anostomidae	*Schizodon fasciatus*	Native	Lisa, piau, uruchila
Characiformes	Prochilodontidae	*Semaprochilodus insignis*	Introduced, invasive	Sábalo
Characiformes	Serrasalmidae	*Serrasalmus rhombeus*	Native	Piraña blanca, palometa
Siluriformes	Pimelodidae	*Sorubim lima*	Native	Paleta, pico de pato, tawalla
Siluriformes	Pimelodidae	*Sorubimichthys planiceps*	Native	Paleta, pantalón, pez leña
Gymnotiformes	Sternopygidae	*Sternopygus macrurus*	Native	Anguila, cuchillo
Synbranchiformes	Synbranchidae	*Synbranchus marmoratus*	Native	Musum, muqui, anguila
Siluriformes	Auchenipetridae	*Tatia aulopygia*	Native	Torito, novia tijera
Characiformes	Characidae	*Tetragonopterus argenteus*	Native	Sardina, panete
Characiformes	Gasteropelecidae	*Thoracocharax stellatus*	Native	Hacha
Siluriformes	Auchenipetridae	*Trachelyopterus galeatus*	Native	Torito
Siluriformes	Trichomycteridae	*Trichomycterus chaberti*	Native	Pez ciego
Characiformes	Triportheidae	*Triportheus angulatus*	Native	Sardina, pechuga, panete
Siluriformes	Pimelodidae	*Zungaro zungaro*	Native	Chanana, chanana amarillo, yaú. Muturo

(Based on Sarmiento et al., 2014)

key (*Ateles chamek*), the Howler Monkeys (*Alouatta* spp.) and a couple of species of the genus *Callicebus* that are in the VU category.

The most numerous group are rodents (order Rodentia) with 13 families (Erethizontidae, Chinchilidae, Dinomyidae, Caviidae, Dasyproctidae, Cuniculidae, Ctenomyidae, Octodontidae, Abrocomidae, Echimyidae, Myocastoridae, Sciuridae, Cricetidae) and 140 species. They represent about 40% of mammalian fauna. Most are small, but there are larger species such as Pacas (*Agouti* spp.) in the Yungas ecosystems and lowlands or the Capybara of the Chaco and Amazonia that can reach up to 40 kg. Among the rodents is the chinchilla (*Chinchilla brevicaudata*, critically endangered species CR) considered extinct until recent of which some wild populations were recently found. Lagomorpha with one family (Leporidae) and a widely distributed species and very common in the Gran Chaco prairies: the Tapeti (*Sylvilagus brasiliensis*). The second group in richness are bats (order Chiroptera) with eight families (Noctilionidae, Vespertilionidae, Molossidae, Emballonuridae, Natalidae, Mormoopidae, Phyllostomidae and Thyropteridae) and 132 species; the best-known members are the hematophagous Vampire (*Desmodus rotundus*), but they are of fundamental importance as pollinators and invertebrate controllers. In recent years they have gained importance for their role in the transmission of diseases and emerging diseases (Aguirre, 2007) (plates 1.1–1.4).

The Carnivora order is represented by six families (Felidae, Canidae, Ursidae, Procyonidae, Mustelidae, Mephytidae), and 28 species. The most emblematic species are as follows: the Jaguar (*Panthera onca*) widespread in the Chaco and Amazonia, the Andean Cat (*Laopardus jacobita*, a highly endangered species of the high Andean region), the Spectacled Bear (*Tremarctos ornatus*) from the upper Yungas, and the maned wolf (*Chrysocyon brachyurus*) from the lowlands of the Amazon.

Among the largest mammals are the Tapir or Anta (*Tapirus terrestris*), the only species of the family Tapiridae (order Perissodactyla), deers, guanacos, and vicunas, as well as peccaries (order Artiodactyla) with three families (Tayassuidae, Camelidae, and Cervidae) and 12 species. Among the most emblematic species are the Vicunas (*Vicugna vicugna*) of the high Andean region and the guanacos (*Lama guanicoe*) with a very small population in the Chaco region. The Chacoan Peccary or Tagua species (*Catagomus wagneri*) from the Chaco, considered extinct until 1971.

Finally, among marsupials there are the American Opossums of the infraclase Metatheria and order Didelphimorphia that include a family (Didelphidae) and 39 species.

A   The Huaso (*Mazama americana*)   B. A Slot (*Bradypus variegatus*)

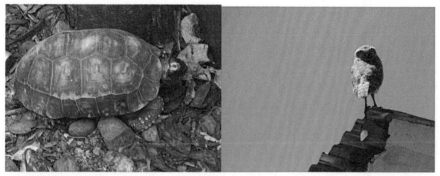

C. A terrestrial turtle (*Geochelone carbonaria*)

D. Among smallest owls (*Athene cunicularia*)

E. Brown Woolly Monkey (*Lagothrix cana*)   F. Amazon Sicurí (*Eunectes murinus*)

Plate 1

Plate 2 (a) *Copernicia alba* palm forest in Chaco vegetation, (b) Tucumanian-Bolivian forest, (c) *Polylepis tomentella* forest, (d) Yungas forest, (e) Amazonian forest, (f) Wetland of Titicaca Lake.

A. The Patujú plant (*Heliconia rostrata*)　　B. Theltapallo (*Cajophora horrida*)

C. The cocoa fruit (*Theobroma cacao*)　　D. *Parajubaea torallyi*, the national Bolivian palm tree.

Plate 3

A. The Taropehyacinth (*Eichhornia azurea*)

B. The Amazonian almond (*Bertholletia excelsa*).

Plate 4

1.3.2.2   Birds

Knowledge about the species richness of Bolivia has increased thanks to the work of several scientists; 1,274 species were known ca. 30 years ago (Remsen and Traylor, 1989) and that list of birds was subsequently permanently updated by the Harmony Association in the country until the recent publication of Herzog et al. (2017). The present avifauna of Bolivia includes approximately 1,435 species of birds (Tables 1.2 and 1.4), of which 24 are endemic (the majority in the dry inter Andean valleys, then in the Yungas, few in the Llanos de Moxos and one Amazonian); 31 are globally threatened with extinction and two were introduced. Therefore, Bolivia is the fifth country with the most bird species in the world and approximately implies 45% of the South American richness.

The national bird of Bolivia is the Andean condor (*Vultur gryphus*). It is mainly a carrion bird, which feeds on carrion and reaches sexual maturity at five- or six-years-old and reaches the longest age of 50 years and nests between 3,000 and 5,000 meters elevation, usually in inaccessible rock formations.

The families with the greatest number of species are the Tyrannidae with 201, Furnariidae with 101, Thraupide with 85 and Trochilidae with 83. The greatest richness is concentrated in the Amazon region and the Yungas. 45% of bird species are restricted to 1–2 ecoregions, especially in mountains closely related to very narrow altitudinal gradients; while in the Chaco, 57% live in more than five regions (Herzog, 2003). For example, the Andean ostriche (*Pterocnemia pennata* which is found at 4,000 m in the southern Puna) and *Fulica cornuta* between 3,000–5,200 m elevation in the center to south Altiplano, in the Tucumanian-Bolivian forests are typical *Cinchlus schulzi* (1,500–2,500 m), *Amazona tucumana* (1,700–2,200 m) and *Penelope dabbenei* (1,800–3,000 m); in the Cerrado and Chiquitanian semideciduos forests is an emblematic parrot: *Anodorhynchus hyacinthinus*, among other cases. Among migratory bird species, there are three species of flamingos and several species of swallows (Scolopacidae).

1.3.2.3 Freshwater Fishes

The richness of fish in Bolivia is primarily due to the fact that they are found in the three watersheds of freshwater that contain distinctive groups of species, although during the floods interconnections between river microbasins (Beni, Mamoré and Parapetí) are generated, maintaining the exclusive ichthyofauna of each basin (Carvajal-Vallejos and Van Damme, 2009; Van Damme et al., 2009). In the endorheic basin of the Altiplano there are 29 species and most of it in the Titicaca Lake (25), dominated by the genus *Orestias* and *Trichomycterus* (of Amazonian origin) among those with native species, as well as two introduced ones; while in the Amazon basin, 802 native and two introduced species are represented; finally, in the Paraguay-Paraná basin with 226 species (Table 1.2, Sarmiento et al., 2014). Findings reported in the NW of Bolivia and in the Amazon basin concentrate contain 62.5% of all known species of the Bolivian Amazon and 48.8% nationwide; so the upper Rio Orthon basin must be considered as a potential hot spot for the species richness of freshwater fishes (Chernoff et al., 2000).

One of the first works that document a Bolivian inventory of native fish is from Terrazas (1970) with 352 species. Bolivian freshwater fishes are in total 908 species and a sample of this richness is addressed by 106 species with three introduced and invasive ones (see Table 1.5, http://www.pecesde-bolivia.com/investigacion), but it is estimated to be a total of 1,400.

As it happens in the Neotropical region, the greater number of species in Bolivia corresponds to both orders Characiformes (354 species) and Siluriformes (336) that represent the 76% of all known species in the country. The families with the greatest number of species are Characidae (Characiformes) with 208 species, Loricariidae (Siluriformes) with 79 species, and Cichlidae (Perciformes) 70 species (Sarmiento et al., 2014).

The Bolivian ichtyofauna includes 12 introduced species among which are the Trout (*Oncorhynchus mykiss*) and the Pejerrey (*Odontesthes bonariensis*) in the Altiplano basin (Lake Titicaca) and mainly Andean lagoons, in addition to an introduced species in the Amazon basin: the Paiche (*Arapaima* sp.), which can reach more than 2 m in length.

There is a predominance of small species with less than 15 cm in length that include widely distributed species such as Tetras or Sardines of the genera *Astyanax, Moenkhausia, Hemigrammus* and *Hyphessobrycon* (Characiformes) belonging to the family Characidae, which are common as aquarium species. Among the Siluriformes several species of reduced size correspond to the group of Cory Catfish (*Corydoras* spp.) and also very frequent in aquariums.

The Amazonian fish fauna is characterized by the presence of emblematic species such as Piranha (*Serrasalmus, Pygocentrus*), known for their ferocity and their habits of attack in large groups. Among other most feared species are the Stingrays (*Potamotrygon motoro, P. tatianae,* and others) that have a sting that produces a very painful sting and difficult to cure; or the famous Candiru (*Vandellia cirrhosa*) with hematophagous habit and which can penetrate the urinary ducts of large vertebrates including man.

In the Amazon and Chaco, among the most abundant species is the Shad (*Prochilodus nigricans* and *P. lineatus*) that are very important for human consumption and are part of commercial fishing. These areas are also characterized by the presence of some large species, which are generally part of commercial fisheries and which are often migratory species, such as large catfishes: Piraiba (*Brachyplatystoma rousseauxii, B filamentosum*), Barred Sorubim (*Pseudoplatystoma fasciatum*) and Spotted Sorubim (*P corruscans*); among the Characiformes stand out the Giant Pacu or Tambaqui (*Colossoma macropomum*) and pirapitinga (*Piaractus brachypomus*).

In the Altiplano, the most common species is the carachi (*Orestias agassii*), which is a very important part of subsistence fishing, mainly on the shores of Lake Titicaca; the basin is also notable for the presence of two catfish species such as the Titicaca Lake (*Trichomycterus rivulatus*).

1.3.2.4 Amphibians and Reptiles

The documentation on the herpetofauna of Bolivia has had an important advance. In the case of amphibians, De la Riva (1990) inventoried 112 species and 251 are currently known (https://amphibiaweb.org); more than 98%

(247 species) are represented in the Anura order with three families with a greater number of species representing more than 70% of the known species: Leptodactylidae: 46 spp.; Strabomantidae: 56 spp.; and Hylidae: 75 spp., whereas only one species of salamander and three caecilians have been recorded (https://amphibiaweb.org). Most of the 41-recorded endemisms are found in the ecoregion of the Yungas. Amphibians represent 43.7% of the total herpetofauna (564 spp.) of Bolivia.

In the case of reptiles, several lists of snakes and saurians were updated from the 80's to the present, so there is an inventory of 313 species from Bolivia, with the Squamata order having the largest number of species with 279 (snakes: 175; lizards: 104) (Uetz et al., 2017). Crocodylia: 6 spp.; Testudines: 14 spp.; Amphisbaenia: 14 spp. Crocodylia and Testudines are characteristic of the alluvial plains of Bolivia, the rest are distributed throughout the country; *Tachymenis* (Colubridae) and *Liolaemus* (Tropiduridae) are typical of Andean zones, while *Bothrops* has species in both the Andes and in the Amazon as well as in dry Andean forests between 250–4,000 m elevation (Gonzáles and Reichle, 2003). Poisonous vipers belong to Viperidae family (17 species) that includes *Bothrops, Lachesis* and *Crotalus*; and also Elapidae (11 spp.), which includes the corals (*Micrurus*): in total 28. Colubridae also has poison glands but they are opistoglifs and are in general, less dangerous. Widely distributed venomous snakes are as follows: the Rattlesnake (*Crotalus durissus*, widely distributed in South America and in Bolivia: in the Amazon and Chaco, up to 1500 m in the Yungas and hilly Chaco montano. *Lachesis muta* is also poisonous but only in the Amazon.

The widest reptile distribution is the Yacare (Caiman yacare), which is found in the Amazonia and Chaco up to 800 m elevation. Among the land turtles, the Red-footed Tortoise (*Chelonoidis carbonarius*) is found in the Amazon and Chaco, in rainforests, dry thorny forests and in savanna areas; whereas aquatic turtles Geoffroy's Side-necked Turtle (*Phrynops geoffroanus*), has a similar distribution.

1.3.2.5 Insects

Among insects of Bolivia, the inventory is still incomplete but there is already an important support on 342 families, 4,334 genera, and 13,719 species. Assuming that this total represents the 75%, it could be estimated that there could be at least 28,000 species of insects (http://insectoid.info/checklist/insecta/bolivia/) (see Table 1.6). The order of Lepidoptera is the best studied and records a total of 3,000 species and it places Bolivia among the four countries in the world with the greatest diversity in the group.

Table 1.6 Number of insect species recorded within different insect orders up to
ca. 17000 and expected existing number (based on several sources)

Order	Published 1993	Published/ unpublished 1993	Registered 2006	Expected
Thysanura	5	5	5	5
Diplura	2	2	2	3
Protura	2	2	2	2
Collembola	280	283	285	300
Ephemeroptera	45	45	48	48
Odonata	44	44	48	48
Plecoptera	35	35	35	35
Orthoptera	28	28	29	31
Dermaptera	3	3	3	3
Blattodea	6	6	10	10
Psocoptera	17	51	53	57
Phthiraptera	41	246	250	525
Hemiptera (Homoptera + Heteroptera)	1059	1143	1221	1320
Thysanoptera	90	96	98	120
Megaloptera	5	5	5	5
Raphidioptera	3	3	3	4
Neuroptera	55	55	57	57
Coleoptera	3293	3375	3495	3800
Strepsiptera	3	6	6	7
Mecoptera	5	5	5	5
Siphonaptera	51	55	60	60
Diptera	2654	3955	4639	6029
Lepidoptera	2092	2092	2194	2400
Trichoptera	189	195	199	200
Hymenoptera	1272	2959	4000	8158
Total	**11279**	**14694**	**16752**	**28,000**

1.4 Species for Which Bolivia Has a Special Responsibility

Based on the issuance of legal tools, definition of strategies and the support
of supreme decrees for the red list books of threatened species of flora and
fauna of Bolivia, governmental authorities have defined special responsibility species (SRS) for which the country has a special responsibility as they

have their main distribution including those which are threatened, vulnerable, rare and also migratory birds. Among Andean species are included the following: *Orestias* spp. fishes, the Titicaca Lake Diver (*Rollandia microptera*) and the Giant Frog (*Telmatobius culeus*) in the Titicaca Lake, camelids of the Altiplano such as Vicunha (*Vigugna vicugna*) and Alpaca (*Vicugna pacos*), as well as an endemic dasypodid of the high plateau, the Andean Quirquincho (*Chaetophractus nationi*), and three migratory flamingos: Parihuana (*Phoenicoparrus andinus*), Small Parina (*Phoenicoparrus jamesi*) and the Chilean (*Phoenicopterus chilensis*) in arid landscapes with various colors lagoons of southwestern Bolivia. In inter Andean valleys, both the chinchilla rodent Vizcacha (*Lagidium viscacia*) and the cathartid Andean Vultur (*Vultur gryphus*). Among invertebrates, the list is as follows: *Cheilonycha auripennis* and *Oxygonia prodiga* (Coleoptera, Cicindelidae) of the Cerrado, the Chiquitano dry forest and Gran Chaco (between 450–600 m) as well as the southwest of the Amazon (pre-Andean Amazonian forest) and Yungas (La Paz and Santa Cruz) between 1000–1700 m, respectively.

Some plant species are also important for their representativeness in the country, such as Andean cushions (*Distichia muscoides* and *D. filamentosus*) related with glaciar wetlands, *Oxychloe andina* (Poaceae) a highly specialized hialine plant in saline deserts of the Andean region, a large bromeliad in inter Andean valleys (*Puya raymondii*), well preserved Amazonian forests with Mahogany trees (*Swietenia macrophylla*), *Couroupita guianensis*, and Amazonian Chestnut (*Bertholletia excelsa*) in terra firme forests, the Asaí Palm (*Euterpe precatoria*) in flooded sites, and *Caryocar brasiliensis* in the Cerrado ecoregion of Santa Cruz and Beni.

1.5 National Laws and International Obligations

Aligned with the Convention of Biological Diversity held in Rio de Janeiro (Brazil), Bolivia approved in 1993 the General Environmental Law that includes the consideration of the environmental conservation of natural heritage (e.g., landscapes, ecosystems, flora, fauna, microorganisms, genetic resources), together with the regulations of protection, prevention and buffering against extraction activities of natural resources. Among these national development sectors of use include civil infrastructure construction, forestry and oil extraction, among others. This law was later complemented with the General Forestry Law of 1996 for the evaluation and monitoring of forests, as well as for the design of management plans for economically important species, mainly of timber plants, as well as adequate strategies for the use and conservation of species and forests.

Bolivia has ratified international environmental agreements in order to contribute to the conservation and sustainable use of biodiversity, like the Convention on Biological Diversity which also works at the level of working groups, related to the rights of peoples, genetic resources, traditional knowledge, global strategies for the conservation of plants, among others. Since 2012, Bolivia has also actively participated in the United Nations' IPBES (Intergovernmental Science-Policy Platform on Biodiversity and Ecosystem Services) to establish a strong link between the scientific community and decision-makers on issues of biodiversity and ecosystem functioning, as well as local knowledge of human communities and sustainable use about topics of modeling and scenarios, regional and global assessments. Based on such an important cultural heritage with 35 indigenous peoples, Bolivia ratified in 2013 the Nagoya Protocol on Access to Genetic Resources and Fair and Equitable Sharing of the Benefits Derived from its Utilization of the Convention on Biological Diversity (https://bolivia.infoleyes.com/norma/6080/protocolo-de-nagoya-sobre-acceso-a-los-recursos-gen%C3%A9ticos-y-participaci%C3%B3n-justa-y-equitativa-en-los-beneficios-que-se-deriven-de-su-utilizaci%C3%B3n-al-convenio-sobre-diversidad-biol%C3%B3gica-onu-protocolo-de-nagoya).

1.6 Biodiversity Threats

The country of Bolivia, at the heart of South America, is a perfect example of where high levels of biodiversity deficient information and high degrees of degradation overlap (Fernandez et al., 2015). The main threats are those that are related to the disturbance and destruction of the last intact and functional forests, especially if they are wet forests; these threats are related to the use of land in these areas, that is, the advance of the agricultural frontier due to the expansion of mechanized agriculture and livestock and colonization (Ibisch and Mérida, 2003).

Among other factors that threatens directly or indirectly the Bolivian biodiversity are as follows: Extraction of parts of plant individuals (food, medicinal use); extraction of (parts of) individuals from plants and animals (food, medicinal, for construction, forestry including use as firewood); disturbance of habitats that leads to a disinhibition of fauna (human presence, vehicular or boat traffic, noise, sawmills, opening of roads, oil wells, pipelines, tourism, etc.); variations in the quality of the habitat of wild species generated by grazing and land use change; fluctuation in the quality of the habitat of wild species generated by changes in the population density of other species alien to the original natural ecosystem (including the introduction of exotic species, captive-bred individuals and pathogens); changes in the community and habitat of native wild species, due to changes in the population density of one or more wild species that interact with them (e.g., local disappearance of key species); changes in the quality of the habitat of wild species due to inorganic and organic contamination (e.g., mercury, pesticides, hydrocarbons); conversion of habitats of wild species into crops and pastures, causing fragmentation and even the disappearance of remaining habitats (including the use of fire to deforest); disappearance of habitats of wild species as a consequence of urbanization; and accelerated displacement or disappearance of habitats of wild species as a consequence of anthropogenic climate changes at the local, regional and/or global level. In Bolivia, 25,000 hectares are deforested every month. According to the Forestry Superintendency's report in 2007, of 300,000 hectares cleared every year, 75% occurs in the Department of Santa Cruz; 20% in Beni and Pando and 5% in Cochabamba and Tarija.

The most degraded ecosystems correspond especially to the northern and southern Puna and the inter-Andean dry forests; in the lowlands, the most degraded area is the zone of agronomic development in the periphery of the city of Santa Cruz (Ibisch and Mérida, 2003). But in the Andes, the prolonged presence of agrocentric cultures and a high population density in an environmentally sensitive environment have caused further degradation; human activity has been concentrated for thousands of years in the wet and semihumid Puna, dry Puna Seca and in the ecoregion of the dry Inter-Andean forests. Therefore, some spots of natural vegetation were almost eliminated. On the other hand, the ecoregions with the highest priority for conservation are the well-conserved forests of the lowlands and the (nor) oriental slopes: Amazonian forests (Southwestern Amazonia) and Yungas, Chiquitano dry Forest, Gran Chaco, hilly Chaco, and Tucuman-Bolivian forests.

Two Red List books of vascular plants have evaluated the conservation status of native species (in both high and lowlands of Bolivia). 607 vascular plant species have shown some degree of threat (among trees, shrubs and some herbaceous plants): 76 are endangered, 70 are vulnerable category and 13 are almost threatened; in the Yungas ecoregion there are 65 threatened species; in the valleys, 41 and in the high Andean 19 (Navarro et al., 2012); in the Amazon, 31, and in the Chaco two; especially in the lowlands there are many threatened species like timber such as Palo Maria (*Calycophyllum spruceanum*, Rubiaceae), the Coloradillo (*Spondias mombin*, Anacardiaceae), among others. As well as a Red List Book of wild relatives (Moraes et al., 2009) with the report of 65 wild relatives of cultivated plants under various categories of threat and therefore the priority for conservation. However, very few conservation strategies and plans were subsequently addressed to reverse the conservation problems of endangered species of flora and fauna. According to Aguirre et al. (2009), the Vertebrates Red List of Bolivia it was evidenced that in total there are 193 threatened and one extinct species: a fish from Lake Titicaca called Humanto (*Orestias cuvieri*); 22 species are critically endangered (CR), 46 endangered (EN), and 125 vulnerable (VU).

1.7 The Nature Index

In Bolivia, 34.5% of all the records fall within a protected area with an IUCN protection status, which indicates that biodiversity observations have been sampled slightly more inside protected areas given that 25% of the total area of the country is under protection (Fernandez et al., 2015). According to these authors, by using ratios in average plants have 10 times less records per species than animals; for animals, the ecoregion with the lowest number of records relative to the number of species was the Prepuna, with an average of two records per species; the larger number of records relative to the number of species were from southwestern Amazon and Gran Chaco with an average of 21.3 and 22.7 records per species, respectively. Whereas for plants, the ecoregions with the lowest number of records relative to species were Titicaca lake, Prepuna and southern Puna, with an average of 1.6, 2.2, and 2.7 records per species, respectively; and the largest number of records relative to species were from Yungas and southwestern Amazon with an average of 10 and 13 records per species, respectively. Finally, from a spatial perspective the ecoregions that were better sampled are: Inter Andean dry Forests (46.1% of the total area sampled), Yungas (43.1%) and Tucuman-Bolivian forests (39.6); but the ecoregions that were least sampled were southern Puna (8.5%), Prepuna (12.3%) and Gran Chaco (15.3%) (Fernandez et al., 2015).

1.8 Protected and Conservation Areas

According to the Environmental Law, protected areas are natural areas with or without human intervention, declared under the protection of the State through legal provisions, with the purpose of protecting and conserving flora and fauna, genetic resources, natural ecosystems, watersheds and areas with values of scientific, aesthetic, historical, economic and social interest. The Bolivian protected areas are categorized into national parks, natural monuments, natural areas of integral management, wildlife reserves, and national sanctuaries as the National System under the General Law on the Environment with 22 sites that cover about 15% (16.7 millions hectares) of the national land mass and two of them are also designated under UNESCO as biosphere reserves (see Table 1.1). The largest protected area is the Madidi national park with ca. 2 millions hectares. However, at the departmental and municipal levels, efforts were also made to protect landscapes and species that are of interest for conservation in the country and that constitute approximately 10 millions hectares (Vides and Reichle, 2003). Finally, to complement the efforts to conserve the natural patrimony, the territories of origin of the indigenous peoples represented in the country are in total 5.5 millions hectares. Therefore, this total conserved areas reach to 25%. However, as a trend for half of them, a 60% of the surface they cover is considered to be of conservation value; several areas are centers of plant diversity with distinctive floristic units, high species richness or endemism centers (Pacheco et al., 1994). Other habitat types, particularly in Andean and Amazonian wetlands, as well as in dry vegetation of inter Andean valleys, and the E-NE side of Bolivia are not yet adequately represented.

A national park, a place that requires strict and permanent protection of natural resources, ecosystems and biogeographical provinces that exist in it, to ensure that they also benefit future generations. *Natural Monuments*, to preserve the outstanding natural features of sites with spectacular landscapes that have geological, physiographic and paleontological deposits, as well as a rich biological diversity. *Wildlife reserves*, to protect, manage and sustainably use wildlife, under official surveillance. *National Sanctuary*, for strict and permanent protection of those sites that harbor endemic species of flora and fauna, threatened or in danger of extinction, a natural community or a unique ecosystem. *Integrated Management Natural Areas*, to make compatible the conservation of biological diversity and the sustainable development of local populations.

Eleven wetlands (including rivers, lakes and others) of Bolivia have been designated as RAMSAR sites. In 2013, the CEPF (Fund for Alliances for Critical Ecosystems) designated the delimitation of the tropical Andes as a biodiversity hotspot according to the concept of Myers et al. (2000); Bolivia is part of all its Andean coverage and includes four protected areas: Madidi, Apolobamba, Pilón Lajas and Amboró (NatureServe & EcoDecision, 2015).

Keywords

- conservation areas
- protected areas

- nature index

References

Aguirre, L. F., (2007). *Historia Natural, Distribución y Conservación de los Murciélagos de Bolivia.* Centro de Ecología y Difusión Simón I. Patiño, Santa Cruz, Bolivia.

Aguirre, L. F., Aguayo, R., Balderrama, J., Cortez, C., & Tarifa, T., (2009). *Libro rojo de la Fauna Silvestre de Vertebrados de Bolivia.* Ministerio de Medio Ambiente y Agua, La Paz, Bolivia.

Ahlfeld, F., (1970). Zur Tektonik des andinen Boliviens. *Geologische Rundschau, 59,* 1124–1140.

Anderson, S., (1997). Mammals of Bolivia: Taxonomy and distribution. *Bulletin of American Museum of Natural History, 231,* 1–652.

Andrade, M., Forno, R., Palenque, E. R., & Zaratti, F., (1998). Estudio preliminar del efectode alturasobre la radiaciónsolar ultravioletaB. *Revista Boliviana de Física, 4,* 14.

Araujo-Murakami, A., Milliken, W. B., Klitgaard, B., Carrion-Cuellar, A. M., Vargas-Lucindo, S., & Parada-Arias, R., (2016). Biomasa y carbono en los bosques amazónicos de tierra firme e inundable (várzea) en el oeste de Pando. *Kempffiana, 12*(1), 3–19.

Argollo, J., (2016). Aspectos geológicos. In: Moraes R., M., Øllgaard, B., Kvist, L. P., Borchsenius, F., & Balslev, H., (eds.). *Botánica Económica de los Andes Centrales.* Universidad Mayor de San Andrés, La Paz, Bolivia (2006), pp. 1–10.

Beck, S. G., & Moraes, R., (1997). M. Llanos de Mojos region, Bolivia. In: Davis, S. D., Heywood, V. H., Herrera-MacBryde, O., Villa-Lobos, J., & Hamilton, A. C., (eds.). *Centers of Plant Diversity: A Guide and Strategy for Their Conservation.* vol. 3. *The Americas.* WWF/UICN, Oxford. pp. 421–425.

Beck, S. G., Killeen, T. J., & García, E., E., (1993). Vegetación de Bolivia. In: Killeen, T. J., García, E., E., & Beck, S. G., (eds.). *Guía de Arboles de Bolivia.* Herbario Nacional de Bolivia – Missouri Botanical Garden, edit. Quipus srl., La Paz, Bolivia. pp. 6–24.

Cadima, F. M. M., Fernández, T. E., & López, Z. L. F., (2005). *Algas de Bolivia con Enfasis en el Fitoplancton, Importancia, Ecología, Aplicaciones y Distribución de Géneros.* Fundación Simón I. Patiño, Cochabamba, Bolivia.

Carvajal-Vallejos, F. M., Bigorne, R., Zeballos Fernández, A. J., Sarmiento, J., Barrera, S., Yunoki, T., et al., (2014). Fish-AMAZBOL: A database on freshwater fishes of the Bolivian Amazon. *Hydrobiologia, 732,* 19–27.

Chernoff, B., Machado-Allison, A., Willink, P., Sarmiento, J., Barrera, S., Menezes, N., & Ortega, H., (2000). Fishes of three Bolivian rivers: Diversity, distribution y conservation. *Interciencia, 25*(6), 273–283.

Churchill, S. P., Sanjinés, A. N. N., & Aldana, M., (2009). C. *Catálogo de Las Briofitas de Bolivia: Diversidad, Distribución y Ecología.* Missouri Botanical Garden – Museo de Historia Natural Noel Kempff Mercado, Santa Cruz, Bolivia.

Dangles, O., Rabatel, A., Kraemer, M., Zeballos, G., Soruco, A., Jacobsen, D., et al., (2017). Ecosystem sentinels for climate change? Evidence of wetland cover changes over the last 30 years in the tropical Andes. *PLoS ONE, 12*(5), e0175814.

De la Riva, I., (1990). Lista preliminar comentada de los anfibios de Bolivia con datos sobre su distribución. *Bulletin Museum Regni Sciences Naturalis Torino, 8,* 261–319.

Emck, P., Moreira-Muñoz, A., & Richter, M., (2006). El clima y sus efectos en la vegetación. In: Moraes R. M., Øllgaard, B., Kvist, L. P., Borchsenius, F., & Balslev, H., (eds.) *Botánica*

Económica de los Andes Centrales. Universidad Mayor de San Andrés, La Paz, Bolivia, pp. 11–36.

Fernández, M., Navarro, L. M., Apaza-Quevedo, A., Gallegos, S. C., Marques, A., Zambrana-Torrelio, C., et al., (2015). Challenges and opportunities for the Bolivian biodiversity observation network. *J. Life on Earth Biodiversity, 16*.

Flakus, A., & Rodriguez, P., (2017). *Lichens and Lichenicolous Fungi of Bolivia*. http://bio.botany.pl/lichens-bolivia/, http://bio.botany.pl/lichens-bolivia/en,strona,catalog,5.html (accessed 25 Jun 2018).

Galli, C., (2017). *Worldwide Mollusc Species Data Base*. [Available: http://www.bagniliggia.it/WMSD/WMSDdownload.htm] (accessed 25 Jun 2018).

Gerold, G., (2003). Formación del suelo en Bolivia. In: Ibisch, P., & Mérida, G., (eds.). *Biodiversidad: La riqueza de Bolivia. Estado de Conocimiento y Conservación*. Fundación Amigos de la Naturaleza, Santa Cruz, Bolivia. pp. 18–31.

Gonzáles, L., & Reichle, S., (2003). Reptiles. In: Ibisch, P., & Mérida, G., (eds.). *Biodiversidad: La Riqueza de Bolivia. Estado de Conocimiento y Conservación*. Fundación Amigos de la Naturaleza, Santa Cruz, Bolivia. pp. 137–141.

Goulding, M., Barthem, R., & Ferreira, E., (2003). *The Smithsonian Atlas of the Amazon*. Smithsonian Books Washington and London, USA and England.

Graf, K., (1981). Palynological investigations of two postglacial peat bogs near the boundary of Bolivia and Perú. *J. Biogeography, 8*, 353–368.

Hanagarth, W., (1993). *Acerca de la Geoecología de las Sabanas del Beni en el Noreste de Bolivia*. Instituto de Ecología, La Paz, Bolivia.

Herzog, S. K., (2003). Aves. In: Ibisch, P., & Mérida, G., (eds.). *Biodiversidad: La riqueza de Bolivia. Estado de Conocimiento y Conservación*. Fundación Amigos de la Naturaleza, Santa Cruz, Bolivia. pp. 141–145.

Herzog, S. K., Terrill, R. S., Jahn, A. E., Remsen, Jr. J. V., Maillard, Z. O., García-Solíz, V. H., et al., (2017). *Aves de Bolivia. Guía de Campo*. La Rosa Editorial, Santa Cruz, Bolivia.

Ibisch, P. L., (1998). Bolivia is a megadiversity country and a developing country. In: Barthlott, W., & Winger, M., (eds.). *Biodiversity: A Challenge for Development Research and Policy*. Springer Verlag, Berlin, Germany. pp. 213–241.

Ibisch, P. L., Beck, S. G., Gerkmann, B., & Carretero, A., (2003). Ecoregiones y ecosistemas. In: Ibisch, P., & Mérida, G., (eds.). *Biodiversidad: La riqueza de Bolivia. Estado de Conocimiento y Conservación*. Fundación Amigos de la Naturaleza, Santa Cruz, Bolivia. pp. 47–88.

Ibisch, P. L., Columba, K., & Reichle, S., (2002). *Plan de Conservación y Desarrollo Sostenible Para el Bosque Seco Chiquitano, Cerrado y Pantanal Boliviano*. Editorial Fundación Amigos de la Naturaleza, Santa Cruz, Bolivia.

Ibisch, P., & Mérida, G., (2003). *Biodiversidad: La Riqueza de Bolivia*. Fundación Amigos de la Naturaleza, Santa Cruz, Bolivia.

Josse, C., Navarro, G., Comer, P., Evans, R., Faber-Lagendoen, D., Fellows, M., et al., (2003). *Ecological Systems of Latin America and the Caribbean: A Working Classification of Terrestrial Systems*. Nature Serve, Arlington, USA.

Jørgensen, P. M., Nee, M. H., & Beck, S. G., (2014). *Catálogo de las Plantas Vasculares de Bolivia*. vol. 2, Missouri Botanical Garden Press, St. Louis, USA.

Killeen, T. J., García, E., & Beck, S. G., (1993). *Guía de Arboles de Bolivia*. Herbario Nacional de Bolivia – Missouri Botanical Garden, La Paz, Bolivia.

Kroll, O., Hershler, R., Albrecht, C., Terrazas, E. M., Apaza, R., Fuentealba, C., Wolff, C., & Wilke, T., (2012). The endemic gastropod fauna of Lake Titicaca: Correlation between molecular evolution and hydrographic history. *Ecology Evolution, 2*(7), 1517–1530.

Langstroth, P. R., & Riding, S., (2011). Biogeography of the Llanos de Moxos: Natural and anthropogenic determinants. *Geographica Helvetica, 66*(3), 183–192.

Lauer, W., & Rafiqpoor, M. D., (1986). Die jungpleistozäne Vergletscherung im Vorland der Apolobamba-Kordillere (Bolivien). *Erdkunde, 40,* 125–145.

Maldonado, M., (2005). Hidroecoregiones *y* ambientes acuáticos. In: Navarro, G., & Maldonado, M., (eds.). *Geografía Ecológica de Bolivia. Vegetación y Ambientes Acuáticos.* Editorial Centro de Ecología *y* Difusión Simón I. Patiño, Santa Cruz, Bolivia. pp. 502–696.

Martinez, C., (1980). *Gèologie de Andes Boliviennes, Structure et Evolution de la Chaine Hercynienne et de la Chaine Andine Dans le Nord de la Cordillère des Andes de Bolivie.* Travaux et Documents de l'ORSTOM, N 119, Paris, France.

MMAyA (Ministerio de Medio Ambiente *y* Agua), (2013). Mapa de bosques de Bolivia. Monitoreo de la Deforestación en la Región Amazónica, Sala de Observación Bolivia, Dirección General de Gestión *y* Desarrollo Forestal, La Paz, Bolivia.

Montes de Oca, I., (2005). *Enciclopedia Geográfica de Bolivia.* Editora Atenea srl., La Paz, Bolivia. http://www.udape.gob.bo/portales_html/portalSIG/atlasUdape1234567/ atlas10_2009/html/La%20 geograf%C3%ADa%20de%20bolivia.pdf.

Moraes, R. M., & Beck, S., (1992). Diversidad florística de Bolivia. In: Marconi, M., (ed.). *Conservación de la Diversidad Biológica en Bolivia.* Centro de Datos para la ConservaciónBolivia/ USAIDBolivia, La Paz, Bolivia. pp. 73–111.

Moraes, R. M., (1999). *Ecología de Palmeras en Valles Interandinos de Bolivia.* Revista Boliviana de Ecología *y* Conservación Ambiental, *5,* 3–12.

Moraes, R. M., (2016). Palmares asociados a los llanos inundados en Bolivia: Ecorregiones de Heath, Moxos, Pantanal *y* Chaco. In: Lasso, C. A., Colonnello, G., & Moraes, R. M., (eds.). *Morichales, Cananguchales y Otros Palmares Inundables de Suramérica, Parte II.* Colombia, Venezuela, Brasil, Perú, Bolivia, Paraguay, Uruguay *y* Argentina. Serie Editorial Recursos Hidrobiológicos *y* Pesqueros Continentales de Colombia. Instituto de Investigación de los Recursos Biológicos Alexander von Humboldt (IAvH), Bogotá, Colombia. pp. 333–344.

Moraes, R. M., Mostacedo, B., & Altamirano, S., (2009). *Libro Rojo de Parientes Silvestres de Cultivos de Bolivia.* Ministerio de Medio Ambiente *y* Agua – Proyecto UNEP/GEF, Plural editores, La Paz.

Moraes, R. M., Zenteno-Ruiz, F. S., & Fuentes, C. A. F., (2017). Árboles de Bolivia: Actualización e implicaciones del conocimiento. *Kempffiana, 13*(1), 1–90.

Myers, N., Mittermeier, R. A., Da Fonseca, C. G., Gustavo, A. B., & Kent, J., (2000). Biodiversity hotspots for conservation priorities. *Nature, 403,* 853–858.

NatureServe & EcoDecision, (2015). *Hotspot de Biodiversidad de los Andes Tropicales.* Critycal Ecosystem Partnership Fund, Washington DC, USA.

Navarro, G. S., Arrázola, M., Atahuachi, N., De la Barra, M., Mercado, W., Ferreira, M., & Moraes, R., (2012). *Libro rojo de la Flora Amenazada de Bolivia. vol. I – Zona andina.* Ministerio de Medio Ambiente *y* Agua – Rumbol srl., La Paz.

Navarro, G., & Ferreira, W., (2009). *Mapa de Vegetación de Bolivia. 1:250. 000.* CD ROM. The Nature Conservancy, Washington DC, USA.

Navarro, G., (2002). Vegetación *y* unidades biogeográficas de Bolivia. In: Navarro, G., & Maldonado, M., (eds.). *Geografía Ecológica de Bolivia. Vegetacióny Ambientes Acuáticas.* Fundación Simón I. Patiño, Cochabamba, Bolivia, 1–500.

Navarro, G., (2011). *Clasificación de la Vegetación de Bolivia.* Fundación Simón I. Patiño, Santa Cruz, Bolivia.

Newman, D., Kaishian, P., Padilla, A. P., & Cuba Pinto, I. *Fungi of Bolivia. Mycology Collections Portal* (http://mycoportal.org/portal/checklists/checklist.php?cl=49&pid=7).

Noblet, C., Lavenu, A., & Marocco, R., (1996). Concept of continuum as opposed to periodic tectonism in the Andes. *Tectonophysics, 255,* 67–78.

Pacheco, L. F., Simonetti, J. A., & Moraes, R. M., (1994). Conservation of Bolivian flora: Representativeness of phytogeographic zones in the national system of protected areas. *Biodiversity and Conservation, 3,* 751–756.

Paz, O. R., Tejada, F., Díaz Cuentas, S., & Arana, I., (2010). *Vulnerabilidad de los Medios de Vida ante Elcambio Climático en Bolivia.* Liga de Defensa del Medio Ambiente/*Agencia Sueca para el Desarrollo Internacional.* La Paz, Bolivia.

PNCC (Programa Nacional de Cambios Climáticos), (2007). *Vulnerabilidad y Adaptación al Cambio Climáticoen Bolivia.* Ministerio de Planificación del Desarrollo – NCAP, La Paz, Bolivia.

Pouilly, M., Beck, S. G., Moraes, R. M., & Ibañez, C., (2004). *Diversidad Biológica en la Llanura de Inundación del Río Mamoré. Importancia ecológica de la Dinámica Fluvial.* Centro de Ecología *y* Difusión Simón I. Patiño, Santa Cruz, Bolivia.

Quevedo, L., & Urioste, J., (2010). El manejo forestal en las tierras bajas de Bolivia. In: *Informe del Estado Ambiental de Bolivia.* Liga de Defensa del Medio Ambiente, La Paz, Bolivia. pp. 323–335.

Rafiqpoor, D., Nowicki, C., Villarpando, R., Jarvis, A., Jones, E. P., Sommer, H., & Ibisch, P. L., (2003). In: Ibisch, P., & Mérida, G., (eds.). *Biodiversidad: La Riqueza de Bolivia. Estado de Conocimiento y Conservación.* Fundación Amigos de la Naturaleza, Santa Cruz, Bolivia. pp. 31–46.

Reis, R. E., Albert, J. S., Di Dario, F., Mincarone, M. M., Petry, P., & Rocha, L. A., (2016). Fish biodiversity and conservation in South America. *J. Fish Biology, 89,* 12–47.

Remsen, Jr. J. V., & Traylor, M. A., (1989). *An Annotated List of the Birds of Bolivia.* Buteo Books, Vermillion, South Dakota, USA.

Ribera, M. O., Libermann, M., Beck, S., & Moraes, R. M., (1996). Vegetación de Bolivia. In: Mihotek, B. K., (ed.). *Comunidades, Territorios Indígenas y Biodiversidad de Bolivia. Memoria Explicativa.* Centro de Investigación *y* Manejo de Recursos Naturales Renovables, Universidad Autónoma Gabriel René Moreno, Santa Cruz, Bolivia. pp. 169–222.

Richter, M., & Moreira-Muñoz, A., (2005). Heterogeneidad climática *y* diversidad de la vegetación en el sur de Ecuador: Un método de fitoindicación. *Revista Peruana de Biología, 12*(2), 217–238.

Rivas, S., (1968). *Geología de la Región Norte del Lago Titicaca.* Servicio Geológico de Bolivia, *2,* La Paz, Bolivia.

Rochat, P., (2000). *Structure et Cinamatique de l'Altiplano Nord Bolivien au Sein des Andes Centrales.* University of Grenoble, Grenoble, France.

Salazar-Bravo, J., & Emmons, L., (2003). Mamíferos. In: Ibisch, P., & Mérida, G., (eds.). *Biodiversidad: La Riqueza de Bolivia. Estado de Conocimiento y Conservación.* Fundación Amigos de la Naturaleza, Santa Cruz, Bolivia. pp. 146–148.

Sarmiento, J., Bigorne, R., Carvajal-Vallejos, F. M., Maldonado, M., Lecia, K. E., & Oberdoff, T., (2014). *Peces de Bolivia/Bolivian Fishes.* IRD-BioFresh (EU), Plural editores, La Paz, Bolivia.

Sarmiento, J., Moraes, R. M., Aguirre, L. F., & Specht, R., (2016). Vertebrados de Espíritu, Llanos de Moxos: Un palmar estacionalmente inundable de Bolivia. In: Lasso, C. A., Colonnello, G., & Moraes, R. M., (eds.). *Morichales, Cananguchales y Otros Palmares Inundables de Suramérica, Parte II.* Colombia, Venezuela, Brasil, Perú, Bolivia, Paraguay, Uruguay y Argentina. Serie Editorial Recursos Hidrobiológicos y Pesqueros Continentales de Colombia. Instituto de Investigación de los Recursos Biológicos Alexander von Humboldt (IAvH), Bogotá, Colombia. pp. 347–372.

Servant, M., (1977). *El Cuadro Estratigráfico del Plio-Cuaternario del Altiplano de los Andes Tropicales de Bolivia.* Revista de Facultad de Ciencias Puras y Naturales, UMSA, La Paz, *1*, 23–29.

SNHN (Servicio Nacional de Hidrografía Naval), (1998). *Hidrografía de Bolivia.* 1ra Edición. La Paz, Bolivia.

Tarifa, T., & Aguirre, L. F., (2009). Mamíferos. In: Aguirre, L. F., Aguayo, R., Balderrama, J., Cortez, C., & Tarifa, T., (eds.). *Libro Rojo de la Fauna Silvestre de Vertebrados de Bolivia.* Ministerio de Medio Ambiente y Agua, La Paz, Bolivia., pp. 419–572.

Tejada, V. E., Arze García, M. E., Moraes, R. M., Bustillos, G. F., Larrazábal, V. O. D. R., Trepp, C. A., et al., (2017). *Food and Nutrition Security. Bolivia, a Country of Incalculable Wealth.* National Germany Academy of Sciences- Leopoldina, IAP- IANAS, Halle, Germany.

Terrazas, U., (1970). *W. Lista de Peces Bolivianos.* Academia Nacional de Ciencias de Bolivia, La Paz, Bolivia.

Uetz, P., Freed, P., & Jirí, H., (2017). *The Reptile Database*, http://www.reptile-database.org, (accessed 25 Jun 2018).

Van Damme, P. A., Carvajal-Vallejos, F., Sarmiento, J., Barrera, S., Osinaga, K., & Miranda-Chumacero, Peces, G., (2009). In: Aguirre, L. F., Aguayo, R., Balderrama, J., Cortez, C., & Tarifa, T., (eds.). *Libro Rojo de la Fauna Silvestre de Vertebrados de Bolivia.* Ministerio de Medio Ambiente y Agua, La Paz, Bolivia. pp. 25–90.

Vides, R., & Reichle, S., (2003). Áreas protegidas departamentales y municipales. In: Ibisch, P., & Mérida, G., (eds.). *Biodiversidad: La Riqueza de Bolivia. Estado de Conocimiento y Conservación.* Fundación Amigos de la Naturaleza, Santa Cruz, Bolivia. pp. 364–379.

Villarroel, D., Munhoz, C. B. R., & Proença, C. E. B., (2006). Campos y sabanas del Cerrado en Bolivia: Delimitación, síntesis terminológica y sus características fisionómicas. *Kempffiana, 12*(1), 47–80.

CHAPTER 2

Biodiversity in Brazil

ADRIANA LUIZA RIBEIRO de OLIVEIRA, ELIEL de JESUS AMARAL, and DANIEL SILVA SANTIAGO

University of Brasília Institute of Biological Sciences – Block E, 1st Floor,
Darcy Ribeiro University Campus – Asa Norte, Brasília-Distrito Federal, 70910-900, Brasília,
E-mail: ribeirosenna1@gmail.com

2.1 Introduction

Brazil is a country with continental size. It is one of the megadiversity countries of the world (Myers et al., 2000; Mittermier et al., 2004) with more than 40,000 species of plants and fungi (Forzza et al., 2010). These plant species together with abiotic features compound six biomes: Amazon forest, Atlantic forest, Caatinga, Cerrado, Pampas, and Pantanal (IBGE, 2004a, b) (Figure. 2.1). The Amazon forest is in northern and central-western Brazil, occupying 49.3% of the territory and extending through to Bolivia, Peru, Ecuador, Colombia, Venezuela, and the Guianas (Kress et al., 1998). The Cerrado is a savannah in the Central area of Brazil occupying 23.9% of the territory, with marginal continuous extensions in north-eastern Paraguay and Bolivia (Ab'Sáber, 1983; Mendonça et al., 2008). It is predominantly a grassland with woody elements and has a diverse mosaic of vegetation that grows on the rocks known as "campos rupestres" (Giulietti and Pirani, 1988). Atlantic forest is a narrow strip from sea level to the eastern highlands of Brazil, becoming broader toward the south occupying 13% of the Brazilian territory (Stehmann et al., 2009), extending marginally into Argentina and Uruguay. Caatinga occurs only in north-eastern Brazil occupying 9.9% of national territory, it is a xerophilous thorny forest of drylands (Andrade-Lima, 1981). Pantanal is a grassland periodically flooded by the rivers Paraná and Paraguay in central-western Brazil, occupying 1.8% of the Brazilian territory, continuing into Bolivia, Paraguay, and Argentina (Pott and Pott, 1997). Pampas is composed of grassland vegetation from southern Brazil. It is occupying 2.1% of the Brazilian territory and it is found also in Argentina, Uruguay, and eastern Paraguay (Boldrini, 2009). Living in these biomes there are more than 4,000 terrestrial vertebrates (MMA, 2016) and 96,660–129,840 invertebrates (Lewinsohn and Prado, 2005). Since the beginning of the colonization in 1500, the flora and fauna of Brazil have been discovered and lost at the same time. The destruction of natural habitats is the biggest

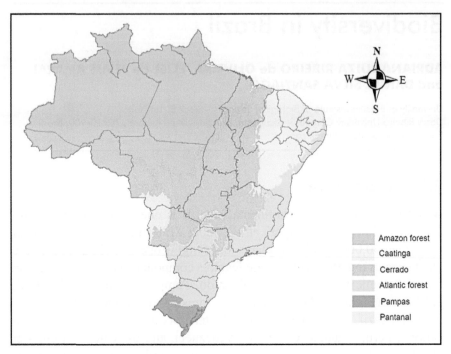

Figure 2.1 Map of the Brazilian biomes.

threat to species. The Atlantic forest has been suffering from continuous deforestation along the centuries. The natural areas of Amazon forest, Cerrado, and Pampas are being replaced by crops or pastures for livestock. In Amazon forest, deforestation to get wood for timber and paper companies also depicts the forests. Even preserved areas in Amazon are empty of life, the hunters continuously have killed the animals that disperse the seeds, leaving no future to these forests (Fernadez, 2005). To prevent widespread misery and catastrophic biodiversity loss, humanity must practice a more environmentally sustainable alternative to business as usual (Ripple et al., 2017).

In this chapter, we will present an overview about the history of the discovery of Brazilian biodiversity. We will depict the biomes (fauna and flora), including numbers of biodiversity, endemism and threatened species of Brazil. Also, we will make a brief comment about the main threats to biodiversity in these biomes and show amazing natural places to visit in Brazil.

2.2 A Brief History of the Brazilian Biodiversity

Since the discovery of Brazil in 1500, until 1800's, there was not much information about flora and fauna of the country. The invasion of Portugal, by the army of Napoleon Bonaparte, brought the royal family and the court in 1808 to Brazil. After the arrival, Dom João began the negotiations for the marriage of his son, Dom Pedro de Alcântara with the Archduchess of Austria, "Dona Carolina Josefa Leopoldina." The Archeduchess was sent to Brazil with the royal entourage that included members of the Royal Academy of Sciences of Austria in a scientific and artistic mission. The king of Bavaria took the occasion to integrate both Martius, from the Academia of Science of Bavaria, as well as Spix, from Munich Zoological Museum to study the Brazilian flora and fauna. Dona Leopoldina and her entourage left Trieste, Italy, on 1817. Soon, the Martius and Spix arrived, they started a trip of three years, from 1817 to 1820, from São Paulo to the Amazon, performing a survey of animals and plants (Henriques, 2008). After they returned to Europe, they started to study the rich material gathered in Brazil. Spix lived only six years after the travel but published studies about monkeys, amphibia, and reptiles (Fittkau, 2001). Martius published, between 1833 and 1906, the first and complete inventory of Brazilian plants, a detailed and comprehensive *Flora brasiliensis* in which 19,958 species of plant, algae, and fungi were reported for Brazil (Martius, 1833; Urban 1906). Nowadays this incredible study is fully digitalized, and it is available for all (http://florabrasiliensis.cria.org.br/). Another incredible result of the expedition of Martius and Spix was the description and a map of the five botanical regions of Brazil, named greek deites (Martius, 1833). Although the limits of these regions have been inaccurate all biomes are represented. Nayades, Hamadryades, Oreades Dryades and Napeias match with Amazon forest, Caatinga, Cerrado, Atlantic forest, and Pampas. The unknown region on the map corresponds to part of Pantanal area. Over the following century, thousands of new species and new distribution records for Brazil were published and a lot of vegetation lost, but no subsequent comprehensive survey of the Brazilian flora was completed until The Brazilian Catalogue of Plants and Fungi (Forzza et al., 2010), continuously updated (Forzza et al., 2012; BFG, 2015). This catalog was the base to build a comprehensive online flora of Brazil (Flora do Brasil, 2020). When it be complete identification keys, descriptions, images of the all known Brazilian species also will be available. The online Brazilian flora will help all people around the world to know the biodiversity of the country as well, it will help the authorities to protect this biodiversity.

2.3   The Biomes

2.3.1   Atlantic Forest

Atlantic forest is a narrow north-south strip from sea level to the eastern highlands of Brazil, becoming broader toward the south, and extending marginally into Argentina and Uruguay (Stehmann et al., 2009). Originally it occupied 1,296,446 km^2 nowadays only 11%–12% of this area remains (IBGE, 2004a,b; Ribeiro et al., 2009; Stehmann et al., 2009). This biome consists mainly of forest phytophysiognomes as dense ombrophilous forest, mixed ombrophilous forest and seasonal semi-deciduous forest (Figure 2.2A–B). We also found, open formations as the shrub and herbaceous vegetation on sandy soils called "restingas" (Figure 2.2C), mangroves and high-altitude grasslands called "campo rupestre and campo de altitude." The singularity of Atlantic forest is due to the drastic geographic and climate variations as the latitudinal range (3–30°S), altitudinal range (sea level 0–2,890 m) and different rainfall regimes with dry and wet climate in the Northeast until humid areas with more than 3,600 mm of annual rain in "Serra do Mar" a mountain range close the sea (Rizzini, 1979; Thomas et al., 1998; Oliveira-Filho and Fontes, 2000; Câmara, 2005; Silva and Casteleti, 2005; Leme and Siqueira-Filho, 2006).

2.3.2   Amazon Forest

The Amazon forest is the largest biome of Brazil, covering an area of 4,196,943 km^2 (IBGE, 2004a,b). The Brazilian Amazon correspond to 60% of total area of the biome, followed by Peru with 13%, Colombia with 10%, and with minor amounts in Venezuela, Ecuador, Bolivia, Guyana, Suriname and French Guiana. The soil of the forest is predominantly composed of quartz and kaolinite (Cáuper et al., 2006). The climate is hot and humid, with an annual average temperature around 25°C, with torrential rains well distributed during the year (IBGE, 2004a,b). The Amazon basin is the largest in the world with 1,100 effluents and covers approximately 6 million km^2. Its main river, the Amazon River (Figure 2.2D–F), flows into the Atlantic Ocean about 175 million liters of water every second. The singularity of Amazon river shapes remarkable phytophysiognomies as the permanently flooded forest (Figure 2.2D) and a seasonally flooded forest that occupies 10% of Amazon. The not flooded "Terra firme" forests (Figure 2.2E) occupy 90% of the region, representing, by far, the most conspicuous area of Tropical moist broadleaf forest. Also, we found some open areas of savannah

Figure 2.2 Biomes of Brazil. A–C. Atlantic forest; A. Ombrophilous forest in the city of Rio de Janeiro; B. *Arauacária* forest remaining in the city of Campos do Jordão; C. Natural lake in the National Park of Restinga de Jurubatiba; D–F. Amazon forest in the state of Amazonas; D. permanently flooded forest; E. "Terra firme" forest; F. Meeting of waters. the place where "Negro" river meets the "Solimões" river to form Amazon river.

called "Campinas," usually composed by grassland, where isolated shrubs stand between 1–5 meters high, with some individuals reaching nine meters (Ferreira, 2009).

2.3.3 Cerrado

The Cerrado is a savannah, with 2 million km², located in the heart of Brazil that extends to Paraguay and Bolivia (IBGE, 2004a,b) (Figure 2.3A–D). The soils are lateric, sandy soils rich and iron and others have metals as

Figure 2.3   Biomes of Brazil. A–D. The Brazilian Cerrado; A–B. Natural fields in the state of Minas Gerais; C–D. Natural places in the state of Goiás.C. Field and forest along the river; D. Veredas; E–F. The Caatinga in the state of Bahia; E. Dunes of São Francisco river. F. "Carnaúbal" a flooded tree formation dominated by *Copernicia prunifera* (Miller) H.E. Moore.

Aluminum and Manganese. The climate is Tropical wet and dry, Aw of Köppen, like the most of Savannas (Richards, 1976). The annual precipitation is around 1,500 mm per year concentrate of October to March with dry winter and rainy summer. In the four dry months, the relative humidity of the air is very low increasing the incidence of fires. The average temperature varies in more hot months to 25–30°C, in margin of forest and 30–35°C, near of margin of deserts and in more cold months to 13–18°C in margin of the forest and 8–18°C near of margin of deserts (Nix, 1983). The well-established seasons have a direct influence on the vegetation, that is highly adapted to fire of dry season (Eiten, 1972). The history of the Cerrado is related to the fire that occurs in this biome since the Pleistocene (Miranda et al., 2002). There are forest, savannah and grasslands formations in the Brazilian Cerrado that compound a vegetational mosaic. Dry forests are not associated with rivers (Figure 2.3C), they have trees of 15–25 m with different grades of deciduous leaves. "Cerradão" is a xeromorphic forest with a sparse arboreal layer, the trees are 8–15 m high and floristic features are like the Cerrado strict sense. The gallery forest has perennial trees associated with small rivers and streams whereas Riparian forest occurs along big rivers (Ribeiro and Walter, 2008). The *cerrado strictu sensu* is a savannah formation with small trees, twisted and scattered shrubs. Many species present an underground structure "xilopodium" to keep water and nutrients. It has the vegetative reproduction ability that helps against fire that is common in the Cerrado. The Cerrado parks are formed by grass vegetation and trees on the small elevations called "murundus." The "palmeirais" and the "veredas"(Figure 2.3D) also are savanna formations, the former exclusively composed by the palm tree *Syagrus oleracea* (Mart.) Becc., whereas "veredas" is dominated by the palm tree *Mauritia flexuosa* L.f., this species is always associated with water bodies. The grasslands formations are called clean and dirty grassland (Figure 2.3 A–B) the former has only or mainly herbaceous species whereas dirty grasslands have shrubs and grass in shallow soil. The "campo rupestre" is a grassland formation with shrubs and herbaceous species on shallow soils and rock outcrops.

2.3.4 *Caatinga*

Caatinga is a biome that occurs only in Brazilian territory mainly in the Northeast region with about 800,000 km² (IBGE 2004a,b). The name "caatinga' means, in indigenous language, white forest. It is due to the appearance of the vegetation in the dry season when the leaves fall, and the white stems of the trees and shrubs remain (Albuquerque and Bandeira, 1995).

The origin of this biome occurred in Cretaceous. Since then, erosion and climatic features built a depression with stone, shallow and acid soils composed of crystalline rocks, gneisses, granites, and xyster. The erosion of the rocks should create a basic soil, but the rain carries the bases and the result is an acid soil. The hydrography is formed by seasonal rivers in the rainy season, when the rain goes the rivers become dry. The São Francisco is the only perennial river that crosses the Caatinga (Figure 2.3E). The climate is extremely with higher solar radiation, average temperature, evapotranspiration, lower humidity, precipitation, and nebulosity (Reis, 1976). The rainy season occurs from November to January in the South region or from February to April in the North region of the Caatinga. The precipitation is about 1000 mm per year, but there are regions with 700 mm or even 500 mm per year (Prado, 2003). The high average temperature is a strong character of Caatinga from 26 to 28°C (Nimer, 1972). There are many phytophysiognomies in Caatinga: forests with big trees of 15–20 m high until rocky outcrops with shrubs, cacti and bromeliads (Leal et al., 2003). There is also residual relief with forests as the inselbergs in the state of Paraíba. A special physiognomy is the"carnaúbal, areas dominated by *Copernicia prunifera* (Mill.) H.E. Moore, a palm tree endemic of Northeast (Giulietti et al., 2003) (Figure 2.3F). These places are temporary flooded and have a rich flora and fauna associated.

2.3.5 Pantanal

The Pantanal is a great extension of the flooded earth with 150, 355 km² in the Central-west region (IBGE, 2004a,b). The climate is moist tropical with rainy summer and dry winter. The high temperature is 34°C and the low is 15°C, with the average of 25°C. The genesis of Pantanal started in the Pleistocene when climate variations change the fluvial systems. The Andes uplift changes the geology. The area received and retained a lot of sediment of tectonic origin. In the subsequent ages the low relief landscape and the high humid as the high number of rivers the uplands around, continuously brought sediment to the region, shaped the Pantanal (Souza and Sousa, 2010). The vegetation that covers the lowlands is heterogeneous. There are flooded and dry fields, swamps, savannah and forest among others. The flora is composed by 50% of broadly distributed species, 30% of species from Cerrado and 20% are of other biomes. The fields are large areas that in the rainy season are flooded, they are formed mainly by grass. The forests are in higher areas, they are much important because are shelter for animals in the rainy season when the fields are flooded. Pantanal has places with a dominant species that give the name to the formation as "Canjiqueiral" a popular name of *Byrsonima cydoniifolia* A. Juss. Many individuals of this species make a homogeneous formation. Similarly, we have Carandazal with the palm tree *Copernicia alba* Morong & Britton and the niceness paratudal with *Tabebuia aurea* (Silva Manso) Benth. & Hook.f. ex S. Moore. It is a big tree, with yellow flowers, that blooms between August and September after the leaves fall. Also, we can find a lot of aquatic and amphibian plant species in the flooded areas that add beauty and richness to the flora.

2.3.6 Pampas

The Pampa is a Steppe that occurs in the south of Brazil, extending through Uruguay and Argentina. It occupies an area of 176,496 km² in the state of Rio Grande do Sul, (IBGE, 2004a,b). The word "pampa" originates from the native indigenous vocabulary "quéchua pampa," which means plain. The climate of the region is subtropical, Cfb of Köppen, with temperatures that varies to 9°C–26°C and well distributed 1,000–2,000 m of annual rainfall (Kuinchtner and Buriol, 2001). Polar fronts bring low temperatures and frost in the winter. The soil is fertile and widely used for agriculture and livestock, the main economic activities in the region. The pampa is an open biome with native fields of herbaceous and shrub species. There are, also, mountain ranges, rocky outcrops and smooth elevations that occur in the plains called "coxilhas."

2.4 Diversity and Endemism

Brazil has 32,629 species of angiosperms (flowering plants) among them 18,421 (57.4%) are endemic (BFG, 2015) (Table 2.1) (Figure 2.4). These plant species compound the biomes together more than 4,000 terrestrial vertebrates (MMA, 2016) (Table 2.2) (Figure 2.5) and 96,660–129,840 invertebrates (Lewinsohn and Prado, 2005).

It's interesting to note that the flowering plants, (>300,000 species) and insects (likely >1 million species) are the most diverse groups in nature.

Table 2.1 Total and Endemic Number of Native and Naturalized Angiosperms and Gymnosperms (BFG, 2015)

Group/species	Total species	Endemic species	Endemism%
Angiosperms	32,629	18,421	57.4
Gymnosperms	30	2	8.7

Figure 2.4 Families of flowering plants with high diversity in Brazil. Species
photographed in the Cerrado; A. Asteraceae. *Trichogonia* sp. B.
Bromeliaceae *Vriesea atropurpurea* Silveira; C. Eriocaulaceae *Eriocaulon
elichrysoides* Bong., D. Fabaceae *Mimosa* sp., E. Poaceae *Ichnanthus
procurrens* (Nees ex Trin.) Swallen, F. Orchidaceae. *Cattleya cinnabarina*
(Bateman ex Lindl.) Van den Berg.

There is evidence that novel defense traits reduce herbivory and that such
evolutionary novelty spurs diversification. Also, the evolution of host-plants
and insects was temporally correlated, in a phylogenetic reconstruction of

Table 2.2 Diversity of Vertebrata and Terrestrial Invertebrate in Brazil (MMA, 2016)

	Class	**Species number**
Terrestrial vertebrates	Amphibious	978
	Reptiles	751
	Birds	1,900
	Mammals	712
Total		4,340
Fishes	Marine	1,380
	Freshwater	3,287
Invertebrates (Lewinsohn & Prado, 2005)		96,669–129,840

ancestral states of characters, corroborating coevolution of these groups (Futuyma and Agrawal, 2009). The Orchidaceae is renowned for its large number of species (>20,000) and its many diverse, even bizarre, specialized pollination systems (Cozzolino and Widmer, 2005; Peakall, 2007, 2010; Scopece et al., 2007, 2008) whereas the Asteraceae are generalists, pollinated by several insect groups such as Coleoptera, Diptera, and Hymenoptera (Noronha and Gottsberger, 1980; Arroyo et al., 1982; Sazima and Machado, 1983; Abbot and Irwin, 1988; Herrera, 1990; Iwata 1990, 1992; Grombone-Guaratini et al., 2004). Although these families have different strategies of reproduction, both are pollinated by insects and reached a great evolutive success emerging among the most diversity families in the world. Also, they figure among the five richest in Brazil (Table 2.3). Asteraceae is the most diverse family in open formations as Cerrado and Pampas whereas Orchidaceae is especially rich in Atlantic forest and Amazon forest (Table 2.4) living mainly on the trees as epiphytes (Figure 2.4).

The Atlantic forest is the richest biome and it has more endemic species. The Cerrado and Amazon forest have almost the same diversity, following Caatinga, Pampas, and Pantanal (Table 2.5). In the Pantanal, the endemism of flora and fauna is rare because the area is a recent lowland without geographic barriers that contribute to the speciation (Alho and Sabino, 2011).

In Atlantic forest, there are 15,001 species of angiosperms, almost half of them are endemic (BFG, 2015). This wealth is greater than that of some continents (12,500 in Europe) (Forzza et al., 2010). Regarding fauna, the surveys already carried out indicate that the Atlantic Forest contains 849 species of birds, 370 species of amphibians, 200 species of reptiles, 270

Figure 2.5   Fauna of Brazil. A–C. Species photographed in the Cerrado – Chapada
dos Veadeiros National Park; A. Blue-and-yellow macaws *Ara ararauna*
(Linnaeus, 1758); B. "Lobo guará" wolf *Chrysocyon brachyurus* (Illiger,
1815); C. "Jararaca" snake *Bothrops moojeni* (Hoge, 1966); D–F.
Species photographed in Amazon forest; D–E. Spider monkey *Ateles
paniscus* (Linnaeus, 1758) and "onça pintada" a jaguar *Panthera onca*
(Linnaeus, 1758) in the zoo of Research Intstitute of Amazon; F. Alligator
Caiman crocodiles crocodilus (Linnaeus, 1758) in the Amazonas river.
A–C photos by Ricardo Couto used with permission.

Table 2.3 Five Most Diverse Families of Angiosperms in Brazil (BFG, 2015)

	Fabaceae	Orchidaceae	Asteraceae	Rubiaceae	Melastomataceae
Species	2,756	2,548	2,013	1,375	1,367
Endemic spp.	1,507	1,636	1,317	726	894

of mammals and about 350 species of fish (Campanili and Wigold, 2010). Moreover, outstanding levels of endemism make the Atlantic Forest one of the most distinctive biogeographic unit in Neotropical region (Prance, 1982). The Cerrado has many open grass fields where predominantly herbaceous families as Asteraceae, Poaceae and Eriocaulaceae reach incredible diversity. Asteraceae has 33 genera and 694 species endemics of this biome (Flora do Brasil, 2020). *Paepalanthus* (Eriocaulaceae) is the third largest genus in Brazil with 338 species, among them 323 are endemic from the Cerrado (BFG, 2015). Amazon forest is a well-defined phytogeographic region dominated by humid tropical rainforest, with high biomass (Pires, 1973; Braga 1979; Daly and Mitchell, 2000). It is difficult to know the richness of plants in the Amazon, but recently, Steege et al. (2013) estimated approximately 16,000 species of Amazonian trees that exceed ten centimeters in diameter at the height of the chest. It represents 30% of all remaining tropical forests in the world and harbors a high biodiversity, with an estimate of 1,300 spe-

Table 2.4 Five Most Diversity Families of Angiosperms in the Brazilian Biomes (BFG, 2015)

Atlantic forest	Cerrado	Amazon forest	Caatinga	Pampa	Pantanal
Orchidaceae (1,574 spp.)	Asteraceae (1,216 spp.)	Fabaceae (1,119 spp.)	Fabaceae (605 spp.)	Asteraceae (299 spp.)	Poaceae (162 spp.)
Fabaceae (964 spp.)	Fabaceae (1,2017 spp.)	Orchidaceae (882 spp.)	Poaceae (282 spp.)	Poaceae (266 spp.)	Fabaceae (153 spp.)
Brome–liaceae (921 spp.)	Orchidaceae (727 spp.)	Rubiaceae (728 spp.)	Asteraceae (284 spp.)	Cyperaceae (141 spp.)	Malvaceae (70 spp.)
Asteraceae (885 spp.)	Poaceae (648 spp.)	Melasto–mataceae (495 spp.)	Euphorbiaceae (232 spp.)	Fabaceae (127 spp.)	Asteraceae (67 spp.)
Poaceae (734 spp.)	Melasto–mataceae (484 spp.)	Poaceae (440 spp.)	Rubiaceae (168 spp.)	Iridaceae/Solanaceae (44 spp.)	Cyperaceae (58 spp.)

Table 2.5 Diversity and Endemism of Angiosperms in Different Brazilian Biomes (BFG 2015)

Biomes/Numbers	Species	Endemic species
Atlantic forest	15,001	7,432
Cerrado	12,097	4,252
Amazon	11,896	1,900
Caatinga	4,657	913
Pampa	1,685	102
Pantanal	1,277	54

cies of birds and 427 species of mammals (Silva et al., 2005; Lewinsohn and Prado, 2005) (Figure 2.5 D–E). The Pampa plays a significant role in the conservation of biodiversity, as it presents a wealth of flora and fauna that are still poorly researched. There are 1,685 angiosperms, with a remarkable diversity of grasses (266 species) and legumes (127 species), in addition to many species of cacti that occur in areas of rocky outcrops (BFG, 2015). The fauna is also diverse, with almost 500 species of birds and more than 100 species of terrestrial mammals.

2.5 Brazil Biodiversity: Threats, Endangered Species, and Protected Areas

Since the discovery of Brazil, the flora was used for commercial purposes. The "Pau-Brasil" (*Paubrasilia echinata* (Lam.) Gagnon, H.C. Lima & G.P. Lewis), a tree of Fabaceae that can reach 30 meters and has a red, precious wood, was very common in the coast. The Portuguese soon began to cut off the trees to commercialize them until almost completely extinction of the species, this commerce was known as the first economic cycle. Nowadays, there are 2113 species in the Official List of Brazilian Flora Species Threatened with Extinction (MMA, 2014), including "Pau Brasil" the species that give the name to the country. The colonial exploitation continued to inland with other economic cycles as sugarcane for sugar production, coffee and gold extraction and brought great loss of diversity. In the following centuries, the anthropogenic impacts became stronger because of the increase in population and consumption. The Atlantic Forest is considered a global Hotspot, one of the richest and most threatened areas on the planet. Originally it occupied 1,296,446 km^2 nowadays only 11%–12% of this area remains (IBGE 2004a,b; Ribeiro et al., 2009; Stehmann et al., 2009). The biome was the first to be exploited by the colonizers and since then it has been suffering a continuous loss of coverage. The agriculture, industrialization and the urban expansion with more than 145 million Brazilians living in its area brought waste and pollution. The Restinga is especially affected, it is suffering rapid habitat destruction because of intense traffic, real estate development, and logging activity along the coast (Rocha et al., 2007; Zamith and Scarano, 2006). Thus, unprotected areas of natural Restinga vegetation are under imminent threat, especially the accessible open shrubby formations found nearest the coast (Hensold et al., 2012). Although this, preserved areas of Atlantic forest can be found even in Rio de Janeiro, a populous state. In the city of Rio de Janeiro you can visit Tijuca National Park to know the Ombrophilous forest. The Serra do Mar mountain range and its high-altitude grasslands can be seen with a quick travel from Rio to Petrópolis or Teresópolis where is located the National Park of the "Serra dos Órgãos." Natural Restinga can be seen in The National "Park of Restinga de Jurubatiba" located in North of state. The inland occupation by the colonizers began three centuries ago but the modern occupation of the Cerrado started in 1970 with the modern agriculture in Brazil. Since then the biome has been losing the coverage due to agriculture expansion, livestock and urbanization but 54% of the original coverage remains (MMA, 2013). Beautiful and natural areas of Cerrado are protected and can be visited in the National Park of the Chapada dos Veadeiros in the state of Goiás. Amazon forest has many natural resources, it suffers a continuous wood exploitation as well with the expansion of livestock and agriculture. Although the natural coverage is 85% (MMA, 2013) and the deforestation has fallen in 2017, an area of 6624 km^2 was devastated in this period (INPE, continuous updated). The cities of Manaus and Belém in the North region are the entrance gates to the Amazon forest. In Manaus, the rivers are the main way to know the forest and see gorgeous flora and fauna. Pantanal is the most preserved biome with 86.7% of original coverage. It has amazing places to visit. In this biome the vegetation in flooded areas as well animals as alligators, macaws and jaguars are easily seen. Pampas is a grassland biome that suffers from the livestock and agriculture as well as urbanization and about 40% of native coverage remains (MMA, 2013). Caatinga suffers from the dry climate being a poverty region in Brazil. It has about 60% of natural coverage. These last biomes have less tourist attraction but also have natural beauty. The Canyons and dunes of São Francisco river are amazing places to know and the National Park of "Serra das Confusões" has natural vegetation and archeological sites. Many species of animals still live in these amazing places and sometimes are seen by local population and tourists.

Keywords

- Brazilian biomes
- Brazilian flora
- endemism

References

Ab'Sáber, A. N., (1983). O domínio dos cerrados: Introdução ao conhecimento. *Revista Servidor Público, 3*, 41–55.

Abbot, R. J., & Irwin, J. A., (1988). Pollinator movements and the polymorphism for outcrossing rate at the ray locus in common groundsel, *Senecio Vulgaris, Heredity, 60*, 295–298.

Albuquerque, S. G., & Bandeira, G. R. L., (1995). Effect of thinning and slashing on forage phytomass from a caatinga of Petrolina, Pernambuco, Brazil. *Pesquisa Agropecuária Brasileira, 30*, 885–891.

Alho, C. J. R., & Sabino, J., (2011). A conservation agenda for the Pantanal's biodiversity. *Braz. J. Biol., 71*(1), 327–335.

Andrade-Lima, D., (1981). The caatinga dominium. *Revista Brasileira de Botânica, 4*, 149–153.

Arroyo, M. T. K., Primack, R., & Armesto, J., (1982). Community studies in pollination ecology in the high temperate Andes of Central Chile, I, Pollination mechanisms and altitudinal variation. *Amer. J. Bot., 69*, 82–97.

BFG (Brazilian Flora group), (2015). Growing knowledge: An overview of seed plant diversity in Brazil. *Rodriguésia, 66*(4), 1085–1113. doi: 10. 1590/2175–7860201566411.

Boldrini, I. I., (2009). A flora dos campos do Rio Grande do Sul. In: Pillar, V. D. P., Muller, S. C., Castilhos, Z. M. S., & Jacques, A. V. A., (eds.). *Campos Sulinos: Conservação e Uso Sustentável da Biodiversidade.* Ministério do Meio Ambiente, pp. 63–77.

Braga, P. I. S., (1979). Subdivisão fitogeográfica, tipos de vegetação, conservação e inventário florístico da floresta Amazônica. *Supl. Acta. Amazonica, 9*, 53–80.

Câmara, I. G., (2005). Breve história da conservação da Mata Atlântica. In: Galindo-Leal, C., & Câmara, I. G., (eds.). *Mata Atlântica: Biodiversidade, Ameaças e Perspectivas.* Fundação SOS Mata Atlântica/Conservação Internacional, São Paulo/Belo Horizonte, pp. 31–42.

Campanili, M., & Wigold, B. S., (2010). *Mata Atlântica: Manual de Adequação Ambiental.– Brasília: MMA/SBF.* Available in: http://www.mma.gov.br/estruturas/202/_arquivos/adequao_ambiental_publicao_web_202. pdf.

Cáuper, G. C., Cáuper, F. R. M., & Brito, L. L., (2006). *Biodiversidade Amazônica.* Manaus. Amazonas. Centro Cultural dos Povos da Amazônia – CCPA.

Cozzolino, S., & Widmer, A., (2005). Orchid diversity: An evolutionary consequence of deception? *Trends in Ecology and Evolution, 20*, 487–494.

Daly, D. C., & Mitchell, J. D., (2000). Lowland vegetation of tropical South America – An overview. In: Lentz, D. L., (ed.). *Imperfect Balance: Landscape Transformations in the PreColumbian Americas.* Columbia University Press, New York, pp. 392–453.

Eiten, G., (1972). The cerrado vegetation of Brazil. *The Botanical Review, 38*(2), 201–341.

Fernandez, F., (2005). Aprendendo a lição de Chaco Canyon: Do "Desenvolvimento Sustentável" a uma Vida Sustentável. *Reflexão, 6*(15), 1–19.

Ferreira, C. A. C., (2009). *Análise Comparativa de Vegetação Lenhosa do Ecossistema de Campina na Amazônia Brasileira.* Tese (Doutorado em Biologia Tropical e Recursos Naturais) – Convênio INPA e UFAM, Manaus, 277.

Fittkau, E. J., (2001). *Johann Baptist Ritter Von Spix: Primeiro Zoólogo de Munique e Pesquisador no Brasil.* História, Ciências, Saúde-Manguinhos, *8*, 1109–1135. Available in: https://dx.doi.org/10.1590/S0104-59702001000500017.

Flora do Brasil, (2020). *Em construção.* Jardim Botânico do Rio de Janeiro. Available in: http://floradobrasil.jbrj.gov.br.

Forzza, R., Baumgratz, J. F. A., Bicudo, C. E. M., et al., (2010). *Catálogo de Plantas e Fungos do Brasil,* vol. 1. Rio de Janeiro, Andrea Jakobsson Estúdio: Instituto de Pesquisas Jardim Botânico do Rio de Janeiro.

Forzza, R., Baumgratz, J. F. A., Bicudo, C. E. M., et al., (2012). New Brazilian floristic list highlights conservation challenges. *Bioscience, 62*(1), 39–45.

Futuyma, D. J., & Agrawal, A. A., (2009). Macroevolution and the biological diversity of plants and herbivores. *Proc. Nation. Acad. Sci., 106*(43), 18054–18061.

Giulietti, A. M., & Pirani, J. R., (1988). Patterns of geographic distribution of some plant species from the Espinhaço range, Minas Gerais and Bahia. In: Heyer, W. R., & Vanzolini, P. E., (eds.). *Proceedings of a Workshop on Neotropical Distribution Patterns.* Academia Brasileira de Ciências. pp. 39–69.

Giulietti, A. M., Neta, A. L. B., Castro, A. A. J. F., et al., (2003). Diagnóstico da vegetação nativa do bioma Caatinga. In: Silva, J. M. C., Tabarelli, M., Fonseca, M. T., & Lins, L. V., (eds.). *Biodiversidade da Caatinga: Areas e Ações Prioritárias Para a Conservação.* Brasília, DF: Ministério do Meio Ambiente, Universidade Federal de Pernambuco.

Grombone-Guaratini, M. T., Solferini, V. N., & Semir, J., (2004). Reproductive biology in species of *Bidens* L. (Asteraceae). *Sci. Agric. (Piracicaba, Braz.), 61*(2), 185–189.

Henriques, R. P. B., (2008). A viagem que revelou a biodiversidade. *Ciência Hoje, 42*(252), 24–29.

Hensold, N., Oliveira, A. L. R., & Giulietti, A. M., (2012). *Syngonanthus restingensis* (Eriocaulaceae): A remarkable new species endemic to Brasilian coastal shrublands. *Phytotaxa, 40*, 1–11.

Herrera, C. M., (1990). Daily patterns of pollinator activity, differential pollinating effectiveness, floral resource availability, in a summer-flowering Mediterranean shrub. *Oikos, 58*, 277–288.

IBGE – Instituto Brasileiro de Geografia e Estatística. (IBGE), (2004). Mapa de biomas. IBGE. (2 November 2011, ftp://geoftp.ibge. gov. br/mapas/tematicos/.

IBGE – Instituto Brasileiro de Geografia e Estatística., (2004a). *Mapa da Cobertura Vegetal.* Available in: http://www.mma.gov.br/estruturas/sbf_chm_rbbio/_arquivos/mapas_cobertura_vegetal.pdf.

IBGE, (2004b). *Mapa de Biomas do Brasil, Primeira Aproximação.* Rio de Janeiro: IBGE. Available in www.ibge.gov.br.

Instituto Nacional de Pesquisas espaciais contínuos updated INPE http://www.obt.inpe.br/prodes/dashboard/prodes-rates.html.

Iwata, M., (1990). Studies on the insects community of flowers: I. The interspecific relations among flower visiting insects in Iriomote Island. *Proceedings of the Faculty of Agriculture of Kyushu Tocai University, 9*, 17–24.

Iwata, M., (1992). Studies on the insects community of flowers: II. The interspecific relations among flower visiting insects in Iriomote Island. *Proceedings of the Faculty of Agriculture of Kyushu Tocai University, 11*, 99–107.

Kress, W. J., Heyer, W. R., Acevedo, P., et al., (1998). Amazonian biodiversity: Assessing conservation priorities with taxonomic data. *Biodiversity and Conservation, 7*, 1577–1587.

Kuinchtner, A., & Buriol, G. A., (2001). Clima do estado do Rio Grande do Sul segundo a classificação climática de Köppen e Thornthwaite. *Disclinarum Scientia Serie Ciencias Exatas, 2*(1), 171–182.

Leal, I. R., Tabarelli, M., & Silva, J. M. C., (2003). *Ecologia e Conservação da Caatinga.* Recife: ed. Universitária da UFPE.

Leme, E. M. C., & Siqueira-Filho, J. A., (2006). A Mata Atlântica – aspectos gerais. In: Siqueira-Filho, J. A., & Leme, E. M. C., (eds.). *Fragmentos de Mata Atlântica do Nordeste – Biodiversidade, Conservação e Suas Bromélias.* Andréa Jakobsson Estúdio, Rio de Janeiro., pp. 47–79.

Lewinsohn, T. L., & Prado, P. I., (2005). Quantas espécies há no Brasil? *Megadiversidade, 1*, 36–42.

Martius, C. P. F., (1833). *Pars Prior: Algæ, Lichenes, Hepaticæ.* Flora Brasiliensis 1. Cottae.

Mendonça, R. C., Felfili, J. M., Walter, B. M. T., Silva Júnior, M. C., Rezende, A. V., Filgueiras, T. S., et al., (2008). Flora vascular do bioma Cerrado: Checklist.com 12. 356 espécies. In: Sano, S. M., & Ribeiro, J. F., (eds). *Cerrado: Ecologia e Flora,* vol. 2. Embrapa Cerrados, Embrapa Informação tecnológica, pp. 421–1279.

Miranda, H. S., Bustamante, M. M. C., & Miranda, A. C., (2002). The Fire Factor. In: Oliveira, P. S., & Marquis, R. J., (eds.). *The Cerrados of Brazil: Ecology and Natural History of a Neotropical Savanna.* New York, Columbia University Press. pp. 51–68.

Mittermier, R. A., Robles, G. P., Hoffmann, M., Pilgrim, J., Brooks, T., Mittermeier, C. G., et al., (2004). *Hotspots Revisited: Earth's Biologically Richest and Most Endangered Terrestrial Ecoregions.* CEMEX/Agrupación Sierra Madre, Mexico City.

MMA, (2013). *Projeto TERRACLASS Cerrado Mapeamento do Uso e Cobertura Vegetal do Cerrado.* Available in: http://www.dpi.inpe.br/tccerrado/TCCerrado_2013.pdf.

MMA, (2014). *Lista Oficial Das Espécies Brasileiras Ameaçadas de Extinção.* Available in: http://pesquisa.in.gov.br/imprensa/jsp/visualiza/index. jsp?data=18/12/2014&jornal=1 &pagina=111&totalArquivos=144.

MMA, (2016). 5° relatório nacional para a Convenção Sobre Diversidade Biológica/Ministério do Meio Ambiente. Secretaria de Biodiversidade e Florestas, Coordenador Carlos Alberto de Mattos Scaramuzza. Brasília: MMA.

MMA-Ministério do Meio Ambiente, *Mata Atlântica manual de Adequação Ambiental.*

Myers, N., Mittermeier, R. A., Mittermeier, C. G., Fonseca, G. A. B., & Kent, J., (2000). Biodiversity hotspots for conservation priorities. *Nature, 403*, 853–858.

Nimer, E., (1972). Climatologia da região Nordeste do Brasil. Introdução à climatologia dinâmica. *Revista Brasileira de Geografia, 34*, 3–51.

Nix, H. A., (1983). Climate of tropical savannas. In: Bourliére, F., (ed.). *Ecosystems of the World 13: Tropical Savannas.* Amsterdan, Oxford, New York, Elsevier Scientific Publishing Company, pp. 37–62.

Noronha, M. R., & Gottsberger, G., (1980). A polinização de *Aspilia floribunda* (Asteraceae) e *Cochlospermum regium* (Cochlospermaceae) e a relação das abelhas visitantes com outras plantas do cerrado de Botucatu, Estado de São Paulo. *Revista Brasileira de Botânica, 3*, 67–77.

Oliveira-Filho, A. T., & Fontes, M. A. L., (2000). Patterns of floristic differentiation among Atlantic forests in Southeastern Brazil and the influence of climate. *Biotropica, 32*, 793–810.

Peakall, R., (2007). Speciation in the Orchidaceae: Confronting the challenges. *Molecular Ecology, 16*, 2834–2837.

Peakall, R., Ebert, D., Poldy, J., et al., (2010). Pollinator specificity, floral odor chemistry and the phylogeny of Australian sexually deceptive *Chiloglottis* orchids: Implications for pollinator-driven speciation. *New Phytologist, 188*, 437–450.

Pires, J. M., (1973). Tipos de Vegetação da Amazônia, In: *O Museu Goeldi no ano do Sesquicentenário*. Belém, MPEG (Publ. Avulsas, 20), pp. 179–202.

Pott, A., & Pott, V., (1994). *Plantas do Pantanal. Brasília*, Embrapa. 320 pp.

Prado, D. E., (2003). As caatingas da América do Sul. In: Leal, I. R., Tabarelli, M., & Silva, J. M. C., (eds.). *Ecologia e Conservação da Caatinga. –* Recife: ed. Universitária da UFPE, Cap. 1, pp. 3–74. Available in: http://www.mma.gov.br/estruturas/203/_arquivos/5_livro_ecologia_e_conservao_da_caatinga_203. pdf.

Prance, G. T., (1982). *Biological Diversification in the Tropics*. Columbia Univ. Press, New York.

Reis, A. C., (1976). Clima da caatinga. *Anais da Academia Brasileira de Ciências, 48*, 325–335.

Ribeiro, J. F., & Walter, B. M. T., (2008). As principais fitofisionomias do Bioma Cerrado. In: Sano, S. P., Almeida, S. M., & Ribeiro, J. F., (eds.). *Cerrado: Ecologia e Flora Embrapa Cerrados, Planaltina.*, pp. 151–212.

Ribeiro, M. C., Metzger, J. P., Martensen, A. C., Ponzoni, F. J., & Hirota, M. M., (2009). The Brazilian Atlantic Forest: How much is left, and how is the remaining forest distributed? Implications for conservation. *Biological Conservation, 142*, 1144–1156.

Richards, P. W., (1976). *The Tropical Rain Forest: An Ecological Study*, 2nd ed. pp 600.

Ripple, W. J., Wolf, C., Newsome, T. M., et al., (2017). World Scientists' Warning to Humanity: A Second Notice. *Bioscience, 67*(12), 1026–1028.

Rizzini, C. T., (1979). *Tratado de Fitogeografia do Brasil*, vol. *2. Aspectos ecológicos.* Hucitec/Edusp, São Paulo.

Rocha, C. F. D., Bergallo, H. G., Van Sluys, M., Alves, M. A. S., & Jamel, C. E., (2007). The remnants of restinga habitats in the Brazilian Atlantic forest of Rio de Janeiro state, Brazil: habitat loss and risk of disappearance. *Brazilian J. Biol., 67*, 263–273.

Sazima, M., & Machado, I. C., (1983). Biologia floral de *Mutisia coccinia* St. Hil. (Asteraceae). *Revista Brasileira de Botânica, 6*, 103–108.

Scopece, G., Musacchio, A., Widmer, A., & Cozzolino, S., (2007). Patterns of reproductive isolation in Mediterranean deceptive orchids. *Evolution, 61*, 2623–42.

Scopece, G., Widmer, A., & Cozzolino, S., (2008). Evolution of postzygotic reproductive isolation in a guild of deceptive orchids. *The American Naturalist, 171*, 315–326.

Silva, J. M. C., & Casteleti, C. H. M., (2005). Estado da biodiversidade da Mata Atlântica brasileira. In: Galindo-Leal, C., & Câmara, I. G., (eds.). *Mata Atlântica: Biodiversidade,*

Ameaças e Perspectivas. Fundação SOS Mata Atlântica/Conservação Internacional, São Paulo/Belo Horizonte, pp. 43–59.

Silva, J. M., Rylands, A. B., & Fonseca, G. A. B., (2005). O destino das áreas de endemismo na Amazônia. *Megadiversidade, 1,* 124–131.

Souza, C. A., & Souza, J. B., (2010). Pantanal mato-grossense: Origem, evolução e as características atuais. *Revista Eletrônica da Associação dos Geógrafos Brasileiros (Seção Três Lagoas*/MS), *11*(7), 34–54.

Stehmann, J. R., Forzza, R. C., Salino, A., et al., (2009). Diversidade Taxonômica na Floresta Atlântica. In: Stehmann, J. R., et al., (eds.). *Plantas da Floresta Atlântica – Rio de Janeiro: Jardim Botânico do Rio de Janeiro, Parte. 1,* pp. 3–12.

Ter Steege, H., Pitman, N. C. A., Sabatier, D., et al., (2013). Hyperdominance in the amazonian tree flora. *Science, 432,* 325–335.

Thomas, W. W., Carvalho, A. M. V., Amorim, A. M., Garisson, J., & Arbeláez, A. L., (1998). Plant endemism in two forests in southern Bahia, Brazil. *Biodiversity and Conservation, 7,* 311–322.

Urban, I., (1906). Index familiarum. In: Von Martius, C. P. F., (ed.). *Pars Prior: Algae, Lichenes, Hepaticae. Flora Brasiliensis* 1. Cottae., pp. 239–268.

Zamith, L. R., & Scarano, F. R., (2006). Restoration of a restinga sandy coastal plain in Brazil: Survival and growth of planted woody species. *Restoration Ecology, 14,* 87–94.

CHAPTER 3

Biodiversity in Canada: An Overview

DAVID A. GALBRAITH

Science Department, Royal Botanical Gardens, 680 Plains Road West, Burlington, Ontario L7T4H4, Canada, E-mail: dgalbraith@rbg.ca

3.1 Introduction

Covering the northern third of the North American continent, Canada encompasses 9,984,670 km² (3,854,085 mi2), the second-largest country on earth if lakes and wetlands are included in its area (Statistics Canada, 2016). Of this area, 8.9% (891,163 km² or 344,080 mi²) is lakes and wetlands (Fisheries and Oceans Canada, 2016a). On the basis of terrestrial area alone, Canada ranks as the third-largest country. It also ranks third if Alaska, Hawai'i and island territories are included in the area of the United States. The human population of Canada was estimated to be 35,749,600 in 2015 (Statistics Canada, 2015)

Canada's northernmost point is Cape Columbia, Ellesmere Island, Nunavut (83° 10' N, 74° 36' W), and its southernmost point is Middle Island, Lake Erie, Ontario (41° 41' N, 82° 41' W). The easterly most point is Cape Spear, Newfoundland (47° 31' N, 52° 37' W), and the most westerly point is the border between the Yukon and Alaska (USA) 2 km west of Windy Peak (60° 18' N, 141° 0' W).

This enormous country thus spans 88 degrees of latitude east-west and 41 degrees of longitude north-south. The geography of Canada is highly varied. A large portion of the country is arctic, but temperate regions span the continent from rainforests in British Columbia in the west to rich deciduous hardwood forests in the east. Politically Canada is organized into ten provinces (west to east: British Columbia, Alberta, Saskatchewan, Manitoba, Ontario, Quebec, Newfoundland and Labrador, New Brunswick, Prince Edward Island, and Nova Scotia) and three territories in the north (west to east: Yukon, Northwest Territory, Nunavut). In this overview, we will consider biodiversity in Canada at the levels of ecosystems and species richness, how this diversity is used, and threats to biological diversity.

3.2 Major Ecozones in Canada

Canada is divided into 18 major terrestrial ecozones (Table 3.1) and 12 maritime ecozones (Table 3.2) by the Canadian Council on Ecological Areas (Figure 3.1). These are further divided into 194 ecoregions of varying size. Ecozones are large areas delineated on the basis of shared geographic and ecological factors, including abiotic and biotic elements. Ecoregions have been delineated as areas within ecozones that share geographic, flora, and fauna characteristics. Neither ecozones nor ecoregions correspond to political jurisdictional boundaries in Canada.

The largest ecozone is the Boreal Shield, spanning the northern portion of Saskatchewan, central Manitoba, the central area of northern Ontario, central Quebec, southern Labrador, and the entire island of Newfoundland. At 1,897,362 km^2, the Boreal Shield itself is larger than the country of Sudan and would be the 16th largest nation on earth, if it was itself a country. The smallest ecozone, the Tundra Cordillera, covers 28,980 km^2 and is larger than the country of Armenia. The Mixedwood Plains of southern Ontario and southern Quebec is home to Canada's greatest human population and largest developed areas, and is larger by area than about half of the individual sovereign countries on earth (source for country areas: Central Intelligence Agency, 2013).

The present patterns of species distribution across Canada have been affected by several important factors. First, most of the landmass of Canada has been subject to the effects of ice sheets during the most recent major glacial episode, the Wisconsinan Ice Age, at the end of the Pleistocene. This episode resulted in as much as two kilometers of ice moving over even southerly areas such as Ontario prior to approximately 15,000 years ago. The glaciers destroyed the biotic fabric of the landscape, pushing topsoil and organic material ahead of themselves as they spread. As they retreated they left behind varying patterns of sedimentary material. As the glaciers retreated it is likely that nutrients and freshwater made available by the melt contributed to rich biotic communities along their edges similar to today's tundra.

Not all of the landmass of Canada was covered by the most recent glaciation. In the far northwest, an area remained free of ice. Stretching across northwestern Siberia, Alaska, and the Yukon was an ice-free refuge named Beringia (Beaudoin and Reintjes, 1994). The area of this refuge is today the home of the largest proportion of endemic plant species in Canada. Along the west coast, isolated mountaintops, and to the east, the exposed continental shelf off of Nova Scotia may also have provided refuges (Banfield, 1974).

Table 3.1   Terrestrial Ecozones in Canada*

Ecozone Name	Location	Area (square kilometers)
Boreal Shield	Northern Manitoba, Saskatchewan, Ontario, southern Quebec, Newfoundland	1,897,362
Northern Arctic	Remaining lands of Arctic Archipelago; northeastern Nunavut	1,481,480
Taiga Shield	Eastern Northwest Territories, southern Nunavut, northern Manitoba, Saskatchewan	1,322,786
Southern Arctic	Northern areas of Nunavut, North-West Territories; Northwestern Quebec	957,139
Boreal Plains	Northeastern British Columbia, Northern Alberta, central Saskatchewan, southern Manitoba	779,471
Boreal Cordillera	Northern British Columbia, southern Yukon Territory, southwestern Northwest Territories	557,937
Taiga Plains	Central Northwest Territories; northern Alberta; northeastern British Columbia	554,014
Prairies	Southern Alberta, Saskatchewan, Manitoba	465,990
Montane Cordillera	Central British Columbia, southwest Alberta	437,761
Hudson Plains	Northeast Manitoba, northern Ontario, western Quebec	350,693
Arctic Cordillera	Northeast edges of Baffin, Devon, Ellesmere Islands in Nunavut; Northwestern Labrador	233,618
Taiga Cordillera	Northern Yukon Territory, western Northwest Territories	231,161
Pacific Maritime	Pacific coastal area of British Columbia, southern Yukon Territory	216,942
Mixedwood Plains	Southern Ontario, southern Quebec	116,206
Atlantic Maritime	Eastern New Brunswick, Prince Edward Island, Nova Scotia	110,590
Atlantic Highlands	Eastern Quebec, western New Brunswick	93,017
Semi-Arid Plateaux	Southcentral British Columbia	56,434
Tundra Cordillera	Northern Yukon Territory, northwestern Northwest Territories	28,980
Total Classified Terrestrial Area:		**9,891,581**

*Ecozones have been defined by the Canadian Council on Ecological Areas. Ordered by largest areas to smallest areas.

Areas source: Environment and Climate Change Canada (2017). Used by permission. © Her Majesty the Queen in Right of Canada, as represented by the Minister of the Environment, 2017.

Following the retreat of the glaciers, plant and animal species recolonized the landscape from various refugia, mostly to the south, constituting a temperate American group. However, many species in Canada are also found in Asia, or northern Europe, including large cursorial mammals like Moose (*Alces alces*) and Grey Wolves (*Canis lupis*). It is thought that these holarctic species colonized Canada after the glaciation by following ice or land bridge routes in the far northwest.

Second, the north-south extent of Canada means that varying latitudes receive different amounts of sunlight at different times of the year, and throughout the country seasonality is strong. The three most northerly ecozones in the Arctic Archipelago, the Northern Arctic, Southern Arctic, and the Arctic Cordillera, and northern sections of the mainland lie north of the Boreal Forest or tree line and constitute the Canadian Arctic Tundra. The tundra exhibits ground that is frozen year-round (permafrost), very short growing seasons, low precipitation, and cold, short winters. Despite these harsh conditions, the tundra abounds with life, both within the marine realm and in terrestrial systems in the summertime.

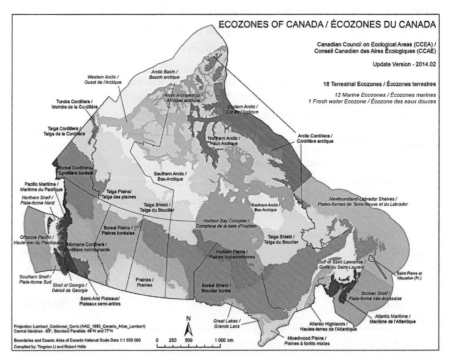

Figure 3.1 Terrestrial ecozones of Canada. *Source:* Canadian Council on Ecological Areas (CCEA). Used by permission.

Table 3.2 Maritime and Aquatic Ecozones in Canada*

Ecozone Name	Location	Area (square kilometers)
Hudson Bay Complex	Hudson Bay, James Bay, and ocean south of Baffin Island	1,244,670
Newfoundland and Labrador Shelves	North Atlantic Ocean north and east of Newfoundland and Labrador	1,041,588
Eastern Arctic	Arctic Ocean east of and between Baffin and Ellesmere Islands	782,636
Arctic Basin	Arctic Ocean northwest of Arctic Archipelago	752,053
Western Arctic	Western Arctic Ocean around Banks, Victoria Islands	539,807
Scotian Shelf	North Atlantic Ocean east of Nova Scotia (includes Bay of Fundy)	416,296
Offshore Pacific	Offshore Pacific Ocean west of British Columbia	315,724
Arctic Archipelago	Arctic Ocean around north and west of Ellesmere Island and smaller islands in Arctic Archipelago	268,792
Gulf of Saint Lawrence	Oceanic gulf between eastern Quebec, Newfoundland, Nova Scotia, Prince Edward Island, and New Brunswick	246,648
Northern Shelf	Pacific Ocean between Haida Gwaii and mainland British Columbia; ocean west of Haida Gwaii	101,663
Great Lakes	Laurentian Great Lakes (freshwater)	88,250
Southern Shelf	Off west coast of Vancouver Island	28,158
Strait of Georgia	Between Vancouver Island and mainland British Columbia	8,969
Total Classified Marine and Aquatic Area:		5,835,254

*Marine ecozones include those areas of the Pacific Arctic, and Atlantic Oceans within the territorial claim of Canada. Ecozones have been defined by the Canadian Council on Ecological Areas. Ordered by largest areas to smallest areas.

Areas source: Environment and Climate Change Canada (2017). Used by permission. © Her Majesty the Queen in Right of Canada, as represented by the Minister of the Environment, 2017.

South of the Canadian Arctic Tundra the Taiga or Boreal Forest stretches across the continent from the Pacific to the Atlantic coasts. The Boreal Forest covers approximately 2.7 million km², covering nearly 30% of Canada's land mass (Natural Resources Canada, 2017). It includes several ecozones: the Taiga Cordillera, Boreal Cordillera, Taiga Plains, Taiga Shield, Hudson Plains, Boreal Shield, Montane Cordillera, and parts of the Boreal Shield. The Boreal Forest is a diverse region biologically, including numerous deciduous and coniferous tree species, large mammalian populations, and 150 species of birds.

The southern portions of Canada are temperate zones, including temperate rainforests in the far west in British Columbia, large grasslands and prairies in Alberta, Saskatchewan, and Manitoba, and temperate mixed forests in southern Ontario, Southern Quebec, and the maritime provinces of New Brunswick, Prince Edward Island, and Nova Scotia. These regions share much in terms of ecology and biodiversity with contiguous areas to the south in the United States. This southern region stretches from Pacific to Atlantic and includes some of the richest areas for species diversity in Canada, as well as the highest concentrations of human settlement and habitat disturbance.

Third, Canada also exhibits considerable variation in relief. The lowest surface point in Canada is generally taken to be sea level around the three ocean coastlines. The highest point, Mount Logan in the Yukon, reaches 5,959 m (19,551 ft) a.s.l. and is the second-highest point in North America. As a result of this geographic relief, biotic regions in Canada also span temperate lowland areas to tundra and glaciers in elevation.

Finally, geographic variation across a major continent provides Canada with a diversity of biomes. Maritime areas on the west and east coast tend to be characterized by moderated temperatures influenced by oceans. In the south of Ontario, the climate is also moderated by the presence of the Laurentian Great Lakes, the largest group of freshwater lakes on earth, containing more than 20% of the world's surface freshwater. The large water mass of these lakes acts as a heat sink, modifying the mid-continental climate, and also serve as sources of precipitation in down-wind areas.

In the west the Rocky Mountains force moist air moving east from the Pacific Ocean and maritime regions up, resulting in a rain shadow effect to the east. This is most pronounced in the South Okanagan region of British Columbia, a semi-arid shrub-steppe sometimes called the Okanagan Desert, in the Semi-Arid Plateaux ecozone. Further east, the southern portions of Alberta, Saskatchewan, and Manitoba of the Prairies Ecozone are temperate grasslands and shrublands.

3.3   Plant and Wildlife Diversity in Canada

As a northern temperate country, Canada's flora is not as rich as that of a tropical region or biodiversity hotspot (Mittermeier et al., 1998), but is nevertheless diverse and fascinating. The flora is comprised of 5,211 species of vascular plants representing 172 families (Table 3.3). Canada's flora shares much with that of the United States to the south, as well as with other circumpolar areas where repopulation was possible following the Wisconsinan glaciation. As noted above, the distribution of species diversity across Canada is related to both present and historical biogeographic factors, including routes of establishment following glaciation, distribution of temperature and energy availability, seasonality, and other biotic and abiotic factors. Among vascular plants, the most diverse family is the Asteraceae, with 597 species or 11.5% of all of the vascular plants of Canada. Although there are 172 families of vascular plants in Canada, just the most speciose ten families account for 53% of all of the species (data from Canadian Endangered Species Conservation Council, 2016). Almost exactly one-third of the vascular plant flora of Canada consists of introduced species (Canadian Endangered Species Conservation Council, 2016).

Among approximately 1,000 tetrapod species of vertebrates found in Canada, birds are by far the largest class, constituting nearly 70% of all species (Table 3.4). The lungless salamanders (Plethodontidae) are the most speciose of the ten families of amphibians, with 10 of the 48 species in Canada. Among the 13 reptilian families, the colubrid snakes (Colubridae) present 21 of the 49 species, or over 40%. The largest among the 73 bird families in Canada is the Anatidae, the ducks, accounting for 59 of 678 species. Thirty-nine different families of mammals are found in Canada. Of these, the 32 species of Cricetidae (Voles, Lemmings, and relatives) account for the largest single proportion, at 14%. This is rather fitting as the largest Order among the mammals of Canada are the rodents, which includes Canada's national mammal and largest rodent, the Beaver (*Castor canadensis*) (species counts from data accompanying Canadian Endangered Species Conservation Council, 2016).

3.4   Access to Biodiversity Information for Canada

Several sources of information on the biodiversity of Canada are available. Canada is a member of the Global Biodiversity Information Facility (GBIF), The Canadian portion of which is termed CBIF (http://www.cbif.gc.ca/) established under the United Nations Convention on Biological Diversity. Canada's own implementation of the biodiversity database facility is named

Table 3.3 Species Diversity of Vascular Plants in Canada

Group		Families	Species
Ferns and Allies	Lycopodiophyta (clubmosses, spikemosses, and quillworts)	3	43
	Monilophyta (ferns and horsetails)	19	136
Gymnosperms	Pinophyta	4	43
Angiosperms	Nymphaeales	2	12
	Magnoliids	5	10
	Monocots	33	1,269
	Eudicots	106	3,698
Total		**172**	**5,211**

Note: Derived from species lists accompanying Canadian Endangered Species Conservation Council (2016).

Canadensys (http://data.canadensys.net/). As of 31 October 2017, the online Canadensys database includes 3,820,574 records spanning thousands of animal and plant taxa; 2,943,309 of these include georeference information.

Table 3.4 Vertebrate Species Diversity in Canada

	Taxon	Families	Species
Fishes	Agnatha (Jawless Fishes)	2	13
	Chondrichthyes (Cartilaginous Fishes)	8	73
	Osteichthyes (Boney Fishes)	37	1,293
	Totals	47	1,379
Tetrapods	Amphibians	10	48
	Reptiles	13	49
	Birds	67	678
	Marine Mammals	13	51
	Terrestrial Mammals	28	171
	Totals	**129**	**997**

Note: Two species of marine mammals are within families that also include terrestrial taxa. The total number of mammalian families in Canada is 39. Derived from species lists accompanying Canadian Endangered Species Conservation Council (2016).

Canada is also a member of the NatureServe network, which includes conservation and biodiversity data sources across Canada and the USA.

Many major biodiversity collections are found at museums and related institutions across Canada. Collections data for the Canadian Museum of Nature, for example, can be accessed online (https://nature.ca/en/research-collections/collections/animals). Increasingly museum and herbarium databases are being made available collectively via Canadensys.

3.5   Biodiversity, Utilization, and Economic Activity in Canada

The economy of Canada benefits from several major industries that make use of wild species and biological resources, in addition to cultivated or domesticated taxa. These industries include forestry, fisheries, and agriculture. In addition, in some cases, extractive industries accessing mineral resources interact with, and threaten, biological diversity as a consequence of mining or fossil fuel extraction. These are noted in the following subsections.

3.5.1   Forestry

The forests of Canada cover 3.47 million km^2, of which 2.70 million km^2 are Boreal Forest. Of the total, approximately 780,000 ha (7,800 km^2 or 0.23%) are logged annually (Natural Resources Canada, 2017a). Forestry in Canada accounts for approximately 1.25% of the country's real gross domestic product, or $19 billion (Natural Resources Canada, 2006). The management of timber cut on public lands (85% of the total harvest) is administered by individual provinces, and generally falls well below levels considered as the maximum for sustainable forest management (National Forestry Database, 2017). Sustainability of forestry practices by private businesses are monitored by three industry initiatives: The Forest Stewardship Council, the Sustainable Forest Initiative, and the Canadian Standards Association Sustainable Forest Management Standard. As of 2014, 1.61 million km^2 of forests in Canada are reported to be managed under sustainable forestry standards by the Forest Products Association of Canada (FPAC, Undated).

3.5.2   Fisheries

Both commercial and recreational fisheries are undertaken in marine areas and in freshwater lakes and rivers across Canada. The annual value of fish landed in Canada is approximately $3.3 billion. Aquaculture production amounts to another $980 million per year (Fisheries and Oceans Canada, 2017). In 2015, for example, marine catches totaled 817,637 tons per year, of which 169,379 metric tons were groundfish (species such as halibut, sole, and flounder), 185,091 metric tons were pelagic fish (such as salmon, herring, anchovy, and tuna) and 448,695 metric tons were shellfish (invertebrates such as lobster, crab, and clams) Approximately 3.5 times as much fish (in mass) are caught on the Atlantic coast of Canada as on the Pacific each year (Fisheries and Oceans Canada, 2017). Canada has established a Sustainable Fisheries Framework to promote the development of sustainable practices in the industry (Fisheries and Oceans Canada, 2016b).

3.5.3   Agriculture

Canada has supported the development of Environmental Farm Plans to encourage land management in accord with the principles of biodiversity conservation. The extensive areas of Canada in use for farming include some areas important for wildlife. Based on 2011 census data, 30% (0.2 million km^2) of agricultural land in Canada is considered also as wildlife habitat, as it includes woodlands and wetlands (mostly in eastern Canada), and also natural lands for pasture (mostly in western Canada) (Jarovek and Grant, 2011).

3.5.4   Extractive Industries

The Mining Association of Canada has established a rating system for Biodiversity Conservation Management, in which each mine is categorized as to three indicators: corporate accountability, conservation planning, and implementation. The association produces an annual report indicating progress on these measures (Mining Association of Canada, 2017). Production of petroleum or other liquid fossil fuels is most intensive in Alberta and Saskatchewan. In northern Alberta, several hundred square kilometers of boreal forest has been affected or destroyed by extraction of bitumen from oil sands.

3.6   Challenges and Major Threats to Biodiversity in Canada

3.6.1   Protected Areas and Biodiversity Conservation in Canada

Protected areas are key to the conservation of biological diversity. According to 2016 data, Canada's Protected Area Network encompasses about

10.5% of targeted lands within ecosystems, and 5% of coastal and marine ecosystems, protected as provincial or national parks (Environment and Climate Change Canada, 2017). Canada has established a target for 2020 of having 17% of each terrestrial ecosystem, and 10% of each coastal and marine area, protected through a network of protected areas and other effective area-based conservation measures, in harmony with the United Nations' Aichi Biodiversity target number 11. Canada expresses this as Target 1 in its Biodiversity Targets for 2020.

To date 18 UNESCO World Biosphere Reserves have been designated in Canada. Biosphere reserves are areas rich in biological diversity and natural beauty in which people also live, with an increasing focus on sustainability. Approximately 2,000,000 Canadians live within UNESCO Biosphere Reserves, roughly 5.6% of the population (Canadian Commission for UNESCO, Undated).

3.6.2 Threats to Species

The pattern of species at risk of extinction in Canada closely follows both the distribution of wildlife species and that of human populations. Large numbers of at-risk plants and animals are found in southern Ontario, southern Quebec, the southern Prairie Provinces and southern British Columbia. These are the areas of greatest habitat fragmentation, change, and loss.

Causes for threats to species within Canada are the same as those that threaten species survival elsewhere: most are linked to habitat loss, degradation and fragmentation, competition with invasive alien species, and overharvesting. As of late 2017, COSEWIC has listed 743 species of plants, invertebrates, and vertebrates as being at-risk (Table 3.4). Since the arrival of Europeans, it is thought that at least 15 species formerly found in Canada have become extinct (Table 3.5). A further 23 species are thought to be extirpated from Canada, species which were formerly found within Canada and while they still survive elsewhere are locally extinct within the country (Tables 3.6 and 3.7).

3.6.3 Legislative Protection and Conservation Measures

Canada has developed a multilevel system for the protection of at-risk populations and species of certain taxa, with legislation for their protection and recovery at both the Federal level and within certain provinces. Not all species within Canada are covered. Vertebrates, vascular plants, mosses, lichens, and specific invertebrate groups such as Lepidopterans and

Mollusks are typically considered for conservation measures, while others such as the majority of Arthropods are not. Federal legislation, the Species at Risk Act (2006), establishes that the goal of conservation efforts is to prevent the extinction of genetically differentiated populations or species within Canada, regardless of whether the species is endemic to Canada (found only within its territory) or whether the species is wider-spread.

Species are added or removed from classification as being at-risk through a species-by-species assessment process. The process, developed by the Committee on the Status of Endangered Wildlife in Canada (COSEWIC), predates the Species at Risk Act (2006) and is applied to plants as well as animals. Some provinces have parallel listing processes, such as Ontario, which has its own at-risk species act and a similar committee called the Committee for the Status of Species at Risk in Ontario (COSSARO).

Table 3.5   Status as of October 2017 of At-Risk Eukaryote Species or Distinct Populations as Determined by COSEWIC in Canada

Taxon	Extinct or Extirpated	Endangered	Threatened	Special Concern	Total
Lichens	0	6	5	10	21
Mosses	2	7	3	5	17
Vascular Plants	3	99	47	46	192
Mollusks	3	19	5	13	40
Arthropods	4	42	8	12	66
Fishes (Marine)	1	26	12	23	62
Fishes (Freshwater)	8	33	27	32	100
Amphibians	2	12	6	9	29
Reptiles	5	19	10	11	45
Birds	5	30	32	22	89
Mammals (Marine)	3	11	6	17	37
Mammals (Terrestrial)	3	16	11	15	45
Total	39	320	172	215	743

Data source: Environment and Climate Change Canada (2017); Data courtesy of Environment and Climate Change Canada.

Table 3.6 Eukaryotic Species Formerly Known From Canada but Considered Extinct by COSEWIC

Taxon	Species	Common Name	Range[1]
Plants (Moss)	*Neomacounia nitida*	Macoun's Shining Moss	ON
Mollusk	*Lottia alveus*	Eelgrass Limpet	ATO
Fish	*Rhinichthys cataractae smithi*	Banff Longnose Dace	AB
	Sander vitreusglaucus	Blue Walleye	ON
	Salmo salar (Lake Ontario subspecies)	Atlantic Salmon	ON, ATO
	Gasterosteus aculeatus (both Limnetic and Benthic taxa)	Hadley Lake Benthic Threespine Stickleback	BC
	Coregonus johannae	Deepwater Cisco	ON
	Coregonuskiyi orientalis	Lake Ontario Kiyi	
Birds	*Pinguinus impennis*	Great Auk	QC, NB, NS, NL
	Camptorhynchus labradorius	Labrador Duck	QC, NB, NS, NL
	Ectopistes migratorius	Passenger Pigeon	SK, MB, ON, QC, NB, PE, NS
Mammals	*Mustela macrodon*	Sea Mink	NB, NS, ATO
	Odobenus rosmarus (Nova Scotia – Newfoundland – Gulf of St Lawrence population)	Atlantic Walrus	QU, NB, PE, NS, NL, ATO
	Rangifer tarandus dawsoni	Caribou (*dawsoni* subspecies)	BC
	Ursus arctos (Ungava population)	Grizzly Bear	QC, NL

Notes: 1 – BC British Columbia; AB Alberta; ON Ontario; ATO Atlantic Ocean; NB New Brunswick; NL Newfoundland and Labrador; NS Nova Scotia; PE Prince Edward Island; QC Quebec; SK Saskatchewan.

Date source: Committee on the Status of Endangered Wildlife in Canada (2017); Data courtesy of Environment and Climate Change Canada.

Table 3.7 Eukaryotic Species Formerly Known From Canada but Considered Extirpated (Extinct Within Canada but Extant Elsewhere) by COSEWIC

Taxon	Species	Common Name	Range[1]	Notes
Moss	*Ptychomitrium incurvum*	Incurved Grizzled Moss	ON	
Vascular Plant	*Desmodium illinoense*	Illinois Tick-trefoil	ON	
Vascular Plant	*Lupinus oreganus*	Oregon Lupine	BC	
Vascular Plant	*Collinsia verna*	Spring Blue-eyed Mary	ON	
Arthropod	*Nicrophorus americanus*	American Burying Beetle	ON, QC	
Arthropod	*Callo phrysirus*	Frosted Elfin	ON	
Arthropod	*Euchloe ausonides insulanus*	Island Marble	BC	
Arthropod	*Lycaeides melissa samuelis*	Karner Blue	ON	
Mollusk	*Alasmidonta heterodon*	Dwarf Wedgemussel	NB	
Mollusk	*Cryptomastix devia*	Puget Oregonian	NC	
Fish	*Erimystax x-punctatus*	Gravel Chub	ON	
Fish	*Polyodon spathula*	Paddlefish	ON	
Amphibian	*Ambystoma tigrinum* (Carolinian population)	Eastern Tiger Salamander	ON	2
Amphibian	*Gyrinophilus porphyriticus* (Carolinian population)	Spring Salamander	ON	2
Reptile	*Terrapene carolina*	Eastern Box Turtle	ON	2
Reptile	*Pituophis catenifer catenifer*	Pacific Gophersnake	BC	

Taxon	Species	Common Name	Range[1]	Notes
Reptile	*Actinemys marmorata*	Pacific Pond Turtle	BC	
Reptile	*Phrynosoma douglasii*	Pygmy Short-horned Lizard	BC	
Reptile	*Crotalus horridus*	Timber Rattlesnake	ON	
Bird	*Tympanuchus cupido*	Greater Prairie-Chicken	AB, SK, MB, ON	
Bird	*Centrocercus urophasianus phaios*	Greater Sage-Grouse *phaios* subspecies	BC	
Mammal	*Mustela nigripes*	Black-footed Ferret	AB, SK	3
Mammal	*Eschrichtius robustus* (Atlantic population)	Grey Whale	ATO	

Notes: 1 – BC – British Columbia; AB – Alberta; ON – Ontario; ATO – Atlantic Ocean; NB – New Brunswick; NL – Newfoundland and Labrador; NS – Nova Scotia; PE – Prince Edward Island; QC – Quebec; SK – Saskatchewan.

2 – Species is listed by COSEWIC as Extirpated but has not been listed by SARA as Extirpated.

3 – *Mustela nigripes* is formally listed as Extirpated in Canada but has been reintroduced to its native range in 2009 through a captive breeding program led by Toronto Zoo (Toronto Zoo, Undated).

Data source: Committee on the Status of Endangered Wildlife in Canada (2017); Data courtesy of Environment and Climate Change Canada.

Every five years, a joint committee consisting of federal and provincial ministers responsible for species at risk, the Canadian Endangered Species Conservation Council, produces an overall report on trends regarding the status of individual species. In the most recent report, dated 2015, the ranks of 29,848 species were assessed in a wide variety of taxonomic groups, including plants, some invertebrate taxa, and vertebrates (Canadian Endangered Species Conservation Council, 2016). In the 2015 assessment, 1,659 species were identified as possibly being at risk in Canada, with over 80% of species considered secure. Fewer than 100 of the species assessed are considered endemic to Canada. These reports and the evolving data they represent are contributing to the COSEWIC listing process by identifying taxa that may need the more detailed assessment of a status report and consideration of the committee. The listing process, criteria, and categories for endangered species in Canada is broadly similar to those developed by the International Union for the Conservation of Nature (IUCN) with some differences.

The main goal of endangered species legislation in Canada is to identify and mitigate risks to distinctive populations or species through a recovery process. Species, which are suspected as being at risk, are assessed through the preparation of a detailed status report and evaluation by COSEWIC which functions as a peer-review panel. Following recommendations by the committee actual listing is decided by the Minister of Environment and Climate Change. Once listed as at-risk, the Species at Risk Act (2006) calls for the formation of a Recovery Team, which prepares recommendations for steps to ameliorate identified risks for each taxon. Subsequent reassessment of status may recommend changes or retention of the previous status.

3.7 Concluding Remarks

Although a north-temperate to arctic country, Canada's biological diversity is complex and varied. The majority of wild species are at this time secure, but as the effects of climate change develop over the coming decades and centuries, even familiar, common species may become precarious as habitat envelopes change. Over 80% of Canada's human population now lives in cities or urban areas, and many are disconnected from natural spaces and exposure to native species. Both the public awareness that leads to the conservation of biological diversity and the well-being of Canada's human population may well depend on reconnecting the majority of the population with the country's rich natural heritage.

Acknowledgments

Thanks to the Canadian Council on Ecological Areas for permission to reproduce the ecozones map in Figure 3.1. Thanks to Corey Burt for assistance in organizing information on the flora of Canada. Thanks to Bruce Bennett, Nadia Cavallin, Peter Kelly, Laurel McIvor, Patrick Moldowan, Tyler Smith, and Peter Sokoloff for allowing reproduction of their photographs.

Keywords

- challenges
- threats

References

Beaudoin, A. B., & Reintjes, F. D., (1994). *Late Quaternary Studies in Beringia and Beyond, 1950–1993: An Annotated Bibliography.* Provincial Museum of Manitoba. Archaeological Survey Occaisional Paper No. 35, online document: URL: http://www.culture-tourism.alberta.ca/documents/Occasional35-QuaternaryStudiesBeringia–1994. pdf. Retrieved 2017–12–02.

Canadian Commission for UNESCO, Biosphere Reserves in Canada. Undated. Online Document. URL: http://unesco.ca/home-accueil/biosphere%20new/biosphere%20reserves%20in%20canada-%20reserves%20de%20la%20biosphere%20au%20canada. Retrieved 2017–12–02.

Canadian Council on Ecological Areas, ecozones introduction, (2014). On-Line Document: URL: http://www.ccea.org/ecozones-introduction/. Retrieved 2017–12–02.

Canadian Endangered Species Conservation Council, (2016), *Wild Species 2015: The General Status of Species in Canada.* National General Status Working Group, online document. URL: http://www.registrelep-sararegistry.gc.ca/virtual_sara/files/reports/Wild%20Species%202015.pdf. Retrieved 2017–12–02.

Central Intelligence Agency, (2013). *The World Factbook 2013–14.* Washington, DC: Central Intelligence Agency, online document. URL: https://www.cia.gov/library/publications/the-world-factbook/index.html Retrieved 2017–12–02.

Committee on the Status of Endangered Wildlife in Canada, (2017). *Committee on the Status of Endangered Wildlife in Canada.* Government of Canada, online document. URL: https://www.canada.ca/en/environment-climate-change/services/committee-status-endangered-wildlife.html. Retrieved 2017–12–02.

Environment and Climate Change Canada, (2017a), Canadian environmental sustainability indicators: Canada's protected areas, online document. URL: http://www.ec.gc.ca/indicateurs-indicators/default. asp?lang=en&n=478A1D3D–1. Retrieved 2017–12–02.

Environment and Climate Change Canada, (2017b), *Species at Risk Public Registry,* online document. UR: http://www.registrelep-sararegistry.gc.ca/sar/index/default_e. cfm Retrieved 2017–10–22.

Fisheries and Oceans Canada, (2016b), *Sustainable Fisheries Framework,* Government of Canada, online document. URL http://www.dfo-mpo.gc.ca/reports-rapports/regs/sff-cpd/overview-cadreeng. htm Retrieved 2017–10–23.

Fisheries and Oceans Canada, (2017) *Canada's Fisheries Fast Facts 2016*. Government of Canada, online document. URL http://www.dfo-mpo.gc.ca/stats/facts-Info–16-eng. htm Retrieved 2017–10–23.

Fisheries and Oceans Canada, Oceans, (2016a), *Government of Canada, Fisheries and Oceans Canada, Communications*, online document. URL: www.dfo-mpo.gc.ca. Retrieved 2016–03–04.

Javorek, S. K., & Grant, M. C., (2011). *Trends in Wildlife Habitat Capacity on Agricultural land in Canada, 1986–2006, Canadian Biodiversity: Ecosystem Status and Trends 2010*. Technical Thematic Report no. 14. Canadian Councils of Resource Ministers. Ottawa, ON., online document: URL: http://www.biodivcanada.ca/default. asp?lang=En&n=137E1147–1 Retrieved 2017–10–23.

Mining Association of Canada, (2017), *Toward Sustainable Mining*. The Mining Association of Canada, online document. URL: http://mining.ca/towards-sustainable-mining. Retrieved 2017–12–02.

Mittermeier, R. A., Myers, N., Thomsen, J. B., Da Fonseca, G. A., & Olivieri, S., (1998). Biodiversity hotspots and major tropical wilderness areas: Approaches to setting conservation priorities. *Conservation Biology, 12*, 516–520.

National Forestry Database, (2017). *Wood Supply vs. Actual Harvest – Industrial Roundwood, 1990–2015*. Canadian Council of Forest Ministers, online document. URL: http:// nfdp.ccfm.org/data/graphs/graph_21_c_e.php (accessed on 23 Oct 2017).

Natural Resources Canada, (2006). *Overview of Canada's Forest Industry*. Natural Resources Canada., online document. URL: https://www.nrcan.gc.ca/forests/industry/overview/13311 Accessed (accessed on 23 Oct 2017).

Natural Resources Canada, (2017a). *Statistical Data*. Government of Canada. Online document. URL https://cfs.nrcan.gc.ca/statsprofile. (accessed on 23 Oct 2017).

Natural Resources Canada, (2017b). *Boreal Forest*. Government of Canada. Online document. URL: http://www.nrcan.gc.ca/forests/boreal/13071 (accessed on 22 Oct 2017).

Statistics Canada, (2015). *Canada's Population Estimates, First Quarter 2015*. Government of Canada, online document. URL: http://www.statcan.gc.ca/dailyquotidien/150617/ dq150617c-eng.htm (accessed on 22 Oct 2017).

Statistics Canada, (2016). *Government of Canada: Statistics*. Land and freshwater area, by province and territory. Statistics Canada. Online document. URL: http://www.statcan. gc.ca/tables-tableaux/sum-som/l01/cst01/phys01-eng.htm. Retrieved 2016–11–02.

Toronto Zoo. Black-footed ferret. Toronto Zoo. Online document. Undated, URL http://www. torontozoo.com/ExploretheZoo/AnimalDetails.asp?pg=367. Retrieved 2017–10–22.

Plate 3.1   Examples of terrestrial ecosystem diversity in Canada. From to left: (a) Muskeg near Route de la Baie James and Eeyou Istchee Cree Territory, northwest Quebec (Photo by Tyler Smith); (b) Lake Hazen, Ellesmere Island, Nunavut. (Photo by Paul Sokoloff); (c) *Eriophorum scheuchzeri* subsp. *arcticum* (Photo by Paul Sokoloff); (d) Boreal Forest in Newfoundland (Photo by Laurel McIvor); (e) Wetlands and mixed forests in southern Ontario (photo by David Galbraith); (f) Pointe Basse, Magdalen Island coast, Gulf of St. Laurence (photo by Tyler Smith).

Plate 3.2   Examples of tetrapod vertebrates of Canada. From top left: (a) Northern
Leopard Frog (*Lithobates pipiens*), southern Ontario (photo by David
Galbraith); (b) Common Snapping Turtle (*Chelydra serpentina*),
Algonquin Park, Ontario (photo by Patrick Moldowan); (c) Common
Loon (*Gavia immer*), Hamilton, Ontario (photo by David Galbraith);
(d) Northern Cardinal (*Cardinalis cardinalis*), Male, Burlington, Ontario
(photo by David Galbraith); (e) Harbor Seal (*Phoca vitulina*), Lunenburg,
Nova Scotia (photo by David Galbraith); (f) Moose (*Alces alces* subsp.
americana), Algonquin Park, Ontario (photo by Patrick Moldowan).

Plate 3.3 Examples of plant species diversity of Canada. From top left: (a) Sheep Laurel (*Kalmia angustifolia*), northern Quebec (photo by Tyler Smith); (b) Lichens in northern Quebec: brighter mounded species: *Cladonia stellaris*; darker nonmounded species: *Cladonia stygia* (photo by Tyler Smith); (c) Three Flowered Avens (*Geum triflorum*) (photo by Peter Kelly); (d) Wood Poppy (*Stylophorum diphyllum*), southern Ontario (Photo by David Galbraith); (e) *Nestotus macleanii*, Eastern Beringia, Yukon Territory (photo by Bruce Bennett); (f) Cinnamon fern (*Osmundastrum cinnamomeum*), Bacchus Woods, Ontario (photo by Nadia Cavallin).

CHAPTER 4

Still Searching the Rich Coast: Biodiversity of Costa Rica, Numbers, Processes, Patterns, and Challenges

GERARDO AVALOS

Full Professor of Tropical Ecology, School of Biology, University of Costa Rica, 11501-2060 San José, Costa Rica, and Director, Center for Sustainable Development Studies, The School for Field Studies, 100 Cummings Center, Suite 534-G Beverly, MA 01915, USA, E-mail: gavalos@fieldstudies.org

4.1 Introduction

Discovered by Christopher Columbus on his fourth voyage in 1502, Costa Rica received the Spanish name of the *rich coast* because of the variety of gold ornaments that the indigenous people displayed during the first contacts with the Spaniards. Despite not being a country rich in precious minerals, Costa Rica lives up to its name because of the enormous diversity of its flora and fauna, much of which continues to be underexplored. Across different groups, Costa Rica has one of the world's largest species concentrations per unit of area (Kappelle, 2016) for a territory that is just 51,000 km² but is among the 20 most diverse countries in the world (Obando, 2002).

Any attempt to summarize the complexity of Costa Rican biodiversity will be too ambitious and is doomed to be incomplete, even for the most abundant and best-known groups. The current number of estimated species is close to 500,000 (Table 4.1) resulting from a complex biogeographic history of biotic interchanges between North and South America. Despite numerous generations of national and foreign scholars who have led biodiversity studies in Costa Rica (Gómez and Savage, 1983; Savage, 2002), the country still remains relatively unexplored, especially for less conspicuous groups (i.e., fungi, bryophytes, nematodes, lichens) and for speciose groups such as insects. Costa Rica continues to serve as a model to understand the critical role of biodiversity inventories in facilitating the comprehension of the patterns that determine the distribution of species diversity, as well as the importance of taxonomic knowledge for the management, conservation, and sustainable use of biological diversity.

It is my objective to present the latest assessment of species richness for the most conspicuous groups of organisms in Costa Rica, highlighting the key ter-

Table 4.1   Reported and Estimated Number of Species Across Conspicuous Groups of Organisms for Costa Rica

Group	Reported species	Estimated species	Estimated percentage of known species
Fungi	3,000[1]	50,000 to 70,000[1]	0.06 to 0.04
Algae	502[2]	5,000–6,000[2]	0.1 to 0.03
Vascular plants	10,712[3]	12,000[4]	89
Mammals	252[5]		100
Freshwater fishes	252[6]		100
Marine fishes	1,794[6]		100
Amphibians	207[7]		100
Reptiles	247[7]		100
Birds	918[8]–920[9]		100
Insects	69,039[4]	365,000–500,000[10]	0.19 to 0.14
Coleoptera	35,000[4]	40,000[11]	87.5
Hymenoptera	20,000[12]	35,000[10]	57
Diptera	36,310[13]		?
Lepidoptera	Butterflies 1,549 [14]		?
	Moths 11,451[14]	14,000[15]	82

Sources: 1: J. Carranza, 2: C. Fernández-García: 3: M. Grayum, 4: Obando (2002), 5: Rodríguez-Herrera, 6: Angulo et al. (2013), 7: F. Bolaños, 8: AOCR, 9:UCOR, 10: P. Hanson, the upper limit includes other arthropods such as mites, 11: A. Solís, 12: Gaston et al. (1996), 13: Brown et al. (2009), 14: L.H. Murillo-Hiller, 15: Chacón et al. (2007).

restrial ecosystems on which they depend, as well as to sketch the most important ecological patterns responsible for generating and maintaining this unique biodiversity. I will end this chapter by identifying the current and future challenges to protect Costa Rican biodiversity. My approach will concentrate on terrestrial ecosystems. The biodiversity of marine habitats has been analyzed in detail by Wehrtmann and Cortés (2008). The patterns of diversity have been explored since the mid-nineteenth century by a continuous flow of researchers, both national and foreign, and hopefully, will inspire future generations of biologists to continue the exploration of the rich coast (Gómez and Savage, 1983).

4.2   Current Status of Taxonomic Knowledge

The estimation of the country's species numbers faces many of the challenges of the world's taxonomists (May, 2010), even for a nation that is among the best-known in the tropics. Still, taxonomic and biosystematic work is often considered an old trade and part of classic natural history science in strong contrast to the hard-core modeling and hypothesis testing of current ecological theories. Consequently, funds to carry out taxonomic work and biological inventories are scarce and receive a low priority, discouraging young scientists from pursuing a career in taxonomy. Thus, there is lack of sufficient taxonomists, especially for groups that are highly speciose and poorly known, although other authors are less pessimistic (see Costello et al., 2013). Taxonomic work represents the foundation of sound ecological work. More inventories and fieldwork are required across all groups, but especially those that are highly diverse and poorly known taxonomically. Some estimate the approximate number of taxonomists worldwide to be close to 30,000–40,000 (Costello et al., 2013). Modern taxonomy has more recruits now and more advanced techniques available to differentiate among species, although limited field collection and field inventories (Wilson, 2017) continue to be a major drawback. On the brighter side, taxonomists are increasing in speciose regions such as South America and Southeast Asia (Joppa et al., 2011; Costello et al., 2013).

Estimates of the world's diversity affect the projections of Costa Rica's biodiversity. More than 2 million species have been described worldwide (Costello et al., 2013; Wilson 2017). Current estimates of the total number of species range from 2 to 3 up to 8 million following Costello et al. (2013)'s conservative assessment, to Erwin (1982)'s 30 million, to over 100 million (Ehrlich and Wilson, 1991). The number of species is a moving target although the current tendency is to adjust the actual number by incorporating taxonomic corrections, which may account for 20% of the already named species in some groups (Costello et al., 2013). Just the species number for the class Insecta could exceed over 6 million (Stork et al., 2015). Most conservative and probably most realistic estimates approach 11 million species (Larsen et al., 2017). These estimates consider only eukaryotic species. If bacteria and other eukaryotic groups such as mites, nematodes, apicomplexan protists, and microsporidian fungi are included, the numbers will sky-rocket. For instance, Larsen et al. (2017) show that the approximate number of species ranges 1 to 6 billion, with bacteria representing 70–90% of the species.

Following Obando (2002), Costa Rica has estimated to harbor 500,000 species combining the numbers from the most conspicuous groups. This represents about 5% of the Earth's diversity (Table 4.1). Insects alone may contribute 250,000 species (P. Hanson, pers. com.). Other arthropods such as mites may have as many species as insects resulting in half million species of only arthropods (P. Hanson, pers. com). In addition, recent estimates of the Earth's biodiversity (Larsen et al., 2017) may bring the numbers for other groups (bacteria, fungi, nematodes) up, affecting the expected numbers for Costa Rica.

The numbers presented in Table 4.1 come from the opinion of experts on the biota of the country as well as from key papers. Better-known groups such as vascular plants and vertebrates have a mostly completed inventory, but even within these groups, it has been difficult to keep an updated tally since the species number constantly changes. For example, 406 new species of vascular plants were described after the publication of the first volume of *Manual de Plantas de Costa Rica* (Hammel et al., 2004). Tracking new species of insects and other arthropods described for Costa Rica is also very difficult. There is no other alternative but to follow the experts' opinion. Even in the case of the best-known groups, the numbers will change in the short-term. Table 4.1 also makes evident the lack of information for the lesser-known groups, as well as groups not shown in the table, such as marine invertebrates that are more difficult to monitor and study. In all cases, making estimates of diversity continues to be a purely speculative task especially when considering that some habitats are not sampled at all or very rarely.

Therefore, current estimates of the total number of species in Costa Rica are, for the most part, incomplete, since there are many groups in urgent need of being inventoried, especially microorganisms such as bacteria, nematodes, and fungi. There are currently no consistent national efforts to maintain an integrated inventory of the country's biodiversity despite of previous attempts to do so (i.e., the database of the National Institute of Biodiversity, INBIO, and the all taxa biological inventory – ATBI– launched by Daniel Janzen in Guanacaste National Park, to mention two examples), including the implementation of gap analysis and the creation of maps analyzing the distribution of major ecosystems and vegetation types (Herrera and Gómez, 1993). The sampling protocol across groups has been sparse, localized, short-lived, and lacks priority for funding, even though an informed database of species distribution would facilitate the management, conservation, and restoration of species diversity now and in the future.

4.3 Summary of Costa Rican Geography and Climate

The high species diversity of Costa Rica is associated with ample topographic and climatic heterogeneity concentrated in a small area. Located in

Southern Central America between Nicaragua and Panama, the layout of the country follows a NW-SE direction between north latitudes of 8° 02'26" and 11° 13'12" N, and west longitudes of 82° 33'48" and 85°57'57", facing the Caribbean Sea to the northeast and the Pacific Ocean to the west and south (Figure 4.1). This small country is the third in size in Central America, with an area comparable to the US state of West Virginia (51,100 km² land mass, but over 589,000 km² of ocean area). The high exposure to the climatic influence of Pacific and Atlantic oceans in combination with a landscape dominated by elevational gradients where 70% of the continental mass is covered by mountains creates a high level of environmental heterogeneity, which has influenced the packing of a large number of species distributed over steep elevational gradients (Burger, 1980).

The extensive exposure to oceans influences many characteristics of the country's climate such as the patterns of rain distribution and erosion including the orogenic location of rivers and watersheds. This exposes the country to localized climatic phenomena that once close enough to the mainland can affect the entire territory. Herrera (2016) calls this feature the *isthmic factor* to reflect the narrowness of the territory and its exposure to the ocean climate. The outline of the territory decreases towards the SW, having a minimum width of 119 km and a maximum of 464 km. Costa Rica does not have many island territories. The only oceanic island is Cocos Island, located 532 km SW to the harbor city of Puntarenas (5° 32' N, 87° 04'W) and is only 26 km² in area. The rest of the islands are concentrated in the Gulf of Nicoya,

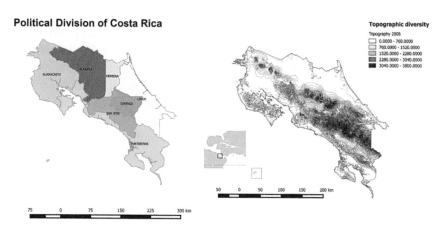

Figure 4.1　Elevational range and political division of Costa Rica.

except Caño Island facing the Osa Peninsula, and Uvita facing the harbor of Limón.

The presence of a mountain system crossing the country in an NW-SE orientation and increasing in elevation, area, and age in that direction, influence the distribution of the main wind fields, the northeasterly trade winds and the equatorial westerlies. This mountain system effectively divides the country in two and consists of the Guanacaste, Tilarán, the Central, and the Talamanca Mountain ranges. The Pacific slope covers 26,585 km^2 and has a more heterogeneous orography with deep canyons and more branched river systems than the Caribbean slope (24,115 km^2), which is characterized by extensive humid plains in the northeast, that narrow towards the SE border with Panama due to the increased area of the Talamanca. Guanacaste Mountain Range consists of seven mountain peaks less than 2,000 m, which are isolated by elevation and harbor cloud forests in their summits. Here, vegetation typical of the highlands happens at significantly lower elevations than in the rest of the Cordilleras (representing an example of the *Massenerhebung effect,* Flenley 1995; Kappelle et al., 1995). The Tilarán Mountain Range is the shortest one with elevations under 2,000 m and runs almost parallel to the Guanacaste and Central Mountain Slopes. The Central Mountain Range is composed by four volcanic massifs whose elevation nears 3,000 m, all active volcanoes. Volcanic eruptions have added to the heterogeneity of the landscape, and have changed the distribution of vegetation in recent times. In this case, and in the Talamanca, the Caribbean slope shows a more gentle and continuous inclination relative to the Pacific slope. This has implications in the degree of environmental gradients whose physical factors tend to vary more strongly in the Pacific relative to the Caribbean slope, affecting the distribution of vegetation types and organisms in general (Hammel et al., 2004). The bulk of the human population (1 out of the 4.8 million inhabitants) is concentrated in the fertile valleys and mid elevations (i.e., 1,200 m) of the Central Mountain Range. The Talamanca start in Cartago and continue across the southern border into Panama. These mountains are massive and do not have active volcanoes on the Costa Rican side although their origin is clearly volcanic. The highest point is Cerro Chirripó with 3,820 m, after which multiple mountain peaks surpass 3,000 m and are dominated by Páramo ecosystems close to their summits. In addition, to Páramo, the highest mountain peaks of the Talamanca show the signs of previous glaciations during the Quaternary when they were covered by ice (Lachniet and Seltzer, 2002), such as glacial lakes, moraines, and glacial valleys shaped in V and U.

The latitudinal position 10°N above the equator defines many of the climate patterns of Costa Rica due to the action of the Intertropical Convergence Zone (ITCZ), an area where the trade winds of the north and south merge to create a low pressure, and thus, intense rains. The ITCZ determines the Costa Rican climate and the rain distribution, especially the timing and intensity of the rainy season from May to early November. The precipitation of Costa Rica is thus characterized by dry conditions in the Pacific Northwest, the central Pacific, and the western slopes of the Central Valley from late November to April, and humid conditions in the rest of the country (from May to late November). The precipitation regime of the Caribbean contrasts with that of the Pacific, being weakly seasonal, with rains for most of the year, and a dry spell in March-April and September-October. The Caribbean slope is exposed to frequent wet spells lasting several days. During the dry season on the Pacific slope there is almost no rain and about 70% of canopy trees become deciduous, especially in highly seasonal forests such as the dry forests of the NW (Janzen, 1967).

Precipitation and temperature follow steep elevational gradients. One of the wettest places in the country is Tapantí with over 7,000 mm of rainfall per year, which is located just 18 km away from some of the driest areas, Valle del Guarco, with nearly 600 mm of rain per year. The summits of the highest mountains have very low temperatures with averages close to 7.2°C and cold conditions during the dry season (down to −11°C but most of the time −5°C) with frost often early in the morning. In some cases, frost covers extensive areas although it lasts only a few hours. Mild temperatures dominate mid elevations (800 to 1,200 m) and warm conditions are prevalent on both coasts in the lowlands, especially along the Pacific slope.

Considering the critical influence of climate on organismal distribution, several attempts have been made to classify ecosystems according to climatic differences. The Holdridge Life Zone System has been traditionally used to characterize the distribution of vegetation, and thus, to map the ecosystem distribution of Costa Rica (Tosi and Holdridge, 1966; Kappelle, 2016). The system uses three climatic variables to predict the presence of a botanical association or "life zone": average annual biotemperature, average annual precipitation, and average potential evapotranspiration. It then combines these factors with latitude, altitude, and precipitation to define 30 life zones. A "life zone" is the expected vegetation type resulting from the combination of these three climatic factors plus elevation and precipitation. Twelve life zones (Holdridge and Grenke, 1971) are found within Costa Rica in habitats ranging from intertidal zones to the top of the mountains above 3,000 m where the northernmost distribution of the Páramo ecosystem is found, to tropical dry forests in the Pacific Northwest and tropical lowland rainforest along the Pacific Southeast and most of the Caribbean slope. The Holdridge system classifies vegetation mostly on physiognomy rather than species composition or indicator species. However, vegetation types follow a more continuous distribution with overlapping species and do not show clear limits between life zones especially along elevation gradients (Lieberman et al., 1996). The Holdridge Life Zone System does not consider seasonality or climatic variation, which is critical to determining the distribution of organisms. For instance, seasonality imposes limits on plant distribution and is responsible for the increase in species richness on the climatically more heterogeneous Pacific slope relative to the Caribbean (Hammel et al., 2004). Thus, the Holdridge Life Zone System, despite of its wide acceptance and application, has limited predictive value in the field. There has been a sustained effort to map and differentiate vegetation types in Costa Rica, and this includes additional mapping efforts such as the "biotic units" of Herrera and Gómez (1993) but still has mixed results.

Hammel et al. (2004) propose a more simplified system based on these authors' extensive field experience distinguishing five vegetation types: dry forest, moist forest, wet forest, rain forest, and páramo. This classification is consistent with Kohlmann et al. (2002). These authors do not discard the Holdridge System altogether but recommend complementing it with a more detailed characterization of vegetation based on indicator species. Indicator species integrate environmental variation better accounting for physical factors that often are difficult to quantify along elevational gradients, such as soil and topographic differences. They treat mangroves, beaches, and islands as discrete categories outside of their proposed classification. Mangroves cover 41,592 ha in Costa Rica and increase in diversity towards the south of the country especially along the Pacific slope, where they are concentrated along river mouths or in areas facing the ocean.

4.4   Historic Origin of the Biodiversity of Costa Rica

4.4.1   Geological History

Geologically, Costa Rica is part of Southern Central America (southern Nicaragua, Costa Rica, and western Panama), which is a complex territory, although much younger than the rest of Central America (this is, Northern Central America, including Chiapas, Guatemala, Honduras and Northern Nicaragua, composing the Chortis and Maya terrains, formed about 570 Ma, compared to the 208 Ma of Southern Central America). Southern Central America emerged as a volcanic island arc resulting from the subduction of the Cocos Plate under the Caribbean Plate. This process caused the closure of the former interoceanic canal separating the Atlantic from the Pacific Oceans (and also separating North America from South America) due to the emergence of a continuous land bridge that was completed only 2.8 Ma during the late Pliocene (O'Dea et al., 2016). The closure of the canal and the emergence of the isthmus constitute the most important geological and biogeographical event of the past 60 million years (Coates, 1997). Before the closure, North and South America had been isolated since the breakup of Pangea during the Early Cretaceous Period (150–140 Ma), which provided sufficient time for both land masses to develop divergent biotas.

The first materials from the volcanic archipelago emerged 80 Ma during the late Cretaceous period. These materials moved towards the East, forming part of what later would become the Greater Antilles. Later on, at the end of the Cretaceous, 65 Ma, the Galápagos hotspot began to generate materials that increased the size of the Caribbean Plate, pushing the Greater Antilles further up to the northeast and properly forming the Central American Interoceanic Archipelago on the western edge of the Caribbean Plate. During the Eocene (40 Ma) the Caribbean plate broadened in size and collided with the Bahamas and Florida platform forming the islands of Cuba and Hispaniola. Eleven million years ago several islands of the archipelago emerged and increased in elevation in what is now western Panama and Costa Rica, restricting the passage of water from the Caribbean to the Pacific, and coinciding with the effects of the last Pleistocene glaciation. At the time the ocean levels and the global temperature dropped, which facilitated the passage of temperate organisms from North and South America. In addition, the Costa Rican territory acquired a formation very similar to the present one 16–15 Ma during the middle of the Miocene when the interoceanic canal reached 2,000 m in depth. At the end of the Miocene, about 7 Ma, the depth of the canal was reduced to 150 m. Under these circumstances, the archipelago facilitated the passage of biotic elements crossing the canal using random methods, such as island hopping and rafting (i.e., fallen trees torn down during storms), or dispersing on their own. During the late Pliocene, 3 Ma, Costa Rica reached its almost final shape, leaving vestiges of the former canal in the San Juan River and Lake Nicaragua (the Nicaraguan depression), as well as very shallow strips of land (50 m deep) along the Tempisque-San Carlos corridor, the depression where the Panama Canal is currently located, and the corridor formed by the Valley of the Atrato River in the present border between Panama and Colombia. The continuous subduction of the Cocos Plate under the Caribbean Plate is responsible for the significant contrast in the geological characteristics of the Pacific coast of Central America (a deep ocean trench down to 2,000 m very close to the coast, active volcanoes, steep slope, frequent earthquakes, and active subduction) in relation to the Atlantic coast (no trenches, no volcanoes, gentle slope, no subduction, and few earthquakes), as well as for the significant increase in size and elevation of the Talamanca mountains.

The Cocos Underwater Ridge, a strip of very dense material (2,000 m high and 200 km wide, from Punta Burica to the west of the Panama fracture zone), which is thick enough to prevent the flowing of lava and the formation of volcanoes, is responsible for uplifting the Talamanca to almost 4,000 m in elevation, the highest point in Southern Central America (Chirripó, 3,820 m). Formed by basalts originated on the ocean floor as result of the activity of the Galapagos hotspot, the Cocos Ridge is responsible for the absence of volcanoes on the Costa Rican side of the Talamanca (the Panamanian side has one single volcano, Barú). The increase in elevation of the Talamanca Mountain Range is fundamental to understanding Costa Rican biogeography. This mountain range covers a significant portion of the country maintaining elevational gradients on both slopes, but especially abrupt gradients on the Pacific slope. The significant elevation of this mountain range generated a great variety of habitats, from sea level to the summit of Chirripó. The gradients resulted in an elevation barrier for many groups, separating organisms whose previous distributions were continuous from the Pacific and the Atlantic slopes. In this Cordillera, greater habitat area and habitat heterogeneity converge to generate high species diversity.

Exotic terrains forming the Complex of Nicoya added the last geologic features of Costa Rica. The Complex of Nicoya is located on the Pacific coast in areas such as the Nicoya Peninsula, the central Pacific, as well as the Osa Peninsula and Punta Burica. This terrain was formed by marine basaltic rocks dragged on top of the Cocos plate, it was too light to undergo subduction, and was located closer to South America before hitting the rest of the Costa Rican mainland. The age of the Nicoya Complex contrasts significantly with the age of the rest of the region. This mechanism helps to explain the particularities of the flora of the Osa Peninsula, more akin to South American elements.

The interoceanic canal was finally closed at the end of the Miocene, only 2.8 Ma. For the first time since the breakup of Pangea, North America and South America were united, resulting in several crucial biogeographical consequences. First, the Central American isthmus provided a continuous route for the species interchange between South and North America. The placental mammals of North America invaded South America triggering the extinction of marsupials. This was compounded by climate change and the end of the last Pleistocene glaciation, which facilitated the extinction of the megafauna (giant mammal herbivores what required ample open spaces like dry and humid savannas). Secondly, the isthmus became a barrier separating the marine fauna of the Caribbean from that of the Pacific. The chemistry of the Caribbean Sea changed radically with respect to that of the Pacific Ocean (this started 15 Ma when the chemistry of the canal changed and the depth decreased), contributing to the geographic barrier effect favoring the development of more coral reefs in the Caribbean in relation to the Pacific Oceans. Finally, the Gulf Stream was created, which changed the climatic conditions of the Caribbean coast of Central America all the way up to Mexico and the Southern United States.

4.5 General Biogeography

The Central American isthmus at the time of the closure of the canal was the stage of the Great American Biotic Interchange (GABI) between North and South America (Marshall, 1988). Thus, the events of the last 3 Ma are critical to understanding the current composition of the Costa Rican flora and fauna. Molecular, geological, and fossil evidence have been used to determine the timing of the interchange as well as to identify the groups and mechanisms involved (Marshall, 1988; Webb, 2006). The evidence is biased towards mammal fossils, which are more abundant and have wider dispersion. However, the migration of large mammals, mainly from South America to North America started way before the closure of the canal, since the Paleocene (65 Ma), giving rise to what is known as the archaic mammal fauna. Organisms able to cross the canal included edentates and giant armadillos of the genus *Glyptotherium* (MacFadden, 2006; Laurito and Valerio, 2013) in addition to other South American hoofed mammals such as giant sloths (*Megatherium* and *Mylodon*) and mammals of open savanna environments (i.e., the megafauna component mentioned above). These components continued crossing even after the emergence of the land bridge and had geographic ranges encompassing the entire American continent. During the Oligocene (30 Ma), caviomorph rodents (capybaras, field mice, and porcupines), as well as platyrrhine monkeys, migrated from north to south. It is presumed that these elements arrived through rafting or island hopping. With the formation of the Central American Isthmus during the late Pliocene, the climate became drier and colder. These conditions favored the invasion of

North American components, such as felids, cervids, canids, cricetid mice, and camelids related to llamas and alpacas, in addition to a diverse contingent of bat species. Other elements such as the giant *Titanis* bird from South America reached Texas between 5 and 4.7 Ma. Giant sloths (*Eremotherium*), giant anteaters (*Myrmecophaga*), capybaras (*Hydrochoerus*) and opossums (*Didelphis*) moved across the Isthmus, as well as a toxodon (*Mixotoxodon*), but of all these only a few reached the southern United States (today, armadillos, porcupines and opossums are abundant in the US).

The isthmian bridge produced a dramatic increase in the number of species of mammals in South America from North American origin, whereas in North America the number of South American elements was much smaller. The North American fauna not only dispersed into South America but also diversified there. This was the case of cricetid rodents, which are considered to be pseudodispersants – elements that disperse into a new area and radiated into new species. The closure of the canal coincided with the end of the Pleistocene glaciation. Conditions became warmer and more humid, favoring the passage of the Neotropical elements of South American rainforests (hummingbirds, toucans, and anteaters). The tropical flavor of the Costa Rican biota is thus a fairly recent acquisition.

4.6   Patterns of Endemism and Diversity Distribution

The current composition of Costa Rican biota follows biogeographic trends of distribution and dispersion influenced by the dynamic geologic history of the Southern Central American Isthmus as well as by environmental gradients. In this case, more temperate elements from North and South America colonized the cold highlands whereas more tropical elements from South America colonized mid-elevations and lowlands.

The levels of endemism for Costa Rica are moderate (1.4%, Obando, 2002), but vary with the group (i.e., Coleoptera, plants, birds, reptiles, Kohlmann et al., 2007). Endemism patterns are relatively well known compared to other speciose tropical regions due to the relatively higher availability of taxonomists, research institutions (Museums and herbaria), and a sustained effort to inventory diversity. Amphibians and reptiles show the highest degree of endemism (16.9%), followed by freshwater fishes (14%), vascular plants (12%), mammals (2.5%) and birds (0.8%). Plants contribute with 1,000 species to the 5,000 endemic species projected for Central America (Hammel et al., 2004). The country is divided into four endemism regions following Obando (2002) and Kohlmann et al. (2007): the Central Mountain Slope, Talamanca, the Central Pacific Region, and the Golfo Dulce area.

Cloud forests, in general, are expected to have high levels of endemism. In addition, Cocos Island can be considered as a fifth region since it maintains a high level of endemism for groups such as vascular plants (70 out of 235 species), ferns (15 out of 76 species), insects (65 out of 362), and an overall level of 11% endemism across groups (Obando, 2002). The distinction of areas of endemism is affected by the sampling bias, since there are areas that have been better studied than others. This would explain the apparent contradiction of higher endemism of beetles in the mountains of Guanacaste in relation to Talamanca, which is more massive, larger in area, and generally more diverse in habitat types (Kohlmann et al., 2007) although more research is needed.

Following Hammel et al. (2004), the plant families with more endemics are Acanthaceae (23%, i.e., the genus *Justicia* with 47% of endemic species), Bromeliaceae (20.2%, the genus *Werauhia* with 46.5% of the species); Lauraceae (18.9%); Marantaceae (22%, especially the genus *Calathea* with 27% of endemic species); and Orchidaceae (26% concentrated in the genera *Epidendrum*, *Lepanthes* and *Pleurothallis* with 41%, 90% and 24%, respectively). The genus *Inga* is highly endemic within the Fabaceae with 12 out of 53 species, with the Caribbean lowlands showing the highest number of species per ha (7.6, Zamora and Pennington 2001).

Out of the 11 conservation areas in which the country is divided (Figure 4.2), the top plant diversity is found within the Central Pacific Slope Conservation Area. The Pacific slope is in general topographically and climatically more heterogeneous. The Central Pacific region, with 4,000 species, has the greatest alpha diversity. This area is unique in Mesoamerica since it includes an ecotone and biogeographic barrier between the tropical dry forests of the NW and the tropical rainforests of the SW (Osa and Golfo Dulce), with the Tárcoles River and Carara National Park as the site of this gradient (Kohlmann and Wilkinson, 2007). Here, climatic and geologic factors converge (i.e., the coincidence of the border of the microplate of Panama and Costa Rica) separating groups such as plants, dung beetles, butterflies, birds, amphibians, and reptiles. This elevational gradient also connects Carara with the still botanically unexplored Turrubares highlands.

The Central Pacific slope is followed in diversity by the elevational gradient of the Central Mountain Slope Conservation Area (3,989 plant species), connecting Braulio Carrillo, one of the largest national parks in Costa Rica (475 km²) with the Caribbean slope forming the longest elevational gradient under legal protection in Central America, which goes from 30 m in elevation at La Selva Protected Zone to the summit of Barva Volcano at 2,906 m.

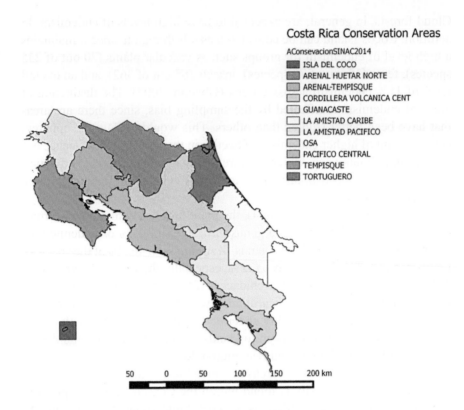

Costa Rica Conservation Areas

AConsevacionSINAC2014
- ISLA DEL COCO
- ARENAL HUETAR NORTE
- ARENAL-TEMPISQUE
- CORDILLERA VOLCANICA CENT
- GUANACASTE
- LA AMISTAD CARIBE
- LA AMISTAD PACIFICO
- OSA
- PACIFICO CENTRAL
- TEMPISQUE
- TORTUGUERO

50   0   50   100   150   200 km

Figure 4.2   The 11 conservation areas of Costa Rica.

The Guanacaste Mountain Range is the smallest in stature but has steep environmental gradients from the dry forest habitat at the base of the Cordillera on the Pacific to the cloud forests of their isolated summits, down to the rainforests of the Caribbean slope. The areas near the mountain peaks (above 800 m) have wetter conditions on both slopes. The presence of volcanoes and other signs of volcanic activity, such as vents in the Rincón de La Vieja volcano, create conditions typical of xeric habitats or dry forests in the middle of rainforests, generating greater environmental heterogeneity. A similar case takes place in the hilly areas of the Tempisque Basin, specifically in Palo Verde National Park, where limestone soils generate environments characteristic of a seasonal desert dominated by the cactus *Stenocereus aragonii*. A peculiar aspect of this mountain range is the abundance and diversity of species of the family Rutaceae, which contains 75% (16 species) of genera of this family and 60% (33) of the total of the species reported for the country (Hammel et al., 2004). The province of Guanacaste deserves special attention because it has historically maintained tropical dry forests. The tropical dry forest is the most threatened and least known tropical ecosystem (Miles et al., 2006) as there are no extensive areas covered by primary dry forest. The strong seasonality of rainfall and the presence of large rivers adding fertile sediments to soils in many areas of the Tempisque basin facilitated human habitation and intensive agriculture since humans invaded these latitudes 11,000 years ago (Janzen and Hallwachs, 2016). More recently, recurring anthropogenic fires removed the original vegetation, facilitated the establishment of large cattle ranches, which fostered anthropic savannas. Fire suppression and control reverses this process facilitating the regeneration of the dry forest. Critical to the restoration of the dry forest is the maintenance of corridors and habitat mosaics connecting the summits of the mountain range with the lowlands, which facilitate the completion of the life cycle of many insect pollinators, especially butterflies and moths (Janzen, 1987). The restoration of the dry forest of Santa Rosa National Park and other areas in the Guanacaste National Park is a classic example of ecological restoration in the tropics. However, the threat is still latent, and these forests are exposed to recurrent fires, illegal hunting, and agricultural and urban expansion. As a result, dry forests have lost a significant proportion of the original biota. Protected forests follow a successional process possibly very different from the one shown by the original forests before the arrival of the human species.

The Tilarán Mountain slope is possibly one of the best-known floristically. In this mountain range, rainfall seasonality parallels the rest of the country, with a very humid Caribbean slope and a drier Pacific slope. Typical dry forest environments predominate below 600 m. Also, on the Pacific side, but at 1,200 m, there are representative samples of the premontane rain forest, an environment with a higher and more humid canopy where coffee farms concentrate in Monteverde. Monteverde maintains prime cloud forest environments in 6 of the 12 life zones reported for the country, characterized by abrupt elevational gradients over short distances. Cloud forests reach high complexity (and possibly higher diversity) above 1,400 m on the Caribbean side and 1,500 m on the Pacific side. Epiphytes in general, and especially orchids, reach peaks of diversity, as do the arborescent ferns. Elfin forests dominate the continental divide.

Altitudinal gradients characterize the Central and Talamanca mountain ranges. The Central Mountain Range shares common elements with the mountain ranges of Tilarán and Talamanca. On Irazú and Turrialba volcanoes there are typical elements of páramo vegetation, which is found above the treeline and is dominated by tussock grasses, and small shrubs with sclerophyllous foliage and cushion plants, although many elements of the adjacent oak forest invade the páramo as ecotypes (this is the case of the genus *Quercus* and *Escallonia*, Luteyn et al., 1999). The area covered by páramos is small and limited to mountain peaks above 3,000 m, although many authors consider páramo to be restricted to the Talamanca Mountain Range. From the Caribbean side in the elevation range between 750 and 1,250 m, the Central Mountain Range presents a transition zone where highland elements mix with typical elements of the lowlands (i.e., the palms *Euterpe precatoria*, *Geonoma edulis*, *Iriartea deltoidea*, and *Prestoea acuminata*). Arborescent ferns are very abundant in this elevational range. In terms of the number of fern species, the greatest diversity lays between 300 and 500 m, with 150 species from 55 families at 300 m and 131 species from 55 families at 500 m. The diversity of species decreases with elevation until reaching the páramo ecosystem.

As mentioned above, amphibians and reptiles show a higher degree of endemism relative to other groups, and for the most part, follow similar trends relative to endemism distribution. The degree of endemism is related to the vagility of the organisms involved. Reptiles such as snakes have many traits associated with a sedentary lifestyle characteristic of sit-and-wait predators (cryptic colors and reduced daily mobility), which could associate the reduced territories and higher phylopatry with restricted dispersion. However, these ecological characteristics need to be analyzed within the context of the climatic and geologic changes that some areas experienced during the last Pleistocene glaciation. There are no Costa Rican endemic families or genera of amphibians and reptiles, but at the species level, there are 52 amphibian species (3 caecilians, 25 salamanders – showing the highest degree of endemism with 58%, and 24 anurans). Endemic reptiles include 9 lizards –12% endemic species- and 10 snakes. At the group level, salamanders show the highest level of endemism (58%). Endemic species tend to concentrate above 1,000 m, and 32% of the endemics are restricted to the highlands of Talamanca as result of the population isolation associated with the large area and topographic and climatic heterogeneity typical of these mountains. In addition, species such as *Oophaga granulifera*, *Phyllobates vittatus*, *Lachesis melanocephala*, *Leptodeira rubricata*, *Porthidium porrasi*, and *P. volcanicum* diverged in the South Pacific region of Costa Rica (the Osa Peninsula) after separating from similar taxa from the Caribbean lowlands (Lamar and Sasa, 2004). This divergence is associated with successive cooling and warming periods of the last Pliocene to Pleistocene glaciations. The uplift of the

Talamanca separated the biota of the Caribbean from that of the Pacific lowlands. Sasa et al. (2010) present a thorough analysis of amphibian and reptile species distribution and determine 159 regional endemics whose distribution is continuous into Panama (88 species), relative to the number of species unique to Costa Rica (71).

4.7 Selected Groups

4.7.1 Insects

Many of the conditions that make Costa Rica a hotspot of species diversity are reflected in the diversity of arthropods. The significant economic importance of arthropods for human health, agricultural production, and the overall economy make it a group likely to receive greater attention relative to others. However, a coordinated effort relying on intensive sampling over a variety of sites and habitats is still necessary to approach realistic numbers regarding the species diversity of specific groups. The network of malaise traps set up for Hymenoptera (Gaston et al., 1996) is an example, showing that the task at hand is not critically overwhelming after producing over a million specimens in more than 50 sites and 150 malaise traps concentrated on the sampling of one of the four most speciose insect orders. Efforts like these were complemented by the extensive collections of parasitoids of folivorous lepidotera larvae from D.H. Janzen and the Lepilab of Santa Rosa National Park (Brown et al., 2014).

In the case of the Hymenoptera, Gaston et al. (1996) estimate the number of species for Costa Rica in 20,000 with Costa Rica having twice as many species per unit of area as other tropical and temperate sites that have been inventoried. The authors concluded that the number of Hymenoptera species will increase but the estimates will not double and that 20,000 species is a realistic figure. The Costa Rican species of Hymenopterans have a higher proportion of egg parasitoids and eusocial species whereas temperate species (i.e., British Isles) are mostly represented by folivorous species on plant families absent or poorly represented in Costa Rica (i.e., Salicaceae, Rosaceae, Betulaceae, and Pinaceae). Intensive sampling over long periods (10–15 years) found families that were very scarce as well as increased the latitudinal ranges of other groups by thousands of kilometers, although more habitats and forest strata still need to be sampled, such as the canopy of tropical lowland rainforests. These almost complete inventories are extrapolated to estimate the global number of species expected for Hymenoptera. Gaston et al. (1996) consider that there are 300,000 to 1 million species in the world if Costa Rica contributes anything between 5–10% of the global biodiversity across groups (suggesting that over 10% of the Hymenoptera species have been described). This calculation helps to estimate the total global number of insect species, which may range from 10 to 30 million, more than the total number of species of all organisms that have been described so far. The picture is changing from one reflecting high endemism and thus dispro-portionally high levels of species diversity in the tropics to a more realistic pattern of species with very wide geographic ranges extending sometimes across thousands of kilometers as the sampling becomes more representative across space and time. As the sampling effort expands to include a wider geographic areas and more time periods, more species accumulate, extend-ing the geographic range of many groups, especially those represented by very rare or very scarce species. Other factors need to be taken into account, such as habitat specificity (bee species show a diversity peak in temperate, arid regions once corrected by sampled area). The number of host species, as influenced by the richness and diversity of host plants and their more diverse chemistry in the tropics, also generates a latitudinal gradient in spe-cies diversity.

Other speciose groups include Lepidoptera (1,549 species of butterflies, L.H. Murillo-Hiller, pers. com., and 11,451 approximate species of moths, although this number could be close to 14,000 species – Chacón et al., 2007). Coleoptera is estimated to contribute 40,000 species distributed in 107 fami-lies (A. Solís, pers. com.).

4.7.2 Vascular Plants

The formal exploration of the Costa Rican biota begins with the works of Emmanuel von Friedrichsthal in 1839 and Anders Oersted in 1846. The second half of the nineteenth century was a period of intense explora-tion led by European naturalists who did the first collections of plants and animals, including Joseph von Warszewicz, Moritz Wagner, Carl von Scherzer, Karl Hoffmann, Alexander von Frantzius, Hermann Wendland and Helmut Polakowsky, to name a few (see Hilje-Quirós, 2013). These naturalists were not constrained to one particular group, but combined dif-ferent subjects and scientific disciplines, from botany and zoology to geol-ogy and soil science. Such interest in the Costa Rican biota was heightened by government efforts to attract European scientists to help develop high school education and agricultural training. These efforts later evolved in the establishment of the first formal University (the University of Santo

Tomás) and the National Museum in 1889. Anastasio Alfaro was the first director of the National Museum and continued developing connections with international collectors, consolidated the herbarium, and organized the plant collections. Alfaro assembled a list of 1,218 plant species (Alfaro 1888) following the ground-breaking work of *Biologia Centrali-Americana*. Alfaro's work was followed by *Primitiae florae costaricensis* by Durand and Pittier (which in addition to 5,000 plant species also included fungi, bryophytes, and lichens) from the late nineteenth and early twentieth century. The contingent of European naturalists brought by the Costa Rican government was later furthered by Henri Pittier, who stimulated the development of Costa Rican sciences in different fields, including meteorology, geography, and botany, and started a systematic exploration of the country. Pittier's work left a very deep mark on natural sciences. From 1887–1904, Pittier and his collaborators made Costa Rica one of the best-known tropical countries. The work of Pittier was continued by Alfaro and Karl Werckle, who expanded plant collections of groups such as ferns and orchids. Werckle wrote the first analysis of biogeographic relationships of Costa Rican plants (Werckle, 1909). Among Werckle's collections there were infertile specimens of what would become a formerly endemic Costa Rican plant family, the Ticodendraceae, and the species *Ticodendrum incognitum* (Gómez-Laurito and Gómez, 1989). Ticodendraceae was recently expanded to Colombia.

At the turn of the twentieth century, official support for the development of the natural sciences weakened and Costa Rica fell into an impasse that lasted until the middle of the twentieth century. During this period, the previous interest and attention of European scientists were replaced by North American scholars, who continued the collection of specimens of many groups in addition to a few but bright Costa Rican botanists, such as Alberto Brenes and Ottón Jiménez.

Paul Standley completed *Flora of Costa Rica* in four volumes in 1938 with 6,088 spp. In 1940, Standley expanded the list to 7,000 species. With the establishment of the School of Biology at the University of Costa Rica in 1957, biodiversity exploration enters a modern era supported by biodiversity collections and catalogs maintained at the Museum of Zoology and the UCR Herbaria. Newly fledged university departments as well as the National Museum, the SJO Herbaria, Botanical Gardens, and NGOs such as the Tropical Science Center and the Organization for Tropical Studies, collaborated in the training of the local human resource as well as the establishment of collaboration agreements with international institutions fostering the continued exploration of the Costa Rican biota. The latest effort of the botanical inventory comes from the publication of *Flora Costaricens,* a project initiated by William Burger in 1971 and still ongoing. This work formed the basis of *Manual de Plantas de Costa Rica*, the official project and main reference to the flora of Costa Rica. *Manual de Plantas* is a joint effort between the Missouri Botanical Garden, the National Museum of Costa Rica, and the Costa Rican Institute of Biodiversity. The Manual represents the complete inventory of the Costa Rican Flora (seed plants) consisting of eight volumes of which seven have been published, starting in 2004. The Manual includes native as well as non-native species that have either naturalized or have been cultivated at a commercial scale. This project brings together the efforts of local and international institutions as well as the expertise of botanists from different fields, who carry out the treatment of different species through the analysis of field specimens and herbaria (more than 300,000 specimens coming from herbaria harboring Costa Rican samples, i.e., the Missouri Botanical Garden, the Field Museum of Natural History, the Smithsonian Institution, and local herbaria such as the National Museum and the University of Costa Rica) constituting a sustained effort expanding botanical exploration to poorly sampled areas.

Keeping an updated list is becoming more difficult due to the lack of a consolidated database correcting for synonymia, taxonomic changes, new species, range expansions, and species extinctions. Thus, the expert opinion becomes very important. The basic trends that the Manual detected in terms of the abundance of plant families are the ones still in place. In this case, the most common families are Orchidaceae, Fabaceae, Poaceae, Asteraceae, and Rubiaceae. Bogarín et al. (2016) report 1,574 orchid species for Costa Rica (30% endemic), representing 8% of the orchid species diversity worldwide, with the genera *Lepanthes, Pleurothallis, Stelis* and *Epidendrum* as the most speciose. The total number is 256 species higher than the value reported in the Manual (1,318). Orchidaceae more than doubles the numbers reported for Fabaceae, Poaceae, Asteraceae, and Rubiaceae (Table 4.2).

4.7.3 Mammals

The latest list of Costa Rican mammals by Rodríguez-Herrera et al. (2014) reports 249 species. Two mice (*Diplomys labilis,* Ramírez-Fernández et al., 2015) and one shrew (*Cryptotis thomasi,* Woodman & Timm, 2017) were recently added to the list. When this chapter was written, the Bush Dog (*Speothos venaticus*) presence in Costa Rica was confirmed at Las

Table 4.2 Number of Plant Species Found in Costa Rica*

	Hammel et al. (2004)	M.H. Grayum (pers. com.)
Ferns and fern allies	1,112	1,125 (or 1,500 following Alexander Rojas, pers. com.)
Bryophytes	8,500	
Gymnosperms	13	13
Angiosperms	8,236	9,587**
Monocots	2,986	
Dicots	5,250	
Total seed plants	8,249	9,600
Total vascular plants	9.361	10,712

*Using data from Hammel et al., (2004) following the updates of M.H. Grayum (pers. com) and others.
**Current numbers of monocots and dicots are difficult to estimate. M.H. Grayum reports that disproportionately more monocots than dicots have been recently described, especially orchids.

Tablas Protected Zone close to the border with Panama (González-Maya et al., 2017), making for a total of 252 species (Rodríguez-Herrera et al., 2014). Cases like this, where rare species are expanding their range could be more common in the near future. The list is updated every 10 years, and more species are added including new records, changes in taxonomic classification (i.e., subspecies elevated to species), range extensions, and newly discovered species (i.e., Woodman and Timm, 2017). This shows that the inventory of mammal species is far from being over. The most speciose orders are Chiroptera with 114, Rodentia with 49, Cetacea with 31, and Carnivora with 25. Terrestrial mammals include 218 species. There are 24 endemic species concentrated above 1,500 m, especially the highlands of the Central Mountain Range, the Talamanca, and the Tilarán Mountain Range, all representing centers of endemism. The Talamancas cover a wide elevational gradient, are highly heterogeneous, and maintain one of the largest forested areas in southern Central America. They will continue to be a source of new species across different groups. The endemic mammal species include six rodents (*Orthogeomys heterodus, O. cherriei, Heteromys oresterus, H. nubicolens, Reithrodontomys rodriguezi,* and *R. musseri*). This number increases to 21 species if the Panamanian portion of the Talamanca Mountain Range and the western Chiriquí area is included. Two other rodent species are shared with Nicaragua, *Reithrodontomys brevirostris* and *R. paradoxus*, totaling 24 endemic species of mammals (this is, one primate -*Cebus imitator*-, 16 rodents, one forest rabbit –*Sylvilagus gabbi*, three shrews, and three bats).

Attention to the most speciose groups, such as bats, has generated greater sampling and monitoring, increasing the list of bat species. The Orders Cetacea, Chiroptera, and Rodentia will likely add more species in the near future. The list of bats went from 109 in 2002 (LaVal and LaVal, 2002) to 114 in 2014 (Rodríguez-Herrera et al., 2014). The impact of this interest comes from the creation of research collaborations, under the leadership of Bernal Rodríguez-Herrera and supported by Tirimbina Biological Reserve and the University of Costa Rica. These latter institutions serve as a hub to attract international researchers, provide research facilities, and train the local human resource necessary to maintain the interest.

Species collections will likely add more species as taxonomic work expands and comparative research is done regarding collections and field research (Rocha et al., 2014), in addition to the application of genetic tools together with standard taxonomic techniques. The Museum of Zoology at the University of Costa Rica (with more than 5 million specimens), as well as other institutions in the country (the Natural History Department at the National Museum), maintain rich collections whose contribution to the future taxonomic work will be fundamental. In addition, sampling and monitoring in still poorly explored habitats need to continue, especially regarding nocturnal and arboreal species difficult to detect not only due to their habits but also due to low densities (i.e., Córdoba-Alfaro, 2016).

4.7.4 Birds

In contrast to the rest of the groups, birds enjoy a privileged place because the species list is updated annually (Garrigues et al., 2016). In addition, the group is the focus of very well organized bird watchers that keep a real-time record of new sightins, range expansions, name changes, and possible extinction. The most updated bird inventories are maintained by the Costa Rican Ornithological Association, AOCR (Garrigues et al., 2016) as well as by the Union of Costa Rican Ornithologists, UCRO (Sandoval and Sánchez, 2017). In addition, other organizations keep annual bird counts (i.e., the La Selva Christmas Bird Count run by the Organization for Tropical Studies since 1985). Bird interest has been constant since early naturalists acknowledged the special value of Costa Rica as an area of high species diversity. The Guide to the Birds of Costa Rica by Stiles et al. (1989), a project that took 17 years to complete, was fundamental in heightening the attention and research focused on the birds in Costa Rica. It also increased tourism revenue since birdwatchers come in the thousands every year to enjoy the advantages of spotting birds from different habitats in a very diverse country. Daniel Janzen, one of the most important tropical ecologists, has pointed out the relevance of Stiles et al. (1989) in boosting interest in the natural sciences and creating revenue to the extent that nature tourism has become one, if not the main, source of economic income for Costa Rica.

The latest list from AOCR indicates 918 species (or 920 species following the UCRO list). These lists distinguish different types of endemic species. Many endemics in the Talamanca highlands extend their range into Panama (51 species), or along the western Pacific slope (18 species). Similarly, the Caribbean slope includes endemic species ranging from Honduras to Panama (25 species). Examples include the Hoffman's Woodpecker (*Melanerpes hoffmannii*) which is the only tropical dry forest endemic found from southern Honduras to the western slopes of the Central Valley in Costa Rica, and the Three-wattled Bellbird (*Procnias tricarunculatus*), whose range extends from Honduras to Panama.

The most speciose families are Tyrannidae (84), Trochilidae (52), Parulidae (51), Thraupidae (49), and Accipitridae (39). Individually, these species make up only 5% of the total number of species, which reflects the high species diversity of the avifauna. About 65% of the avifauna is resident, 20% are long-distance migrants, and the rest are elevational migrants. Approximate values apply since the status of some species still needs to be fully established.

Although the distribution and abundance of most species are well known, studies on the ecological factors determining distributional patterns following elevational gradients are still scarce (see Blake and Loiselle, 2000). Long-term bird inventories, aside from the Christmas bird counts, are absent for most of the country, although organizations such as AOCR are actively fostering citizen's science (May, 2013). There are a few studies including elevational gradients (Loiselle and Blake, 1992; Young et al., 1998), but these are the exceptions. Long-term projects are required to standardized methodologies to separate natural fluctuations in bird populations from environmental variation, such as habitat loss, the disturbance of remaining habitats, and responses to climate change.

More studies at the biogeographic and ecological level are required to determine the patterns that affect both the historical and ecological distribution of birds. Nevertheless, there are some studies that have analyzed the historical variation in species composition (Barrantes et al., 2009). Highland birds of western Talamanca (including Panama) may have the highest ende-mism of Central America.

4.7.5 Amphibians and Reptiles

The earlier works on herpetofaunal diversity are congruent with the contri-butions of the first naturalists that arrived into the country during the last quarter of the nineteenth century. Among these early contributors, Edward D. Cope (end of the nineteenth century) and Edward H. Taylor (mid-twenti-eth century) deserve special mention. However, it is the monumental work of Savage (2002) that establishes the baseline relative to what is known about the Costa Rican amphibians and reptiles. Savage (2002) lists 396 spe-cies of reptiles and amphibians. Sasa et al. (2010) updated these numbers and report 420 species (189 amphibians, 231 reptiles), of which 71 species (19) are endemic to the country.

Federico Bolaños, one of the leading experts on Costa Rican herpetol-ogy, is in the process of updating the list along with other collaborators. The current numbers include 454 species: 207 species of amphibians (148 frogs and toads, 51 salamanders, and 8 cecilids), and 247 species of reptiles (2 crocodiles, 230 squamata – 89 lizards and 141 snakes, and 14 to 15 turtles depending on whether the Green Turtle is considered as one or two species). This list has grown in 58 species since Savage (2002), and has experienced major changes due to the refinement of molecular biology techniques that have separated formerly cryptic species. Also, many species have expanded their range due to habitat loss and fragmentation, which favors the dispersal of some groups. Considering that Costa Rica represents only 0.03% of the Earth's area, it is striking that its amphibian and reptile diversity is compara-ble to that of Mexico, Colombia, or Brazil, countries that surpass Costa Rica several times in the area. This reflects the considerable diversity of Costa Rica at the global level, which is the highest in Mesoamerica. The number of amphibian species (189) represents 3% of the total number of species found worldwide, whereas the number of snake species (138) corresponds to 4.4% of the global total (Sasa et al., 2010). The list shows high diversity at the family level (16 families of amphibians and 30 of reptiles), and includes 47 and 108 genera of amphibians and reptiles, respectively, with anurans being the most diverse at the family level.

Sasa et al. (2010) identified 47 species of reptiles and amphibians occur-ring in five marine environments (including islands), which constitute 12% of the species reported for Costa Rica. On the Caribbean coast, herpeto-logical studies concentrate on sea turtle biology. On the Pacific coast, the research has focused on marine turtle conservation, dynamics of crocodile populations, and species use of various mangroves and islands (Sasa et al., 2009). Six sea turtles and one pelagic sea snake are considered truly marine, but many other species use salt marshes, mangrove swamps, and coastal beaches and dunes.

Interested readers should refer to Sasa et al. (2010) for a thorough analysis of the recent status of the Costa Rican herpetofauna. These authors analyzed the biogeographic origin, current distribution, current threats, conservation status, and taxonomic changes, providing an updated assessment of the work of Savage (2002). What follows highlights the major features of the herpetofauna, but readers looking for the fine details should refer to Sasa et al. (2010).

Most of the country has been thoroughly sampled producing extensive biological collections, such as the one at the Museum of Zoology at the University of Costa Rica. However, certain habitats still remain underexplored, such as insular environments, including Cocos Island (2 endemic lizards, *Anolis townsendi* and *Sphaerodactylus pacificus*), as well as the summits of the Talamanca Mountain Range. Information from the Museum of Zoology allowed Sasa et al. (2010) to map the diversity of reptiles and amphibians for most of the country and found the highest species richness associated to the Caribbean lowlands (Cahuita, Guayacán de Siquirres, and Puerto Viejo de Sarapiquí, with more than 81 species). High species richness is dispersed in contrasting areas of the country (40–80 species in Bajo La Hondura, Dominical, Golfito, Las Cruces de San Vito, the Manuel Alberto Brenes Biological Reserve in San Ramón, Rincón de Osa, and Sirena). There is an obvious sampling bias since places within the Central Valley have been collected with more intensity relative to other areas such as the northern Caribbean lowlands, the western slope of the Nicoya Peninsula and the Talamanca highlands. Much of the herpetological diversity is concentrated in places of high species richness separated by areas of low species diversity. Many factors come into play to explain this pattern, such as the habitat characteristics favoring the occurrence of certain stages of the amphibian life cycle, resource distribution, habitat fragmentation, and habitat loss. More research is required to separate natural fluctuations in the abundance of amphibians and reptiles from factors associated with habitat quality and resource distribution.

The distribution of the herpetofauna varies with life zones. General patterns follow the trends described by Duellman (1966) regarding vegetation types and climatic parameters. In this respect, temperature and precipitation, as well as vegetation complexity (reflected in the variety of microhabitats – types and number of perches, suitable breeding sites, leaf litter, forest strata) affect the distribution of amphibians and reptiles (Duellman, 1966). Less diverse life zones include the Lower Montane Moist Forest, which has only 4 species (Savage, 2002). This life zone corresponds to major populated areas in the Central Valley, which have removed the original vegetation since the beginning of the twentieth century. The Montane Wet Forest (6 species), Montane Rainforest (15), and Sub-Alpine Pluvial Páramo (5) have low species richness due to low temperatures, and extreme climatic conditions (daily and seasonal drops in temperature close to −11°C). The Páramo environment has a smaller area in Costa Rica relative to the South American Páramos, and thus the fauna associated with this habitat is poor. Therefore, there are only a few reptiles and amphibians in the Páramo including a few lizards (*Sceloporus malachiticus* and *Mesaspis monticola*) and one salamander (*Bolitoglossa pesrubra*), which have restricted daily habits due to low temperatures. The greatest species richness is found in Lowland Moist Forest (233 species) and Lowland Wet Forest (237 species), as well as in Premontane Wet Forest and Premontane Rainforest (205 and 203 species, respectively). These life zones have high levels of humidity, a short dry season, and a relatively high mean biotemperature, factors associated with high herpetological diversity (Duellman, 1966). As can be deduced from these patterns, there is a gradient in species richness following elevation, rain, and temperature distribution. Species richness decreases with elevation, but the drop is steepest after a 1,000 m. Lowland Moist Forest (233 species) and Lowland Wet Forest (237 species) on the Caribbean slope, and the Premontane Wet and Premontane Rainforest (205 and 203 species, respectively) show the highest species richness. These habitats are complex, warm, weakly seasonal, and maintain a relatively large area compared to other life zones. In addition, Premontane Wet and Premontane Rainforests are located at mid elevations (700–1,500 m), and have a mix of species from the highlands and the lowlands. Mid-elevations around 800 m represent one of the peaks in species diversity across groups (see Kohlmann et al. (2007) for the case of Scarabaeinae beetles), and amphibians and reptiles are not the exceptions.

The five ecoregions described in Savage (2002) and Sasa et al. (2010) summarize many of the differences mentioned above and highlight the effects of rain and temperature determining the distribution of faunistic and floristic groups.

1. *Pacific Northwest.* This region includes most of the lowlands in the Guanacaste province, such as the Nicoya Peninsula and the western

side of the Grande de Tárcoles watershed. Kohlmann and Wilkinson (2007) consider this latter area as a geographic barrier between biogeographic areas. Here, there is a higher prevalence of reptiles relative to amphibians, and some groups, which are typically dry forest species, reach their southernmost limits here (i.e., *Rhinophrynus dorsalis*).

2. *Pacific Southwest*. This region includes the Central Pacific lowlands starting in Carara National Park and south to the Osa Peninsula. The affinities are found with the Caribbean Slope as mentioned above with assemblages that got separated after the uplift of the Talamanca.

3. *Caribbean Lowlands*. These form a monolithic unit starting in the lower slopes of the Guanacaste, Central, and the Talamanca Mountain ranges, and is the largest in area. This is the second most diverse area for reptiles and amphibians in Costa Rica (141 species of reptiles and 95 of amphibians).

4. *Central Mountain Range*. This includes the region between 500 and 1,500 in elevation (the premontane belt). Constant rains and complex habitats give rise to the most diverse area of herpetofauna in the country (162 species of reptiles and 146 of amphibians).

5. *Talamanca Mountain Range*. This massive area has elevational gradients on the Pacific and Caribbean sides and a variety of habitats and life zones. Here, amphibian diversity surpasses reptile diversity (62 vs. 42 species), and is especially rich in salamanders. It represents a priority area for biological exploration.

4.8   Costa Rican Habitats Requiring Special Protection

What criteria intervene to define conservation priorities? An integrative approach should be implemented to allocate scarce resources and further research efforts to unexplored ecosystems that are in danger of being heavily disturbed or disappearing altogether. Species richness, the degree of functional diversity, biogeographic importance, and contribution to endemism, are all common criteria applied to identify conservation priorities based on an increased return of conservation benefits relative to the resource investment (Bottrill et al., 2008). The following habitats deserve more attention:

1. *Dry Tropical Forests:* This ecosystem has been historically degraded by human populations through the use of recurring fires to control succession. In addition, the dry forest has suffered intensive logging, hunting, as well as livestock and agricultural expansion, and more

recently, an urban expansion which has had considerable impacts on water resources in an ecosystem very sensitive to water shortages. Climate change will certainly stress dry forest areas with greater intensity. It would be reasonable to consider changing the economic activities that take place in these areas, especially livestock, agriculture, and the expansion of urban areas.

2. *Mountain Bogs:* These are environments with a very limited distribution covering a small geographic area and located mainly in the Talamanca mountain range. Although they contribute little to the number of species, mountain bogs maintain a unique flora and show high diversity at the family level. They present a window into the past: peat moss soils accumulate a pollen record that goes back to the Pleistocene, showing ample changes in vegetation since the last glaciation. Mountain bogs are considered wetlands and are very humid environments during the wet season but during the dry season they can lose most of the superficial water and are exposed to fires. Many mountain bogs are outside the system of conservation areas, and in addition to being fragile they also suffer plant extraction, and artificial draining to make room for livestock.

3. *Páramos:* These are ecosystems of relatively low species diversity, but adapted to climatic extremes. Páramos experience the conditions of a temperate summer and winter in the same day, being humid during the rainy season, but very dry environments during the dry season, and thus, exposed to anthropogenic fires. The flora of the páramos is well-known although many ethnobotanical uses are at risk of being lost. Mountain bogs and páramos take a long time to recover from fires.

4. *Cloud Forests:* These ecosystems are particularly diverse in the epiphytic component. They include unique environmental extremes, such deeply shaded environments and microhabitats sheltering miniature orchids. The complexity of the canopy and the epiphytic richness ensures the addition of new species in the near future. Cloud forests are being disturbed by climate change, as they depend on cloud cover and high humidity and such conditions are altered during prolonged dry seasons. Drier and longer dry seasons will also affect their phenological behavior, plant production, and foodwebs. Relatively little is known about the effects of global warming on the cloud forest biota.

5. *Elevational Gradients:* Costa Rica is a mountainous country where elevational gradients predominate. Maintaining continuous altitudinal corridors and habitat mosaics is critical for elevational migrants

as well as for resident species that may move along the corridor on a seasonal basis. This condition applies to both rainforests and dry forests. However, many altitudinal gradients are outside protected areas. More attention needs to be devoted to the Central Pacific region, which, in addition to elevational gradients, represents a transition zone between the tropical dry forest and the tropical lowland rainforests. This region of Costa Rica is experiencing a major boom in the urban expansion, especially close to the coastline.

6. *The Rainforests of Corcovado National Park:* This area harbors a rich biota with strong Amazonian affinities. This national park has suffered recurrent impacts from gold miners, hunters, loggers, and drug traffickers, and is now under a high visitation pressure. Recently, the park system established agreements with neighboring communities so that they bring visitors to the national park. Corcovado represents an area where many of the pressures affecting protected areas act. In the same way, the solutions and the management of the biodiversity of Corcovado will serve as an example for other protected areas.

4.9 Conclusions

The history of conservation in Costa Rica has shown a constant contrast of success and conflict. The country has achieved important goals, such as the official protecting of 26% of its territory in conservation areas, reducing the historic rates of deforestation of the 1980s, reaching 53% of the country's surface covered by forest and generating almost all of its energy from renewable sources in recent years. However, population demands on natural resources are increasing due to urban expansion, the growth of the road infrastructure, the expansion of high impact monocultures such as pineapple, a greater demand for water, increased pollution, continuous dependence on fossil fuels for transportation, and the lack of adaptation to climate change. Clearly, Costa Rica is reaching a turning point. Biodiversity conservation and management cannot be separated from increased demands for ecosystem services. Human communities must benefit from conservation if the protected areas are to survive.

Conventional causes of biodiversity loss (hunting, extraction, habitat loss, introduction of invasive species) are now compounded with the fast pace of climate change, which will certainly accelerate the rate of species extinction. As the process of global warming continues, it is more evident that the country is exposed to extreme climatic events. As climatic phenomena become more frequent and intense, it becomes more urgent to implement firm policies to balance economic development with natural resources. Costa Rica serves again as a small laboratory to implement solutions, or to observe the dire consequences of inaction.

Biodiversity inventories require more field biologists doing species monitoring and working along with national parks and local communities. Taxonomic work represents the foundation of successful management and conservation science. Consistent biological inventories translate into better proxies to understand human impacts on ecosystems. Nowadays, a career in biology represents one of the fields where it is more difficult to secure a job. Unfortunately, the country has a crisis of priorities. Biodiversity inventories, as well as the analysis of the distribution of species, are crucial conditions to understand and maintain the functionality of ecosystems and of the services they provide. However, there are few job positions to address these processes.

Currently, there is no consolidated system to keep track of the number of species across groups. The National Institute of Biodiversity (INBIO) helped to perform this role and maintained databases for a diverse array of plant and animal groups. INBIO's efforts consolidated the expertise of ecologists and taxonomists who recur to INBIO to exchange and consult data. However, this integrated effort was short-lived, and although some of the online databases are still functional they are not up to date. The information on species diversity is dispersed and back into the domain of experts, scattered institutions, and university departments. INBIO, through its publishing house, made biological information available to the general public and popularized the biological knowledge of many groups through field guides. The field guides were used not only in Costa Rica but also in the less-known neighboring countries. In its heyday, INBIO developed innovative technological platforms such as the generation of a barcode associated with each specimen. In addition, INBIO shared the information and cataloging methods with similar institutions in other countries, demonstrating the benefits of biological diversity. The progressive disappearance of this institution has been a great loss for Costa Rica.

The National System of Conservation Areas (SINAC) needs to be strengthened. National Parks and other conservation areas suffer a high degree of isolation and edge effects even though deforestation inside protected areas has decreased significantly since the 1970s (Sánchez-Azofeifa et al., 2003). This is a country where diversity is associated with topographic heterogeneity, which requires that protected areas are laid out strategically to include elevational gradients. However, protected areas covering elevational gradients are the exception.

SINAC was born with the aim of providing autonomy to protected areas while keeping a landscape ecology vision by grouping national parks according to geographic location so that they could meet their needs and plan their budgets without waiting for decisions made in San José. However, most of the decision power remained in the central office. Recently, (as of 2017), the country's highest court of law gave SINAC about 5 years to improve the management and infrastructure of protected areas. This includes protection (stopping hunting and extraction as well as improving park connectivity) and visitor services. National parks possibly contribute 60% of the Costa Rican GDP indirectly, and generate a few million dollars annually only in entrance fees. These economic benefits dilute in the bureaucratic maze of SINAC. Aggressive actions are required to reduce SINAC's bureaucratic apparatus and get more boots on the ground.

A concerted, multidisciplinary effort, from all sectors of the society, is required to understand and respond to the challenges of balancing economic growth with biodiversity protection. Achieving this goal while confronting climate change represents the most serious test Costa Rica will face in the near future.

Acknowledgments

Many people contributed with papers, unpublished data, and expressed their opinions. I would like to thank Paul Hanson, Roy H. May, Alvaro Herrera, Bernal Rodríguez-Herrera, Alexander Rojas, Mario Blanco Coto, Federico Bolaños, Angel Solís, Julieta Carranza, Arturo Angulo, Michael H. Grayum, Odalisca Breedy, Diego Bogarín, Mahmood Sasa, Mitzi Campos, Luis Ricardo Murillo-Hiller, and Cindy Fernández. Catie Morris proofread the manuscript and added comments that significantly improved the writing.

Keywords

- amphibians
- mammals
- taxonomic knowledge
- reptiles

References

Angulo, A., Garita-Alvarado, C. A., Bussing, W. A., & López, M. I., (2013). Annotated checklist of the freshwater fishes of continental and insular Costa Rica: Additions and nomenclatural revisions. *Check List, 9*(5), 987–1019.

Barrantes, G., (2009). The role of historical and local factors in determining species composition of the highland avifauna of Costa Rica and western Panamá. *Revista de Biología Tropical, 57*, 333–346.

Blake, J. G., & Loiselle, B. A., (2000). Diversity of birds along an elevational gradient in the Cordillera Central, Costa Rica. *The Auk, 117*(3), 663–686.

Bogarín, D., Pupulin, F., Smets, E., & Gravendeel, B., (2016). Evolutionary diversification and historical biogeography of the orchidaceae in Central America with Emphasison Costa Rica and Panama. *Lankesteriana, 16*(2), 189–200.

Bottrill, M. C., Joseph, L. N., Carwardine, J., Bode, M., Cook, C., Game, E. T., & Pressey, R. L., (2008). Is conservation triage just smart decision making? *Trends in Ecology & Evolution, 23*(12), 649–654.

Brown, B. V., Borkent, A., Cumming, J. M., Wood, D. M., Woodley, N. E., & Zumbado, M. A., (2009). *Manual of Central American Diptera, vol. 1.* National Research Council, Ottawa, Ontario, pp. 1–714.

Brown, J. W., Janzen, D. H., Hallwachs, W., Zahiri, R., Hajibabaei, M., & Hebert, P. D., (2014). Cracking complex taxonomy of Costa Ricanmoths: Anacrusis Zeller (Lepidoptera: Tortricidae: Tortricinae). *J. Lepidopterists'Soc., 68*(4), 248–263.

Burger, W. C., (1980). Why are there so many kinds of flowering plants in Costa Rica? Por qué hay tanta variedad de plantas angiospermas en Costa Rica? *Brenesia, 17,* 371–388.

Chacón, I., Ramirez, J. J. M., & Herrera, J. A., (2007). *Mariposas de Costa Rica: Orden Lepidoptera.* Inst. Nacional de Biodiversidad.

Coates, A. G., (1997). The Forging of Central America. In: *Central America: A natural and cultural history,* (pp. 1–37) Yale University Press, New Haven, CT.

Córdoba-Alfaro, J., (2016). Presence and distribution of *Sphiggurus mexicanus* (Rodentia: Erethizontidae) for the South Pacific lowland of Costa Rica. *Brenesia, 85–86,* 72–74.

Costello, M. J., May, R. M., & Stork, N. E., (2013). Can we name Earth's species before they go extinct? *Science, 339*(6118), 413–416.

Duellman, W. E., (1966). The Central American herpetofauna: An ecological perspective. *Copeia, 4,* 700–719.

Ehrlich, P. R., & Wilson, E. O., (1991). Biodiversity studies: Science and policy. *Science, 253*(5021), 758.

Erwin, T. L., (1982). Tropical forests: Their richness in Coleoptera and other arthropod species. *Coleopterists Bulletin, 36*(1), 74–75.

Flenley, J. R., (1995). Cloud forest, the Massenerhebung effect, ultraviolet insolation. In: *Tropical Montane Cloud Forests.*(pp. 150–155) Springer, New York.

Garrigues, R., Araya-Salas, M., Camacho-Varela, P., Montoya, M., Obando-Calderón, G., & Ramírez Alán, O., (2016). *Official List of the Birds of Costa Rica – Updated in 2016.* Scientific Committee, Costa Rican Ornithological Association. Zeledonia, 20, 2.

Gaston, K., Gauld, I., & Hanson, P., (1996). The size and composition of the Hymenoptera fauna of Costa Rica. *J. Biogeography, 23*(1), 105–113.

Gómez, L. D., & Savage, J. M., (1983). Searchers on that rich coast: Costa Rican field biology, 1400–1980. *In: Costa Rican Natural History,* (pp. 1-11) University of Chicago Press, Chicago.

Gómez-Laurito, J., & Gómez, L. D., (1989). *Ticodendron:* A new tree from Central America. *Annals of the Missouri Botanical Garden, 76*(4), 1148–1151.

González-Maya, J. F., Gómez-Hoyos, D. A., & Schipper, J., (2017). First confirmed records of the bush dog (Carnivora: Canidae) for Costa Rica. *Neotropical Biology and Conservation, 12*(3), 238–241.

Hammel, B. E., Grayum, M. H., Herrera, C., & Zamora, N., (2004). *Manual de Plantas de Costa Rica,* volumen I. Introduction, Monographs in Systematic Botany from the Missouri Botanical Garden, *92,* 1–299.

Herrera, W., & Goméz, L. D. P., (1993). *Mapa de Unidades Bióticas de Costa Rica.* Instituto Nacional de Biodiversidad, INBIO, Heredia.

Herrera, W., (2016). *Climate of Costa Rica. Costa Rican Ecosystems, (pp. 19–29)* University of Chicago Press, Chicago.

Hilje-Quirós, L., (2013). Los primeros exploradores de la entomofauna costarricense. *Brenesia, 80,* 65–88.

Holdridge, L. R., & Grenke, W. C., (1971). *Forest Environments in Tropical Life Zones: A Pilot Study.* Pergamon Press.

Janzen, D. H., & Hallwachs, W., (2016). *Biodiversity Conservation History and Future in Costa Rica: The Case of Área de Conservación Guanacaste (ACG). In: Costa Rican Ecosystems,* (pp. 290-341). University of Chicago Press, Chicago.

Janzen, D. H., (1967). Synchronization of sexual reproduction of trees within the dry season in Central America. *Evolution, 21*(3), 620–637.

Janzen, D. H., (1987). Insect diversity of a Costa Rican dry forest: Why keep it, and how? *Biol. J. Linn. Soc., 30*(4), 343–356.

Joppa, L. N., Roberts, D. L., & Pimm, S. L., (2011). The population ecology and social behavior of taxonomists. *Trends in Ecology & Evolution, 26*(11), 551–553.

Kappelle, M., (2016). *Costa Rican Ecosystems.* University of Chicago Press, Chicago.

Kappelle, M., Van Uffelen, J. G., & Cleef, A. M., (1995). Altitudinal zonation of montane *Quercus* forests along two transects in Chirripó National Park, Costa Rica. *Vegetatio, 119*(2), 119–153.

Kohlmann, B., & Wilkinson, M. J., (2007). *The Tárcoles Line: Biogeographic Effects of the Talamanca Range in Lower Central America.* La línea de Tárcoles: Efectos biogeográficos de la Cordillera de Talamanca en la parte baja de Centroamérica. Giornale Italiano di Entomologia, *12*(54), 1–30.

Kohlmann, B., Solis, A., Russo, R., Elle, O., & Soto, X., (2007). Biodiversity, Conservation, and Hotspot Atlas of Costa Rica: A Dung Beetle Perspective (Coleoptera: Scarabaeidae: Scarabaeinae), *Zootaxa,* 1457, 1–34.

Kohlmann, B., Wilkinson, J., & Lulla, K., (2002). *Costa Rica From Space.* EARTH University, San José, Costa Rica.

Lachniet, M. S., & Seltzer, G. O., (2002). Late Quaternary glaciation of Costa Rica. *Geological Society of America Bulletin, 114*(5), 547–558.

Lamar, W. W., & Sasa, M., (2004). A new species of hognose pit viper genus *Porthidium* from the southwestern Pacific of Costa Rica (Serpentes: Viperidae). *Rev. Biol. Trop., 51,* 797–804.

Larsen, B. B., Miller, E. C., Rhodes, M. K., & Wiens, J. J., (2017). Inordinate fondness multiplied and redistributed: The number of species on earth and the new pie of life. *The Quarterly Review of Biology, 92*(3), 229–265.

Laurito, C. A., & Valerio, A. L., (2013). *Scirrotherium antelucanus*, a new species of Pampatheriidae (Mammalia, Xenarthra, Cingulata) from the Upper Miocene of Costa Rica, Central America. *Revista Geológica de América Central, 49*, 45–62.

LaVal, L. K. H., & Bernal, R. K. L., (2002). Murciélagos de Costa Rica. Costa Ricanbats (No. Sirsi) i9789968702638).

Lieberman, D., Lieberman, M., Peralta, R., & Hartshorn, G. S., (1996). Tropical forest structure and composition on a large-scale altitudinal gradient in Costa Rica. *J. Ecology,* 137–152.

Loiselle, B. A., & Blake, J. G., (1992). Population variation in a tropical bird community. *BioScience, 42*(11), 838–845.

Luteyn, J. L., Churchill, S. P., Griffin, III, D., Gradstein, S. R., Sipman, H. J. M., & Gavilanes, A., (1999). A checklist of plant diversity, geographical distribution, and botanical literature. *New York Bot Gard, 84*, 1–278.

MacFadden, B. J., (2006). Extinct mammalian biodiversity of the ancient New World tropics. *Trends in Ecology & Evolution, 21*(3), 157–165.

Marshall, L. G., (1988). Land mammals and the great American interchange. *American Scientist, 76*(4), 380–388.

May, R. H., (2013). *En los pasos de Zeledón: Historia de la ornitología nacional y la Asociación Ornitológica de Costa Rica (No. 598. 072347286 M466)*. Asociación Ornitológica de Costa Rica, San José (Costa Rica).

May, R. M., (2010). Tropical arthropod species, more or less? *Science, 329*(5987), 41–42.

O'Dea, A., Lessios, H. A., Coates, A. G., Eytan, R. I., Restrepo-Moreno, S. A., Cione, A. L., & Stallard, R. F., (2016). Formation of the Isthmus of Panama. *Science Advances, 2*(8), e1600883.

Obando, V., (2002). *Biodiversidad en Costa Rica: Estado del Conocimiento y Gestión*. Editorial INBio, Santo Domingo de Heredia, Costa Rica.

Ramírez-Fernández, J. D., Córdoba-Alfaro, J., Salas-Solano, D., & Rodríguez-Herrera, B., (2015). Extension of the known geographic distribution of *Diplomyslabilis* (Mammalia: Rodentia: Echimyidae): First record for Costa Rica. *Check List, 11*(5), 1745.

Rocha, L. A., Aleixo, A., Allen, G., Almeda, F., Baldwin, C. C., Barclay, M. V., & Berumen, M. L., (2014). Specimen collection: An essential tool. *Science, 344*(6186), 814–815.

Rodríguez-Herrera, B., Ramírez-Fernández, J. D., Villalobos-Chaves, D., & Sánchez, R., (2014). Actualización de la lista de especies de mamíferos vivientes de Costa Rica. *Mastozoología Neotropical, 21*(2), 275–289.

Sánchez-Azofeifa, G. A., Daily, G. C., Pfaff, A. S., & Busch, C., (2003). Integrity and isolation of Costa Rica's national parks and biological reserves: Examining the dynamics of land-cover change. *Biological Conservation, 109*(1), 123–135.

Sandoval, L., & Sánchez, C., (2017). *Lista de Aves de Costa Rica: Vigésimo Quinta Actualización. Unión de Ornitólogos de Costa Rica*, (http://uniondeornitologos.com/wp-content/uploads/2017/07/Lista-de-Aves-de-Costa-Rica-XXV.pdf).

Sasa, M., Chaves, G. A., & Patrick, L. D., (2009). Marine reptiles and amphibians. *Marine Biodiversity of Costa Rica, Central America*, pp. 459–468.

Sasa, M., Chaves, G., & Porras, L. W., (2010). The Costa Rican herpetofauna: Conservation status and future perspectives. *Conservation of Mesoamerican Amphibians and Reptiles*, pp. 510–603.

Savage, J. M., (2002). *The Amphibians and Reptiles of Costa Rica: A Herpetofauna Between Two Continents, Between Two Seas.* University of Chicago Press.

Stiles, F. G., & Skutch, A. F., (1989). *A Guide to the Birds of Costa Rica.* Cornell University Press.

Stork, N. E., McBroom, J., Gely, C., & Hamilton, A. J., (2015). Approaches narrow global species estimates for beetles, insects, and terrestrial arthropods. *Proc. Nat. Acad. Sci., 112*(24), 7519–7523.

Tosi, J. A., & Holdridge, L. R., (1966). *Ecological Map of Costa Rica.* Tropical Science Center, San Jose, San Pedro.

Webb, S. D., (2006). The great American biotic interchange: Patterns and processes. *Annals of the Missouri Botanical Garden, 93*(2), 245–257.

Wehrtmann, I. S., & Cortés, J., (2008). *Marine Biodiversity of Costa Rica, Central America,* vol. 86. Springer Science & Business Media.

Werckle, C., (1909). *La sub region fitogeográfica costarricense.* Soc. Nac. Agric, Switzerland.

Wilson, E. O., (2017). Biodiversity research requires more boots on the ground. *Nature Ecology & Evolution, 1*(11), 1590.

Woodman, N., & Timm, R. M., (2017). A new species of small-eared shrew in the *Cryptotis thomasi* species group from Costa Rica (Mammalia: Eulipotyphla: Soricidae). *Mammal Research, 62*(1), 89–101.

Young, B. E., DeRosier, D., & Powell, G. V., (1998). Diversity and conservation of understory birds in the Tilarán Mountains, Costa Rica. *The Auk,* 998–1016.

Zamora, N., & Pennington, T. D., (2001). *Guabas y cuajiniquiles de Costa Rica (Inga spp.).* Editorial INBio.

Plate 4.1   (a) Summit of Cerro de La Muerte, páramo vegetation; (b) Botos Lagoon in Poás National Park; (c) View towards Peñas Blancas Valley from Monteverde Cloud Forest Reserve; (d) Oak forest at Cerro de La Muerte; (e) Mid-elevation forest (500 m) in Braulio Carrillo National Park; (f) Giant oak tree at Cerro de La Muerte; (g) Playa Naranjo, Santa Rosa National Park.

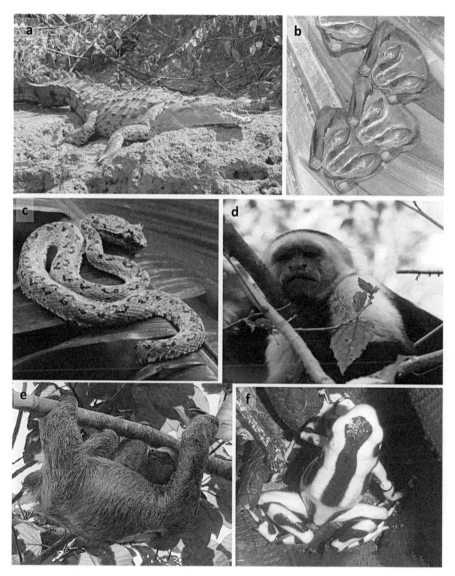

Plate 4.2 (a) American Crocodile (*Crocodylus acutus*); (b) Tent-making Bat (*Uroderma bilobatum*); (c) Eye-lash Viper (*Bothriechis schlegelii*); (d) White-headed Capuchin (*Cebus imitator*); (e) Brown-throated Three-toed Sloth (*Bradypus variegatus*); (f) Green-and-black Poison Dart Frog (*Dentrobates auratus*).

Plate 4.3 (a) Resplendent Quetzal (*Pharomachrus mocinno*); (b) Violet Sabrewing (*Campylopterus hemileucurus*); (c) Coppery-headed Emerald (*Elvira cupreiceps*); (d) Fiery-throated Hummingbird (*Panterpe insignis*); (e) Green Violetear (*Colibri thalassinus*); (f) Gartered Trogon (*Trogon caligatus*); (g) Volcano Hummingbird (*Selasphorus flammula*); (h) Royal Flycatcher (*Onychorhynchus coronatus*); (i) Double-toothed Kite (*Harpagus bidentatus*).

CHAPTER 5

Biodiversity in Cuba

DENNIS DENIS,[1] DARYL D. CRUZ-FLORES,[2] and ERNESTO TESTÉ[3]

[1]Professor at the University of Havana, Cuba
[2]Researcher at Ecology and Systematic Institute, Cuba
[3]Researcher at National Botanical Garden of Cuba, E-mail: dda@fbio.uh.cu

5.1 Introduction

The Cuban archipelago is located in one of the 35 hotspots of biodiversity on the planet. These places represent regions of an exceptionally high concentration of natural ecosystems, species and endemisms (Zachos and Habel, 2011). Despite this, the Insular Caribbean contains one of the most threatened biota due to the high density of the human population and other pressures of socioeconomic origin (Shi et al., 2005). Within the region, Cuba is the largest island, which together with its proximity to the continent, diversity of ecosystems and biogeographic history, makes it a key nucleus for the conservation of the region's biodiversity.

In the Cuban archipelago, inhabits the largest number of plant and vertebrate species of the Antilles and hosts a high percentage of exclusive species. In particular, it is considered among the most diverse insular territories in plants worldwide and is the first island in the number of species of this group per square kilometer (Gonzalez-Torres et al., 2016). The terrestrial biota of Cuba exhibits, excluding protozoa, algae and bacteria, around 25, 733 known autochthonous taxa. Insects are the largest percentage, followed by fungi and plants (angiosperms and gymnosperms), which together represent 76% of the known Cuban terrestrial biodiversity. These figures underestimate the complete diversity of life, since some groups have received little or no attention from taxonomy specialists to date. Additionally, molecular biology techniques are currently allowing the identification of numerous cryptic and species complexes (i.e., Alonso et al., 2012; Agnarsson et al., 2016; Doadrio et al., 2009).

Cuba possesses 58.5% of the vascular plants and 52.2% of the birds described or registered in the Antilles, and in other groups, it hosts up to a quarter of the species. The vertebrates are the best-known group and on which a large part of the resources dedicated to conservation fall; however, these represent only 2.6% of the Cuban terrestrial biota. The Cuban terrestrial fauna is composed of 16,553 species, highlighting as the most diverse groups insects and mollusks, within invertebrates, and birds and reptiles, within vertebrates.

The Cuban fauna is typically insular and is characterized by a high endemism – produced by the *in situ* evolution of most species, the strong prevalence of invertebrates over vertebrates (93.5% of fauna), the absence of megafauna, and the predominance of flying forms over terrestrial. An important feature is the emergence of groups with strong adaptive radiation processes, through which the represented groups diversify extraordinarily, producing many species related to each other, but which occupy very different habitats and resources. Such is the case of the genus *Anolis*, widely spread throughout the Caribbean islands; frogs of the genus *Eleutherodactylus*, ants of the genus *Leptothorax* and mollusks of the family Urocoptidae.

The Cuban fauna is free of animals considered to be a serious danger to humans, as there is no poisonous species, wild beasts or large carnivores (except, perhaps, oceanic sharks or larger crocodiles). On the other hand, extreme variations in size can be found in the Cuban fauna with the presence of some world records: one of the smallest frog *Eleutherodactylus iberia* (less than 10 mm); the smallest bird, bee hummingbird (*Mellisuga helenae*), locally named "zunzuncito" with 6 cm long and less than 2 g in weight; the smallest bat, known as the Butterfly bat (*Natalus lepidus*), with 22 cm wingspan and body weight of about 4–5 g and the largest of the insectivores, which is the Cuban solenodon("almiquí[1]") (*Solenodon cubanus*).

The most distinctive character of the Cuban terrestrial fauna is, without doubt, its marked endemism and extreme localization of animal forms. The relative antiquity (upper Eocene) of a good portion of the Cuban territory and the independent evolution of the main components of its biota due to geographic isolation conditioned by insularity and the repeated rise and fall of sea level throughout its geological history that produced sequences of isolations and reconnections of vast territories that emerged over thousands of years, gave rise to the high endemism that characterizes the Cuban archipelago (Iturralde-Vinent and MacPhee, 1999; Hedges, 2006).

The extreme localization of many species arose from this *in situ* evolutionary process. Thus, many species of mollusks reside only in a single mogote (a special and unique type of karstic elevation from western Cuba), differing even in races according to the slope they occupy. Several species of lizards and frogs live exclusively on one mountain or in a specific locality. Certain species of ticks are only known from the hot rooms of some particular caves. Mollusks are the invertebrates with the highest percentage of endemism, while for vertebrates this distinction corresponds to amphibians, followed by reptiles.

---

1   Many Cuban nature elements, such as geographic places, animals, and plants, received common names from Native American pre-Columbian's language.

In Cuba, five terrestrial Ecoregions of importance for conservation are recognized, according to the categorization of the World Wide Fund for Nature, these are: wet and dry forests, pine forests, wetlands, xeromorphic thickets, and mangroves. These are areas with particular ecological, climatic and geomorphological characteristics, which are considered important for global conservation because, among other values, they host numerous flora and fauna endemisms, species of relict distribution and high species richness.

The knowledge of the Cuban biodiversity could be considered among the largest of the islands of the world, due to the great efforts developed in its inventory and the long history of animal and plants researches of Cuban specialists. Although some specific groups of biota have been little or nothing studied, most of the major groups have exhaustive inventories and are updated by the National Biodiversity Center, a governmental institution created in 1998, which edited the first "National Study on Biological Diversity in the Republic of Cuba."

5.1.1 *Geography, Geology, and Climate Conditions*

Cuba is a tropical archipelago formed by the island of Cuba with 104,556 km^2, the Island of Youth, of 2,204 km^2, and around 4,195 cays and adjacent islets grouped into four archipelagos: Los Colorados, Sabana-Camagüey, Canarreos and Jardines de la Reina, totaling an area of 3,126 km^2. It also includes a submarine platform of 67,831 km^2, which yields a total area of 178,753 km^2, whose biodiversity can be considered as Cuban.

The relief is mostly flat, in more than 80% of the area and the rest are areas of moderate height and mountains. The four most important orographic groups are: Guaniguanico, Guamuhaya, the Nipe-Sagua-Baracoa, and Sierra Maestra mountain ranges, which are located in the western, central and eastern regions of the island, respectively. In the Sierra Maestra, the Turquino Peak and the Cuba Peak, with 1974 and 1874 m a.s.l., are the highest elevations of the Cuban archipelago. The mountainous regions of Cuba represent centers of speciation, climatic refuges and exclusive sites of many endemic and endangered lineages of plants and animals, although the natural vegetation in these places shows high to moderate values of fragmentation. The secular trends of vertical displacement of the different parts of the island produce a longitudinal twist that makes the south coast mainly low and flooded and the north coast high and rocky, except for the southeast region that tends to rise around 15 mm by year.

More than 66% of the earth's surface is formed by karstified rocks and there is a high density of caverns and caves. Due to the shape of the island, the rivers in Cuba are not large and have a short route, from the center of the island to the north coast or to the south coast. There are 573 river basins, most are dammed and the rivers of greater extension are the Cauto, Zaza, Sagua la Grande, Caonao, and the Toa.

The climate of Cuba is determined by the long and narrow configuration of the island, which makes the sea temper the tropical conditions, in addition, by the proximity to North America, the little-rugged relief, and insularity. The archipelago is exposed to the trade winds that act from the northeast in winter and from the east-northeast in summer. In general, the climate is moderate tropical and maintains a marked seasonal cycle of rainfall and with the temporary influence of continental masses of cold air. The rainy season extends from May to October and the least rainy season from November to April. The rainfall regime reaches an average of 1,375 mm per year, but its spatial distribution is not homogeneous. In the extreme northeast, it can reach 3,000 mm and in the south of Guantanamo 600 mm a year. The temperatures are generally high and the annual average values go from 22°C to 28°C or higher in the southern coast of the eastern region, magnitudes lower than 20°C are reported in the highest parts of the mountainous areas. The records of the maximum average temperature are between 27°C and 32°C and the minimum average temperature between 17°C and 23°C.

The topical hurricanes or cyclones that characterize the region may also have had an important catalytic effect within the *in situ* evolution of Cuban wild populations of plants and animals, which are naturally small and may be affected by them. This relative annual climate variability, however, has a long-term stability, since it is assumed that there have been no notable changes during the last 40 million years, which has also contributed to the high biodiversity of the biota.

Cuba, as an archipelago, is of relatively old origin with a very complex historical biogeography and geological structure, since it is at the junction of two tectonic blocks. The western and central region of Cuba belongs to the North American plate that continues with the Bahamas and Florida. The eastern region is part of the Caribbean plate, a system of submerged valleys and ridges that emerge in Cozumel and Yucatan. But the Sierra Maestra and the entire southeastern portion is part of the Cayman Dorsal, through whose base a geological link is established with northern Central America.

In the Jurassic–Cretaceous stage, in the place of Cuba, there was a platform that bordered the continental slope, with a chain of active volcanic islands. It is not until the Eocene, approximately 40 million years ago, that the folded substrate of Cuba emerges, shaping three archipelagos that correspond to the current mountain ranges. Finally, to the Holocene, the rocky massifs that constitute the substrate of Cuba occupy their current position in relation to North and Central America, with mainly vertical movements. From the Upper Miocene – Pliocene, the silhouette of Cuba and the Isle of Youth began to take current shape, but the actual shape of the archipelago dates back only 1 million years. Throughout these historical periods the islands were covered with colonizing flora and fauna from the continent that began to evolve *in situ* at a very rapid rate due to speciation processes exacerbated by the long isolation stages with intermediate periods of geographical union and biological mixture, with consecutive separations due to changes in sea levels, which gave rise to the high endemism that characterizes its biodiversity.

The complexity of the processes of origin of the Cuban archipelago, its insularity condition, the relative stability of some of its regions and climate, the high geological variety of the soils, and the tropical and intercontinental location have acted synergistically with the evolutionary process of life to give rise to the Cuban biota, with a high number of endemism that reaches the density of 1 endemism for every 12 km^2.

5.2 Main Ecosystem Characteristics

5.2.1 Coastal Ecosystems

Due to the shape of the island of Cuba and its character as an archipelago, coastal ecosystems contain an important part of its biodiversity. The coasts and the marine platform of Cuba are formed by a varied set of habitats, among them the most important are wetlands (marshes and muddy areas), coastal lagoons, mangroves, marine seagrass, coral reefs, rocky coastline, and sandy beaches.

The cumulative coasts of sandy beaches are coastal ecosystems of low biological diversity due to their physical homogeneity, low productivity, and high turbulence, although they are home to many invertebrates species associated with sand. In general, they have vegetation composed of numerous small plant species, with xerophytic, squat, and succulent leaves, due to salt stress. They usually contain naturally a complex of herbaceous plants adapted to salinity; towards the interior, on rocky substratum there are abundant uverales or seagrape forests (*Coccoloba uvifera*) associated with small palms (*Thrinax* spp.) and Gum Tree (*Bursera simaruba*). Frequently, the sandbanks are associated with dense marine grasses and patch reefs. However, its greatest importance in Cuba is linked to beach and sun tourism, which is today one of the main economic activities of the country and constitutes a danger to biodiversity if it is not developed in a controlled manner.

The abrasive or karstic coasts can be high or low, and alternate with the extensive sandy beaches. They are covered with vegetation, higher from the strip subjected to the direct action of the waves. Generally, they are plant formations called coastal xeromorphic scrub, with microphylla shrub species that present a marked xerophyticism, in the part closest to the sea the vegetation is constituted, mainly, by succulent plants, shrubs, and herbs.

Towards the interior of the sea, the seagrass ecosystems, known in Cuba as *seibadales*, stand out: unconsolidated sediment bottoms (mud and sand) with the development of seagrasses and algae. This biotope is the most widespread in the Cuban platform (more than 50% of its surface), especially in the Gulf of Batabano, the Sabana-Camagüey Archipelago and the northern platform of Pinar del Río. The extraordinary primary production in seagrasses makes them the main source of energy input that sustains the biological and fishing productivity of the platform, which is exported to the reefs through the species that feed on them. They constitute an important habitat for the settlement and breeding of the juvenile stages of many commercial species and an important fishing substrate. They also act as bottom stabilizers, prevent the erosion of reefs and beaches, regulate the concentration of oxygen and carbon dioxide in the sea, and in many areas, they produce a large part of the beach sands. In the Cuban archipelago, there are six species of marine herbs among which *Thalassia testudinum* and *Syringodium filiforme* are the most abundant. The leaves of *T. testudinum* are generally covered by epiphytes such as hydrozoans, small anemones, serpulid worms, the sponge *Chondrilla nucula*, as well as some macroalgae. Colonies of the coral *Manicina areolata* are also very common in the seagrass beds. The invertebrate fauna associated with the *T. testudinum* ecosystem are sponges, some of them commercially exploited (genus *Spongia* and *Hippiospongia*, and *Ircinia strobilina*), polychaetes, bivalves (*Pinna carnea, Atrinarigida, Arcophagia fausta, Ark zebra,* and *Barbatia cancellaria*), gastropods (*Strombus gigas, Cassis madagacarensis, C. tuberosa, C. flammea, Turbo castanea, Cerithium algicola,* and *C. litteratum*), spiny lobsters (*Panulirus argus*), decapods (*Petrochirus diogenes, Dardanus venosus* and *Clibanarius* spp.), echinoids (*Lytechinus variegates* and *Meoma ventricosa*), starfish (*Oreaster reticulatus, Echinaster echinophorus,* and *E. spinulosus*) and sea cucumbers (*Stychopus badionotus* and *Holothuria mexicana*). The vertebrates are characterized by many fish species, mostly juveniles of the Lutjanidae, Serranidae and Pomadacidae families, which feed on small polychaete worms and nondecapod crustaceans among other food items.

Farther from the coast, coral reefs appear: massive structures of biological origin, solid and with diverse forms, created mainly by around 40 species of scleractine corals and other sessile organisms such as gorgonians, sponges, algae and tubular polychaetes. The coral reefs of Cuba are considered among the most beautiful in the Caribbean and are an important tourist attraction. They are inhabited by a great diversity of crustaceans, sea urchins, holothurians, asteroids, ophiuroids, mollusks, bryozoan colonies and numerous fish of striking shapes and colors, being the most diverse habitat and rich in species and forms among marine habitats. Coral reefs have great ecological value, representing a vital area of refuge, feeding or reproduction of a large number of species. They also have economic importance because of their ecosystem services of protection against wave action, protecting the coast from erosion. More than 50% of Cuba's reefs are separated from the main island by extensive seagrasses or soft bottoms and cays.

5.2.2 Freshwater Ecosystems

The rivers and lakes are underrepresented in the country, due to their elongated and narrow shape. There are, however, more than 200 small rivers of little course and little flow. The longest is the Cauto River that runs from east to west in the eastern zone, with an enormous system of 32 tributaries that add up to 343 km of channels and cover an area of 8,969 km². The river of highest flow is the Toa, also in the eastern region, with 58 tributaries that pour their waters into the central channel that forms a basin of 1,052 km². There are some lagoons with significant sizes such as La Leche lagoon in Ciego de Ávila, with an area of 67 km² or the Tesoro lagoon, in Matanzas, with 9 km². The scarcity of lakes is compensated by more than 2,226 dams that, together with the associated channel systems, make up 40.5% of the inland water surface of the island of Cuba (310,676 ha). These bodies of water are used by many birds, particularly migratory species, which use them as places to rest or feed.

There is a large number of coastal lagoons inserted in mangroves and estuarine complexes, with an approximate area of 9,500 km². These are bodies of shallow water connected to the sea by narrow channels or estuaries, with little saltwater exchange but still under the influence of the tides. In the southern coastal marshes, for example, there are 6,393 ha of lagoons and estuaries; while in the mangrove system of the Birama swamp, the surface of lagoons, estuaries, and inlets exceeds 6,500 ha.

5.2.3 *Wetland Ecosystems*

In the wetland inventory of the neotropical region, the main listed Cuban wetlands represent about 15.7% of the national territory. This value, however, is underestimated since it does not include large parts of the coastal marshy strip that borders, practically, the entire south coast of the island. The most important wetland systems in Cuba are the Zapata swamps, the Lanier swamp, on the Isla de la Juventud, and the Birama swamp (Mugica et al., 2006). These wetlands contain a very diverse fauna, which includes 186 species of birds, 57 of freshwater fish, and innumerable invertebrates and marine fish.

The Zapata marshes are the largest system of wetlands in the Caribbean, located in the province of Matanzas. It has a length of 175 km and a mean width of 14 to 16 km, with a maximum span of 58 km. The territory consists of marine surfaces with the presence of carbonated rocks (carso) in two well-defined blocks: the western swamp and the eastern swamp both separated by the Bay of Pigs. The territory contains one of the most extensive speleo lacustrine systems of the Antilles, characterized by a layer of underground water under an extensive system of karstic rocks, with numerous geological accidents such as casimbas, cenotes, and rocky lagoons. It contains important hydrological resources, areas of reproduction, spawning, and development of marine and terrestrial species of high economic value, systems of submerged marine terraces and highly unique coral reefs. In the Zapata marshes, there are more than 212 species of vertebrates (17.9% endemic), and a high variety of invertebrates. This region is one of the most important in Cuba because of the diversity of birds (250 species) and because it has a large number of endemic and threatened species (21 and 16, respectively). In the area of Santo Tomas, there are two local endemic bird species, unique because of their restricted distribution: the Zapata Wren and the Zapata Rail that face a serious risk of extinction. In addition, to birds, this extensive wetland maintains the largest wild populations of the Cuban crocodile, in addition to the American crocodile. Among the mammals, the greater fishing bat, the manatee and the dwarf hutia stand out. In addition, to the fauna there are more than 900 species of plants with 13% of endemism.

The Birama swamp is the second largest wetland in Cuba and the Caribbean, with an area of 57 048 ha. Located at the mouth of the Cauto River, it is the largest water basin in the country. It is a deltaic swamp that, when it joins the rest of the mangroves of the Guacanayabo Gulf, covers part of the southern coastal areas of the provinces of Granma and Las Tunas. In this place are the healthiest mangrove formations in the country with trees reaching 20 m of height and many of their areas have remained relatively undisturbed due to the difficulty of access. It is an intricate system of estuaries, lagoons and swamps of singular beauty and high degree of wilderness. It contains large populations of waterfowl, which move between its mangroves, estuaries, rice fields, and nearby shrimp farming ponds. The swamp also hosts the largest and densest world breeding population of *Crocodylus acutus*. It is an important wintering site for migratory ducks and shorebirds, which use the enormous lagoons of fresh and brackish water of the system. The wetland is declared a Ramsar site and contains two wildlife refuges: Delta del Cauto and Monte Cabaniguán (Mugica et al., 2006).

Coastal mangroves cover about 240,000 km^2 of the coasts of the Cuban archipelago, which constitutes about 4.8% of the territory emerged, and represent 26% of the forest area, which places Cuba at the head of the Caribbean in this natural resources. This complex ecosystem is located on coasts of biological origin, cumulative, swampy and in the estuaries with freshwater runoff, although also in typical saline environments. They are plant formations of very low diversity, dominated mainly by four tree species: *Rhizophora mangle* (red mangrove), *Avicennia germinans* (mangrove prieto), *Laguncularia racemosa* (pataban) and *Conocarpus erectus* (yana). But they also host up to 115 species of plants, belonging to 85 genera and 46 families, of which 28 are trees, 17 shrubs, 44 herbs, 15 lianas, 10 epiphytes, and 1 hemiparasite. At least 10 species of endemic plants are directly related to mangrove ecosystems. A large number of species of algae is associated with the submerged part, of which 22 species of Chlorophyceae, 18 Rodophyceae, and 7 Phaeophyceae have been described.

The submerged roots of the mangroves serve as a substrate and refuge for a high diversity of marine species, some of them of commercial value. These forests contribute energy to the aquatic habitat, through its leaves, branches, and roots, which become part of the detritus accumulated in the sediments. Mangroves protect coasts and other habitats from the erosion platform that causes waves, winds and coastal currents, filter pollutants and prevent them from reaching coral reefs.

5.2.4 *Forest*

In Cuba, several vegetal formations have been described, but in a general way, they can be grouped in forests, scrub, herbaceous vegetation, vegetation complexes, and secondary vegetation. There is a high degree of fragmentation and isolation of the natural vegetation nuclei and the areas that still retain a certain degree of naturalness and representativeness of the terrestrial biota constitute only about 10% of the Cuban archipelago.

5.2.5 Caves

The Cuban archipelago has about 70% of its surface covered by calcareous rocks and has one of the highest densities of caves per unit of area in the insular Caribbean, which has favored an important specialization in the use of these shelters by various zoological groups (Nuñez Jiménez et al., 1988). The hypogeous or subterranean ecosystems constitute an ecological entity with very particular characteristics and high levels of endemism. The Cuban archipelago has the richest and most diverse speleofauna of the Antilles, composed by 807-recorded species, mostly invertebrates. The most representative groups in cave environments are arachnids, insects, myriapods, and crustaceans. Warm or hot caves, with environmental conditions created by bats of the families Phyllostomidae and Mormoopidae, represent 1% of the total caves of the country, but some of them can sustain populations of more than 100,000 bats in each.

5.3 Organisms and Taxonomic Groups

5.3.1 Cuban Plant Biodiversity

The complexity and uniqueness of the flora of Cuba has captivated more than two centuries the attention of numerous botanists, naturalists and collectors of plants, both Cuban and foreign. Among the main explorers that the island of Cuba has received during this time are the German botanists Alexander von Humboldt and Johanes Bisse, the Swedish Eric L. Ekman, the North Americans Charles Wright and Nathaniel Britton and the French Priests Leon and Alain, just to mention some of the great connoisseurs of the flora of Cuba. Among the Cuban botanists can be noted for their great contributions to José A. de la Ossa, Antonio Ponce de León and Julián Acuña.

There are many criteria about the number of species that make up the flora of Cuba, although most authors give estimations between 7,000 and 7,500 plant species. Within these estimates are included both native species and exotic species present in Cuba. In agreement with González-Torres et al. (2016) Cuba's native flora is composed of 6,950 taxa, and by 337 exotic taxa (Oviedo and González-Oliva, 2015), which gives a total of 7,287 plants present in Cuba. These estimates place Cuba as the botanical richest island territory worldwide (Whittaker and Fernández-Palacios, 2007) and the first island in terms of plant species per square kilometer.

One of the most important characteristics of the flora of Cuba is its high degree of endemism, and of the total number of native species reported about 50% (3,075 taxa) are endemic, many of which are local endemics, whose distribution is restricted to small areas of the Island. Another important aspect in the endemic flora of Cuba is the existence of 63 endemic genera, many of which are monotypic. This high degree of endemism, both local and regional, and the condition of an island, makes the Cuban flora highly vulnerable to different environmental and anthropic pressures, which makes the flora of Cuba highly threatened. At present, 66% of the native taxa present in Cuba have already been evaluated, of which 46% are included in some threat category and more than half of them represent exclusive taxa of Cuba (González-Torres et al., 2016). The extensive analyzes on the state of conservation of the flora of Cuba, has allowed us to realize the extinction of 22 taxa.

5.3.1.1 Bryophytes

The bryophytes include all the nonvascular plants, which are taxonomically grouped into three divisions: Hepatophyta, Bryophyta, and Anthocerotophyta. In Cuba, these divisions together comprise about 929 taxa, most of which are distributed to the eastern region of the Island. The best-represented division for Cuban flora is Hepatophyta, which has 498 taxa, grouped into 107 genera and 32 families The Bryophyta division is composed of 416 species, with only 18 endemics, 165 genera and 48 families, and the Anthocerotophyta division, which is the scarcest, has only 15 species, 6 genera and 3 families (Duarte, 1997; Motito, 2007; Rivera, 2011). The greatest diversity of mosses (Bryophyta) is concentrated in the mountain ranges Sierra Maestra and Nipe-Sagua-Baracoa, in eastern Cuba. Most Cuban mosses are terrestrial, rupicolous and epiphytic. Its distribution is strongly influenced by humidity and light intensity. One of the aspects that most influences the diversity of mosses in Cuba is the height, living the majority of the taxa between 500 and 1,000 msm (Motito, 2007). The majority of the anthocerotes are of terrestrial habitats and they inhabit humid, open places, disturbed and mineralized soils of rivers and mountains; although it is also common to find them in cultivated areas and on rocks in streams of water (Rivera, 2011). In the case of liverworts, many of the ecological aspects remain unknown. As you can see the bryophytes have been and are one of the least studied groups of flora in Cuba, and perhaps worldwide.

5.3.1.2 Lycophytes and Ferns

The latest update on lycophyte and fern species from Cuba (Sánchez, 2017) states that for Cuba there are 599 taxa reported, grouped into 37 families and 131 genera. Lycophytes are the least diverse of the two groups, only composed of 53 species, 17 of which are endemic to Cuba. These 17 endemic taxa belong to the genus *Selaginella,* and represent more than half of the family, which suggests that Cuba represents an important point of diversification of the genus (Sánchez, 2017). The ferns represent 546 species, with only 76 endemics. The degree of endemism of these groups is quite low, as their spores are capable of traveling distances of up to 3 200 km and a single spore can potentially establish a species in a new site (Regalado et al., 2015). Regarding the degree of threat of this group, 54.5% of its taxa are in some category of threat (González-Torres et al., 2016).

5.3.1.3 Gymnosperms

The gymnosperms are a little diverse group in Cuba, which has 18 species grouped into 4 families and 5 genera. Of the total species reported for Cuba, 11 are endemic (61%) and 11 species are in some category of threat. The most diverse family of this group is Zamiaceae, with 8 species, among which stands out *Microcycas calocoma*, considered as National Monument of Cuba and one of the 50 most threatened plants of Cuba (González-Torres et al., 2013).

5.3.1.4 Angiosperms

Flowering plants make up the majority of Cuban plant species. It is composed of 5,404 taxa, grouped into about 1,129 genera and 271 botanical families. The diversity of flowering plants in Cuba is enormous, and these can be found throughout the Island, from high fertility soils to areas with high concentrations of heavy metals. In Cuba, this group can have a terrestrial, epiphytic, rupicolous, aquatic and marine habit, and the highest density of species is found in the main mountainous groups (Guaniguanico, Guamuhaya, Sierra Maestra, and Nipe-Sagua-Baracoa) and towards the east of Cuba.

The most diverse botanical family in Cuba is Rubiaceae, with 478 taxa, followed by Asteraceae (322 taxa) and Orchidaceae (296 taxa). There are four other families that exceed 200 taxa (Myrtaceae, Poaceae, Cyperaceae and Euphorbiaceae). The majority of these families have high percentages of endemic species. As for the large genera, the main genera are *Eugenia*

(Myrtaceae) with 107 taxa, *Rhynchospora* (Cyperaceae) with 87 taxa and *Miconia* (Melastomataceae) with 86 taxa. Some of these data may rise in the future product of the continuous taxonomic updates of these families, which have been working for many years due to the existence of the "Flora of Cuba Project"; that although it also includes bryophytes, lycophytes and ferns and gymnosperms, the flowering plants are the group that has received the most attention from botanists.

Up to this point, we had seen that the flora of Cuba is characterized by the high diversity of plant groups and above all its high values of endemism and the fragility of its species to different external pressures. But this is not the only thing that makes the flora of Cuba so special and unique. The flora of the Island includes several ancient lineages, among which stands out *Leugenbergia zinniflora,* which is an arboreal cactus endemic to Cuba. We can also find unique species such as *Pinguicula lignicola* and *P. jackii* (Lentibulariaceae); unique insectivorous plants that live on limestone; the rest of the Cuban species grows on nutrient-poor substrates such as silicic sands and serpentinites (González-Torres et al., 2013).

Within the Caribbean region, Cuba is recognized as an important center for plant diversification and radiation, which has greatly influenced plant diversity and the high values of endemism. One of the most relevant examples of Cuba as a plant diversification center is that of the *Buxus* genus (Buxaceae). Worldwide, this genus has about 100 species, 37 of which are in Cuba, 35 of which are endemic to the Island (Kohler, 2014). Other examples are the genera of *Copernicia* palms, which of 21 species worldwide 16 are endemic to Cuba and *Coccothrinax,* which has 43 of the 50 species reported for the genus. But these are not the only ones, other genera such as *Leptocereus* (Cactaceae), *Roystonea* (Arecaceae) and *Antillanthus (*Asteraceae).

5.3.1.5 Exotic Plants

For several centuries in Cuba, many exotic species have been introduced for ornamental purposes, cultivars or by the unconscious action of man. Many of these introduced species have acclimated in an extraordinary way to the environmental conditions of the Island, which has led to their establishment and expansion. For this reason, Cuba currently has 337 invading exotic taxa, 53% of which are herbs. Most of these taxa come from Asia and Continental America (Oviedo and González-Oliva, 2015). Of this group of invasive species, 100 taxa stand out as the most harmful, among which are *Dichrostachys cinerea, Hedychium coronarium, Hyparrhenia rufa, Melaleuca quinquenervia,* among others.

5.3.1.6   Plant Formations

The high diversity of plants present in Cuba is mainly due to the high diversity of types of plant formations or ecoregions, which depend directly on the type of soil where they are grown. Depending on the classification there are more or less vegetable formations; the most used classification gives Cuba 29 different types of plant formations (Capote and Berazaín, 1984). This high diversity of vegetation is largely due to the complex geological formation of Cuba, which is recognized as a large mosaic of different types of soils. Each zone of Cuba has large variations of these types of soils, which directly influence the types of plants that grow on them.

The most biodiverse formations are the mountain rainforests, the xeromorphic bushes on serpentines, the coastal and subcoastal bushes and the pine forests. All these types of vegetation have high concentrations of endemic species. Perhaps the most extreme of the plant formations are those that develop on serpentines (xeromorphic spiny thorny scrub). These formations conditions are so extreme that many plants species are endemic and with unique characteristics. The most extreme case is found in the serpentines of Yamanigüey in the north of eastern Cuba, where about 90% of the flora is endemic to the region; making it the most biodiverse place in the entire Caribbean Hotspot.

5.3.2   *Terrestrial Invertebrates*

The greatest diversity of animals on the island corresponds to terrestrial invertebrates, whose fauna is estimated at approximately 11,700 species (Vales et al., 1998). However, this great diversity is one of the aspects that make the knowledge of the group in Cuba not so deep. In addition, the small size often presented by individuals, which makes identification difficult, and the lack of experienced taxonomists also influence the lack of studies focused on many of its members. Among the groups of invertebrates that inhabit the island, greater knowledge is available for mollusks, insects, and arachnids. As for other taxa of the Cuban fauna, the areas of higher diversification and representativeness of the invertebrates on the island are the major mountain systems such as Guaniguanico, Guamuhaya, Sierra Maestra, and Nipe-Sagua-Baracoa. These regions also constitute the areas of greatest endemism in Cuba.

Cuban invertebrates have been included in few conservation studies, due to the lack of information that exists on distribution, population parameters and trends. Currently, 130 under genus taxa (98 species and 32 subspecies) of terrestrial invertebrates are categorized under the IUCN system and they belong to Gastropoda (82), Insecta (38) and Arachnida (10). Of the total of categorized species, 46 falls into the category of Critically Endangered, 24 Threatened and 27 as Endangered (Hidalgo-Gato et al., 2016).

5.3.2.1 Insects

Insects stand out among the most diverse invertebrate groups in Cuba. As far as currently registered, the Cuban entomofauna is composed of 25 orders and 8,459 species. However, some estimates suggest that the number of species on the island may be around 15,000. Compared to other Caribbean islands, knowledge about the taxonomy of some insect orders in Cuba is relatively high. However, the information related to the ecology, population dynamics and composition of the assemblages, is still scarce and there are large territories of the island where entomological studies have not been carried out at all. The most diverse orders are Coleoptera (beetles) with 32% of the total species numbers, followed by Lepidoptera (butterflies and moths, 18% of the known species), Hymenoptera (wasps, bees, and ants, 13%) and Hemiptera (bugs and aphids, 15%).

In the Cuban archipelago, there are 2,673 species of beetles, belonging to 87 families, and about 56% are endemic (Peck, 2005). However, the number of species must be greater considering that they can be found in many diverse habitats in regions not yet explored. Among these, we can mention aquatic or semiaquatic habitats, habitats associated with leaf litter, soil, decomposing organic matter and semi-decomposed wood in forested areas. Lepidoptera, the second best order of insects represented in Cuba, contains 1,557 species (around 289 are exclusive) included in 863 genera, 56 families and 27 superfamilies (Núñez and Barro, 2012). At present, they are the best-known taxonomically group of Cuban insects. The third order in representativeness is Hymenoptera, with 1,069 reported species, grouped into 14 superfamilies, 49 families, and 474 genera. Within this, bees are the group most studied so far, in terms of their taxonomy, although some studies have been conducted on ecological and biogeographical aspects.

Endemism varies significantly between and within the different orders of insects, in correspondence with differences in the dispersion abilities and biogeographic history of the species (i.e. Matos-Maraví et al., 2014). Some researchers have estimated that around a third of the known insects of the Cuban archipelago are endemic to this territory. According to others, the figure could be between 40% and 60%. But it is not easy to calculate the real number of species composing this fauna, so estimations vary from three times higher than the current one to more than 2,000 species to be discovered. There are extreme groups where the endemism exceeds 90%, as in the wood insects (Phasmatodea) and the mutilids (Hymenoptera) (Genaro and Tejuca, 1999).

5.3.2.2 Mollusks

Mollusks are one of the most diverse and endemic invertebrate group in Cuba. Despite the relatively recent origin of the Cuban archipelago, the mollusk fauna of Cuban archipelago is one of the most diverse in the world (Espinosa and Ortea, 1999). To date, 1391 species have been inventoried, of which 910 are stylomatophores, 476 prosobranchs and 5 pulmonary systematomaphores (Hernández et al., 2017). The values of endemism reported for this group range between 95% and 96%, equivalent to state that by every 100 species of terrestrial mollusks found in Cuba only four or five may be found elsewhere. The high endemism of this group, together with the microlocalization and very low population abundance, make up a very distinctive faunistic scenario at the Antillean and world level, which has led to Cuba being recognized as "the paradise of terrestrial snails." The restricted localization of some of these species is so marked that it even surpasses many of plants of our native flora. Terrestrial mollusks are one of the intensively studied and best-known zoological groups in Cuba and are represented in all terrestrial ecosystems on the island.

The Cuban terrestrial prosobranch species are grouped into 52 genera and 6 families. However, these figures are not definitive, since they may vary according to the taxonomic criteria followed and the frequent presence of synonyms in many species names. The endemism within this group of mollusks in Cuba is also very high (99.1%), only four species are not endemic. The family Annulariidae (= Pomatiidae), of the Littorinoidea superfamily, is the most numerous with 352 species and 772 subspecies, arranged in 4 subfamilies, 35 genera (31 endemic) and 74 subgenera (66 endemics). Of the remaining 5 families of prosobranchs present in Cuba, Helicinidae is the next richest with 75 species located in 3 subfamilies, 11 genera (4 endemic species). The other families that make up this group are Megalomastomatidae (33), Truncatellidae (7), Proserpinidae (2), and Neocyclotidae (Hernández et al., 2017).

The Urocoptidae family is the largest in Pulmonata, with 585 species (and 781 subspecies). Other families well represented on the island are Cerionidae (90), Cepolidae (59), Oleacinidae (38), Subulinidae (33) and Pleurodontidae (28).

The genus *Polymita* Beck, 1837 with six polychromic species is recognized as one of the most notable land snail for the beauty, brilliance and colorful variability of the shells. All the species of this genus are critically threatened (Hidalgo-Gato et al., 2016). The anthropogenic land uses and other environmental impacts have caused a marked reduction and fragmentation of their natural habitats. In addition, the uncontrolled recollection and illegal trade of shells for crafts or souvenirs for a century have caused serious declines in polymitas' population currently decimated to critical levels.

Another land snail genus also emblematic of the Cuba island is *Liguus*, represented by four species: *L. fasciatus* (Müller, 1774), *L. blainnianus* (Poey, 1851), *L. flammellus* Clench, 1934 and *L. vittatus*, the first shared with Florida and the last three endemic to Cuba. In the Caribbean only La Hispaniola host another species of *Liguus*. All four species are distinguished by a remarkable high chromatic diversity, which is evident in the beauty of their shells. This also has made them focus of indiscriminate recollections.

5.3.2.3 Arachnids

The Arachnida class constitutes one of the groups of terrestrial invertebrates best-represented in the Cuban archipelago. The fauna of Cuban arachnids stands out for presenting all the orders described worldwide and has 1,328 species (75% endemic) (Hidalgo-Gato et al., 2016). Within the insular Caribbean, Cuba is the country of greatest diversity and highest information about its spiders, mites, and ticks. According to literature published from 1897 to 2005, there are about 682 species, 340 genera and 129 families registered for mites and 32 species of ticks, 12 of which are endemic (Barros-Batesti et al., 2006).

In Cuban nature, there are 600 species of spiders and 247 are endemic (Mancina et al., 2017). Most of the Cuban spiders are grouped in the Araneidae family (13.2% of the total species), in Salticidae (12.8%) and in Theridiidae (9.5%), which, taken together, contain one-third of all those described or registered in this country. At the other extreme, there are 16 families that are only represented in our fauna by one or two species.

Another very diversified order of arachnids in Cuba are harvestmen (Opiliones), scorpions (Scorpiones) and the schizomids (Schizomida), with 68, 56, and 55 species reported. Also, 38 species of false scorpions (Pseudoscorpiones) are currently known, with 16 endemics. The remaining orders are represented in Cuban fauna by only few species: Amblypygicontains 16 known species (12 endemic), Solifugae 7 endemic species, Ricinulei, 10 endemic species, and Palpigradi one species, also endemic.

The orders of arachnids best studied in Cuba to date are: Scorpions, Schizomida, Araneae, Thelyphonida, Amblypygi and Solifugae. Since 1970, it is rare to find a year in which several new species of Cuban arachnids was not described and the process continuous (see Plate 5.2 for an example).

5.3.3 Terrestrial Vertebrates

In Cuba, about 655 species of terrestrial and freshwater vertebrates have been recorded. Stand out for the number of species the groups of birds and reptiles, and for the high percentage of endemism amphibians, reptiles and freshwater fish. Fifty-seven species of freshwater fish, 62 species of amphibians, 155 species of reptiles, 368 species of living birds and 35 species of mammals have been recorded in Cuba, but the numbers slightly varies in relation to taxonomic disputes.

Fishes with fluvial habits in Cuba are grouped into 36 genera and 19 families, and 23 of the species (40.35%) are endemic. These numbers make Cuba the Caribbean country with the highest species richness and endemism, within this group of vertebrates. Among the main values of the Cuban ichthyofauna are three endemic genera: *Girardinus* with seven species and the monotypic *Alepidomus* and *Quintana*. The families Bitithidae and Lepisosteidae include very charismatic species such as cavefishes and the manjuarí (Cuban gar) (*Atractosteus tristoechus*). Cavefishes include four species of the genus *Lucifuga*, adapted to live in groundwater, where light almost does not penetrate or in absolute darkness, which is why they have undergone strong adaptive changes (slim bodies, depigmentation, loss of vision, etc.). The Cuban gar belongs to a group of fish of the Carboniferous period with the body protected by enameled or armored scales. Another family of extraordinary scientific interest is Poeciliidae, which includes the genus *Gambusia*, fishes of small sizes that ingest large quantities of mosquito larvae. Some invasive fish have been introduced and become serious threats to other aquatic species populations, such as African catfish (*Claria* sp.) and trout (*Micropterus salmoides*).

With 62 known to date species and 95% of endemism, Cuba is home to almost a third of all Antillean amphibians, all members of the Anura order, in four families, two of which have a single representative: Family Ranidae with the Bull Frog (*Lithobates catesbeiana*) and the Family Osteopilidae with the Cuban tree frog (*Osteopilus septentrionalis*). The list of autochthonous species includes eight toads of the genus *Peltophryne* and 52 small frogs of the genus *Eleutherodactylus*. Only three species are not exclusive to Cuba: *Eleutherodactylus planirostris*, *O. septentrionalis* and *L.catesbeianus* (the bull frog, introduced by man) (Díaz and Cádiz, 2008). Some of the smallest frog species in the world are found in Cuba and belong to genus *Eleutherodactylus* (*E. limbatus* and *E. iberia*). *Eleutherodactylus iberia*, lives in the area of the Toa river, Guantanamo province, and has approximately 10 mm snout-to-cloaca length.

The Cuban herpetological fauna has 142 species of reptiles, grouped in 29 genera, 17 families and three of the four living orders, of which 110 are endemics (Rodríguez Schettino, 2003). These numbers also place Cuba, within the Antilles, in second place in terms of species richness, outreached only by Hispaniola. The fauna of reptiles of Cuba does not possess poisonous or toxic species, nor that attack man if they are not stressed in refuges or resting places. Among the most charismatic species is the largest of lizards: the iguana (*Cyclura nubila*), and the largest of the ophidians, known as Majá de Santa María (Cuban boa) (*Epicrates angulifer*), which can reach 3 m in length, but it is totally harmless, both endemic genera of the Antilles. Among the very small sized reptiles is *Sphaerodactylus schwartzi*, considered the second smallest reptile in the world, with a snout-to-cloaca length between 18 and 20 mm. Some reptiles are used as food for human consumption (loggerhead, crocodile, freshwater turtle, etc.), to use the skin or carapace for the production of handicrafts (crocodile, ophidians and turtles). The fat of some species has a wide medicinal use for the cure of some respiratory and osteomuscular affections (crocodile and Cuban boa) and the eggs of the hawksbill turtle are attributed aphrodisiac properties.

Three reptile genera are endemic of Cuba (*Cricosauria* – 1 species, *Chamaleolis* – 4 species and *Cadea* – 2 species). Among the lizards there is a supernumerary genus, *Anolis*, with 55 species, and possibly many more to be discovered. The genus *Leiocephalus*, in spite of having only 6 species, 5 of them are endemic (*L. carinatus, L. cubensis, L. stictigaster, L. macropus, L. onaneyi* and *L. raviceps*), and shows an elevated genetic diversity expressed in 39 subspecies. Another species with this genetic variability among reptiles is *Ameiva auberi*, a unique species with 22 subspecies throughout the national territory.

In addition, in Cuba, there are 25 species of ophidians, 18 endemic, and four species of sea turtles (family Chelonidae) frequent the Cuban waters. There is also a freshwater turtle species *Trachemis decusata*, an endemic subspecies under a permanent legal protection and classified as Nearly Threatened due to the remarkable decline in natural populations for its use as food or pet.

Finally, there are three species of crocodiles inhabiting Cuba, as apical predators in some of the main wetland ecosystems. The most important for animal conservation is the Cuban crocodile (*Crocodylus rhombifer*), considered Critically Threatened and of which there is still very little information about the status of wild populations. This species is included in category I of the Convention on International Trade in Endangered Species of Wild Fauna and Flora (CITES). It has the smallest geographical distribution of the group since it is restricted to the Zapata swamp, in densities between 11 and 105 individuals per square kilometer, although in the past it was widely distributed throughout the island. It is a type of freshwater crocodile of medium size, which reaches between 3.5 m and 4.9 m long. Its main threats come from hunting, which at the beginning of the last century resulted in more than 90,000 individuals in a period of 10 years until it was legally prohibited in 1967. Also, it is affected by habitat destruction, by introduced exotic species and by hybridization with *Crocodylus acutus*, which endangers the genetic integrity of the species. The conservation of this crocodile has been focused on captivity breeding programs, for which there are five large hatcheries in the country, from which controlled reintroductions are carried out in natural places.

The American crocodile (*Crocodylus acutus*) has also suffered a sharp reduction in its populations, although it remains locally common. Its largest populations are concentrated in the Birama swamp, around the mouth of the Cauto River, where one of the largest nesting sites in the region is located. In the swamp of Lanier is very abundant the alligator known as babilla (*Caiman crocodylus*), an introduced species that has reached a wild population estimated in more than 40,000 individuals.

Within vertebrates, birds are the most diverse group, having registered for Cuba 369 species, of them 149 residents and 220 are migratory. Of the resident species 22 are endemic, including three endemic genera: *Xiphidiopicus, Priotelus, and Teretristis*.

'The Cuban Trogon or Tocororo (*Priotelus temnurus*), national bird of Cuba, and the Bee hummingbird (*Mellisuga helenae*) stand out for their color and charisma. Bee hummingbird is the smallest bird in the world with 5.5 cm from the tips of the beak to the tail. Also, remarkable bird species are the Cuba Tody or Cartacuba (*Todus multicolor*) that nest underground and the Cuban Parrot or Cotorra (*Amazona leucocephala*) that has been very exploited as pet by its ability to imitate phrases of the human language. The last three bird species discovered by science in Cuba have a very restricted distribution to a locality in the center of the Zapata Swamp and were the Zapata Wren (*F. cerverai*), Zapata rail (*C. cerverai*) and Zapata Sparrow (*T. inexpectata*).

There are also many introduced species of birds: *Gallus gallus, Meleagris gallopavo, Pavo cristatus, Columba livia, Cairina moschata, Passer domesti-*

cus, *Numida meleagris, Phasianu scolchicus*, among others. Some of the native birds are subjected to a strong hunting pressure for captivity as songbirds, such as the Cuban Finch (*Tiaris canora*), yellow-faced grassquit (*Tiaris olivacea*), the Cuban Bullfinch (*Melopyrrha nigra*), northern mockingbird (*Mimus polyglottos*) and the Indigo bunting (*Passer inacyanea*). In the last century, several species such as the Cattle Egret (*Bubulcus ibis*), the Fulvous Whistling-duck (*Dendrocygna bicolor*), the White-cheeked pintail (*Anas bahamensis*) and the Cowbird (*Molothru sbonariensis*), arrived and naturally colonized the Cuban territory. More recently, Cuba was colonized by tricolored munia (*Lonchura malacca*) and the Eurasian collared dove (*Streptopelia decaopto*).

The Cuban fauna of mammals, like in all West Indies, are characterized by a low diversity of species. In the Cuban archipelago, there are only 38 species of native terrestrial mammals. These are included in the orders Chiroptera (bats), Rodentia (jutías) and Insectivora (almiquí). In prehistoric times this fauna was more extensive, existing many species of sloths (*Xenarthra*), monkeys (Primates) and other rodents and insectivores already extinct (Cooke et al., 2017).

Chiroptera are the most diverse mammal order in Cuba, with 27 bat species, 17 of which use the caves as shelters and 12 are strictly cave dwellings. The latter includes *Phyllonycteris poeyi*, known as the "hot cave bat," for almost exclusively inhabiting this type of refuge, forming colonies of at least 15,000 individuals, the highest concentrations of mammals observed in the region (Silva Taboada, 1979). Also, it is remarkable the butterfly bat, also one of the smallest bat in the world.

The Hutias are a group of rodents of the family Capromyidae, exclusive of the Antilles, of the genera *Capromys, Mespocapromys,* and *Mysateles*. These herbivorous mammals, with essentially arboreal and nocturnal habits, have had great adaptive success and the 10 species in Cuba are endemic. Some of them are widely distributed, such as the Conga Hutia (*Capromys pilorides*) and the CarabalíHutia (*Mysateles prehensilis*) but others have very restricted distribution, inhabiting only some small mangrove cays. The rat hutia (*Capromys auritus*) is a local endemic of Fragoso cay, categorized as Critically Endangered due to the small size of its unique population, the restricted nature of its habitat and the threat of tourism development. The Cuban solenodon is an insectivore species considered described as a "living fossil" due to inhabit the island for millions of years, but at the arrival of the Spanish first sailors it was already a rare species. Their status, according to the latest edition of IUCN, is Critically Endangered/Extinct, but some individuals are supposed to survive only in very remote corners of the Sierra Maestra and the mountains of Sagua – Baracoa.

5.3.4 Cuban Marine Biodiversity

The Cuba platform is one of the most biologically diverse among the islands of the Western Hemisphere. Nearly 550 species of marine bacteria are known, but microorganisms, in general, have been relatively little studied. There are about 950 plant species (micro and macroalgae and vascular plants). The number of known invertebrates surpasses the figure of 4,500 species and that of vertebrates about 1,040 (mainly fish). In the Cuban seas, about 50 species of solitary or colonial ascidians and 255 sponges are known, of which 4 are economic resources: *Hippospongia lachne, Spongia graminea, S. obscura,* and *S. barbara.* In Cuba, 55 species of gorgonians are known, all of them hermatypical. The most important families are Gorgoniidae, and Plexauridae. In Cuba, 60 coral entities are known, including species, subspecies and ecological forms.

Marine mollusks are one of the most numerous groups of invertebrates. Oysters, clams and mussels contribute 2,000–2,400 tons annually to the economy in Cuba. There are more than 1,000 species of crustaceans present in Cuban waters, being one of the groups of greatest diversity, density and biomass in the marine ecosystem. Among the decapod crustaceans (crabs, shrimps, lobsters) have been reported about 500 species that mainly inhabit the benthos of all coastal and deep ecosystems. Among them are the most important fishing resources of Cuba: spiny lobster and shrimp. Among the nondecapod crustaceans there is a remarkable variety of taxonomic groups among which the amphipods and isopods stand out for their diversity, of which about 450 species have been reported.

In Cuba, 375 species of echinoderms have been recorded: Crinoids: 34, Asteroids: 75, Ophiuroids: 158, Echinoids: 63 and Holothuroids: 45. The green hedgehog, *Lytechinus variegates variegatus,* which lives in seagrass, is a fundamental element in the mobilization of the energy accumulated by the vegetation: they consume large amounts of turtle grass or seiba, but they digest it very little, so it is semidigested defecated and becomes part of the detritus store of the ecosystem. Other species of hedgehogs, such as *Meoma ventricosa* and *Clypeaster rosaceus,* are detritophages that also contribute to further disintegrate organic matter, facilitating bacterial action on it. The black hedgehog *Diadema antillarum,* is a great consumer of algae in the coral reefs, thus contributing to avoid its excessive development.

Fish are among the most diverse organisms in the aquatic environment. The ichthyofauna of Cuba is probably the richest of the Antilles, 1030 species have been reported to date, of which 948 are Teleosts, 80 Chondrichthyes (sharks and rays) and a single species of the subclass Holocephalii (chimera). Of this total, some 40 species inhabit all or part of the freshwater, although many of them also use the estuarine areas. Some 20 species have only been reported for Cuban waters, which could be a consequence of a higher level of knowledge about the fish fauna of Cuba than in other regions of the Greater Caribbean, since a distinctive characteristic of marine organisms is their poor endemism.

Approximately 130 orders of fish are subject to fishing, but only about 40 are of remarkable importance as a resource. Marine fish in Cuba contribute more than 55% of the edible catch. Some have been overexploited, such as the Lane snapper in the Batabano Gulf and the Sabana-Camagüey archipelago, the smooth ones in the coastal lagoons, the creole grouper, the mangrove snapper, etc. Fish are an essential element in the underwater landscape, and it is one of the main attractions for international tourism. However, the abundance of medium and large size fish is poor in many dive sites, as a result of both commercial fishing and recreational and poaching.

The marine reptiles are represented by five species of chelonians: the green turtle, *Chelonia mydas*, the hawksbill, *Eretmochelys imbricata*, the loggerhead, *Caretta caretta*, the Olive Ridley Turtle *Lepidochelys olivacea* and in lesser proportion the leatherback sea turtle, *Dermochelys coriacea* (Rodríguez Schettino, 2003). The excessive exploitation of these or the fragmentation of their nesting habitats have caused a significant decrease in their populations and as a result are included in Appendix I of CITES (species banned for international trade), because they are considered to be in danger of extinction. Among the marine mammals are the Antillean manatee, *Trichechus manatus* and the dolphins, *Tursiops truncatus*.

5.4   Endemic and Threatened Species

The relative antiquity of some regions of the Cuban island territory, the geological diversity, the climatic stability and the complex biogeographic history of the archipelago are responsible for an extreme level of endemism of Cuban biodiversity, to the point of almost half of the living beings in Cuba are endemic. Endemic species are especially concentrated in the mountainous systems of the eastern, western and central regions and in those regions with extreme conditions derived from the nature of the soils or the climate, such as serpentine and sandy siliceous soils.

Among the endemic Cuban mammals are the Cuban solenodont, which is the largest insectivorous mammal that exists, categorized as Critically Endangered by the IUCN, and the 10 species of jutias of the genera *Capromys*, *Mesocapromys* and *Mysateles*, many of them which are restricted to mangrove cays and/or are formed by a single known local population. Bats, even with their remarkable mobility and distributed throughout the national territory, have 14 endemic forms.

Of the approximately 350 species of living birds reported for Cuba, 22 are endemic to the archipelago, and of these, 8 belong to endemic genera. When adding that of 27 other species there are 36 subspecies or geographic races also endemic, it can be concluded that the percentage of endemism in this group is also relatively high (14.16%), taking into account the flight and displacement capacity of the birds that it allows them to disperse and colonize nearby territories. The Cuban Tody (*Todus multicolor*) is one of the endemic species of smaller size and more colorful in its plumage.

In the Cuban archipelago, there are 91 endemic species of reptiles, for 75.2% of endemism, including 3 endemic genera: *Cricosaura, Chamaeleolis,* and *Cadea.* Other genera, although not exclusive to the Cuban territory (*Leiocephalus, Arrhyton, Antillophis, Tropidophis, Tretanorhinus, Epicrates,* and *Cyclura*), are from the Antilles. Of the total endemic reptiles 33 species (36%) are known from a single locality (Rodríguez Schettino, 2003).

Of the 46 existing amphibian species, 93.4% are endemic to the Cuban territory, which gives it great value from the conservationist point of view. The 7 species of toads of the genus *Peltophryne* (endemic Antillean) are exclusive to Cuba and the island is considered the center of dispersal of the group in the Antilles (Rodríguez et al., 2010). Among the leptodactylids, belonging to the genus *Eleutherodactylus* there are 37 species described for Cuba and of them, only one is not endemic (*E. planirostris*) (Díaz and Cádiz, 2008).

Among the Cuban invertebrates, terrestrial mollusks are another group of invertebrates of singular importance due to the high number of endemic species described, which has allowed them to be compared, due to their richness and diversity of forms, with the malacofauna of the Philippines or Hawaii (Espinosa and Ortea, 1999). This is the group that presents the highest endemism (95–96%), standing out the presence of 79 genera and 130 endemic subgenera. A large number of mollusks are local endemics.

Within the Arachnida class, in Cuba, all known species of Opilioni, Palpigradi, Ricinulei and Uropygi are endemic. In the case of insects it is very difficult to specify an exact number of endemics, but in some well-known orders the approximate percentage of endemic species is also relatively high: Trichoptera (61% of the species), Coleoptera (56%), Hymenoptera (45%), Diptera (23.8%) and Homoptera (21%). A considerable part of the ant fauna is endemic and almost half of the species belong to the genus *Leptothorax*. In Cuba, there are 79 endemic species of zoonematodes, for 39% of endemism.

Some Cuban endemic species have been introduced by humans in other regions, such as the *Anolisequestris* lizards, introduced and established in Florida, *Anolisporcatus*, in the Dominican Republic and Florida, the iguana (*Cyclura nubile nubila*), endemic subspecies introduced on Isla Magueyes, southwest of Puerto Rico, the Cuban grassquit (*Tiaris canorus*), introduced in the New Providence Island, Bahamas. Among the terrestrial gastropod mollusks, endemic to Cuba, introduced and established in other countries are *Zachrysia auricoma* (in Mexico, Panama, and Puerto Rico), and *Zachrysia trinitaria* (in Florida) (Espinosa and Ortea, 1999).

5.5   Extinctions and Threats

At present, there are at least 74 species of fungi (Mena et al., 2013), 995 of the flora (González-Torres et al., 2016), 130 of invertebrates (Hidalgo-Gato et al., 2016) and 165 of vertebrates (González et al., 2012) have been classified in some of the threat categories of the International Union for the Conservation of Nature (IUCN, 2012).

Approximately 50% of the native flora is in danger of extinction, 18% is considered Critically Endangered and 25 species have already been declared Extinct. The Red List of Cuban Plants developed by the Group of Specialist in Cuban Plants, following the criteria of the IUCN, has evaluated 4,627 species of plants, which represents about 67% of the estimated total of Cuban plant species. In the recent evaluations of 2016, 31% of angiosperms, 54.5% of pteridophytes and related plants and 78.5% of gymnosperms were threatened with extinction. 61.78% of the species of the endemic genera of Cuba are also in these categories.

There are few islands in the world that have made such an exhaustive analysis of the endangered species of their flora, therefore, it would be correct to say that Cuba is the island with the largest number of threatened plant species of which we have reference today (46.31%). The second island in number of threatened species is Madagascar with 42.1% of its plants at risk of extinction. Fortunately, 73.68% of threatened flora finds protection, to a greater or lesser extent, in the National System of Cuban Protected Areas. There are still about 1,600 species to be categorized, mainly from the families Apocynaceae, Convolvulaceae, Fabaceae, Lamiaceae, Myrtaceae, Orchidaceae and Poaceae. 20% of the species analyzed do not have enough information to evaluate their conservation status; hence the importance of continuing basic studies of Cuban flora.

Reforestation is one of the multiple actions developed in Cuba towards the conservation of biodiversity, which has allowed the increase of forest cover, from 14% in 1959 to 24.7% in the year 2005. In the last decade of the last century natural forests and forest plantations decreased between 2 and 11% in a large part of the world, but in Cuba, there was a total growth of 13% (Earth Trends, 2003), although only 2% was in natural forests. However, the value of the forest indexes should not be overestimated as an indicator of the state of conservation of the flora.

The total number of known species of Cuban fauna exceeds 19,040, of which 5.5% have been categorized according to their degree of threat. The Red List of Cuban fauna includes a total of 1,027 species and 86 subspecies categorized. Among Cuban invertebrates, there are fewer categorized species (264) than vertebrates (763) (Amaro-Valdés, 2012; Mancina et al., 2017). Only 126 indigenous invertebrate species from Cuba appear in the IUCN Red List, representing 0.9% of the species registered for this archipelago.

According to the Vertebrate Red List of Cuba, 52 species of vertebrates are Critically Endangered (31.5%), 42 are Endangered (25.5%), 63 fall in the Vulnerable category (38.2%) and 8 are Near Threatened (4.8%), for a total of 165 species. Of some groups, all their species have been categorized, as is the case of cartilaginous fishes (82 species, 28% threatened), birds (351 species, 6% threatened) and mammals (53 species, 28.3% threatened). In the insular Caribbean area, Cuba has the highest number of threatened species, followed by Haiti, the Dominican Republic and Jamaica.

The Global Amphibian Assessment in 2004 reported 80% of Cuban amphibians classified as threatened and declining, leaving Cuba in fourth place among the countries with the highest percentage of threatened species in America (preceded by Haiti, the Dominican Republic and Jamaica) (Young et al., 2004). Fortunately, to date, the extinction of any amphibian species has not been reported in Cuba and all have been observed in very recent dates (Díaz and Cádiz, 2008).

The Red List includes 87 species of Cuban reptiles (56.1% of those registered) that, for various reasons, are the most endangered in the present and in the near future. Of these, 34 are Critically Endangered; 27, Endangered; 22 are Vulnerable and four, Nearly Threatened.

In the case of birds, due to the disappearance of forests, indiscriminate hunting, collection and trade of different birds, 40 Cuban species have been included in the Red Book as being in danger of extinction; such is the case of the Cuban parakeet (*Aratingaeuops*), the Gundlach's hawk (*Accipiter gundlachi*), the Cuban Crane (*Grus Canadensis nesiotes*), etc., which need urgent protection and management measures for their populations. Among the categorized birds, three species are Critically Endangered, nine Endangered, 18 in the Vulnerable category and two are Near Threatened. In total, 30 species of birds are threatened with extinction and two are almost threatened, which means 8.7% of the living species registered to date. Two species of birds in Cuba are given as recently extinct: the Cuban Macaw (*Ara tricolor*) and the Cuban ivory-billed Woodpecker (*Campephilus principalis bairdi*), and the subspecies of Cuban solitaire of the Isle of Youth (*Myadestes elisabeth retrusus*).

Almost one-third of Cuban mammals are threatened with extinction, if we exclude bats, 75% of mammal species are in danger of extinction – six species are among the 100 most threatened in the world (Cooke et al., 2017; Borroto-Páez and Mancina, 2017). In general, of the 25 species of non-flying land mammals recognized for Cuban territory, only 8 survive today, for 65% extinction. Two species, the dwarf hutia (*Mesocapromys nanus*) and the San Felipe's hutia (*M. sanfelipensis*), may have been extinct at the end of the last century, although in the first of them there is evidence of individuals that could survive in intricate areas of the Zapata Swamp (Silva Taboada et al., 2007).

5.6 National Laws and International Obligations

The government of Cuba has always had an advanced position in relation to international laws and agreements related to the protection of the environment and biological diversity (CITMA, 2014). At the United Nations Conference on Environment and Development (UNCED), better known as the Rio Summit, the then President of the Councils of State and Ministers of the Republic of Cuba, Fidel Castro Ruz, said: "We must use all the science necessary for a sustained development without contamination. We have to pay the ecological debt and not the external debt. The hunger should disappear and not the man."

Since its inception, Cuba has subscribed to Agenda 21, the Conventions on Biological Diversity and Climate Change Convention. In Cuba, there are more than 200 scientific institutions focused on the study of biodiversity, among which the Academy of Sciences of Cuba, and the Ministry of Science, Technology and Environment.

With a well-developed environmental legal system, at least on a theoretical level, there is the political will to include the conservation of biodiversity in all the country's actions. In accordance with Law No. 81 of 1997 on the Environment, there are numerous environmental management instruments such as: the National Environmental Strategy, the National Program for Environment and Development, technical standards on environmental protection, environmental regulations, environmental licenses and environmental impact assessments, the Environmental Information System, the State Environmental Inspection System, environmental education, scientific research and technological innovation, economic regulation and the National Environment Fund.

The environmental management is aimed at ensuring the sustainable development of a territory, based on a multidisciplinary approach for harmoniously coexistence between economic and social developmental projects with the conservation of biological diversity. The Environmental license is the document issued by the administrative authority authorizing the execution of an investment, after assessing that the work to be carried out complies with the necessary technical requirements that allow the conservation of the biological diversity present in the locality. The Environmental impact assessment is the procedure that on the basis of the research of impacts on environment, carried out by the administrative authority, grants the environmental license and allows to assess the positive or undesirable environmental effects that a specific work could cause, and propose the measures aimed at avoiding or mitigating them as much as possible.

There is an Environmental Information System with the purpose of guaranteeing to the government and to the general population, the information required on the state of the local and national Biological Diversity, in such a way that the necessary measures can be taken to ensure their conservation. An Environmental Inspection System, regulated by Resolution No. 130 of 1995 of the Ministry of Science, Technology and the Environment, is aimed at the control and enforcement of compliance with the legal provisions and norms in force in the field of environmental conservation.

A National Environmental Education Strategy involves all state agencies that affect one way or another or contribute to the formation of ethical values and behaviors consistent with the conservation of the environment. The National Environmental Fund was established by Law No. 81 of 1997 and its objective is to finance totally or partially projects or activities aimed at the conservation of the environment, therefore Biological Diversity is among its beneficiaries. The Forestry Law, the Regulations on Biological Diversity, the Law Decree on Protected Areas, the Decree Law on Management of the Coastal Zone and many other legal elements have been set up in an effort to take the protection of the environment to a priority level in the governmental action.

The National Strategy for the Conservation of Biological Diversity responded to the commitments assumed by Cuba, at the Earth Summit, and to the pronouncements contained in Agenda 21, referring to the need to formulate national strategies and action plans. The Strategy was the result of an extensive process of consultations carried out with all the organisms and institutions that in our country affect in one way or another on the Biological Diversity and included 11 basic objectives, referred to topics such as, conservation "*in situ*" and "*ex situ*"; legal system; rehabilitation and restoration of degraded ecosystems; the strengthening of the National System of Protected Areas; the territorial ordering; environmental education; social instruments and incentives; the environmentally safe use of biotechnology, among others.

Cuba is also attached to the Convention to Combat Desertification and Drought, the Convention on Climate Change, the Convention on Biological Diversity, the Convention on Wetlands of International Importance, especially as Waterfowl Habitat (Ramsar) and the Convention on International Trade in Endangered Species of Wild Fauna and Flora (CITES) (CITMA, 2014).

5.7 Biodiversity Threats

The greatest threats to Cuban biodiversity arise from direct anthropogenic activities or from environmental risks. The direct threats come mainly from agriculture (deforestation, use of inappropriate methods), fishing (cultivation of exotic species, overexploitation), tourism industry (construction, uncontrolled activities), mining (in areas in conflict with conservation), civil constructions and urban development, environmental pollution (agrochemicals, vector control, waste, mineralization of water), and activities of illegal extraction of resources (hunting, fishing and poaching).

There are also natural risks, such as the exacerbation of dry periods, heavy rains, sea penetrations and the increase in the intensity and frequency of cyclonic disturbances. The rise in water temperature as a result of global changes, specifically during events related to El Niño, is causing whitening and other coral diseases and facilitating the overgrowth of algae on them, which affects the entire ecosystem, its diversity of species and biological productivity.

The main effects of these threats are alterations, fragmentation or loss of habitats, ecosystems and landscapes, overexploitation of species, contamination of soil, water and air, invasion or introductions of species and erosion. Agroforestry activities have caused the destruction and fragmentation of natural habitats, since the arrival of the first Europeans to the island. In the sixteenth century between 88–92% of the island was covered with forests, but at the beginning of the twentieth century, there was only 41% coverage. The development of cane cultivation in the seventies of the last century produced a maximum deforestation peak that reached 85% of the surface of the island (delRisco, 1995). In recent decades the trend has reversed and there is a slow but sustained increase in forest cover, which currently reaches approximately 30% of the island's land area.

Currently, the landscapes of Cuba constitute a heterogeneous mosaic of agroforestry ecosystems interspersed with fragments of natural vegetation. According to the Ministry of Science, Technology and Environment in 2014, 95% of the fragments of natural vegetation in Cuba were less than 10 km^2 and only 30 areas have a continuous surface greater than 100 km^2. This fragmentation of habitats is listed among the greatest threats to a high number of Cuban species of plants and animals in danger of extinction, but the response of biota to fragmentation has been little studied in Cuba.

The development of mining also undermines the conservation of biodiversity. In Cuba, the most serious case is that of the nickel industry, in the northern part of the eastern territory, whose excavations to extract minerals have devastated large areas of forest. In the same way, the industrial nickel process has become the main environmental pollutant in the region. Precisely, the most important mineral deposits are in the most biodiverse region of the country and this force to take effective measures to minimize the environmental impact of this activity.

Other external factors enhance the threats to Cuban biodiversity, such as the economic blockade by the US, which results in rapid decision making in the face of urgent needs such as the shortage of fuel and freely convertible currency, the necessary socioeconomic development, but not always duly controlled and the economic modifications of the nineties.

In the marine environment, exploitation of fishery resources has been seen which affected the viability and stability of the populations of important resources (fishes, shrimp and lobster). The use of harmful fishing gear, such as hammocks, causes serious damage to seagrasses and patch reefs. Oil pollution, industrial and agricultural waste, as well as sewage, has seriously affected many coastal lagoons, seagrasses and some coral reefs. Tourism also brings associated uncontrolled activities, such as the anchorage of boats on the reefs, underwater fishing, the deposition of garbage and other activities.

The damming of river waters for agricultural and human consumption purposes has led to the salinization of coastal areas and decreases the supply of nutrients, limiting their biological productivity. The interruption of the hydrological regime by the construction of roads through the sea to connect by land with keys of tourist importance contributes to the increase of the salinity and other environmental alterations in the macrolagoon of the archipelago of Sabana – Camagüey.

In Cuba throughout the last centuries hundreds of species of plants and animals have been introduced, whether for crops, ornamentals, biological controls, pets, for food and hunting purposes, although not all have been established and dispersed towards natural ecosystems (Borroto-Páez et al., 2015). It is considered that one of the causes of extinction of many species has been the pressure by exotic animals and plants. In the case of plants, at least 322 taxa have been identified that behave as invaders in Cuba, with a marked predominance of herbs (173 taxa), followed by shrubs, climbers and trees. The species of America (135 taxa) and Asia (127) predominate, followed by those native to Africa (50), Australia-Oceania (17), and Europe (14). Hundreds of square kilometers of the surface of the island are invaded by the African shrub *Dichrostachys cinerea*. Of the list of invasive plants in Cuba, 94% are in the subgroup of species of maximum aggressiveness (46 of 49 taxa), figures that indicate these species as priorities in terms of management at national, regional and local levels. Another 230 plant taxa have been identified as potentially invasive.

Of invasive mammals, 29 species have been reported in 40 islands of the Cuban archipelago. There are large populations of rats (*Rattus rattus* and *R. norvergicus*), feral dogs and cats and mongoose (*Herpestes javanicus*) scattered across the fields. The mongoose was introduced in 1886 to control the pests of rodents and became, in turn, a pest. In addition, you can mention horses, cows, pigs, dogs, cats, farm birds, peccaries, agouties, pacas, camels, llamas, rabbits, deer, Indian and African antelopes, deer, buffalo, green monkeys, to name a few. Numerous endemic species, not abundant and of restricted distributions to a few localities, have been affected by invasive species and have even become extinct (*Nesophontes, Boromys, Mesocapromys sanfelipensis*, etc.) (Borroto-Páez and Mancina, 2017).

The most recent severe invasions have been due to the erroneous introduction and mismanagement of the African catfish (*Clarias gariepinus*) for human consumption and the general invasion of the Lionfish (*Pterois volitans*) in the Caribbean. At present, both species are believed to be a serious threat to the survival of Cuban aquaculture fauna.

Cuba, similar to other islands in the Caribbean, can be strongly impacted by the effects of climate change and has predicted the increase in annual average temperature, intensification and expansion of periods of drought, rise in the mean sea level, and the increase in the frequency and intensity of hurricanes. Some studies are evaluating the possible effects of these environmental alterations on biota.

Agriculture is intensive with the excessive use of resources and low crop rotation. There is also a weak integration between conservation and sustainable use strategies and economic development activities. Problems common to the region are also evident: lack of control over compliance with current legislation, inadequate management of scientific or economic projects, which have led to the departure of important genetic resources and lack of awareness and education environmental.

However, for several decades, many actions in favor of biodiversity have been carried out in Cuba. Reforestation was one of the first that was carried out, from the 60 s of the twentieth century, a plan was drawn up and executed to increase the forest mass that encompassed all the territories and, in particular, the mountains. This has allowed the increase of forest cover, from 14% in 1959 to 24.7% in the year 2005. In the 1990–2000 decade, in a large part of the world and in particular in Central America and the Caribbean, the forests natural and plantations decreased between 2 and 11%; however, in Cuba, there was a total growth of 13% (Earth Trends, 2003).

5.8   Protected and Conservation Areas

The first protected area in Cuba dates back to 1930, the El Cristal National Park, in the province of Holguín, to conserve pine forests and other timber trees. The number of areas was increasing, until in 1997 Law 81 of the Environment was approved, which establishes the objectives and basic principles of the National System of Protected Areas, and its conceptual and regulatory basis, was promulgated through the Decree – Law 201 of 1999. The National Center of Protected Areas, institution belonging to the Ministry of Science, Technology and Environment of Cuba, consolidates this National System of Protected Areas in which three levels of classification are identified: of national significance, local significance, or special sustainable development regions, depending on the connotation or magnitude of their values, representativeness, degree of conservation, uniqueness, extent, complexity, fragility of their ecosystems, economic and social importance, and other relevant elements. The protection categories correspond to the IUCN system: Natural Reserve, National Park, Ecological Reserve, Outstanding Natural Element, Managed Floristic Reserve, Wildlife Refuge, Protected Natural Landscape and Protected Area of Managed Resources.

In Cuba, the National System of Protected Areas is composed of 211 units, 77 of national significance and 134 of local significance. The system covers an area that represents 20.2% of the national territory, including the insular marine platform up to the depth of 200 meters, with 17.16% of the land surface and 24.96% of the land under the system coverage. In addition, the MAB program "Man and the Biosphere" of UNESCO has declared six Biosphere Reserves for Cuba: "Sierra del Rosario" (1985), "Peninsula de Guanahacabibes" (1987), "Baconao" (1987), "Cuchillas del Toa" (1987),

"Ciénaga de Zapata" (1999) and "Bahía de Buenavista" (1999). In all these areas, 95% of the flora species (98% of the endemic and threatened species) are represented, as well as all the endemic birds, the most endemic and endangered vertebrate species, and the sites with the greatest abundance of terrestrial fauna.

The natural values that contain the Cuban protected areas have conferred, also, different types of international recognition by world entities such as the United Nations Educational, Scientific and Cultural Organization (UNESCO), the RAMSAR Convention for the rational use and conservation of wetlands and BirdLife International, among others, declaring 6 Biosphere Reserves, 2 Natural World Heritage Sites and 1 Cultural Landscape, 6 RAMSAR Sites and 28 Important Bird Conservation Areas (IBA, for its acronym in English).

The Viñales National Park was declared a Natural World Heritage Site due to its exceptional natural, historical, social and economic values. It is located in the Sierra de los Organos, in the western sector of the Guaniguanico mountain range, province of Pinar del Río and it is dominated by karstic elevations known as mogotes, with almost vertical walls and flat or rounded peaks, with numerous caves, sinkholes and holes where there is an abundance of vegetation of the Mogote vegetation complex type, which is characterized by its richness, complexity and endemism. Some plants stand out like the palm of saw (*Thrinax microcarpa*), the ceibon (*Bombacopsis cubensis*), the cayman oak (*Ekmanianthe actinophylla*) and great diversity of curujeyes, orchids and ferns, but mainly, the presence of a unique element, the cork palm (*Microcycas calocoma*), considered a living fossil because of its primitive characteristics. Viñales has been considered by many scholars as the Paradise of the Mollusks and is among the regions with the highest local endemism in the country.

In Cuba, 34 types of natural and seminatural plant formations have been identified (Borhidi, 1996; Capote and Berazain, 1984), of which 32 are in the national system with different degrees of representativeness. The mangroves, swamp grasslands, as well as the evergreen, semi-deciduous and pine forests, have representative fragments of vegetation and high priority, as conservation nuclei for the SNAP. According to the Red List of the Flora of Cuba, the national system provides coverage to 3,210 plant species (including 1 386 endemics), of which 1,579 are threatened (González-Torres et al., 2016).

In the SNAP, 49 endemic monotypic genera of the flora are represented, which are located within the limits of 47 protected areas, with national parks being the category of management that hosts the largest number (22). On the other hand, of the 72 species included in the 13 nonmonotypic endemic genera, 62 are present in the system (Castañeira et al., 2013). The SNAP covers in a general way 96.7% of the autochthonous species of terrestrial vertebrates, 91.6% of endemic species and 90.5% of endangered species. The best-represented groups are birds and amphibians with all their percentages above 90%. Within freshwater fish, only 2 endemic and threatened taxa (*Girardinus cubensis* and *Lucifuga subterranea*) have been left without coverage in protected areas (González et al., 2013), as well as 2 species of amphibians, 15 species of reptiles and 2 species of mammals. With respect to birds, SNAP covers 91% of native species, as well as 100% of endemic species and 96.9% of endangered species. The only threatened species not represented in the system is *Pterodroma hasitata*, only known from the town of Las Brujas, south of the slopes of the Sierra Maestra (Rodríguez et al., 2012). The main aquatic bird congregation sites, based on the IBAs of Cuba (Aguilar, 2010), have an acceptable level of protection, if we take into account that in some cases these zones coincide with the boundaries of protected areas identified and in others they contain protected areas with different management categories (González et al., 2013).

Even though the SNAP encompasses a high representativeness of species and ecosystems, there are still problems that limit greater success in the conservation of Cuban biodiversity (CNAP, 2013). For example, some of the units identified in the system do not yet have an administration and in others, the instability of the technical staff, scarce equipment and finan-

Table 5.1 Number of Currently Known Species by Main Groups in Cuban Biota, Percentage of Endemism and Conservation Status by IUCN Criteria

Order	Species	Endemics (%)	Status*		
			CR	**EN**	**VU**
Mollusks	1,391	95–96	29	10	41
Insects	8,459	40–60	13	13	12
Arachnidae	1,328	75	4	1	4
Other invertebrates	452	42			
Freshwater fishes	38	40.3	1	4	1
Amphibians	68	95	1	3	23
Reptiles	153	68	41	23	19
Birds	397	14.7	3	10	17
Mammals	34	50	6	2	3
Plants	7287	50	570	249	151

* Status: CR: Critically endangered; EN: Endangered; VU: Vulnerable.

Table 5.2 Number and Extension of Cuban Protected Areas Bycategory (Data Extracted From CNAP 2013)

Category of Protection	Number of Areas	Total Area (ha)	Terrestrial Area	Marine Area
Managed Resources Protected Area	11	3,466,713	1,379,734	2,076,979
Naturally Outstanding Element	7	24,441	19,329	5,112
Natural Park	14	1,187,123	517,129	669,996
Natural Protected Landscape	2	5,813	5,813	–
Ecological Reserve	23	229,808	140,976	88,832
Faunal Refugee	9	283,018	111,239	171,779
Managed Floristic Reserve	10	42,039	40,467	1,572
National Reserve	4	11,112	8,968	2,144
Biosphere Reserve*	6	446,991	444,492	2,499
Ramsar sites*	4	1,188,411	–	–
Important Bird Areas*	28	2,316,568	–	–

* International categories, not included by SNAP.

cial resources, influence in an adequate management. Additionally, there are gaps and imbalances in the knowledge of the biodiversity that inhabits the areas and in some cases it is not known if the limits of the areas are adequate to guarantee the survival of the populations of flora and fauna that they protect, as well as for the maintenance long-term ecological processes (Tables 5.1 and 5.2).

Keywords

• National laws • International obligations

References

Agnarsson, I., Lequier, S. M., Kuntner, M., Cheng, R. C., Coddington, J. A., & Binford, G., (2016). Phylogeography of a good Caribbean disperser: *Argiope argentata* (Araneae, Araneidae) and a new "cryptic" species from Cuba. *ZooKeys, 625*, 25–44.

Aguilar, S., (2010). *Áreas Importantes Para la Conservación de Las Aves en Cuba*. Editorial Academia, La Habana.

Alonso, R., Crawford, A. J., & Birmingham, E., (2012). Molecular phylogeny of an endemic radiation of Cuban toads (Bufonidae: Peltophryne) based on mitochondrial and nuclear genes. *J. Biogeography, 39*, 434–451.

Amaro-Valdés, S., (2012). *Lista Roja de la Fauna Cubana*. Editorial AMA, La Habana.

Barros-Batesti, D. M., Arzua, M., & Bechara, H. G., (2006). *Garrapatas de Importancia Médicoveterinaria de la Región Neotropical. Unaguíailustradapara identificación de especies*. Sao Paulo, Vox/ICTTD-3/Butantan.

Borhidi, A., (1996). *Phytogeography and Vegetation Ecology of Cuba*. Akademiai Kiadó, Budapest. Second revised and enlarged edition.

Borroto-Páez, R., & Mancina, C. A., (2017). Biodiversity and conservation of Cuban mammals: Past, present, and invasive species. *J. Mammalogy, 98*(4), 964–985.

Borroto-Páez, R., Alonso Bosch, R., Fabres, B. A., & Alvarez, O., (2015). Introduced amphibians and reptiles in the Cuban archipelago. *Herpetological Conservation and Biology, 10*, 985–1012.

Capote, J., & Berazaín, R., (1984). Clasificación de las formaciones vegetales de Cuba. *Revista del Jardín Botánico Nacional, 5*(2), 27–75.

Castañeira, M. A., Valdés, J. A., Hernández, J. A., Rankin, R., & Palmarola, A., (2013). Análisis de vacío de géneros endémicos unitípicos. In: *Plan del Sistema Nacional de Áreas Protegidas de Cuba: Período 2014–2020*. Ministerio de Ciencia, Tecnología y Medio Ambiente, La Habana, Cuba. pp. 158–162.

CITMA (Ministry of Science, Technology and Environment of the Republic of Cuba), (2014). *V Informe Nacional al Convenio sobre la Diversidad Biológica*. La Habana, p. 253.

CNAP (National Center of Protected Areas), (2013). *Plan del Sistema Nacional de Áreas Protegidas de Cuba: Período 2014–2020*. Ministerio de Ciencia, Tecnología y Medio Ambiente, La Habana, Cuba.

Cooke, S. B., Dávalos, L. M., Mychajliw, S., Turvey, T., & Upham, N. S., (2017). Anthropogenic extinction dominates Holocene declines of West Indian mammals. *Annual Review of Ecology, Evolution, and Systematics, 48*, 301–327.

Del Risco, E., (1995). *Los Bosques de Cuba: Su Historia y Características*. Editorial CientíficoTécnica, La Habana.

Díaz, L. M., & Cádiz, A., (2008). Guía taxonómica de los anfibios de Cuba. Abc Taxa, vol. 4. Brussels, Belgium.

Doadrio, I., Perea, S., Alcaraz, L., & Hernández, N., (2009). Molecular phylogeny and biogeography of the Cuban genus *Girardinus* Poey (1854), Relationships within the tribe Girardinini (*Actinopterygii, Poeciliidae*). *Molecular Phylogenetics and Evolution, 50*, 16–30.

Duarte, P., (1997). *Musgos de Cuba*. Fontqueria XLVII.

Earth Trends, (2003). Forest, Grasslands and drylands – Cuba. Country Profiles online at http://earthtrends.wri.org.

Espinosa, J., & Ortea, J., (1999). Moluscos terrestres del archipiélago cubano. *Avicennia, 2*, 1–137.

Genaro, J. A., & Tejuca, A., (1999). Datos cuantitativos, endemismo y estado actual del conocimiento de los insectos cubanos. *Cocuyo, 8*, 23–28.

González, A., Fernández de Arcila, R., & Aguilar, S., (2013). Análisis de vacíos de fauna terrestre. Vertebrados. En: *Plan del Sistema Nacional de Áreas Protegidas de Cuba: Período 2014–2020*. Ministerio de Ciencia, Tecnología y Medio Ambiente, La Habana, Cuba. pp. 166–175.

González, A., Rodríguez Schettino, L., Rodríguez, A., Mancina, C. A., & Ramos, G. I., (2012). *Libro Rojo de los Vertebrados de Cuba*. Editorial Academia, La Habana.

González-Torres, L. G., Palmarola, A., Bécquer, E. R., Berazaín, R., Barrios, D., & Gómez, J. L., (2013). Las 50 plantas más amenazadas de Cuba. *Bissea*, *7*(NE1). p. 107.

González-Torres, L. R., Palmarola, A., González, O. L., Bécquer, E., Testé, E., & Barrios, D., (2016). Lista roja de la flora de Cuba. *Bissea*, *10*(1), 1–352.

González-Torres, L. R., Palmarola, A., González-Oliva, L., Bécquer, E. R., Testé, E., Castañeira-Colomé, M. A., et al., (2016). Lista Roja de la flora de Cuba. *Bissea*, *10*(1), 33–283.

Hedges, S. B., (2006). Paleogeography of the Antilles and origin of West Indian terrestrial vertebrates. *Annals of the Missouri Botanical Garden*, *93*, 231–244.

Hernández, M., Alvarez-Lajonchere, L., Martínez, D., Maceira, D., Fernández, A., & Espinosa, J., (2017). Moluscos terrestres *y* dulceacuícolas. In: Mancina, C. D., & Cruz, D., (eds.). *Diversidad Biológica de Cuba: Métodos de Inventario, Monitoreo y Colecciones Biológicas*. Editorial AMA, La Habana.

Hidalgo-Gato, M., Espinosa, J., & Rodríguez-León, R., (2016). *Libro Rojo de Invertebrados Terrestres de Cuba*. Editorial Academia, La Habana.

Iturralde-Vinent, M. A., & MacPhee, R. D. E., (1999). Paleogeography of the Caribbean region: Implications for Cenozoic biogeography. *Bulletin of the American Museum of Natural History*, *238*, 1–95.

IUCN, (2012). *Red List Categories and Criteria: Version 3. 1. Second edition*, Gland, Switzerland and Cambridge, UK.

Kohler, E., (2014). *Buxaceae*. En: Greuter, W., & Rankin, R., (eds.). *Flora de la República de Cuba. 19*(1), 1–60.

Mancina, C. A., Fernández, R., Castañeira, M. A., Cruz, D. D., & González, A., (2017). Diversidad biológica terrestre de Cuba. Cap. 2. In: Mancina, C. A., & Cruz Flores, D. D., (eds.). *Diversidad Biológica de Cuba: Métodos de Inventario, Monitoreo y Colecciones Biológicas*. Editorial AMA, La Habana.

Matos-Maraví, P., Núñez, Á. R., Peña, C., Miller, J. Y., Sourakov, A., & Wahlberg, N., (2014). Causes of endemic radiation in the Caribbean: Evidence from the historical biogeography and diversification of the butterfly genus *Calisto* (Nymphalidae: Satyrinae: Satyrini). *BMC Evolutionary Biology*, *14*, 199.

Mena, J., Blanco, N., Herrera, S., Ortiz, J. L., Camino, M. C., Cabarroi, M., et al., (2013). Lista roja de hongos *y* Myxomycetes de Cuba. In: *Plan del Sistema Nacional de Áreas Protegidas de Cuba: Período 2014–2020*. Ministerio de Ciencia, Tecnología *y* Medio Ambiente, La Habana, Cuba. pp. 165–166.

Motito, A., (2007). *Los Musgos en Cuba Oriental: Aspectos Sobre su Distribución, Ecología y Conservación*. Tesis de Doctorado, Santiago de Cuba.

Mugica, L., Denis, D., Acosta, M., Jiménez, A., & Rodríguez, A., (2006). *Aves Acuáticas en los Humedales de Cuba*. Editorial Científico Técnica. p. 193.

Nuñez, J. A., Viñas, B. N., Acevedo, G. M., Mateo, R. J., Iturralde, V. M., & Graña, G. A., (1988). *Cuevas y Carsos*. Editorial Científico Técnica, La Habana.

Núñez, R., & Barro, A., (2012). A list of Cuban Lepidoptera (Arthropoda: Insecta). *Zootaxa*, *3384*, 1–59.

Oviedo, R., & González-Oliva, L., (2015). Lista Nacional de Plantas Invasoras *y* potencialmente invasoras en la República de Cuba – 2015. *Bissea*, *9*(NE2), 5–93.

Peck, S. B., (2005). A checklist of the beetles of Cuba with data on distributions and bionomics (Insecta: Coleoptera). *Arthropods of Florida and Neighboring Land Areas, 18*, 1–241.

Regalado, L., Sánchez, C., & González-Oliva, L., (2015). Categorización de helechos y licófitos de la flora de Cuba – 2015. *Bissea*, *9*(3), 1–146.

Rivera, Y., (2011). *La División Anthocerotophyta Rothm. ex Stotl. & Crand- Stotl. en Cuba.* Tesis de Maestría, Jardín Botánico Nacional, Universidad de La Habana.

Rodríguez, A., Vences, M., Nevado, B., Machordom, A., & Verheyen, E., (2010). Biogeographic origin and radiation of Cuban *Eleutherodactylus* frogs of the auriculatus species group, inferred from mitochondrial and nuclear gene sequences. *Molecular Phylogenetics and Evolution*, *54*, 179–186.

Rodríguez, F., Viña Bayés, N., & Viña Dávila, N., (2012). *Pterodroma hasitata*. In: González, H., Alonso, H., Rodríguez, S. L., Rodríguez, A., Mancina, C. A., & Ramos, G. I., (eds.). *Libro Rojo de los Vertebrados de Cuba*. Editorial Academia, La Habana. pp. 209–210.

Rodríguez, S. L., (2003). *Anfibios y Reptiles de Cuba*. UPC Print, Vaasa, Finlandia.

Sánchez, C., (2017). Lista de los helechos y licófitos de Cuba. *Brittonia*. doi: 10.1007/s12228-017-9485-1.

Shi, H., Singh, A., Kant, S., Zhu, Z., & Waller, E., (2005). Integrating habitat status, human population pressure, and protection status into biodiversity conservation priority setting. *Conservation Biology*, *19*, 1273–1285.

Silva, T. G., (1979). *Los Murciélagos de Cuba*. Editorial Academia, La Habana.

Silva, T. G., Suárez, D. W., & Díaz, F. S., (2007). *Compendio de los Mamíferos Autóctonos de Cuba Vivientes y Extinguidos*. EdicionesBoloña, La Habana.

Vales, M. A., Álvarez, A., Montes, L., & Ávila, A., (1998). *Estudio Nacional Sobre la Diversidad Biológica en Cuba*. UNEP.

Whittaker, R. J., & Fernández, P. J. M., (2007). *Island Biogeography, Ecology, Evolution, and Conservation*. Oxford University Press: Oxford.

Young, B. E., Stuart, S. N., Chanson, J. S., Cox, N. A., & Boucher, T. M., (2004). *Joyas que Están Desapareciendo: El Estado de los Anfibios en el Nuevo Mundo*. NatureServe, Arlington, Virginia.

Zachos, F. E., & Habel, J. C., (2011). *Biodiversity Hotspots: Distribution and Protection of Conservation Priority Areas*. Springer: New York, NY.

Plate 5.1 Landscapes of Cuban biodiversity., Picture 1: Landscape of Viñales valley, showing the typical mogotes (Source: Wikimedia)., Picture 2: Cuban mountain región (Photo: Daryl D. Cruz)., Picture 3: Cuban coral reefs (Source: Wikimedia)., Picture 4: Sandy beach in Cuba (Source: Wikimedia)., Picture 5: Aerial view of a typical agricultural area in Cuba (Photo: Dennis Denis)., Picture 6: Mangrove forest in Cuba (Photo: Dennis Denis).

Plate 5.2A Representative pictures of Cuban plant biodiversity., Picture 1: Pluvisilvas del Salto La Sabina, Sierra de Nipe (Photo: Jose L. Gómez)., Picture 2: *Antillanthus pachypodous*, Mina Iberia (Photo: Jose L. Gómez)., Picture 3: Dwarf orchid *Lepantes* sp (Photo: Daryl D. Cruz)., Picture 4: *Copernicia* sp. Isla de la Juventud (Photo: Dennis Denis).,

Plate 5.2B Representative pictures of Cuban plant biodiversity., Picture 5: *Buxus* sp. Yamaniguey (Photo: Mikhail S. Romanov)., Picture 6: Green Orchid (Photo: Dennis Denis)., Picture 7: *Melocactus harlowii Boca de la Bahía de Baitiquirí* (Photo: Sandy Toledo)., Picture 8: *Magnolia virginiana* subsp. *oviedoae* (Photo: Ernesto Teste).

Plate 5.3A Representative invertebrates of Cuban biodiversity., Picture 1: Blue
Scorpion (*Rhopalurus junceus*) (Photo: unknown)., Picture 2: Cristal
Butterfly (*Greta cubana*) (Photo: Thalía Pérez)., Picture 3: Cuban
Polimita (*Polymita picta*) (Photo: Alejandro Barro)., Picture 4: Butterfly
(*Calisto* sp.) (Photo: Daryl D. Cruz).

Plate 5.3B Representative invertebrates of Cuban biodiversity., Picture 5: Cave whip spiders (*Phrynus* sp.) (Photo: Dennis Denis)., Picture 6: Cuban wild cricket (undetermined species) (Photo: Dennis Denis)., Picture 7: Nocturnal Moth (Photo: Dennis Denis)., Picture 8: Terrestrial crab (*Gecarcinus ruricola*) (Photo: Dennis Denis).

Plate 5.4A   Vertebrates representative of Cuban biodiversity., Picture 1: Cuban endemic frog (*Eleutherodactylus* sp.) (Photo: Daryl D. Cruz)., Picture 2: Cuban lizard (*Chamaeleolis* sp.) (Photo: Dennis Denis)., Picture 3: Cuban boa (*Chilabothrus angulifer*) (Photo: Dennis Denis)., Picture 4: Cuban trogon (*Priotelus temnurus*) (Photo: Aslan I. Castellón).

Plate 5.4B Vertebrates representative of Cuban biodiversity., Picture 5: Bee hummingbird (*Mellisuga helenae*) (Photo: Aslan I. Castellón)., Picture 6: Cuban solenodon (*Solenodon cubensis*) (Photo: unknown)., Picture 7: Hutia (*Capromys pilorides*) (Photo: unknown)., Picture 8: Poey's Bat (*Phyllonycteris poeyi*) (Photo: Carlos Mancina).

CHAPTER 6

Plant Biodiversity of Ecuador: A Neotropical Megadiverse Country

PABLO LOZANO CARPIO

Universidad Estatal Amazónica, Paso Lateral km 2 ½ via Napo, Pastaza-Ecuador,
E-mail: plozano@uea.edu.ec, pablo_lozanoc@yahoo.com

6.1 Introduction

Chinchaysuyo, which was later identified as the Kingdom of Quito, was understood as one of the four "suyo" or empires – regions of the Inca, and afterwards known as "Ecuador." Ecuador is a small country, which is located in the northwestern part of South America. It is situated just at the equator and so reasoning the origin of its name (see Figure 6.1). Subsequently, the French Geodesic Mission arrived in 1736 to realize the measurements of the circumference of the earth which was lead by Charles-Marie de la Condamine and his entourage committee of various natural scientists including Joseph Jussiu, who was invited to Ecuador's south because of *Cinchona officinalis* (Quinine); between others, which is a species to heal the yellow fever in Europe and other parts of the planet. It is important to emphasize that this mission contained not only ground measurements, but also more various data of altitude, vegetation, and zoological samples, and ecological texts during many decades in different places in America.

There were diverse natural scientists Ecuador who got attracted to Ecuador to get to know its good features; among others: the Spanish expedition of Juan Tafalla and Juan Manzanilla (1801). After that, the natural scientist Alexander von Humboldt came in January 1802, who traveled across Ecuador with the Ecuadorian Carlos Montufar and who assisted the work on the descriptions of vegetation in the region and in the whole country. Humboldt began his research with trips to the volcano Pichincha for three instances, and on the last one, he was able to register 15 tremors in a time of 36 minutes. This observation fascinated him, while the rumor arised that the German intentionally caused the tremors like a heretic, throwing dust into the crater. He described its interior: "The gorge of the volcano formed a circular hole with a circumference of circa 2 kilometers and all its rough black edges fell downwards into an inaccessible abyss" (Tufiño, 1999).

Figure 6.1 Ecuadorian map and geographycal position.

The well-known natural scientist Edward Whymper, who realized his observations in the region after Humboldt, performed various excursions and characterizations of the region and the country. On his climb on the volcano Pichincha of over 10,000 feet, he found 21 beetle species in a height between 12,000 and 15,600 feet, which were new in science. Additionally, he accomplished data records of the hummingbirds of the volcano Pichincha. Other expeditions were with the German Karl Hartweg (1841) and the Englishman Richard Spruce (1849). The German Friedrich Lehmann preserved plants for the herbarium of Berlin in 1876, when also the German Theodor Wolf arrived, who accentuated the Geomorphology of Ecuador in his work. There is no doubt that the appearance of "Quinina" pushed various expeditions, mainly English and American ones like there was the visit of Julio Steyermark (1944) of the museum of history of Chicago, and like the mission of Holmes Camp (Botanical Garden of New York) to herborize and collect seeds of "Quinina" all over Ecuador, and also the Missouri Botanical Garden was in the place with the floristic inventories of Alwin Gentry in the 1990 s and David Neill who decided to stay in Ecuador, among others.

As a conclusion of this synthesis in which many natural scientists are not mentioned, are added the Swedish Gunnar Harling and Lernnar Anderson who promoted the editorial series "Flora of Ecuador" showing the country's big floristic potential, and furthermore the Danish Henrik Balslev, Peter Jørgensen, Henrik B. Pedersen, Bente Klitgaard, Simon Laegaard, Jens Madsen, and Benjamin Ollgaard are important among others, who contributed in circa 30 years to botanical collections of continental Ecuador, publishing many flora books and articles sharing the knowledge about the national floristic biodiversity. It would be impossible to finish this introduction without talking about the German contributions by the mountain research working group in the South of Ecuador with Erwin Beck, Rainer Bussmann, Jürgen Homeier, Sven Guenter, Michael Richter, inter alia of six German Universities, which conducted more than studies in mountain ecology, having success with various publications and taking part in the professional formation of national researchers.

6.2 Vegetation Biodiversity in Ecuador

Ecuador is divided geographically into three continental regions: Amazon, Mountains, Coast, and the "Galápagos Islands." When we talk about biodiversity in Ecuador which is a very heterogeneous field, we see big contributions of the country to the knowledge of its biota and in particular plant species in the last decades. The statistics define 17,058 species of vascular plants (Ulloa Ulloa and Neill, 2004), which elevates Ecuador as a "megadiverse" country, concentrated on just 275,000 km² or 2% South America (Sierra, 1999). After Valencia et al. (2000) it is indicated that in Ecuador exist 4,011 endemic species which represent 26% of the native flora, later in 2012 there was a verification of 4,500 endemic species in Ecuador (León-Yanez et al., 2012), many of them in critical threat. This high biodiversity primarily is located in the Andean region with 9,865 species (Jørgensen and León Yánes, 1999), while the Coast has 4463 species and the Amazon counts 4,857 or 31.7% of the total diversity. The floristic records did not finish yet, especially in the oriental foothills, where a big part of the diversity is found in niches and endemic species, among other unique ecological characteristics in the country. The records in Galápagos consist of 699 species and according to the provenance of endemic species in the country we have the following:

- Andean Region > 1000 m   –   3038 species   –   68%
- Coast Region < 1000 m   –   799 species   –   18%
- Amazon Region < 1000 m   –   516 species   –   12%
- The Galápagos Islands 0–1.707 m   –   187 species   –   4%

Total – 4,540 evaluated endemic species (León-Yánez et al., 2011).

Further, Ecuador is considered as a member of the 17 megadiverse countries because of its high species concentration and its endemism, where three biodiversity hotspots were identified with the priority of conservation on a global level. These biodiversity "hotspots" are: The humid and very humid tropical rainforests of the Coast region, the forests of the sloped Andean flanks, and the tropical rainforests of the Amazon.

There is an enormous height gradient along the Andean mountain range and the lowlands of the Coast and the Amazon in short distances, influencing a series of abiotic factors (temperature, precipitation, inclination, exposition of the forests, soils, and geomorphology, inter alia), which cause a rich and diverse flora with high endemism. The Ecuadorian Environmental Ministry lists total 90 ecosystems of continental Ecuador in its publication about the classification system of ecosystems (MAE, 2013). In that, many of them described and mentioned for the first time, like these: the formations on sandstones of the Cóndor and Kutukú mountain range; the isolated Páramos of the volcano Sumaco, among other ecosystems which enclose a high biodiversity, nowadays fragile and fragmented.

Two new additions to the flora of Ecuador were done, one in 1999–2004, followed by one of 2005–2010 recording a high floristic biota, numbered in 18,198 vascular plant species, of which circa 5,400 are categorized as endemic on national level (Neill and Ulloa Ulloa, 2011), which equals around 30% of the total biota. This biodiversity is the result of all past geologic and climatologic occurrences, main families are placed in table (6.1) when happened a huge explosion of species and endemism like it is emphasized currently in the country.

The genetically variability of Ecuador is as high as the formerly cited statistics, in which 5,172 useful plants distributed over 238 botanical families were registered, with 11 different benefits and with an amount of 60% of medical plants (De la Torre et al., 2008). Ecuador has been and is the owner

Table 6.1 Major Families of Continental Ecuador (Neill and Ulloa Ulloa, 2011)

	New species	New registrations	Taxonomic changes	Total additions and changes
Orchidaceae	340	25	275	640
Rubiaceae	25	57	12	94
Araceae	61	7	1	69
Sapotaceae	6	47	0	53
Fabaceae	9	15	16	40
Annonaceae	14	7	11	32
Myrtaceae	25	6	0	31
Gesneriaceae	7	1	21	29
Polypodiaceae	9	5	14	28
Asteraceae	13	13	2	28
Melastomataceae	7	11	10	28
Solanaceae	14	13	0	27

of a wide range of phytogenetical resources with an enormous knowledge of medicinal, food plants, among a multitude of applications. Possibly one of the biggest intakes to the international pharmacopeia was to introduce a non-timber forest product (NTFP) of incalculable value until today. The tree bark of "quinine or cascarilla" *Cinchona officinalis* L. and others like the exportation of cinnamon, cocoa, latex rubber, spices, tree dyes, medicinal plants between other products of vegetal origin approve the richness in NTFPs in the equatorial tropics. All this natural landscape of megadiverse characteristics is mixed with the presence of 15 Native ethnicities, therefore one Afro-American, one of white race, and one of the mestizos (combination of immigrants, Italians, Europeans, Americans, Asians, Arabs, inter alia with Indigenous of the country), which includes in this scenario the mixture of knowledge and usage of natural resources (Figure 6.2).

6.3 How Did the Megadiversity Evolve in Ecuador?

The geological and phytogeographical history of Ecuador indicates the presence of different Geological and Climatologic Historians who declare the flora and fauna to be ruled by a series of migrations, adaptations, noting adaptive and radiative specifications of some genus, like in the case of *Teagueia* (orchids) in Pastaza with 26 species who stem from the same

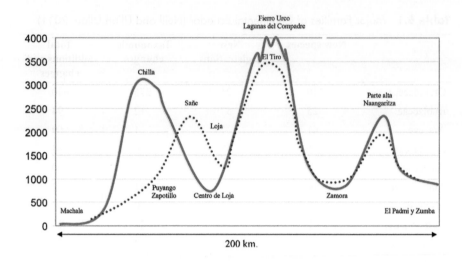

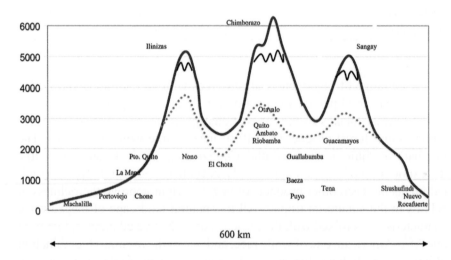

Figure 6.2 The altitudinal section along the countryside.

ancestor Lou Jost (pers. Comm.). It is important to limit the displacement of forests towards West and other fragments of medium coverage forests, called "refuges." Temperatures of 2–6°C below the ones of our time influenced the floristic composition of the Amazon until today. Nevertheless, palynological records of more than 40,000 years of a lake in central Amazon imply that the tropical rainforest continuously covered the region and that the Savannah was absent in the zone during the last maximum ice age, suggesting that the Amazon forest was not fragmented in refuges in the glaciations (Athens, 1997; Bush et al., 1990; Colinvaux, 1993).

The refuge theory has favored an interdisciplinary focus to understand the high diversity of species in Amazon forests. This theory implies a dryer climate during the Pleistocene and a development of the Savannahs in the entire basin with the consequent fragmentation and reduction of forests into refuges. The Amazon forests would have reached its current extent in the course of the Holocene (Colinvaux, 1989). After Liu and Colinvaux (1985), the vegetation zones of the occidental slope in the Andean mountain range of Ecuador during the last ice age could have been at least 700 m lower than today, corresponding to a decrease of 4.5°C. It is known that there were effectively marked precipitation changes in the Pleistocene (Van der Hammen, 1988). In particular, the strong rain decreased in the period between 21,000 years ago and 10,000 years ago (Van der Hammen, 1988).

In the Pleistocene and before (7–2.5 million years ago), while and after the Andean orogenesis there were migration movements from North to South America and reversed. This happened across the isthmus of Panama (Van der Hammen, 1988). Some examples of these movements to South are some Laurasia taxa like the families Juglandaceae and Theaceae inter alia within the Andes mountains, where they turned to be ecologically dominating, however, they coped with a low speciation, at least the wooden taxa, while the invasion into the North from Gondwana are high trees and lianas like *Tecoma* and *Bignonia* in Central America (Gentry, 1982).

In the Oro Province (Southern Ecuadorian coast) there is the petrified forest of Puyango, a relict of the "Araucarioxylon" group, which are dated to the Cretaceous of 120–96 million years ago (Romeijn-Peeters and Beeckman, 2006). The existence of these fossils denotes the South-North migration and adaptation to more humid and hot climates as in the South, wherein any moment the occurrence of low temperatures should have rewarded and forced that the species migrate.

The approvals of the Ecuadorian Amazon show marine fossil embeddings next to the Nangaritza and Bombuscaro rivers (Southeast), indicating that this part of the continent was inundated by the ocean before the appearance of the real Andean mountain chain. The forests were affected by dry climates with seasonal rainfalls (Bush et al., 1990), which was proved by Athens (1997). The recovery of the rainforests begins at the Holocene 10,000–9,000 years ago, the evidence of Carajas (Southwestern Amazon) suggest a fast rainfall increase, equal to the Argentinian Pampas where dry conditions changed to more humidity (Prieto, 1996). None the less, there are other proofs of temperature change to cold and dry conditions 8,000–5,000 years ago (Iriondo and Garcia, 1993), with desertic climate in the Chaco region. Similar conditions are reported from Panama, Guatemala and Costa Rica (Isbele and Hooghiemstra, 1997). Furthermore, the beginning of the temperature gradient of superficial marine water in Ecuador is dated to 5,000 years ago, this is caused by strong wind movements (Rodbell et al., 1999) as a result of the humid conditions which returned to this zone.

In the Ecuadorian Amazon 4,200–3,150 years ago, the climate was characterized by dry periods (Colinvaux, 1989), despite this these results don't match with the realized studies by Athens (1997) for the same place. After other studies, the dry cold conditions and fluctuations were common 390–3,600 years ago, absent between 3,600–2,500 years ago, and scarce 2,800–2,500 years ago (Martin et al., 1993). The final registered events in the Holocene 1,300–800 years ago show that the Western Amazon was distinguished by an excessive rainfall and inundations (Athens, 1997). This fits to the studies of Colinvaux (1989), who describes a humid period in this time range. In relation to our proceeding to the present times, the climate's "stability" turns more homogeneous.

Presented data of the Ecuadorian Amazon by Athens (1997) are controversy comparing with the idea that the flora remained without changes in the late Pleistocene. In all cases, the refuge in the Pleistocene has been accepted by biogeographs like a general explanation for the tropical speciation. Not only the forest species survived successfully in the dry periods in these "islands" of remaining woods, but there also have been repeating cycles of multiple fragmentation ranges which could have augmented the number of species. This could reason why the neotropics have so many species (Gentry, 1982).

The vertical movements of forests were registered when the lower forest limit descended to 700 m a.s.l., which was approved in Mera in the Amazon province of Pastaza, Ecuador. This requires a lower temperature than 4.5°C in the Amazon (Heine, 1994). This data is conformity with the results of Van der Hammen and Cleef (1986), who found the same vertical movement schemes of forests in the Colombian Andes mountains.

This biogeographic perspective of the region demonstrates us big climatic changes with strong dry and cold periods, combined with other humid and hot periods, leading to species migration, floristic adaptations and speciation. Looking at plants, there is a notable relation with Australia/New Zealand (Van der Hammen, 1988). *Colobalanus quitensis*, found in the Andes, is an oligotypical genus which is distributed in the area of central Australia-Antarctica. Other migrants of the South (Austral-Antarctic) at the eastern mountain range of Colombia during the Pleistocene from 2.5 million years ago to 10,000 years ago, while some elements like *Podocarpus* are found in the Eocene (54–38 million years ago). Many more allaying elements arrived in the Plio-Pleistocene, e.g., *Gunnera* (Van der Hammen, 1988).

Palynological data shows that the tropical zone was not stable in the Pleistocene and experimented with severe changes which influenced mostly the altitudinal position of forests (Van der Hammen, 1988; Hooghiemstra and Cleef, 1995). The tentative reconstruction of the Western mountain range of Colombia was determined by the climate changes. To give a general description of the last ice age, the treeline was 1,200 to 1,500 m lower than the actual height, and the temperatures fluctuated 6–7°C colder than today (Van der Hammen, 1988; Van der Hammen and Cleef 1986, cited in Ulloa Ulloa and Jørgensen, 1995).

There is a migration history of flora from North America to South America and reversed. In the Pleistocene Northern Andes, the first mountain elements which arrived in the Eastern range in the Savannah of Bogota were *Hedyosmum* and *Symplocos*. In the middle Pliocene, the genus *Myrica* (*Morella pubescens*) appeared and there existed a gradual enrichment of Andean flora with elements like *Styloceras* and *Juglans* (walnut) in the late Pliocene (Van der Hanmen, 1987). Since the beginning of the Pleistocene there is found *Alnus* (alders), and soon it took part as a very important part of the palynological flora (Van der Hammen and Cleef, 1986), and probably arrived also *Ribes, Berberis,* and *Vaccinium*. At the same time, *Quercus* (oak) came into sight, which was one of the last wooden elements of North America (Hooghiemstra and Cleef, 1995).

Of the current originating families of the Northern hemisphere there are represented in tropical America for example, Myricaceae, Juglandaceae, Betulaceae, Fagaceae, Magnoliaceae, Berberidaceae, Hippocastanaceae, Cyrillidaceae, Clethraceae, Cornaceae, Oleaceae, and Caprifoliaceae. Between the Andes, one observes a decrease of the appearance of these families towards the South. As for instance, *Quercus* (oak) only is found in the Northern Andes, and other wooden families like Sabiaceae, Staphyleaceae, Theaceae, Celastraceae are only represented by 1–2 genera (Gentry, 1982).

It is considerable and important that there was an adaptation in the case of some Savannah species and which were able to ascend to the sub-Andes and the proper Andes. An interesting example is detailed in the study of Cleef et al. (1993), where 12 genera of the current Paramo flora in Costa Rica and Peru were described, which came from the Savannah. Furthermore, Steyermark and Huber (1978) found 18 genera (15%) of the Avila flora (Eastern mountain range of Colombia), which stem from the Savannah (Cleef et al., 1993).

The valley of Jirón-Paute (Austro-Ecuadorian) is understood as a natural barrier (Jørgensen et al., 1995), which was established in the ice age period as an interglacial time, which currently limited the migration of species of the montane forest and Paramo from the Northern part of the country to the South and inversely.

Studies of the family Ericaceae point out differences in genera and species of both parts of the country, however, it was ensured that only a few species cross this barrier and habit in the same way in the North as in the South (Lozano et al., 2003). Examples of distribution in the family Loasaceae indicate a major affinity of genera and species with the North of Peru, more than with the North of the country (Lozano et al., 2003). Besides many more examples which reveal that areas like the South have a center of distribution for some families (Ericaceae, Symplocaceae, Cunoniaceae, Clethraceae) with a high speciation.

In the Sumaco volcano, Napo province of the Amazon region, was identified evolution of the isolation type, otherwise also in Nangaritza-Zamora Chinchipe (a flora related to the Venezuelan Tepuy), like as in the province Loja at the summit of Guachanama in the canton Celica, whose elevated formation among the dry valleys of "Casanga" and "Puyango," denote characteristics of endemism and vegetation associations. Similar features were spotted in paramo of Podocarpus National Park-Loja. Presumably, this richness is attributed because of being the highest peak in the Southern region, where climate conditions and soils are different from the surrounding.

The Southern region of Ecuador is marked by the anticline of "Huacabamba," which influences basically in the lower Andes and also prospectively has affected the speciation and the endemism of this region. The formation of tablelands in the mountain chain of Condor permitted the growth of vegetation with dwarf size like in the Paramo. A few studies relate to the zone, however, there was detected a high endemism in the mountain chain of Condor (Neill, personal communication), with affine species on the Venezuelan Tepuys and in the area of Darien in Panama (Palacios, 1995).

In continuation, four ecosystems in continental Ecuador were described with general attributes to understand its importance and relevant features.

6.4   Importance of Paramos

In aspects of geomorphology, the reliefs and structural geoforms are linked with the orogenesis and development of the Andean chain (ECORAE, 1997). The presence in its center part of the big mountainous axis which constitutes the Andean mountain range, let begin the separation of the country in three regions or continental segments: In the center, the mountainous reliefs of the Andean mountains; in the West the plains and colline reliefs of the coastal region; and in the East the ranges, colline and flatlands of the Amazon (Winckel, 1992). Witnesses of the tectonic evolution in the Ecuadorian Amazon region are the rising of the mountain range of Kutuku and Condor like as various sedimentary basins.

Paramo cover 5% of the territory, some estimations about the surface of the paramo in Ecuador reach up to 12,000 km² (Proyecto-Páramo, 1999). They are of huge importance for the Andean countries: Socio-cultural, biological, hydrologic, economic, among others; like as the distribution of flora and fauna species, and certainly the conservation of genes.

In general, we talk about recent soils of circa 10,000 years. They are dark, acid, rich in organic material and nutrient-poor. The low temperatures slow down the decomposition of organic material and prevent the availability of some essential elements. An eminent quality of some organic soils of the paramo with low drainage is the constant water saturation, regulated by thick layers of vegetation, formed mainly by mosses (*Sphagnum*), which are vegetal sponges with 20 times more weight of deposited water giving infinity to creeks. The soils of the volcanic type with ashes are located above metamorphic rocks. This structure is the base for fundamental environmental services of the paramo: the deposit and distribution of clean and constant water to lower areas where it is used for irrigation, drinking water and hydroelectricity. The emphasis of these soils and the vegetation which protects them, also like their fragility, are topics which were regarded and handled from a different point of view, considering the conversion of the paramo into a, particularly, fragile and strategic ecosystem. In some places, soils of the paramo can extend to several meters of thickness.

The paramo flora show a series of adaptations which allow them to survive in a very hostile environment. Many of them have pubescence to save warmth and hard leaves which inhibit the loss of water by evapotranspiration. Normally, they are stocky species to be protected from the cold and the wind, even if *Espeletia* and *Puya*, and also different tree species can reach up to several meters in height. In some cases, the adaptations can be very sophisticated; in others, they are practically absent and these species have to benefit from the generated microclimas for the rest of the vegetation.

The endemism could amount to 60% of the entire paramo; that means that among 10 species, 6 are endemic (Luteyn, 1999). Leon-Yánez (1999) suggests 270 endemic species in the country. The families of the paramo with the major number of endemic species in Ecuador are Orchidaceae and Asteraceae. This authoress also indicates *Gentianella* (Gentianaceae),

Epidendrum (Orchidaceae), *Lysipomia* (Campanulaceae), *Draba* (Brassi-caceae), and *Lepanthes* (Orchidaceae) as the five richest genera of Ecuadorian endemic species. The only case of endemism in the country and in the ecosystem at a genus level is *Cotopaxia* (Apiaceae) (Jørgensen and León-Yánez, 1999).

In the South of Ecuador (3 °S), paramos have developed above 2,800 m a.s.l. while the upper level can be defined when herbaceous plants and lichens appear dispersely and when perpetual snowfall starts, some almost reach up to 4,800 m a.s.l. In the lowland forests of the Pacific, 260 found species are possible in the tenth part of a hectare; in the montane forests until 1,500 m a.s.l. there can be 130 species, until 2,500 m a.s.l. exactly 100 species are reported and above 3,500 m a.s.l. remain only 35 species.

The isolated paramo of Ecuador correspond to the paramos which are in the inner Ecuadorian Amazon, in particular, in the volcanos cited from North to South: Sumaco and Sangay, therefore the Plateado hill southward Amazon. They are distinguished to be outside of the oriental Andean range, with precipitations higher than 6,000–8,000 mm, with soils which contain volcanic attributes and material like lava and ashes, and with a low organic substrate and low anthropogenic influence. They are situated in heights from circa 3000 m a.s.l. to 3,800 m a.s.l. or more. Other types of paramos of Ecuador to cite are: Dry paramos, shrub paramos, espeletia paramo, and wetland paramos (Figure 6.3).

6.5 Importance of Montane Forests

Ecuador has unique montane ecosystems in the three regions of the country including the coast, mountains, and Amazon side. The montane forest or cloud forest is known for its constant presence of fog, further, there is a big diversity of native flora, with a high endemism. This vegetal belt is located in the center of all flora species of Ecuador (Webster, 1995), which is verified by Jørgensen and León-Yánez (1999), who report the existence of around 9,865 vascular plant species in this ecosystem.

The montane forests of Ecuador are marked by epiphytes and mosses which grow above the trees. The majority of epiphyte species are orchids (Orchidaceae), which count in Ecuador more than 4,000 species (J. del Hierro, personal communication). The families Araceae and Bromeliaceae are also abundant (Webster, 1995) and serve as a habitat for a sum of amphibians, reptiles, and insects like beetles and arachnids. Bussmann (2001), registers 627 epiphyte species in the altitudinal gradient between 1,800 and 3,150 m a.s.l.

Figure 6.3 (A) Paramo landscape (Antisana volcano).

Beyond 1,500 m a.s.l., the montane forests and its floristic composition is sharply distinct from the lowlands, with a predominance of species and genera of Laurasian origin. The family Lauraceae is one of the richest in

Figure 6.3   (B) Paramo landscape (Antisana volcano).

wooden species (wider than 2.5 cm of DBH) in all Andean montane forests which are situated between 1,500 and 2,900 m of altitude, followed by the families Rubiaceae and Melastomataceae. In higher elevations, the families Asteraceae and Ericaceae are the species-richest elements of wooden flora (Gentry, 1995).

Endemism patterns of montane forests on national scale reveals consisting exceptional values, Balslev (1988) yet estimated that half of Ecuador's flora is concentrated in 10% of the national surface, an area which is represented by the regions between 900 and 3,000 m of altitude. The montane forests are also a natural habitat of many varieties of the sylvan species of the cultivated Andean crops. Among some of these wild genera which are associated to Andean crops, we have papaya (*Carica papaya*), tomato (*Lycopersicon esculentum*), tree tomato (*Cyphomandra betacea*), several sylvan species relative to passion fruit (*Passiflora* sp.), avocado (*Persea americana*), grains of the genera of wild beans (*Phaseolus* spp.), blackberry (*Rubus* spp.), gherkin (*Solanum muricatum*), and the potato (*Solanum* spp.). Among the main montane wooden elements we can enumerate: Cedar (*Cedrela montana*), walnut (*Juglans neotropica*), alder (*Alnus acuminata*), andean palm (*Xeroxylon andinum*), quishuar (*Buddleja incana*), almiscle (*Clethra fimbriata*), duco (*Clusia alata*), helecho macho (*Cyathea caracasana*), (*Vallea stipularis*), canelo amarillo (*Nectandra laurel*), payagzi (*Graffenrieda harlingii*), romerillo (*Podocarpu soleifolius*), (*Polylepis pauta*), quinine (*Cinchona oficinalis*), among others.

There are certain ecological differences between the oriental and the occidental mountain range, it can be said that the flanks of the Amazon are principally humid because of the winds from the Atlantic Forest in Brazil, while the montane forests towards West and the Cordillera Real in Ecuador receive a seasonal humidity from the influence of the cold Humboldt Current in the months from June to November and/or hot currents of El Niño from December to May.

The premontane ecosystems delineate an abundant biological diversity, especially floristic in the Andean region. In the Amazon foothills, these forests are continuous and very humid, while the ones in the occidental zones of Colombia and the North of Ecuador (Chocó-Andes) are slightly extended and not continuous, with high rainfalls (>12,000 mm). In the South the belt of the occidental evergreen forest is less humid and narrower, ending in the department of Tumbes in Peru near to the frontier with Ecuador (Lozano et al., 2003). So the lower premontane forests of Ecuador's South/Peru's North constitute the most Southern extension of the humid forests of the region Choco. Nevertheless, next to the Southern limit, the dry valleys separate the areas of the evergreen forests, which are relatively low and more or less isolated (Lozano et al., 2003). It is highlighted that in comparison to the forests of the Andean foothills, the oriental flank is more diverse, even if the endemism is higher in the occident (Valencia et al., 2000). Thus, many species can be endemic in fairly little areas (Lozano et al., 2003), for that reason emerges the necessity to research the premontane forests and lower montane forests of Southwestern Ecuador.

The upper montane ecosystem in Ecuador extends from 3,000 to 3,400 m a.s.l., it is the transition between the cloud forest and the paramo. This forest is also named "Ceja Andina" and is very similar to the cloud forest in its physiognomy and in the number of mosses and epiphyte plants (Valencia et al., 2000), but it differs in the structure and size. Currently, the forest of the "Ceja Andina" has the form of islands of natural forests (fragments or patches), which are relegated to the gorges and/or the soils with steep slopes (Luteyn, 1999). This isolation of the forest comes from various factors, like as provocated by landslides, crumbling or other natural disasters (Neill and Jørgensen, 1999), and from the occasions, which are caused by humans (fires and transformation to agricultural soils). This situation puts in high risk the survival of these forests and the correlated biodiversity.

In Ecuador's North in the hillside zone of declivity in the Ecological Reserve El Ángel, the Ceja Andina forest is composed by islands inside of a espeletia paramo matrix (*Espeletia pycnophyla* Cuatrec). Many of these fragments are very small and thereby they are considered as no functional parts of an ecosystem and under the threat of disappearance, some of them formed by species like *Polylepis* sp. One of the strategies which are suggested by landscape ecology to maintain the viability of ecosystems is the emergence of corridors, which are classified by its size and objectives in: biological corridors, conservation corridors, of habitat, of landscape, among others. The habitat corridors pretend to merge fragments of the same habitat at local scale (less than 1 km). The used strategy for the development of a habitat corridor, between two separated forest fragments of different extents, is significantly important to implement in the Ecuadorian Andes.

The upper and lower montane ecosystems are acknowledged as areas of rich biodiversity with vegetation formations which are unique in the country and with high potential of phylogenetic resources. The intense pressure, under which they are currently subdued by all the progress of humans in the prospection of potential resources, posed them in a critical conservation status. It is worth to note that among the potential resources of great interest are not explicitly plants, but the subsoil where gold was mined since centuries and currently finds of copper among other metals and nonmetals, which disadvantage its conservation (Figure 6.4).

6.6   Importance of Amazonian Forests

The Ecuadorian Amazon region, perceived as a space of convergence of various factors of geographic, historic, geopolitical, sociocultural, economic and environmental sort, constitutes the biggest geographical area with an estimated extent of 116,613 km^2, which represents 45.48% of the country's surface, but it is the least populated region with 739,814 inhabitants. This region holds a rich biodiversity, even that in the year 1999 a number of 200–240 tree species were evidenced (Neill and Jørgensen, 1999), and a total amount of 4,857 trees, shrubs, and herbs like epiphytes, lianas, and others (Jørgensen and León-Yanez, 1999), among them are 522 endemic species at regional scale (León-Yanez et al., 2011). Other studies point out the presence of 1,100 tree species in 25 ha forest in the National Park Yasuni, which means the highest tree species density ever registered in the world (Dangles and Nowicki, 2009).

The forms, rather located in submeridian belts, provide the Ecuadorian Amazon a general relief in shape of stair treads. Similarly, four big types of

Figure 6.4 (A) Cloud montane forest landscape (Sumaco volcano at the Amazon region).

relief forms can be distinguished easily: (i) Sub-Andean reliefs and ridges, (ii) periandean foothills, (iii) periandean collines, and (iv) valleys and river plains.

In general terms, the soils of the humid tropics which include the Ecuadorian Amazon are *ferralitic* soils, caused by their development in an environment of ferralitic meteorization, characterized by climatic conditions with precipitation of more than 1,200 mm and a monthly mean temperature of over 22°C.

The classes to which belong the majority of the soils in the Ecuadorian Amazon region are: Andisols, Entisols, Histosols, Inceptisols, Oxisols, and Ultisols. There are some Spodosols, but its extent is not significant; by contrast, Alfisols, Aridisols, Vertisols, and Mollisols are absent (Valarezo, 2004).

Due to the high humidity in the Amazonian basin, there are dominant floristic groups, among them the bryophytes stand out. Other present groups are pteridophytes like: *Blechnum* sp., further *Lycopodium* and *Sellaginela*. Besides there are other floristic elements of simple wind distribution like: *Baccharis*, like as other species of the genera: *Piptocoma, Cecropia,* and *Vismea*, which appear rapidly to occupy areas devoid of vegetation.

The Amazonia of Ecuador means a wide region with elevated diversity in an altitudinal gradient, like the lower Amazon between 50 m a.s.l. and 400 m a.s.l., where we find the Yasuni and Cuyabeno forests as some examples to cite with rich diversity in plants, insects and fauna in general; we can say that above this height of 400 m a.s.l. to 700 m a.s.l. are Amazonian plains in corrugated slopes with species of high commercial value, passing over to the ridges or foothills above 700 m a.s.l. The existence and dominance of species varies according to the location and the properties of the terrain like we have seen in the lower blackwater wetlands covered by *Mauritia flexuosa* formations and others in the foothills of the Southeastern region with forests of the conifers *Retrophyllum rospligiosii* and *Prumnopitys harmsiana*. The prevalence of the gymnosperms are high with 171 registered tree species (Yaguana et al., 2012).

The Amazonian forests of Ecuador suffer practices of land use change for the establishment of pastureland for beef cattle livestock, others for the benefits of valuable timber like: caoba (*Swietenia macrophylla*), cedro (*Cedrela odorata*), chuncho (*Cedrelinga cateniformis*), guayacan (*Tabebuia serratifolia*), yunyun (*Terminalia amazonia*), and colorado (*Guarea macrophylla*), which have provoked a degradation of the forest, its genes and within a time lapse of not more than 10 years a soil erosion which become poor in presence of aluminum caused by a lack of a natural nutrient recycle, desertifying in large areas (Figure 6.5).

6.7 The Dry Forest Importance

The distribution of dry tropical forests of Latin America is depicted in two principal blocks, the first found in Northern Mexico, Middle America, and the Caribbean Islands, and second, in Southern Brazil, Paraguay, Bolivia, and Argentina (Linares-Palomino et al., 2011). These forests have less than 1,800 mm of annual precipitation, with a period of 3–6 months of the dry season with less than 100 mm of monthly precipitation.

Figure 6.4 (B,C) Cloud montane forest landscape (*Polylepis* forest and Fog Vegetation).

Figure 6.5   Amazon vegetation landscape (Colonso Reserve – Tena province).

The dry forests are vegetation formations, whose substantive feature is the leaf fall, so to say deciduous, circa 75% of its species lose seasonally its leaves. Another found attribute in the dry forests is that the genera (and also many families) have only a few sympatric species, comparing to humid forests where genera like *Ficus*, *Inga*, *Miconia*, *Psychotria*, *Piper*, and *Solanum* can have more than 10 sympatric species.

The occidental dry forests in the province Loja, known as the *center of endemism Tumbesino* (Kessler, 1992), maintain its high amount of endemic bird species. This ecosystem, also known as the center of endemism of arid plants of Guayas (Madsen et al., 2002), are characterized by possessing vegetation of the coastal type. It is extended from the central North to the South of the country with floristic elements, which are adapted and specialized

Figure 6.6   Dry forest vegetation landscape (Puyango Forest and Galapagos Island).

to this ecosystems with extreme arid conditions. The indicating vegetation is: *Croton rivinifolius* (Euphorbiaceae); *Ipomoea carnea* (Convolvulaceae); *Cordia lutea* (Boraginaceae); *Cereus diffusus, Armatocereus cartwrightianus, Hylocereus polyrhizus* (Cactaceae); *Chloroleucon mangense, Pithe-*

cellobium excelsum, Piptadenia flava, Prosopis juliflora (Mimosaceae); *Achatocarpus pubescens* (Achatocarpaceae); *Erythroxylum glaucum* (Erythroxylaceae); *Maytenus octogona* (Celastraceae), *Clitoria brachystegia, Geoffroea spinosa, Lonchocarpus atropurpureus, Piscidia carthagenensis,* and *Pterocapus* sp. (Fabaceae), among others.

The dry tropical forests of the region Tumbesina of the Southwest of Ecuador and the Northwest of Peru are the home of a great number of plants and unique animal species, and they mean a priority for conservation of biodiversity at global scale. Distressingly, more than 95% of these exceptional ecosystems were lost due to human intervention. The forests of the Coast belong to the region Tumbesina, which covers 135,000 km², divided between Ecuador and Peru, from the province Esmeraldas in Ecuador's North to the department of La Libertad in the Northwest of Peru. Along the Coast, the forests are continuous, but in the inter-Andean passage exist dispersed valleys, from Imbabura to Cañar (Northern dry inter-Andean forests), in Azuay and Loja (Southern dry inter-Andean forest), and in Zamora Chinchipe (Eastern dry inter-Andean forest) (Aguirre and Kvist, 2006) (Figure 6.6).

6.8 Effects to Vegetation Biodiversity

Deforestation in Ecuador is a phenomenon which is complex to analyze due to the multiplicity of influencing factors. Among them, the politics of the ex-IERAC (Ecuadorian Institute for Agrarian Reform and Colonization) in the 1960 s and 1970 s can be cited, first when agricultural settlements were the reason of 60% of the annual deforested area, second because of the demand for timber and forest products for generalized use of the population and in industrial processes, and third caused by lack of planning in the implementation of the infrastructure work (petroleum exploitation activity, infrastructure for the generation of electricity, construction of roads, etc.) (Guerrero, 2006).

The transference of areas with forest potential to the agricultural use implied a deforestation of 18,160 km², which is the difference between the potential and the current land usage in the category of forest use, which determines an under-utilization of forest soil (MAE – Ministry of Environment of Ecuador, 2005). It should be added that the conducted deforestation over the time of Ecuador's history and in particular in the last times represent a reduction of around 6,000,000 ha since 1962 until today.

The deforestation of the tropical rainforest during the period understood between 1991 and 2000 is 1,782,832 ha which corresponds to 1.47% of the annual value, so to say 198,092.4 ha annual after data emitted by Clirsen (2005) and formalized by the environmental authority.

The humid tropical rainforest has a deforestation rate of 1.49% of the annual value, and in the *Mauritia flexuosa* palm formations 0.16%, while the dry forest has 2.18% annual, and where population with a high index of poverty and natality colonizes, could confirm the prevailing assumption of many studies about the existing relation between the deforestation, the amount of inhabitants, and poverty, which constitutes one of the facilitator conditions to produce a major deforestation rate.

The deforestation of the humid forest is strongly related to direct causes like the expansion of the agricultural frontier, the formation of pastureland, the enlargement of plantations for the oil palm, petroleum and mining activities, roadways and silviculture among others, connected with indirect causes like tax politics, legalization of property ownership and market pressures of forest products. We have an example of the Ecuadorian Amazon region. In 2008, the program SocioBosque was created, which influenced straightly the conservation of forests, decreased the payments of incentives, which evinced the annual deforestation rate of continental Ecuador for the period of 1990–2000 with –0.65% and for the period of 2000–2008 with –0.58% (MAE, 2013). This denotes a change in the former statistics, which exceeded over 1.4 in the decade of the 1990 s. Without a doubt, it is a problem not only local but regional, in which countries with a great diversity of species take part for different reasons, among them the most relevant should be the loss of hundreds of species without studies every day.

It is worth to highlight that in 2014 the statistics of MAE showed that 12,000,000 ha exist from which 50% were under conservation status with an annual net deforestation of 47,497 ha/year in the period of 2008 to 2014 and a mean annual regeneration from 2008 to 2014 of 50,421 ha/year. Based on this numbers, the Ecuadorian State thought about an annual net deforestation rate of 0% until the year 2020.

6.9 Concluding Remarks

The present compilation of information describes the geologic-climatic-historic reasons in the equatorial zone which have influenced directly the patterns of migration and adaptation of species, succeeding in the lapse of years to establish a rich and diverse flora, of which several originate of different currents of North-South or South-North migration, others Austral-Antarctic, like as the appearance of proper Neotropical flora.

The geographic features of the region and the presence of the Andean mountains allowed the speciation of various genera in the altitudinal gradient at both mountain range flanks, developing an elevated endemism with dominant and highly represented families like it is the case of the orchids, to cite an example.

It remains a territory to explore and to emphasize in this chapter, nevertheless the given statistics from diverse studies yet permits to acknowledge Ecuador as a megadiverse place, unique, with a great endemism and a raised usage of plants by different ethnicities who inhabit the area.

The efforts to maintain this biodiversity in Ecuador have been numerous and in 2008 the "SocioBosque" Program was created, which later in December 2013 changed its original concept into "National Program of Incentives to the Conservation and Sustainable Use of Natural Heritage SocioBosque." At the national level, the deforestation decreased in Ecuador (considering the period 2000–2008 versus 2009–2014) by 39.47%, based on the implementation of related public politics. Promptly the program contributed with 15.75% of the total reduction of deforestation. The reduction of the annual deforestation rate was –0.71% in the period 1990–2000, then –0.66% in the period 2000–2008, and at last 50,421 ha/year in the period 2008–2014. Nonetheless, the lack of planned reforestation programs impinges on directly in the unique source of supply which is the native forest, consisting not of cubic meters of timber but of a great diversity, which we even don't know yet.

Keywords

- ecosystems
- geomorphologic history
- megadiverse
- plant biodiversity

References

Aguirre, Z., & Kvist, P., (2006). Especies leñosas *y* formaciones vegetales en los bosques estacionalmente secos de Ecuador *y* Perú. *Arnaldoa*, *13*(2), 324–350.

Athens, J. S., (1997). Palaeoambiente del oriente Ecuadoriano: Resultados preliminares de columnas de sedimentos procedentes de humedales. *Fronteras de Investigación*, *1*(1), 15–32.

Balslev, H., (1988). Distribution patterns of Ecuadorian plant species. *Taxon*, *37*, 567–577.

Bush, M. B., Colinveaux, P. A., Wiemann, M. C., Piperno, D. R., & Liu, K. B., (1990). Late Pleistocene temperature depression and vegetation change in Ecuadorian Amazonia. *Quat. Res.*, *34*, 330–345.

Bussmann, R. W., (2001). Epiphyte diversity in a Tropical Andean Forest-Reserva Biológica San Francisco, Zamora- Chinchipe, Ecuador. *Ecotropica*, *7*, 43–59.

Cleef, A. M., Van der Hammen, T., & Hooghiemstra, H., (1993). The savanna relationship in the Andean paramo flora. *Opera Botanica*, *121*, 285–290.

Clirsen, (2005). *Deforestación en el Ecuador*. www.clirsen.com.

Colinvaux, P. A., (1989). The past and future Amazon. *Scientific American, 260,* 102–108.

Colinvaux, P., (1993). Pleistocene Biogeography and Diversity in Tropical Forests of South America. In: Goldblatt, P., (ed.). *Biological Relationships Between Africa and South America.* Yale University Press.

Dangles, O., & Nowicki, F., (2009). *Biota Maxima.* Pontificia Universidad Católica del Ecuador. Imprenta Mariscal. Quito-Ecuador.

De la Torre, L., Navarrete, H., Muriel, P., Macía, M. J., & Balslev, H., (2008). *Enciclopedia de las Plantas útiles del Ecuador,* Herbario QCA de la Escuela de Ciencias Biológicas de la Pontificia Universidad Católica del Ecuador *y* Herbario AAU del Departamento de Ciencias Biológicas de la Universidad de Aarhus, Quito *y* Aarhus.

ECORAE, (1997). Plan Maestro para el Ecodesarrollo de la Región Amazónica Ecuadoriana. Diagnóstico Integral de la –RAE-. Quito. En: Valarezo, C., (ed.). *Características, Distribución, Clasificación y Capacidad de Uso de los Suelos en la Región Amazónica Ecuadoriana-RAE.* Universidad nacional de Loja, Centro de estudios *y* desarrollo de la Amazonía, programa de modernización de los servicios agropecuarios. Loja, Ecuador, (2004).

Gentry, A. H., (1982). Phytogeographic patterns as evidence for a Choco Refuge. In: Prance, G. T., (ed.). *Biological Diversification in the Tropics,* New York. pp. 112–136.

Gentry, A., (1995). Patterns of diversity and floristic composition in neotropical montane forest. In: Churchil, S. P., Balslev, H., Forero, E., & Luteyn, J. L., (eds.), *Biodiversity and Conservation of Neotropical Montane Forest.* Bronx, New York.

Guerrero, R., (2006). *La Deforestación en el Ecuador.* Universidad Técnica del Norte.

Heine, K., (1994). The Mera site revisited: Ice-age Amazon in the light of new evidence. *Quaternary International, 21,* 113–119.

Hooghiemstra, H., & Cleef, A. M., (1995). Pleistocene climatic change and environmental and generic dynamics in the North Andean montane forest and páramo. In: Churchill, S., Balslev, H., Forero, E., & Luteyn, J. L., (eds.). *Biodiversity and Conservation of Neotropical Montane Forest.* The New York Botanical Garden, pp. 35–49.

Iriondo, M. H., & Garcia, N. O., (1993). Climatic variations in the argentine plains during the last 18,000 years. *Palaeogeography, Palaeoclimatology, Palaeoecology, 101,* 209–220.

Isbele, G. A., & Hooghiemstra, H., (1997). Vegetation and climate history of montane Costa Rica since the last glacial. *Quaternary Science Reviews, 16,* 589–604.

Jørgensen, P. M., & León-Yánez, S., (1999). *Catálogo de Plantas Vasculares del Ecuador.* Missouri Botanical Garden. St Louis, USA.

Jørgensen, P. M., Ulloa Ulloa, C., Madsen, J., & Valencia, R., (1995). A floristic analysis of the high Andes Ecuador. In: Churchill, S., Balslev, H., Forero, E., & Luteyn, J. L., (eds.). *Biodiversity and Conservation of Neotropical montane Forest.* New York, pp. 221–223.

Kessler, M., (1992). The vegetation of South-West Ecuador, In: Best, B. J., (ed.). *The Threatened Forest of South-West Ecuador.* Biosphere Publications, Leeds. pp. 79–100.

León-Yánez, S., Valencia, R., Pitman, N., Endara, L., Ulloa Ulloa, C., & Navarrete, H., (2011). *Libro Rojo de las Especies Endémicas de Ecuador, 2 Edición.* Publicaciones del herbario QCA, Pontífica Universidad Católica del Ecuador, Quito.

Linares-Palomino, R., Oliveira-Filho, A. T., & Pennington, R. T., (2011). In: Dirzo, R., Young, H. S., Mooney, H. A., & Ceballos, G., (eds.). *Seasonally Dry Tropical Forest: Ecology and Conservation.* Island Press, pp. 3–21.

Liu, K., & Colinveaux, A., (1985). Forest changes in the Amazon Basin during the last Glacial Maximum. *Nature, 318*, 556–557.

Lozano, P., Delgado, T., & Aguirre, Z. M., (2003). *Estado Actual de la Flora Endémica Exclusiva y su Distribución en el Occidente del Parque Nacional Podocarpus*. Publicaciones de la Fundación Ecuadoriana para la Investigación y Desarrollo de la Botánica. Loja, Ecuador.

Luteyn, J., (1999). Introduction to the Páramo Ecosystem. In: Luteyn, J., (ed.). *Páramos: A Checklist of Plant Diversity, Geographical Distribution, and Botanical Literature*. New York, The New York Botanical Garden Press, pp. 1–39.

Madsen, J. E., (2002). Cactus en el sur del Ecuador. In: Aguirre, Z. M., Madsen, J. E., Cotton, E., & Balslev, H., (eds.). *Botánica AustroEcuadoriana – Estudios sobre los Recursos Vegetales en las Provincias de El Oro, Loja y Zamora-Chinchipe*. Ediciones Abya Yala, Quito, pp. 89–303.

Martin, L., Fournier, M., Mourguiart, P., Sifeddine, A., Turcq, B., Absy, M. L., & Flexor, J. M., (1993). Southern oscilations signal in South American palaeoclimatic data of the, (7000). years. *Quaternary Research, 39*, 338–346.

Ministerio del Ambiente del Ecuador, (2013). *Sistema de Clasificación de los Ecosistemas del Ecuador Continental*. Subsecretaria de Patrimonio Natural. Quito.

Neill, D. A., & Jørgensen, P. M., (1999). Climates. In: Jørgensen, P. M., & León-Yánez, S., (eds.). *Catalogue of the Vascular Plants of Ecuador*. Missouri Botanical Garden Press. Saint Louis, Estados Unidos de Norteamérica. 8–13.

Neill, D., & Ulloa Ulloa, C., (2011). *Adiciones a la Flora del Ecuador: Segundo Suplemento, 2005–2010*. Fundación Jatun Sacha. Ministerio del Ambiente. Missouri Botanical Garden. Quito.

Palacios, W., (1995). *Cuenca del Río Nangaritza (Cordillera del Cóndor) una Zona Para Conservar*. Herbario Nacional del Ecuador, QCNE, Quito.

Prieto, A. R., (1996). Late quaternary vegetational and climatic changes in the Pampa grassland of Argentina. *Quaternary Research, 45*, 73–88.

Proyecto Páramo, (1999). *Mapa Preliminar de los Tipos de Páramo del Ecuador*. Mimeógrafo no publicado. Universidad de Amsterdam/Ecociencia/Instituto de Montaña. Quito, Ecuador.

Rodbell, D. T., Seltzer, G. O., Anderson, D. M., Abbott, M. B., Enfield, D. B., & Newman, J. H., (1999). An 15,000-year record of el niño-driven alluviation in Southwestern ecuador. *Science, 283*, 516–519.

Romeijn, -P., & Beeckman, I., (2006). *Los Valores Naturales del Bosque Petrificado de Puyango*. Ecuador Sur del Pacifico al Amazonas. Biotours.

Sierra, R., (1999). *Propuesta Preliminar de un Sistema de Clasificación de Vegetación para el Ecuador Continental*. Proyecto INEFAN/GEF-BIRF y EcoCiencia. Quito, Ecuador.

Steyermark, J. A., & Huber, O., (1978). *Flora del Avila, Venezuela*. Vollmer Foudation, Caracas.

Tufiño, P., (1999). *Alexander Von Humboldt Ecuador Terra Incógnita*. Edimpres S. A., *1*(2), 14–18.

Ulloa Ulloa, C., & Jørgensen, P. M., (1995). Árboles y Arbustos de los Andes del Ecuador. Ediciones ABYA-YALA, Quito-Ecuador, 38.

Ulloa Ulloa, C., & Neill, D., (2004). *Cinco Años de Adiciones a la Flora de Ecuador 1999–2004*. Universidad Técnica Particular de Loja, 75.

Valarezo, C., (2004). *Características, Distribución, Clasificación y Capacidad de Uso de los Suelos en la Región Amazónica Ecuadoriana-RAE,* Universidad Nacional de Loja, Centro de estudios y desarrollo de la Amazonía, programa de modernización de los servicios agropecuarios. Loja, Ecuador, pp. 49–52.

Valencia, R., Pitman, N., León-Yánez, S., & Jørgensen, P. M., (2000). *Libro Rojo de las Plantas Endémicas del Ecuador.* Herbario QCA, Pontificia Universidad Católica del Ecuador, Quito, Ecuador.

Van der Hammen, T., (1988). The tropical flora in historical perspective. *Taxon, 37*(3), 515–518.

Van der, H. T., & Cleef, A., (1986). Development of high andean páramo flora and vegetation. In: Vuilleumier, F., & Monasterio, M., (eds.). *High Altitude Tropical Biogeography.* Oxford, 153–201.

Webster, G. L., (1995). The Panorama of Neotropical Cloud Forests. *Biodiversity and Conservation of NeotropicalMotatne Forests.* s. l., pp. 53–77.

Winckel, A., (1992). *Los Grandes Rasgos del Relieve en el Ecuador.* Los paisajes naturales del Ecuador. volumen 1. Las condiciones generales del medio natural. Centro Ecuadoriano de Investigaciones Geográficas. IRD. IGM. Quito, pp. 19–21.

Yaguana, C., Lozano, D., Neill, D., & Asanza, M., (2012). Diversidad florística y estructura del bosque nublado del río Numbala, Zamora-Chinchipe, Ecuador: El "bosque gigante" de podocarpaceae adyacente al parque nacional Podocarpus. *RevistaAmazónica: Ciencia y Tecnología, 1*(3), 226–247.

CHAPTER 7

Biodiversity in Honduras: The Environment, Flora, Bats, Medium and Large-Sized Mammals, Birds, Freshwater Fishes, and the Amphibians and Reptiles

JAMES R. MCCRANIE,[1] FRANKLIN E. CASTAÑEDA,[2] NEREYDA ESTRADA,[3] LILIAN FERRUFINO,[4] DANIEL GERMER,[5] WILFREDO MATAMOROS,[6] and KEVIN O. SAGASTUME-ESPINOZA[7]

[1]10770 SW 164 Street, Miami, FL, 33157–2933, USA, E-mail: jmccrani@bellsouth.net

[2]Panthera, Tegucigalpa, Honduras

[3]School of Biology, Faculty of Sciences, National Autonomous University of Honduras, Tegucigalpa, Honduras

[4]Herbario Cyril Hardy Nelson Sutherland, School of Biology, Faculty of Sciences, National Autonomous University of Honduras, Tegucigalpa, Honduras

[5]Barrio el Bosque, Calle Principal, Casa 2022, Tegucigalpa, Honduras

[6]Universidadde Ciencias y Artes de Chiapas, Instituto de Ciencias Biológicas, Museo de Zoología, Colección de Ictiología, Tuxtla Gutierrez, Chiapas, México

[7]School of Biology, Faculty of Sciences, National Autonomous University of Honduras, Tegucigalpa, Honduras

7.1 Introduction

7.1.1 General Description of Honduras

Honduras is one of seven political units (countries) comprising Central America. Of those seven countries, only Nicaragua is larger than Honduras. Honduras, with its approximate size of 112,082 km, is located in the north-central portion of Central America and is bordered by Guatemala, El Salvador, Nicaragua, the Caribbean Sea, and the Golfo de Fonseca of the Pacific Ocean. The Caribbean coastline is about 650 km in length, and that of the Golfo de Fonseca is about 150 km in length (Atlas Geográfico de Honduras, 2006). Honduras also contains several island groups. Caribbean islands include the Islas de la Bahía (three main islands and numerous keys), the Cayos Cochinos (two small islands and several tiny cays), the Islas del Cisne or Swan Islands (two islands), the Cayos Miskitos (several small keys), and the Cayos Zapotillos (several small keys). The Golfo de Fonseca of the Pacific Ocean contains a group of 12 small islands or cayos. Honduras is divided into 18 political departments. McCranie (2011) included a map showing the geographical boundaries of those departments. Maps showing the location and a discussion of the general descriptions of the Islas de la Bahía (including the Cayos Cochinos), the islands in the Golfo de Fonseca, and the Islas del Cisne are included in McCranie et al. (2005), McCranie and Gutsche (2016), and McCranie et al. (2017), in that order.

7.1.2 Physiography

The mainland of Honduras contains numerous mountain ranges and groups of smaller mountains, all of which are isolated from each other by interconnecting river valleys and lower elevation terrain. Elevations in the country range from sea level to 2849 m. That highest elevation is on a peak in the Cordillera de Celaque in southwestern Honduras. McCranie (2011) included a map that shows the general locations of the most significant of those mountains. The lower elevations (<600 m) in the country generally extend along both coasts and on the flanks of nearby mountains, on the numerous islands, and in the extensive area of the Mosquitia in northeastern Honduras. Those Mosquitia lowlands are larger than all other lowland areas in the country combined. A map showing the general location of those Mosquitia lowlands is in McCranie et al. (2006).

The complex topology of Honduras, created by its extensive mountain ranges and interconnected river valleys, serve to force the Continental Divide to occur as low as 870 m elevation at one locality and do not exceed 1050 m elevation at two other points (Monroe, 1968). The Caribbean (Atlantic) versant rivers are considerably more extensive than are those of the Pacific versant. Thus, the Continental Divide passes along about the southern quarter portion of the country.

7.1.3 Climate

Honduras is centered at about 15° latitude, thus lies entirely in the tropics. The two most important climatic elements in the tropics, are air temperature and the amount and pattern of precipitation. In the tropics, air temperature is largely dependent on elevation, with the usual lapse rate or vertical temperature gradient being about 0.65°C per 100 m (Koeppe and De Long, 1958). The majority of precipitation in Honduras is a result of the northeastern trade winds bringing moist air from the Caribbean Sea, but other effects are also in place (i.e., a deflected southeast trade wind from the Pacific Ocean;

occasional cold fronts from the north during November to February). Generally, the rainy season occurs from May to November, but climatic changes occurring in the most recent 10–15 years seem to have shifted both the beginning and ending months of the rainy season to about one month later, and with those beginning and endings becoming more erratic (JRM personal observation). McCranie (2011) discussed some of the more important events affecting rainfall in Honduras. Zúniga (1989) provided a general classification of climate regimes based on rainfall patterns, air temperature, and relative humidity, but did not include all of those values for all of his included regimes. Although the Zúniga study is far from complete, it remains the best climatic study available for Honduras. Zúniga (1989) included 11 general primary regimes, with six of those regimes having 1–3 variants. The Zúniga classification regime map and his precipitation map were reproduced in Atlas Geográfico de Honduras (2006) and McCranie (2011). A brief description of those 11 Zúniga regimes including rainfall patterns, air temperatures, and relative humidity (where available) are summarized in Table 7.1.

7.2 Ecosystem Characteristics

7.2.1 Forest Formations

Holdridge (1967) developed a system for classifying forest formations based largely on climatic, edaphic, and atmospheric conditions. McCranie has been using a slightly modified version of that Holdridge system for the time he has been investigating the amphibian and reptilian faunas of Honduras. Nine generalized forest formations are recognized for the Honduran mainland and most nearby islands. A tenth forest formation is also included for the Islas del Cisne (Swan Islands) and the Cayos Miskitos (Miskito Cays) in the Caribbean Sea. That last mentioned formation also likely occurs on the Cayos Zapotillos off the coast northwest of Puerto Cortés, Cortés. Brief discussions are presented below for each of those formations. See McCranie (2011) for a somewhat more detailed discussion of those formations.

7.2.1.1 Lowland Moist Forest Formation (LMF)

The majority of this formation (0–600 m elevation) consists largely of broadleaf evergreen rainforest that receives a relatively high amount of annual precipitation (>2,000 mm), has a relatively high mean annual temperature (>24°C), and has a relatively high humidity (>80%). This broadleaf evergreen rainforest typically has a tall, multistoried, closed-canopy.

Table 7.1 The Climatic Patterns and Regimes of Honduras as Defined by Zúniga (1989)

Zúniga Classification	Rainfall Pattern	Annual Rainfall	Air Temperature Mean (°C)	Mean Annual Relative Humidity
Very rainy without significant dry season	(m) Nov. & Dec. (l) April & May	2880 mm	27.5°C	84%
Very rainy with regular distribution of rain	(m) Oct. & Nov.* (l) April & May	2600–3400 mm	26°C	No data
Tropical CL with very rainy season	(m) Oct. & Nov.* (l) March & April	2200–3000 mm	26.5°C	86%–87%
Transitional very rainy	(m) June & Oct* (l) March & April	2000–2600 mm	26.6–27.3°C	70%–86%
Very rainy with semi stationary trade winds	(m) June& Sept.* (l) March & April	1400–2400 mm	No data	72%–84%
Rainy interior montane	(m) June & Sept.* (l) Feb. & March	1600–2000 mm	21° to 10°C at higher elevations	72%–74%
Light rainy with distinct dry season	(m) June & Oct. (l) Feb. & March*	No data	22° to 18°C at higher elevations	80%

Zúniga Classification	Rainfall Pattern	Annual Rainfall	Air Temperature Mean (°C)	Mean Annual Relative Humidity
Light rainfall with some rainfall in dry season	(m) May& Sept. (l) Feb. & March	400–1000 mm	24°C–26°C	62%–72%
Transitional light rainfall	(m) May& Sept. (l) Jan. & Feb.	No data	No data	60%–82%
Pacific trade wind montane	(m) June& Sept. (l) Dec. & Jan.	No data	No data	80% in area of CD
Short rainy season with strong dry season	(m) June& Sept. (l) Jan. &Feb.	1500–2000 mm	28°C along PC, 22°–24°C in remainder	60% along PC 72%–78% in remainder

All values given for rainfall, temperature, and humidity are approximates as given by Zúniga. Abbreviations included are: CD = continental divide; CL = Caribbean lowlands; l = least rainiest months; m = most rainiest months; mm = millimeters; and PC = Pacific Coast. An asterisk (*) included in rainfall pattern signifies that more monthly variation was reported by Zuniga than could be included in this table.

Unfortunately, the increasing human population has largely destroyed this broadleaf evergreen rainforest. The largest remaining expanse of broadleaf forest now lies in isolated pockets in La Mosquitia of northeastern Honduras. Twenty years ago those Mosquitia forests were a continuous expanse of closed canopy forest extending from the border with Nicaragua to west of the Río Plátano Biosphere Reserve. Even that Río Plátano Reserve, also a World Heritage Site, is now under increased human attack. Other forest variations within the LMF formation include pine savannas (in northeastern Honduras), freshwater swamps and marshes, broadleaf swamp forest (apparently limited to a small area in northeastern Honduras), mangrove swamps, and cocotelas and coastal strand. As a result of the high rainfall in these forests, numerous rivers and smaller watercourses are present. The broadleaf evergreen rainforest of this formation sometimes carries the misnomer "jungle."

7.2.1.2 Lowland Dry Forest Formation (LDF)

This is the dry scrub forest (0–600 m elevation) typically found along the Pacific coast of Central America to northwestern Costa Rica. This forest type has a relatively short rainy season and a long, severe (3–4 months) dry season. As a result, LDF has an open canopy and receives relatively little annual precipitation (1,000–2,000 mm) to go with their high temperatures (>24°C) and relatively low relative humidity (<80%). Most trees are deciduous, losing their leaves during the dry season. Spiny plants and trees are also common. Most LDF has been converted to pastures and other uses for man.

7.2.1.3 Lowland Arid Forest Formation (LAF)

This thorn forest (150–600 m elevation) is another open, hot zone that occurs in several river valleys in southern and central Honduras. Temperatures are high (>24°C) and rainfall is low (500–1,000 mm per year); as a result the terrain is open. The dry season is severe, with several months in which scant, or no precipitation occurs. Succulents provide much of the ground cover. Most LAF has been converted to pastures, with no original forest remaining.

7.2.1.4 Premontane Wet Forest (PWF)

This moderate elevation forest (ca. 601–1,500 m elevation) is characterized by a relatively low mean annual temperature (18°C–24°C) and a relatively high amount of annual rainfall (>2,000 mm). The trees are tall (25–30 m in height) and the canopy is closed. The PWF lies mostly on Caribbean (Atlantic) slopes facing the sea. Numerous small watercourses and several notable rivers are formed in these forests and from higher elevations. The PWF has also suffered heavily at the hands of man.

7.2.1.5   Premontane Moist Forest Formation (PMF)

This moderate elevation forest (ca. 601–1,500 m elevation) is characterized by a relatively slightly lower mean annual temperature (18°C–24°C) and a relatively low amount of annual rainfall (1000–2000 mm). This is largely a region of open pine forests, much of it burned regularly. Gallary forest containing broadleaf trees follows the few watercourses that drain these slopes.

7.2.1.6   Premontane Dry Forest Formation (PDF)

This moderate elevation scrub forest (ca. 601–1250 m elevation) is characterized by a relatively low mean annual temperature (18°C–24°C) and a relatively low amount of annual rainfall (500–1,000 mm). That low amount of rainfall is partially because of a long severe dry season. The PDF is limited to slopes of dry valleys having LDF and LAF. Watercourses are few and close to all of this forest type has been altered by man.

7.2.1.7   Lower Montane Wet Forest Formation (LMWF)

This is the forest formation typically referred to as cloud forest and is generally present at elevations of about 1501 to 2700 m and usually facing the Caribbean Sea. The LMWF receives a relatively high amount of annual precipitation (>2,000 mm) and has relatively cool annual temperatures (mean 12°C–18°C). Pristine LMWF is characterized by having a continuous canopy of tall broadleaf evergreen trees that are often cloaked with epiphytes. These forests become stunted and form elfin forest on wind-swept slopes on Cerro Picucha, Olancho, and on the highest peak in Cerro Celaque, Lempira. Streams and smaller watercourses are numerous. Man has also seriously altered much of this habitat.

7.2.1.8   Lower Montane Moist Forest Formation (LMMF)

This is another "cloud forest" formation, but receives less annual precipitation (1,000–2,000 mm) than does LMWF. It remains similar to LMWF in having relatively cool annual temperatures (mean 12°C–18°C) and occurring at about 1501 to 2700 m elevation. LMMF occurs to the south of the LMWF formation on Pacific versant slopes and on the southernmost Caribbean versant slopes. Streams and smaller watercourses are numerous. Man has also seriously altered much of this habitat.

7.2.1.9 Montane Rainforest Formation (MR)

The MR is restricted to the highest elevations (>2,700 m) of the highest mountains in the country. Mean annual temperatures are relatively low (6°C–12°C) and mean annual precipitation is relatively high (>2,000 mm). The MR is known only on Cerro de Las Minas, Cerro Celaque, Lempira; Cerro El Pital, Ocotepeque; and Cerro Santa Bárbara, Santa Bárbara. Because of their small sizes and difficulties in reaching them, MR is poorly studied in Honduras. Cypress (*Cupressus*) and fir (*Abies*) trees have been reported from the top of Cerro Santa Bárbara (see McCranie 2011 for a brief review of this formation in Honduras).

7.2.1.10 Lowland Dry Forest Formation, West Indian Subregion (LDF[WI])

This formation occurs on the West Indian-like Islas del Cisne (Swan Islands) and on the tiny Cayos Miskitos. The Islas del Cisne has a rather lengthy dry season receiving relatively little precipitation (<2,000 mm) and have relatively high mean annual temperatures (>24°C). These forests are an open scrub forest that contains many cacti. The two areas in Honduras where this formation is known to occur lack surface water of any type; not even pools formed by heavy rains, with the exception of one artificial concrete tank on Big Swan Island (Isla del Cisne Grande). Seawater is present shortly below the upper leaf litter level on at least one island in the Cayos Miskitos (JRM personal observation).

7.3 Honduran Flora Biodiversity

7.3.1 *The Spermatophytes*

Honduras has several floral ecosystems, among the most important of which include: Humid Tropical Forest, Low Montane Forest, Tropical Dry Forest, with the last mentioned considered one of the most threatened ecosystems because of anthropogenic activities. Land use is dominated by broadleaf forest and dense and sparse pine forest (Secretaría de Energía, Recursos Naturales, Ambiente y Minas 2014).

The tropical regions contribute to two-thirds of the estimated 300,000 plant species (Kreft and Jetz, 2007). Honduras has 7,524 known species of vascular plants, of which 214 (2.8%) are considered endemic (Nelson, 2008). This chapter includes 6178 native species of Spermatophytes, with 22 gymnosperm species (0.4%), 295 basal angiosperms (5%), 1809 mono-cotyledons (30%), and 4093 eudicots (66%). Nelson (2008) also listed 651 species of ferns. The families with the greatest number of species are: Orchi-daceae (620), Fabaceae (520), Poaceae (470), Asteraceae (379), and Rubia-ceae (237), all of which are considered among the most diverse angiosperm families in the Neotropics.

Nearly 82 species of Honduran flora are categorized as endemic or meso-endemic with different states of conservation status according to the Red List of The International Union for Conservation of Nature (IUCN) and the Appendices to the Convention on International Trade in Endangered Species of Wild Fauna and Wild Flora (CITES) (Mejía-Ordoñez and House, 2008). Appendix II of CITES demonstrates most of these species belong to the families Orchidaceae and Bromeliaceae.

According to the IUCN list, the following native species are critically endangered: *Desmopsis dolichopetala* R.E. Fr., *Platymiscium albertinae* Standl. & L.O. Williams (Fabaceae), *Ilex williamsii* Standl., *Oreopanax lem-piranus* Hazlett, *Dendropanax hondurensis* M.J. Cannon & Cannon, *Cho-danthus montecillensis* Ant. Molina, *Varronia urticacea* (Standl.) Friesen, *Viburnum molinae* Lundell, *V. subpubescens* Lundell, *Tontelea hondurensis* A.C. Sm., *Connarus popenoei* Standl., *Sloanea shankii* Standl. & L.O. Wil-liams, *Dalbergia intibucana* Standl. & L.O. Williams, *Lonchocarpus moli-nae* Standl. & L.O. Williams, *L. phaseolifolius* Benth., *L. sanctuarii* Standl. & L.O. Williams, *L. trifolius* Standl. & L.O. Williams, *L. yoroensis* Standl., *Terua vallicola* Standl. & F.J. Herm., *Casearia williamsiana* Sleumer, *Molinadendron hondurense* (Standl.) P.K. Endress, *Pleurothyrium roberto-andinoi* C. Nelson, *Quararibea yunckeri* Standl., *Gentlea molinae* Lundell, *Mollinedia butleriana* Standl., *M. ruae* L.O. Williams & Ant. Molina, *Euge-nia coyolensis* Standl., *E. lancetillae* Standl., *Forestiera hondurensis* Standl. & L.O. Williams, *Fraxinus hondurensis* Standl., *Coccoloba cholutecensis* R.A. Howard, *C. lindaviana* R.A. Howard, among others.

However, some of those just mentioned species are in critical states of conservation need, whereas others lack knowledge of their current popu-lation status. Species in danger of extinction include: *Guaiacum sanctum* L., *Anaxagorea phaeocarpa* Mart., *Amphitecna molinae* L.O. Williams, *Terminalia bucidoides* Standl. & L.O. Williams, *Lacunaria panamensis* (Standl.) Standl., *Connarus brachybotryosus* Donn. Sm., *Tetrorchidium brevifolium* Standl. & Steyerm., *Lonchocarpus phlebophyllus* Standl. & Steyerm., *L.retiferus* Standl. & L.O.Williams, *Machaerium nicaraguense* Rudd, *Pithecellobium johansenii* Standl., *P. saxosum* Standl. & Steyerm., *P. stevensonii* (Standl.) Standl. & Steyerm., *Juglans olanchana* Standl. & L.O. Williams, *Ocotea jorge-escobarii* C. Nelson, *Hampea sphaerocarpa* Fryxell, *Blakea brunnea* Gleason, *Trichilia breviflora* S.F. Blake & Standl., *Parathesis vulgata* Lundell, *Eugenia salamensis* Donn. Sm., *Neea acuminatissima* Standl., *Ouratea insulae* L. Riley, *Quiina schippii* Standl., among others.

Studies on the medicinal flora in Honduras have recorded 734 species of medicinal plants of Honduras (House et al., 1999). Previously, House et al. (1995) reported 250 medicinal species, with 157 natural and 93 cultivated. Approximately, 183 plant species with nutritional value have also been identified, with 43 species having commercial value and 98 having medicinal uses (Zea and Cruz, 2006).

A list of the Spermatophyte plants reported in Honduras is shown in Table 7.2. The scientific names were consulted in Nelson (2008) and the databases TROPICOS and The Plant List, both using the APG IV system (2016).

7.4 Vertebrate Taxonomic Groups

7.4.1 Bats of Honduras: A List of Bat Species with Taxonomic Considerations

Bats have a long ongoing history with humans in Honduras, even though it does not begin with modern science. For years bats were worshiped as gods by the Mayan culture, as seen in huge remaining representations of bats in ruins, glyphs, and texts. The Mayans named bats "Zotz," and associated them with the dark, with death, and the Mayan underworld (Xibalba). The "PopolVuh" portraits a journey *Hunahpu* and *Ixbalanque* (mythological figures in the PopolVuh) had to take through Xibalba and the different rooms they visited in order to undertake different tests required of them. One room is named *"Zotzi-ha,"* House of Bats (Cajas, 2009). Fash and Fash (1989) noted the qualities of "Structure 20" (now destroyed by natural factors) demonstrated strong evidence to consider parts of it as an analogy of *"Zotzi-ha"* as mentioned in the PopolVuh.

The Mayan city of Copán, Honduras, has a strong symbolism for bats, a clear example is the emblem glyph for the city, which shows a bat with a strong nasal sheet (*"Zotz"* a member of the family Phyllostomidae) as a rep-

Table 7.2 Spermatophyte Plants List Reported in Honduras

Order	Family	Number of genera	Numbers of species
Alismatales	Alismataceae	4	11
	Araceae	17	90
	Cymodoceaceae	2	2
	Hydrocharitaceae	4	5
	Potamogetonaceae	1	2
	Ruppiaceae	1	1
	Zosteraceae	1	1
Apiales	Apiaceae	12	17
	Araliaceae	5	23
Aquifoliales	Aquifoliaceae	1	13
	Phyllonomaceae	1	3
Arecales	Arecaceae	24	69
Asparagales	Amaryllidaceae	5	9
	Asparagaceae	12	29
	Hypoxidaceae	2	3
	Iridaceae	11	19
	Orchidaceae	131	620
Asterales	Asteraceae	113	379
	Campanulaceae	7	18
	Menyanthaceae	1	1
Boraginales	Boraginaceae	10	61
Brassicales	Bataceae	1	1
	Brassicaceae	4	11
	Capparaceae	9	14
	Caricaceae	2	4
	Cleomaceae	2	11
	Tovariaceae	1	1
	Tropaeolaceae	1	1
Buxales	Buxaceae	1	1
		1	2
Canellales	Winteraceae	1	2
Caryophyllales	Achatocarpaceae	1	1
	Aizoaceae	2	2
	Amaranthaceae	15	45

Table 7.2   (Continued)

Order	Family	Number of genera	Numbers of species
	Basellaceae	1	3
	Cactaceae	21	50
	Caryophyllaceae	3	10
	Droseraceae	1	1
	Molluginaceae	2	2
	Nyctaginaceae	8	18
	Petiveriaceae	3	3
	Phytolaccaceae	2	5
	Polygonaceae	8	32
	Portulacaceae	1	4
	Stegnospermataceae	1	1
	Talinaceae	1	2
Celastrales	Celastraceae	17	25
Ceratophyllales	Ceratophyllaceae	1	1
Chloranthales	Chloranthaceae	1	3
Commelinales	Commelinaceae	8	37
	Haemodoraceae	2	2
	Pontederiaceae	3	8
Cornales	Cornaceae	1	2
	Hydrangeaceae	1	1
	Loasaceae	3	3
Crossosomatales	Staphyleaceae	1	1
Cucurbitales	Begoniaceae	1	29
	Cucurbitaceae	17	29
Cupressales	Cupressaceae	1	2
Cycadales	Zamiaceae	3	8
Dilleniales	Dilleniaceae	4	10
Dioscoreales	Burmanniaceae	4	7
	Dioscoreaceae	1	27
Dipsacales	Adoxaceae	2	9
	Caprifoliaceae	2	7
	Sambucaceae	1	1
Ericales	Actinidiaceae	1	19
	Clethraceae	2	11

Table 7.2 (Continued)

Order	Family	Number of genera	Numbers of species
	Cyrillaceae	1	1
	Ebenaceae	1	6
	Ericaceae	16	30
	Lecythidaceae	2	3
	Marcgraviaceae	4	10
	Mitrastemonaceae	1	1
	Pentaphylacaceae	4	10
	Polemoniaceae	2	5
	Primulaceae	13	48
	Sapotaceae	4	27
	Styraceae	1	4
	Symplocaceae	1	11
	Theaceae	1	2
Fabales	Fabaceae	101	520
	Polygalaceae	4	33
	Surianaceae	1	1
Fagales	Betulaceae	3	5
	Fagaceae	1	23
	Juglandaceae	2	3
	Myricaceae	1	2
	Ticodendraceae	1	1
Gentianales	Apocynaceae	37	111
	Gelsemiaceae	1	1
	Gentianaceae	8	18
	Loganiaceae	3	8
	Rubiaceae	62	237
Geraniales	Geraniaceae	1	1
Gunnerales	Gunneraceae	1	1
Huerteales	Dipentodontaceae	1	1
	Tapisciaceae	1	1
Lamiales	Acanthaceae	27	82
	Bignoniaceae	27	60
	Calceolariaceae	1	1
	Gesneriaceae	18	49

Table 7.2 (Continued)

Order	Family	Number of genera	Numbers of species
	Lamiaceae	17	76
	Lentibulariaceae	3	17
	Linderniaceae	2	7
	Martyniaceae	1	1
	Oleaceae	4	5
	Orobanchaceae	7	17
	Phrymaceae	2	2
	Plantaginaceae	15	38
	Plocospermataceae	1	1
	Schlegeliaceae	1	1
	Scrophulariaceae	2	3
	Verbenaceae	14	51
Laurales	Hernandiaceae	3	6
	Lauraceae	10	68
	Monimiaceae	1	2
	Siparunaceae	1	2
Liliales	Alstroemeriaceae	1	4
	Melanthiaceae	1	1
	Smilacaceae	1	10
Magnoliales	Annonaceae	11	36
	Magnoliaceae	2	5
	Myristicaceae	2	6
Malpighiales	Achariaceae	3	3
	Calophyllaceae	1	1
	Chrysobalanaceae	4	13
	Clusiaceae	3	19
	Cytinaceae	1	1
	Dichapetalaceae	1	2
	Erythroxylaceae	1	7
	Euphorbiaceae	25	174
	Goupiaceae	1	1
	Hypericaceae	2	13
	Lacistemataceae	1	1
	Linaceae	1	3

Table 7.2 (Continued)

Order	Family	Number of genera	Numbers of species
	Malpighiaceae	12	47
	Ochnaceae	6	13
	Passifloraceae	4	49
	Peraceae	1	2
	Phyllanthaceae	6	20
	Picrodendraceae	1	1
	Podostemaceae	4	8
	Putranjivaceae	1	1
	Rhizophoraceae	2	4
	Salicaceae	14	39
	Trigoniaceae	1	2
	Violaceae	5	13
Malvales	Bixaceae	3	4
	Cistaceae	2	5
	Malvaceae	50	162
	Muntingiaceae	2	2
	Thymelaeaceae	2	4
Metteniusales	Metteniusaceae	1	3
Myrtales	Combretaceae	5	14
	Lythraceae	5	27
	Melastomataceae	24	148
	Myrtaceae	6	62
	Onagraceae	5	29
	Vochysiaceae	1	3
Nymphaeales	Cabombaceae	1	2
	Nymphaeaceae	1	8
Oxalidales	Brunelliaceae	1	1
	Connaraceae	3	10
	Cunoniaceae	1	6
	Elaeocarpaceae	1	11
	Oxalidaceae	2	12
Pandanales	Cyclanthaceae	4	9
	Triuridaceae	1	1
Picramniales	Picramniaceae	2	8

Table 7.2 (Continued)

Order	Family	Number of genera	Numbers of species
Pinales	Pinaceae	2	8
	Taxaceae	1	1
Piperales	Aristolochiaceae	1	13
	Piperaceae	3	150
Poales		12	124
	Cyperaceae	20	208
	Eriocaulaceae	3	7
	Juncaceae	1	7
	Mayacaceae	1	1
	Poaceae	94	470
	Typhaceae	1	2
	Xyridaceae	1	9
Podocarpales	Podocarpaceae	1	3
Proteales	Proteaceae	1	1
	Sabiaceae	1	4
Ranunculales	Berberidaceae	1	1
	Menispermaceae	5	13
	Papaveraceae	2	5
	Ranunculaceae	3	6
Rosales	Cannabaceae	4	5
	Moraceae	13	45
	Rhamnaceae	10	20
	Rosaceae	7	22
	Ulmaceae	3	5
	Urticaceae	8	36
Santalales	Balanophoraceae	1	2
	Loranthaceae	5	20
	Olacaceae	3	4
	Opiliaceae	1	2
	Santalaceae	4	30
	Schoepfiaceae	1	1
	Ximeniaceae	1	1
Sapindales	Anacardiaceae	11	16
	Burseraceae	3	14

Table 7.2   (Continued)

Order	Family	Number of genera	Numbers of species
	Meliaceae	5	23
	Rutaceae	10	34
	Sapindaceae	15	67
	Simaroubaceae	2	2
Saxifragales	Altingiaceae	1	1
	Crassulaceae	1	2
	Hamamelidaceae	2	2
Solanales	Convolvulaceae	16	73
	Hydrophyllaceae	1	1
	Solanaceae	20	148
	Sphenocleaceae	1	1
Vitales	Vitaceae	3	12
Zingiberales	Cannaceae	1	4
	Costaceae	1	8
	Heliconiaceae	1	23
	Marantaceae	6	25
	Zingiberaceae	1	4
Zygophyllales	Krameriaceae	1	2
	Zygophyllaceae	2	3
	Total	**1532**	**6178**

resentation of the city itself. "*Zotz*" is not the only glyph found in the Mayan iconography but also "*Qaaw'aZotz*," or Mr. Bat; a 60 cm high three-dimensional sculpture found in Copán, Honduras, and the PopolVuh Museum, Guatemala City (Cajas, 2009), plus other appearances in Mayan culture, like the Mayan calendar.

In the modern era, the study of bats in Honduras started with Goodwin (1942), who reported eight families and 73 bat species for the country. That report is the first large effort at documenting the bat fauna in Honduras. Subsequent studies concerning bat diversity in Honduras and for the Mesoamerican region include Handley (1959, 1960), Davis and Carter (1962), Davis et al. (1964), Jones (1964a, 1964b, 1966), Carter et al. (1966), LaVal (1969), and Valdez and LaVal (1971). Similarly, Dolan and Carter (1979) and Knox Jones et al. (1988) reported some new bat records for Mexico and

Middle America. McCarthy et al. (1993), in a masterpiece, discussed the history of bat studies for northern Central America and reported 119 species for the region, and 98 species specifically for Honduras.

Recent studies contributing to the number of bat species known for Honduras include: Mora (2012), Mora and Lopez (2012), Mora et al. (2014), Espinal and Mora (2016), and Espinal et al. (2016). Several morphological and molecular analyzes of species complexes have resulted in the description of new species, synonymy of others, and species-level elevation of some subspecies (see Baird et al., 2012, 2015; López-Wilchis et al., 2012; Velazco and Patterson, 2013; Mantilla-Meluk, 2014; Medina-Fitoria et al., 2015). As a result, to date there are 110 bat species officially reported for the country (Table 7.3; taken largely from Mora, 2016) and four species of probable occurrence (Table 7.3; see Hernández, 2015; Mora, 2016) in Honduras.

Bats play an important economic role in natural pest control, for which a single colony (125 individuals) of insectivorous bats can eat up to 1.3 million insect pests each year, a value estimated to be around $22.9 billion/year (Boyles et al., 2011). Not only are they enormous arthropod predators, they also play important roles as pollinators and seed dispersers in most tropical forests, with most of these ecosystem services provided by one bat family (Phyllostomidae) in the New World. Bat pollination occurs in around 549 species of angiosperms in the Neotropics (Kunz et al., 2011). This as a whole serves indirectly to humans by facilitating the reproductive success of food plants, reduced cost for pest control, and forest regeneration by natural seed dispersers.

Nevertheless, bats have only recently received some of the attention they deserve in the country (and the World), where there is little to no bat importance conscience in the educational system, an issue that needs to be assessed so that future generations can give bats the much-deserved importance in the natural world that they should demand. With this, possible protection strategies can be planned to provide action for bat protection in Honduras by working hand-by-hand with the ongoing Bat Conservation Strategy (Hernández, 2015).

7.4.2 Medium and Large-Sized Mammals of Honduras: A Brief Discussion and List

Medium and large terrestrial mammals are probably the most significant group of species in terms of their cinegetic and cultural value for humans. The earliest known human-other mammals interactions in the landscape that now belongs to Honduras comes from the "Cueva del Gigante" a large rock shelter located near the city of Marcala, La Paz. Archeologists have found

Table 7.3 Bat Species Known From Honduras. This List Follows Hernández (2015) and Mora (2016)

Family	Subfamily	Genus/Species	Common Name	Taxonomic Notes
Thyropteridae		*Thyroptera tricolor*	Spix's Disk-Winged Bat	
Vespertilionidae		*Bauerus dubiaquercus*	Van Gelder's Bat	
Vespertilionidae		*Perimyotis subflavus*	Eastern Pipistrelle	*Ex-Pipistrellus* Hoofer and Bussche (2003)
Vespertilionidae		*Lasiurus cinereus*	Hoary Bat	
Vespertilionidae		*Lasiurus ega*	Southern Yellow Bat	
Vespertilionidae		*Lasiurus intermedius*	Northern Yellow Bat	
Vespertilionidae		*Lasiurus frantzii*	Southern Red Bat	*Ex-Pipistrellus* Baird et al. (2015)
Vespertilionidae		*Lasiurus egregius*	Big Red Bat	
Vespertilionidae		*Rhogees samenchuae*		
Vespertilionidae		*Rhogees sabickhami*		
Vespertilionidae		*Eptesicus furinalis*	Argentine Brown Bat	
Vespertilionidae		*Eptesicus fuscus*	Big Brown Bat	
Vespertilionidae		*Eptesicus brasiliensis*	Brazilian Brown Bat	
Vespertilionidae		*Myotis keaysi*	Hairy-Legged Myotis	
Vespertilionidae		*Myotis albescens*	Silver-Tipped Myotis	
Vespertilionidae		*Myotis velifer*	Cave Myotis	
Vespertilionidae		*Myotis nigricans*	Black Myotis	
Vespertilionidae		*Myotis elegans*	Elegant Myotis	

Table 7.3 (Continued)

Family	Subfamily	Genus/Species	Common Name	Taxonomic Notes
Vespertilionidae		*Myotis riparius*	Riparian Myotis	
Molossidae		*Tadarida brasiliensis*	Brazilian Free-Tailed Bat	
Molossidae		*Nyctinomops laticaudatus*	Broad-Eared Free-Tailed Bat	
Molossidae		*Nyctinomops aurispinosus*	Peale's-Eared Free-Tailed Bat	
Molossidae		*Nyctinomops macrotis*	Big Free-Tailed Bat	
Molossidae		*Eumops nanus*		
Molossidae		*Eumops hansae*	Sanborn's Bonneted Bat	
Molossidae		*Eumops auripendulus*	Shaw's Mastiff Bat	
Molossidae		*Eumops glaucinus*	Wagner's Bonneted Bat	
Molossidae		*Eumops underwoodi*	Underwood's Bonneted Bat	
Molossidae		*Cynomops mexicanus*	Mexican Dog-Faced Bat	
Molossidae		*Promops centralis*	Big Crested Mastiff Bat	
Molossidae		*Molossus sinaloae*	Sinaloan Mastiff Bat	
Molossidae		*Molossus rufus*	Black Mastiff Bat	
Molossidae		*Molossus bondae*		*Ex-M.currentium*; Eger (2008)
Molossidae		*Molossus molossus*	Palla's Mastiff Bat	
Molossidae		*Molossus coibensis*	Coiban Mastiff Bat	
Molossidae		*Molossus aztecus*	Aztec Mastiff Bat	

Family	Subfamily	Genus/Species	Common Name	Taxonomic Notes
Emballonuridae		*Diclidurus albus*	Northern Ghost Bat	
Emballonuridae		*Rhynchonycteris naso*	Proboscis Bat	
Emballonuridae		*Centronycteris centralis*	Thomas' Shaggy Bat	
Emballonuridae		*Saccopteryx biliniata*	Greater Sac-Winged Bat	
Emballonuridae		*Saccopteryx leptura*	Lesser Sac-Winged Bat	
Emballonuridae		*Balantiopteryx plicata*	Gray Sac-Winged Bat	
Emballonuridae		*Balantiopteryx io*	Thomas' Sac-Winged Bat	
Emballonuridae		*Peropteryx kappleri*	Greater Dog-Like Bat	
Emballonuridae		*Peropteryx macrotis*	Lesser Dog-Like Bat	
Noctilionidae		*Noctilio leporinus*	Greater Bulldog Bat	
Noctilionidae		*Noctilio albiventris*	Lesser Bulldog Bat	
Natalidae		*Natalus mexicanus*	Mexican Greater Funnel-Eared Bat	
Mormoopidae		*Mormoops megalophylla*	Peters' Ghost-Faced Bat	
Mormoopidae		*Pteronotus personatus*	Wagner's Mustached Bat	
Mormoopidae		*Pteronotus mesoamericanus*		A valid species; see Clare et al. (2013)
Mormoopidae		*Pteronotus davyi*	Davy's Naked-Backed Bat	
Mormoopidae		*Pteronotus gymnonotus*	BigNaked-Backed Bat	
Phyllostomidae	Desmodontinae	*Diphyllae caudata*	Hairy-Legged Vampire Bat	

Table 7.3 (Continued)

Family	Subfamily	Genus/Species	Common Name	Taxonomic Notes
Phyllostomidae	Desmodontinae	*Desmodus rotundus*	Vampire Bat	
Phyllostomidae	Desmodontinae	*Diaemus youngi*	White-Winged Vampire Bat	
Phyllostomidae	Glossophaginae	*Choeronycteris mexicana*	Mexican Long-Tongued Bat	
Phyllostomidae	Glossophaginae	*Leptonycteris yerbabuenae*	Lesser Long-Nosed Bat	
Phyllostomidae	Glossophaginae	*Glossophaga soricina*	Palla's Long-Tongued Bat	
Phyllostomidae	Glossophaginae	*Glossophaga commissarisi*	Commissaris' Long-Tongued Bat	
Phyllostomidae	Glossophaginae	*Glossophaga leachii*	Gray's Long-Tongued Bat	
Phyllostomidae	Glossophaginae	*Anoura geoffroyi*	Geoffroy's Tailless Bat	
Phyllostomidae	Glossophaginae	*Lichonycteris obscura*	Dark Long-Tongued Bat	
Phyllostomidae	Glossophaginae	*Choeroniscus godmani*	Godman's Long-Tongued Bat	
Phyllostomidae	Glossophaginae	*Hylonycteris underwoodi*	Underwood's Long-Tongued Bat	
Phyllostomidae	Stenodermatinae	*Sturnira parvidens*	Little Yellow-Shouldered Bat	*Ex-S. lillium*; Velazco and Patterson (2013)
Phyllostomidae	Stenodermatinae	*Sturnira hondurensis*	Highland Yellow-Shouldered Bat	*Ex-S. ludovici*; Velazco and Patterson (2013)
Phyllostomidae	Stenodermatinae	*Centurios enex*	Wrinkle-Faced Bat	

Family	Subfamily	Genus/Species	Common Name	Taxonomic Notes
Phyllostomidae	Stenodermatinae	*Ectophylla alba*	Honduran White Bat	
Phyllostomidae	Stenodermatinae	*Artibeus jamaicensis*	Jamaican Fruit-Eating Bat	
Phyllostomidae	Stenodermatinae	*Artibeus inopinatus*	Honduran Fruit-Eating Bat	
Phyllostomidae	Stenodermatinae	*Artibeus lituratus*	Great Fruit-Eating Bat	
Phyllostomidae	Stenodermatinae	*Dermanura phaeotis*	Pygmy Fruit-Eating Bat	*Ex-Artibeus phaeotis* Solari et al. (2009)
Phyllostomidae	Stenodermatinae	*Dermanura watsoni*	Thomas' Fruit-Eating Bat	*Ex-Artibeus watsoni* Solari et al. (2009)
Phyllostomidae	Stenodermatinae	*Dermanura azteca*	Aztec Fruit-Eating Bat	*Ex-Artibeus aztecus* Solari et al. (2009)
Phyllostomidae	Stenodermatinae	*Dermanura tolteca*	Toltec Fruit-Eating Bat	*Ex-Artibeus toltecus* Solari et al. (2009)
Phyllostomidae	Stenodermatinae	*Chiroderma villosum*	Hairy Big-Eyed Bat	
Phyllostomidae	Stenodermatinae	*Chiroderma salvini*	Salvin's Big-Eyed Bat	
Phyllostomidae	Stenodermatinae	*Enchisthenes hartii*	Velvety Fruit-Eating Bat	
Phyllostomidae	Stenodermatinae	*Vampyriscus nymphaea*	Striped Yellow-Eared Bat	
Phyllostomidae	Stenodermatinae	*Vampyriscus major*	Stripe-Faced Bat	*Ex-Vampyrodes caraccioli* Velazco and Simmons (2011)

Table 7.3 (Continued)

Family	Subfamily	Genus/Species	Common Name	Taxonomic Notes
Phyllostomidae	Stenodermatinae	*Vampyriscus thyone*	Northern Little Yellow-Eared Bat	
Phyllostomidae	Stenodermatinae	*Platyrrhinus helleri*	Heller's Broad-Nosed Bat	
Phyllostomidae	Stenodermatinae	*Urodermama magnirostrum*	Brown Tent-Making Bat	
Phyllostomidae	Stenodermatinae	*Uroderma davisi*	Tent-Making Bat	*Ex-U. bilobulatum* Mantilla-Meluk (2014)
Phyllostomidae	Stenodermatinae	*Uroderma convexum*	Tent-Making Bat	*Ex-U. bilobulatum* Mantilla-Meluk (2014)
Phyllostomidae	Carollinae	*Carollia perspicillata*	Seba's Short-Tailed Bat	
Phyllostomidae	Carollinae	*Carollia sowelli*	Sowell's Short-Tailed Bat	*Ex-C. brevicauda* Baker et al. (2002)
Phyllostomidae	Carollinae	*Carollia castanea*	Chestnut Short-Tailed Bat	
Phyllostomidae	Carollinae	*Carollia subrufa*	Gray Short-Tailed Bat	
Phyllostomidae	Lonchorhininae	*Lonchorhina aurita*	Tomes' Sword-Nosed Bat	
Phyllostomidae	Phyllostominae	*Vampyrum spectrum*	Spectral Bat	
Phyllostomidae	Phyllostominae	*Chrotopterus auritus*	Wooly False Vampire Bat	
Phyllostomidae	Phyllostominae	*Phyllostomus hastatus*	Greater Spear-Nosed Bat	

Family	Subfamily	Genus/Species	Common Name	Taxonomic Notes
Phyllostomidae	Phyllostominae	*Phyllostomus discolor*	Pale Spear-Nosed Bat	
Phyllostomidae	Phyllostominae	*Phyllostomus stenops*	Pale-Faced Bat	
Phyllostomidae	Phyllostominae	*Macrophyllum macrophyllum*	Long-Legged Bat	
Phyllostomidae	Phyllostominae	*Trachops cirrhosus*	Fringe-Lipped Bat	
Phyllostomidae	Phyllostominae	*Tonatia aurophila*	Stripe-Headed Round-Eared Bat	
Phyllostomidae	Phyllostominae	*Gardnerycteris crenulatum*	Striped Hairy-Nosed Bat	*Ex-Mimon crenulatum* Hurtado and Pacheco (2014)
Phyllostomidae	Phyllostominae	*Mimon cozumelae*	Cozumel Golden Bat	
Phyllostomidae	Phyllostominae	*Lophostoma brasiliense*	Pygmy Round-Eared Bat	
Phyllostomidae	Phyllostominae	*Lophostoma silvicolum*	White-Throated Round-Eared Bat	
Phyllostomidae	Phyllostominae	*Lophostoma evotis*	Davis' Round-Eared Bat	
Phyllostomidae	Micronycterinae	*Micronycteris hirsuta*	Hairy Big-Eared Bat	
Phyllostomidae	Micronycterinae	*Micronycteris microtis*	Common Big-Eared Bat	
Phyllostomidae	Micronycterinae	*Micronycteris schmidtorum*	Schmidt's Big-Eared Bat	
Phyllostomidae	Micronycterinae	*Micronycteris minuta*	White-bellied Big-Eared Bat	
Phyllostomidae	Glyphonycterinae	*Glyphonycteris daviesi*	Davies' Big-Eared Bat	
Phyllostomidae	Glyphonycterinae	*Glyphonycteris sylvestris*	Tri-colored Big-Eared Bat	

that this cave was used as a hunting camp by middle Holocene human for-
agers some 10,000 years ago. Remains found in the cave seem to indicate
that large now extinct mammals such as *Euceratherium* were hunted by
humans in the area, as well as other still extant mammal species such as deer
(Scheffler et al., 2012). Large mammals also played an important role for
the Mayans. The jaguar, for example, was considered to be a powerful divin-
ity and remains of this large cat have been found in sacred Mayan burials
in the Mayan ruins of Copán, western Honduras (Rabinowitz, 2014). Only
50 years ago, in the mountains of north-central Honduras, the indigenous
Tolupans still believed they had to request permission to hunt a peccary or a
deer from the Master of the Animals, the Master would allow them to hunt
only what they needed. In return, the Tolupán had to allow the animals to
occasionally feed on their crops. They believed each Master had a real-world
assistant, the assistant of the Master of the peccaries, for example, was the
jaguar, and that for the deer was the puma (Chapman, 1978).

Cecil Underwood made one of the first major mammal collections in
Honduras. Those collections were made in the central and western part of
the country during the 1930s; his work set the beginnings for the "Mammals
of Honduras" published by Goodwin (1942). Goodwin (1942) reported a
total of 95 species and subspecies, including small and marine mammals (he
also reported 73 bat species; see above). Fourteen of those non-bat species
were new to science. Goodwin (1942) also reported several new genera for
the North American Continent.

More recently, Gamero (1978) and Marineros and Martinez (1998) wrote
books on the mammals of Honduras. Their works offered information on the
distribution, cinegetic value, and conservation status of these species. Marineros and Martinez (1998) estimated that a total of 228 species of mammals
are present in Honduran territory, including marine species and bats. Camera
trap surveys became popular in Honduras during the first decade of 2000.
The use of this technology has significantly increased our knowledge on the
distribution, ecology, and conservation status of the mammals in the country.
Several inventories, range extensions, and density estimates of medium and
large mammals have been carried out for specific natural protected areas
in the country (Estrada, 2004, 2006; Castañeda et al., 2013; Gonthier and
Castañeda, 2013; Portillo and Elvir, 2013, 2015, 2016; Merida and Cruz,
2014; McCann, 2015; Schank et al., 2015; Dunn et al., 2016; Petracca et al.,
2017). Most of this research has focused on large mammals such as jaguars
(*Panthera onca*), tapirs (*Tapirus bairdii*), and peccaries (*Pecari tajacu*).

The international organization Panthera has been carrying out camera
trap surveys in Honduras since 2010, accumulating more than 40,000 cam-

era trap nights in more than 50 localities across the country. Results from these surveys indicate white-lipped peccaries (*Tayassu pecari*) have been extirpated from most of the territory, with populations only persisting in the core area of La Moskitia and a locality just outside the Moskitia in the Río Sico area. In 2001, those surveys recorded 70 individuals in one herd in the Rus-Rus area. A camera trap survey done in 2011 in the Sico *y* Paulaya area captured images of at least nine white-lipped peccaries in one frame and a recent survey in southwest Río Plátano captured 15 individuals in one frame. Unfortunately, even in La Moskitia peccaries are rapidly declining due to habitat destruction and poaching. White-lipped peccaries are the most pre-ferred wild game meat by locals and one to two weeks hunting excursions are common practices in order to find and hunt this species. White-lipped peccaries are the most threatened large mammal species in Honduras. Col-lared peccaries (*Tayassu tajacu*), on the contrary, still show a wide distribu-tion, with herds of 2 to 7 individuals documented in cameras across the north coast as well as in central Honduras. Collared peccaries are locally known as "sainos or quequeos" and are frequently registered in cameras at the cloud forest of Cerro Azul Meamber, Pico Pijol and Cusuco National Parks, as well as in the Texíguat Wildlife Refuge. Nonetheless, collared peccaries have also been extirpated from specific localities due to poaching. An example of this is Jeannette Kawas National Park where the last collared peccary was seen in 2004; intensive camera trap surveys and interviews with local hunters have failed to find proof of its current presence in the park. Both species of deer seem to have a wide distribution with brocket deer (*Mazama temama*) being absent from the lowlands and appearing more commonly in the high mountain cloud forest; we have found white tail deer (*Odocoileus virginia-nus*) to be present throughout most of the habitats in the country, but more frequently captured in cameras in the lowlands and mainly in the dry valleys such as Olanchito and Comayagua. Tapirs (*Tapirus bairdii*) can be the most frequently captured species in cameras settled in the Río Plátano Biosphere Reserve and are not uncommon in Pico Bonito and Cusuco National Parks and the Texíguat Wildlife Refuge. Tapirs were extirpated from Jeannette Kawas National Park approximately 40 years ago and it is still possible to find elderlies in the area who hunted this large ungulate back in the 1960s and 1970s. A recent survey in Pico Pijol and Yoro National Parks failed to register the presence of tapirs and locals seem to agree that the species might have already been extirpated from these two areas. We found abundant tapir tracks in a recent expedition to the core of Pico Bonito, although only 15 km from La Ceiba (the third largest city in Honduras); the steep slopes of this locality are difficult for most humans to climb (we were flown in and out by helicopters), thus tapirs seem to prosper there. There are five species of wild cats documented in Honduras: jaguar (*Panthera onca*), puma (*Puma concolor*), jaguarondi (*Puma yagouaroundi*), ocelot (*Leopardus pardalis*), and margay (*L. wiedii*). All five species are present in La Moskitia in eastern Honduras as well as in areas along the north coast such as Pico Bonito and Texíguat. Jaguars, the largest extant feline of America are found from Cabo Gracias a Dios in extreme eastern Honduras and all along the north coast (lowlands and highlands) all the way to the Río Motagua in the Honduran-Guatemala border region. This includes the Merendón Mountain range west of the city of San Pedro Sula; this area was neglected from jaguar distribution maps until 2015 when we collected the first camera trap record of a jaguar in Cusuco National Park. There is also one confirmed record of a subadult male jaguar that was shot and killed by ranchers near the town of Orica, Francisco Morazán, only 70 km northeast of the capital city of Tegucigalpa. Pumas seem to be widely distributed in the country, with individuals recently documented along the north coast, around the Yojoa Lake area, and in the mountains of La Tigra and Uyuca near the capital city of Tegucigalpa. The status of all five species of felines in the southern area of Honduras (e.g., Choluteca, Valle) is unknown. Most of the medium sized mammals such as pacas (*Cuniculus paca*), agouties (*Dasyprocta* sp), coatis (*Nasua narica*), tayras (*Eira barbara*), armadillos (*Dasypus novemcinctus* and *Cabassous centralis*), porcupines (*Coendou mexicanus*), gray fox (*Urocyon cinereoargenteus*), and greater grison (*Galictis vittata*), have been widely documented in the country in several kinds of habitats including agricultural landscape. In conclusion, although many species of medium and large mammals are still present in the Honduran territory, many have been extirpated from specific localities and most of them seem to be declining due to habitat destruction and poaching, with white-lipped peccaries, giant anteater (*Myrmecophaga tridactyla)*, tapirs and jaguars being the most threatened.

We include in Table 7.4 a list of all medium and large-sized mammals known for Honduras. We register a total of 54 species in 10 orders and 18 families (Goodwin, 1942; Martinez and Marineros, 1988; IUCN, 2017). The species *Bassaricyon gabbii* is included in Table 7.4, nonetheless, there are no museum specimens, thus confirmation of its occurrence in Honduras is pending (see Helgen et al., 2013). The order with the highest number of species is Carnivore with 20, followed by Rodentia with nine species. According to the IUCN (2017), three are endangered: *Ateles geoffroyi* (Geoffroy's Spider Monkey), *Dasyprocta ruatanica* (Roatán Island Agouti), and *Tapirus bairdii* (Baird's Tapir). Four species are considered to be Vulnerable: *Trichechus manatus* (West Indian Manatee), *Myrmecophaga tridactyla* (Giant

Table 7.4 Medium and Large-Sized Mammals of Honduras

Order	Family	Genus + Species	Vernacular name	IUCN status
Sirenia	Trichechidae	*Trichechus manatus*	West Indian Manatee	VU
Cingulata	Dasypodidae	*Dasypus novemcinctus*	Nine-banded Armadillo	LC
Cingulata	Dasypodidae	*Cabassous centralis*	Northern Naked-Tailed Armadillo	DD
Pilosa	Bradypodidae	*Bradypus variegatus*	Brown-Throated Three-Toed Sloth	LC
Pilosa	Megalonychidae	*Choloepus hoffmanni*	Hoffmann's Two-Toed Sloth	LC
Pilosa	Cyclopedidae	*Cyclopes didactylus*	Silky Anteater	LC
Pilosa	Myrmecophagidae	*Myrmecophaga tridactyla*	Giant Anteater	VU
Pilosa	Myrmecophagidae	*Tamandua mexicana*	Northern Tamandua	LC
Primates	Cebidae	*Cebus capucinus*	White-Headed Capuchin	DD
Primates	Atelidae	*Alouatta palliata*	Mantled Howler Monkey	LC
Primates	Atelidae	*Ateles geoffroyi*	Geoffroy's Spider Monkey	EN
Rodentia	Erethizontidae	*Coendou mexicanus*	Mexican Hairy Dwarf Porcupine	LC
Rodentia	Dasyproctidae	*Dasyprocta punctata*	Central American Agouti	LC
Rodentia	Dasyproctidae	*Dasyprocta ruatanica*	Roatan Island Agouti	EN
Rodentia	Cuniculidae	*Cuniculus paca*	Lowland Paca	LC
Rodentia	Sciuridae	*Sciurus deppei*	Deppe's Squirrel	LC
Rodentia	Sciuridae	*Sciurus variegatoides*	Variegated Squirrel	LC
Rodentia	Geomyidae	*Orthogeomys grandis*	Giant Pocket Gopher	LC
Rodentia	Geomyidae	*Orthogeomys hispidus*	Hispid Pocket Gopher	LC
Rodentia	Geomyidae	*Orthogeomys matagalpae*	Nicaraguan Pocket Gopher	DD

Table 7.4 (Continued)

Order	Family	Genus + Species	Vernacular name	IUCN status
Lagomorpha	Leporidae	*Sylvilagus brasiliensis*	Tapeti	LC
Lagomorpha	Leporidae	*Sylvilagus floridanus*	Eastern Cottontail Rabbit	LC
Carnivora	Felidae	*Leopardus pardalis*	Ocelot	LC
Carnivora	Felidae	*Leopardus wiedii*	Margay	NT
Carnivora	Felidae	*Puma concolor*	Cougar	LC
Carnivora	Felidae	*Puma yagouaroundi*	Jaguarundi	LC
Carnivora	Felidae	*Panthera onca*	Jaguar	NT
Carnivora	Canidae	*Urocyon cinereoargenteus*	Gray Fox	LC
Carnivora	Canidae	*Canis latrans*	Coyote	LC
Carnivora	Procyonidae	*Bassariscus sumichrasti*	Cacomistle	LC
Carnivora	Procyonidae	*Procyon lotor*	Common Raccoon	LC
Carnivora	Procyonidae	*Nasua narica*	White-Nosed Coati	LC
Carnivora	Procyonidae	*Potos flavus*	Kinkajou	LC
Carnivora	Procyonidae	*Bassaricyon gabbii*	Northern Olingo	LC
Carnivora	Mustelidae	*Mustela frenata*	Long-Tailed Weasel	LC
Carnivora	Mustelidae	*Eira barbara*	Tayra	LC
Carnivora	Mustelidae	*Galictis vittata*	Greater Grison	LC
Carnivora	Mustelidae	*Lontra longicaudis*	Neotropical River Otter	NT
Carnivora	Mephitidae	*Spilogale putorius*	Eastern Spotted Skunk	VU
Carnivora	Mephitidae	*Mephitis macroura*	Hooded Skunk	LC

Order	Family	Genus + Species	Vernacular name	IUCN status
Carnivora	Mephitidae	*Conepatus leuconotus*	American Hog-Nosed Skunk	LC
Carnivora	Mephitidae	*Conepatus semistriatus*	Striped Hog-Nosed Skunk	LC
Perissodactyla	Tapiridae	*Tapirus bairdii*	Baird's Tapir	EN
Artiodactyla	Tayassuidae	*Pecari tajacu*	Collared Peccary	LC
Artiodactyla	Tayassuidae	*Tayassu pecari*	White-Lipped Peccary	VU
Artiodactyla	Cervidae	*Mazama temama*	Red Brocket Deer	DD
Artiodactyla	Cervidae	*Odocoileus virginianus*	White-Tailed Deer	LC
Didelphimorphia	Didelphidae	*Caluromys derbianus*	Derby's Woolly Opossum	LC
Didelphimorphia	Didelphidae	*Chironectes minimus*	Water Opossum	LC
Didelphimorphia	Didelphidae	*Didelphis marsupialis*	Common Opossum	LC
Didelphimorphia	Didelphidae	*Didelphis virginiana*	Virginia Opossum	LC
Didelphimorphia	Didelphidae	*Marmosa alstoni*	Alston's Mouse Opossum	LC
Didelphimorphia	Didelphidae	*Marmosa mexicana*	Mexican Mouse Opossum	LC
Didelphimorphia	Didelphidae	*Philander opossum*	Gray Four-Eyed Opossum	LC

Abbreviations used are: DD = data deficient; EN = endangered; LC = species of least concern; NT = near threatened; and VU = vulnerable.

Anteater), *Spilogale putorius* (Eastern Spotted Skunk), and *Tayassu pecari* (White-Lipped Peccary). Three species are in the Near Threatened category including the Jaguar (*Panthera onca*).

7.4.3 Birds of Honduras: A Taxonomic List of Genera and Species Numbers

Monroe (1968) provided the first detailed distributional and taxonomic study of the Honduran birds published. In the early twenty-first century, Bonta and Anderson (2002) published an updated species checklist of the Honduran birds using the Monroe book as a starting point and their own observations countrywide. In more recent years, two publications arose with detailed information and distribution of the avian diversity in Honduras (Gallardo, 2014; Fagan and Komar, 2016), thus expanding what was known for the country and predicting the possible stability at around 800 species. An updated list of the birds, known to use, or have used, Honduran territory as part of their geographical distribution, including breeding residents, migrants, or only infrequent visitors is presented in Table 7.5. That table includes 47 orders, 75 families, and 432 genera. Those genera contain 770 species known for the country. Honduras sits on a major dividing line between regions of endemism which Gallardo (2014) divided into three sectors: northern, southern, and regional endemics encompassing 41 species, only one of which (*Amazilia luciae*; Honduran Emerald Hummingbird) is endemic to Honduras. That single endemic bird species (0.001% of the total bird species) is in stark contrast to the 128 species (30.8%) of the less mobile amphibian and reptilian species known to occur in Honduras.

7.4.4 Freshwater Fish of Honduras: A Taxonomic Update and List of Species

The Honduran freshwater fish fauna assemblage is formed by 161 native species, distributed in two Classes, 17 orders and 41 families (Table 7.6). Additionally, six exotic fish species are introduced into natural systems in Honduras (Table 7.6). Those exotics also represent one order and one family not otherwise known from the country. Based on salinity tolerance, eight species (5.0% of total native assemblage) are primary, 44 (27.3%) are secondary, and 109 (67.7%) belong to the peripheral category. Clearly, the Honduran freshwater ecosystems are dominated by fish species in families with marine affinities, or are salinity tolerant, with only a small percentage of the country's freshwater fish belonging to nonsalinity-tolerant families (Table 7.6).

Table 7.5 The Birds of Honduras Listed by Order, Family, and Genera

Order	Family	Genera (sp)	Vernacular Name
Tinamiformes	Tinamidae	*Tinamus, Crypturellus* (4)	Tinamous
Anseriformes	Anatidae	*Dendrocygna, Anser, Cairina, Anas, Spatula, Mareca, Aythya, Nomonyx, Oxyura, Branta* (17)	Whistling-Ducks, Geese, Muscovy Ducks, Dabbling Ducks, Diving Ducks, Stiff-tailed Ducks
Galliformes	Cracidae	*Ortalis, Penelope, Penelopina, Crax* (6)	Chachalacas, Guans, Curassow
Galliformes	Odontophoridae	*Dendrortyx, Colinus, Odontophorus, Dactylortyx, Cyrtonyx, Rhynchortyx* (8)	Wood-Partridge, Quails
Podicipediformes	Podicipedidae	*Tachybaptus, Podilymbus* (2)	Grebes
Phoenicopteriformes	Phoenicopteridae	*Phoenicopterus* (1)	Flamingo
Procellariiformes	Hydrobatiidae	*Oceanodroma* (4)	Storm-Petrels
Phaetontiformes	Phaetontidae	*Phaeton* (2)	Tropicbirds
Ciconiiformes	Ciconiidae	*Jabiru, Mycteria* (2)	Stork, Jabiru
Suliformes	Fregatidae	*Fregata* (1)	Frigatebirds
Suliformes	Sulidae	*Sula* (4)	Boobies
Suliformes	Phalacrocoracidae	*Phalacrocorax* (2)	Cormorants
Suliformes	Anhingidae	*Anhinga* (1)	Anhinga
Pelecaniformes	Pelecanidae	*Pelecanus* (2)	Pelicans
Pelecaniformes	Ardeidae	*Botaurus, Ixobrychus, Tigrisoma, Ardea, Egretta, Bubulcus, Butorides, Agamia, Nycticorax, Nyctanassa, Cochlearius* (18)	Bitterns, Herons, Egrets
Pelecaniformes	Threskiornithidae	*Eudocimus, Plegadis, Mesembrinibis, Platalea* (5)	Ibises
Accipitriformes	Cathartidae	*Coragyps, Cathartes, Sarcoramphus* (4)	Vultures

Table 7.5 (Continued)

Order	Family	Genera (sp)	Vernacular Name
Accipitriformes	Pandionidae	*Pandion* (1)	Osprey
Accipitriformes	Accipitridae	*Gampsonyx, Elanus, Chondrohierax, Leptodon, Elanoides, Morphnus, Harpia, Spizaetus, Busarellus, Rostrhamus, Harpagus, Ictinia, Circus, Accipiter, Geranospiza, Buteogallus, Rupornis, Parabuteo, Geranoaetus, Pseudastur, Leucopternis, Buteo* (35)	Eagles, Hawk-Eagles, Kites, Hawks
Eurypygiformes	Eurypigidae	*Eurypyga* (1)	Sunbittern
Gruiformes	Rallidae	*Aramides, Amaurolimnas, Laterallus, Hapalocrex, Porzana, Rallus, Pardirallus, Porphyrio, Gallinula, Fulica* (14)	Wood-Rails, Crakes, Gallinule, Coots
Gruiformes	Heliornithidae	*Heliornis* (1)	Sungrebe
Gruiformes	Aramidae	*Aramus* (1)	Limpkin
Charadriiformes	Burhinidae	*Burhinus* (1)	Thick-knee
Charadriiformes	Recurvirostridae	*Himantopus, Recurvirostra* (2)	
Charadriiformes	Haematopodidae	*Haematopus* (1)	Oystercatcher
Charadriiformes	Charadriidae	*Pluvialis, Vanellus, Charadrius* (8)	Lapwing, Plovers
Charadriiformes	Jacanidae	*Jacana* (1)	Jacana
Charadriiformes	Scolopacidae	*Bartramia, Numenius, Tringa, Limosa, Arenaria, Calidris, Actitis, Limnodromus, Gallinago, Phalaropus* (29)	Godwit, Sandpipers, Phalaropes
Charadriiformes	Stercorariidae	*Stercorarius* (3)	Jaegers
Charadriiformes	Laridae	*Xema, Leucophaeus, Larus, Anous, Onychoprion, Sternula, Phaetusa, Gelochelidon, Hydroprogne, Chlidonias, Sterna, Thalasseus, Rynchops* (25)	Gulls, Terns

Order	Family	Genera (sp)	Vernacular Name
Columbiformes	Columbidae	*Columba, Patagioenas, Zentrygon, Geotrygon, Claravis Columbina, Leptotila, Zenaida, Streptopelia* (22)	Doves, Pigeons
Cuculiformes	Cuculidae	*Coccyzus, Piaya, Tapera, Neomorphus, Dromococcyx, Morococcyx, Geococcyx, Crotophaga* (11)	Cuckoos, Anis
Strigiformes	Tytonidae	*Tyto* (1)	Barn Owl
Strigiformes	Strigidae	*Athene, Pseudoscops, Lophostrix, Megascops, Glaucidium, Ciccaba, Strix, Pulsatrix, Bubo, Asio* (15)	Owls
Caprimulgiformes	Caprimulgidae	*Chordeiles, Lurocalis, Nyctidromus, Nyctiphrynus, Caprimulgus* (12)	Nighthawks, Pauraque, Nightjar
Caprimulgiformes	Nyctibiidae	*Nyctibius* (2)	Potoos
Apodiformes	Apodidae	*Cypseloides, Streptoprocne, Chaetura, Aeronautes, Panyptila* (10)	Swifts
Apodiformes	Trochilidae	*Florisuga, Glaucis, Threnetes, Phaethornis, Colibri, Heliothryx, Anthracothorax, Lophornis, Eugenes, Heliomaster, Lampornis, Lamprolaima, Doricha, Tilmatura, Archilochus, Atthis, Chlorostilbon, Klais, Abeillia, Phaeochroa, Campylopterus, Chalybura, Thalurania, Eupherusa, Microchera, Amazilia, Hylocharis* (43)	Hummingbirds
Trogoniformes	Trogonidae	*Trogon, Pharomachrus* (8)	Trogons, Quetzal
Coraciiformes	Momotidae	*Hylomanes, Aspatha, Momotus, Baryphthengus, Electron, Eumomota* (7)	Motmots
Coraciiformes	Alcedinidae	*Megaceryle, Ceryle, Chloroceryle* (6)	Kingfishers
Galbuliformes	Bucconidae	*Notharchus, Malacoptila, Monasa* (3)	Puffbirds

Table 7.5 (Continued)

Order	Family	Genera (sp)	Vernacular Name
Galbuliformes	Galbulidae	*Galbula, Jacamerops* (2)	Jacamars
Piciformes	Ramphastidae	*Aulacorhynchus, Pteroglossus, Selenidera, Ramphastos* (5)	Toucans
Piciformes	Picidae	*Melanerpes, Sphyrapicus, Picoides, Picumnus, Veniliornis, Piculus, Colaptes, Celeus, Dryocopus, Campephilus* (16)	Woodpeckers
Falconiformes	Falconidae	*Micrastur, Ibycter, Caracara, Milvago, Herpetotheres, Falco* (13)	Falcons, Caracara
Psittaciformes	Psittacidae	*Pyrilia, Pionus, Amazona, Ara, Bolborhynchus, Brotogeris, Eupsittula, Psittacara* (17)	Parrots, Macaws, Parakeets
Passeriformes	Thamnophilidae	*Cymbilaimus, Taraba, Thamnophilus, Thamnistes, Dysithamnus, Epinecrophylla, Cercomacra Myrmotherula, Microrhopias, Gymnocichla, Myrmeciza, Myrmornis, Gymnopithys, Hylophylax, Phaenostictus* (18)	Antshrikes
Passeriformes	Grallaridae	*Grallaria, Hylopezus* (3)	Antpittas
Passeriformes	Formicariidae	*Formicarius* (1)	Antthrush
Passeriformes	Furnariidae	*Sclerurus, Sittasomus, Deconychura, Dendrocincla, Dendrocolaptes, Xiphocolaptes, Xiphorhynchus, Lepidocolaptes, Xenops, Anabacerthia, Hyloctistes, Automolus, Synallaxis* (23)	Leaftossers, Xenops, Woodcreepers, Woodhaunter
Passeriformes	Tyrannidae	*Ornithion, Camptostoma, Myiopagis, Elaenia, Mionectes, Leptopogon, Zimmerius, Lephotriccus, Oncostoma, Poecilotriccus, Todirostrum, Rhynchocyclus, Tolmomyias, Platyrinchus, Onychorhynchus, Terenotriccus, Myiobius, Aphanotriccus, Xenotriccus, Mitrephanes, Contopus, Empidonax, Sayornis, Pyrocephalus, Colonia, Attila, Rhytiperna, Myiarchus, Pitangus, Megarynchus, Myiozetetes, Conopias, Myiodynastes, Legatus, Tyrannus* (63)	Tyrannulets, Elaenias, Spadebills, Flatbills, Flycatchers, Kingbirds

Order	Family	Genera (sp)	Vernacular Name
Passeriformes	Cotingidae	*Cotinga, Lipaugus, Carpodectes, Procnias* (4)	Cotingas
Passeriformes	Pipridae	*Manacus, Corapipo, Chiroxiphia, Ceratopipra, Piprites* (5)	Manakins
Passeriformes	Tityridae	*Tityra, Schiffornis, Laniocera, Pachyramphus* (8)	Tityras, Becards
Passeriformes	Vireonidae	*Vireo, Tunchiornis, Pachysylvia, Hylophilus, Vireolanius, Cyclarhis* (18)	Shrike-Vireo, Vireos
Passeriformes	Corvidae	*Cyanocitta, Calocitta, Cyanocorax, Cyanolyca, Aphelocoma, Corvus* (9)	Jays, Raven
Passeriformes	Hirundinidae	*Atticora, Stelgidopteryx, Progne, Tachycineta, Riparia, Hirundo, Notiochelidon, Petrochelidon* (12)	Swallows
Passeriformes	Certhiidae	*Certhia* (1)	Creeper
Passeriformes	Troglodytidae	*Microcerculus, Cyphorhinus, Salpinectes, Troglodytes, Cistothorus, Pheugopedius, Thryophilus, Uropsila, Cantorchilus, Thryothorus, Henicorhina, Campylorhynchus* (19)	Wrens
Passeriformes	Polioptilidae	*Ramphocaenus, Polioptila* (4)	Gnatwren, Gnatcatcher
Passeriformes	Cinclidae	*Cinclus* (1)	Dipper
Passeriformes	Turdidae	*Sialia, Myadestes, Catharus, Hylocichla, Turdus* (18)	Nightingale-Thrush, Veery, Thrushes, Solitaire
Passeriformes	Mimidae	*Melanotis, Melanoptila, Dumetella, Mimus* (5)	Catbirds, Mockingbirds
Passeriformes	Motacillidae	*Anthus* (1)	Pipit
Passeriformes	Bombycillidae	*Bombycilla* (1)	Waxwing
Passeriformes	Peucedramidae	*Peucedramus* (1)	Olive Warbler

Table 7.5 (Continued)

Order	Family	Genera (sp)	Vernacular Name
Passeriformes	Parulidae	*Seiurus, Helmitheros, Parkesia, Vermivora, Parula, Mniotilta, Protonotaria, Limnothlypis, Oreothlypis, Geothlypis, Setophaga, Basileuterus, Phaeothlypis, Cardellina, Myioborus, Icteria* (54)	Waterthrushes, Warblers
Passeriformes	Thraupidae	*Eucometis, Tachyphonus, Lanio, Ramphocelus, Thraupis, Tangara, Dacnis, Cyanerpes, Chlorophanes, Diglossa, Haplospiza, Sicalis, Volatinia, Sporophila, Oryzoborus, Coereba, Tiaris, Saltator* (30)	Tanagers, Honeycreepers, Bananaquit, Finch, Grassquits, Seedeaters, Saltators
Passeriformes	Emberizidae	*Arremon, Arremonops, Atlapetes, Amaurospiza, Aimophila, Melozone, Peucaea, Spizella, Chondestes, Passerculus, Ammodramus, Melospiza, Zonotrichia, Chlorospingus* (17)	Sparrows, Brushfinches
Passeriformes	Cardinalidae	*Piranga, Habia, Chlorothraupis, Caryothraustes, Cardinalis, Pheucticus, Amaurospiza, Cyanocompsa, Spiza, Passerina* (20)	Cardinal Tanagers, Cardinal Grosbeaks, Cardinals, Dickcissel
Passeriformes	Icteridae	*Dolichonyx, Agelaius, Sturnella, Dives, Quiscalus, Molothrus, Icterus, Amblycercus, Cacicus, Psarocolius* (21)	Blackbirds, Orioles, Caciques, Oropendolas
Passeriformes	Fringillidae	*Euphonia, Chlorophonia, Loxia, Carduelis, Coccothraustes* (11)	Crossbill, Grosbeak, Euphonias, Chlorophonia, Siskins
Passeriformes	Passeridae	*Passer* (1)	House Sparrow
Passeriformes	Estrilididae	*Lonchura* (2)	Munias

Table 7.6 A List of the Freshwater Fishes Known to Occur in Honduras

Order (18)	Family (42)	Species (167)	Salinity	Status
Carcharhiniformes	Carcharhinidae	*Carcharhinus leucas*	Peripheral	Native
Carcharhiniformes	Carcharhinidae	*Rhizoprionodon porosus*	Peripheral	Native
Pristiformes	Pristidae	*Pristis pectinata*	Peripheral	Native
Elopiformes	Megalopidae	*Megalops atlanticus*	Peripheral	Native
Anguilliformes	Anguillidae	*Anguilla rostrata*	Peripheral	Native
Anguilliformes	Ophicthidae	*Myrophis punctatus*	Peripheral	Native
Clupeiformes	Clupeidae	*Harengula clupeola*	Peripheral	Native
Clupeiformes	Clupeidae	*Harengula humeralis*	Peripheral	Native
Clupeiformes	Clupeidae	*Jenkinsia lamprotaenia*	Peripheral	Native
Clupeiformes	Clupeidae	*Ophistonema oglinum*	Peripheral	Native
Clupeiformes	Engraulidae	*Anchoa colonensis*	Peripheral	Native
Clupeiformes	Engraulidae	*Anchoa filifera*	Peripheral	Native
Clupeiformes	Engraulidae	*Anchoa parva*	Peripheral	Native
Clupeiformes	Engraulidae	*Anchovia clupeoides*	Peripheral	Native
Clupeiformes	Engraulidae	*Anchoviella elongata*	Peripheral	Native
Cypriniformes	Cyprinidae	*Ctenopharyngodon idella*	Primary	Introduced
Cypriniformes	Cyprinidae	*Hypophthalmichthys molitrix*	Primary	Introduced
Characiformes	Characidae	*Astyanax aeneus*	Primary	Native
Characiformes	Characidae	*Brycon guatemalensis*	Primary	Native
Characiformes	Characidae	*Hyphessobrycon tortuguerae*	Primary	Native
Characiformes	Characidae	*Roeboides bouchellei*	Primary	Native

Table 7.6 (Continued)

Order (18)	Family (42)	Species (167)	Salinity	Status
Siluriformes	Ariidae	Cathorops higuchii	Peripheral	Native
Siluriformes	Ariidae	Cathorops melanopus	Peripheral	Native
Siluriformes	Ariidae	Cathorops sp	Peripheral	Native
Siluriformes	Ariidae	Cathorops steindachneri	Peripheral	Native
Siluriformes	Ariidae	Cathorops taylori	Peripheral	Native
Siluriformes	Ariidae	Sciades assimilis	Peripheral	Native
Siluriformes	Ariidae	Sciades guatemalensis	Peripheral	Native
Siluriformes	Ariidae	Sciades seemanni	Peripheral	Native
Siluriformes	Ictaluridae	Ictalurus punctatus	Primary	Introduced
Siluriformes	Heptapteridae	Rhamdia guatemalensis	Primary	Native
Siluriformes	Heptapteridae	Rhamdia laticauda	Primary	Native
Gymnotiformes	Gymnotidae	Gymnotus cylindricus	Primary	Native
Gymnotiformes	Gymnotidae	Gymnotus maculosus	Primary	Native
Batrachoidiformes	Batrachoididae	Batrachoides gilberti	Peripheral	Native
Gobiesociformes	Gobiesocidae	Gobiesox strumosus	Peripheral	Native
Atheriniformes	Atherinopsidae	Atherinella argentea	Peripheral	Native
Atheriniformes	Atherinopsidae	Atherinella blackburni	Peripheral	Native
Atheriniformes	Atherinopsidae	Atherinella guija	Peripheral	Native
Atheriniformes	Atherinopsidae	Atherinella meeki	Peripheral	Native
Atheriniformes	Atherinopsidae	Atherinella milleri	Peripheral	Native
Atheriniformes	Atherinopsidae	Atherinella pachylepis	Peripheral	Native

Order (18)	Family (42)	Species (167)	Salinity	Status
Cyprinodontiformes	Rivulidae	*Cynodonichthys tenuis*	Secondary	Native
Cyprinodontiformes	Rivulidae	*Kryptolebias marmoratus*	Secondary	Native
Cyprinodontiformes	Profundulidae	*Profundulus kreiseri*	Secondary	Native
Cyprinodontiformes	Profundulidae	*Tlaloc portillorum*	Secondary	Endemic
Cyprinodontiformes	Poeciliidae	*Alfaro cultratus*	Secondary	Native
Cyprinodontiformes	Poeciliidae	*Alfaro huberi*	Secondary	Native
Cyprinodontiformes	Poeciliidae	*Belonesox belizanus*	Secondary	Native
Cyprinodontiformes	Poeciliidae	*Gambusia nicaraguensis*	Secondary	Native
Cyprinodontiformes	Poeciliidae	*Heterandria anzuetoi*	Secondary	Native
Cyprinodontiformes	Poeciliidae	*Phallichthys amates*	Secondary	Native
Cyprinodontiformes	Poeciliidae	*Poecilia hondurensis*	Secondary	Endemic
Cyprinodontiformes	Poeciliidae	*Poecilia mexicana*	Secondary	Native
Cyprinodontiformes	Poeciliidae	*Poecilia sp. patuca*	Secondary	Native
Cyprinodontiformes	Poeciliidae	*Poeciliopsis pleurospilus*	Secondary	Native
Cyprinodontiformes	Poeciliidae	*Poeciliopsis turrubarensis*	Secondary	Native
Cyprinodontiformes	Poeciliidae	*Pseudoxiphophorus anzuetoi*	Secondary	Native
Cyprinodontiformes	Poeciliidae	*Pseudoxiphophorus bimaculatus*	Secondary	Native
Cyprinodontiformes	Poeciliidae	*Xiphophorus helleri*	Secondary	Native
Cyprinodontiformes	Poeciliidae	*Xiphophorus mayae*	Secondary	Native
Cyprinodontiformes	Anablepidae	*Anableps dowei*	Secondary	Native
Beloniformes	Belonidae	*Strongylura marina*	Peripheral	Native
Beloniformes	Belonidae	*Strongylura notata*	Peripheral	Native

Table 7.6 (Continued)

Order (18)	Family (42)	Species (167)	Salinity	Status
Beloniformes	Belonidae	*Strongylura timucu*	Peripheral	Native
Beloniformes	Hemyrhamphidae	*Hyporhamphus roberti hildebrandi*	Peripheral	Native
Beloniformes	Hemyrhamphidae	*Hyporhamphus unifasciatus*	Peripheral	Native
Syngnathiformes	Syngnathidae	*Microphis brachyurus lineatus*	Peripheral	Native
Syngnathiformes	Syngnathidae	*Pseudophallus mindii*	Peripheral	Native
Syngnathiformes	Syngnathidae	*Pseudophallus starksi*	Peripheral	Native
Syngnathiformes	Syngnathidae	*Syngnathus pelagicus*	Peripheral	Native
Syngnathiformes	Syngnathidae	*Syngnathus scovelli*	Peripheral	Native
Synbranchiformes	Synbranchidae	*Ophisternon aenigmaticum*	Secondary	Native
Synbranchiformes	Synbranchidae	*Synbranchus marmoratus*	Secondary	Native
Perciformes	Centropomidae	*Centropomus ensiferus*	Peripheral	Native
Perciformes	Centropomidae	*Centropomus nigrescens*	Peripheral	Native
Perciformes	Centropomidae	*Centropomus parallelus*	Peripheral	Native
Perciformes	Centropomidae	*Centropomus pectinatus*	Peripheral	Native
Perciformes	Centropomidae	*Centropomus undecimalis*	Peripheral	Native
Perciformes	Centropomidae	*Centropomus unionensis*	Peripheral	Native
Perciformes	Centrarchidae	*Micropterus salmoides*	Secondary	Introduced
Perciformes	Carangidae	*Caranx bartholomaei*	Peripheral	Native
Perciformes	Carangidae	*Caranx latus*	Peripheral	Native
Perciformes	Carangidae	*Olygoplites saurus*	Peripheral	Native
Perciformes	Carangidae	*Trachinotus goodei*	Peripheral	Native

Order (18)	Family (42)	Species (167)	Salinity	Status
Perciformes	Lutjanidae	*Lutjanus apodus*	Peripheral	Native
Perciformes	Lutjanidae	*Lutjanus jocu*	Peripheral	Native
Perciformes	Gerreidae	*Diapterus auratus*	Peripheral	Native
Perciformes	Gerreidae	*Eucinostomus argenteus*	Peripheral	Native
Perciformes	Gerreidae	*Eucinostomus harengulus*	Peripheral	Native
Perciformes	Gerreidae	*Eucinostomus jonesi*	Peripheral	Native
Perciformes	Gerreidae	*Eucinostomus melanopterus*	Peripheral	Native
Perciformes	Gerreidae	*Eugerres plumieri*	Peripheral	Native
Perciformes	Gerreidae	*Gerres cinereus*	Peripheral	Native
Perciformes	Haemulidae	*Pomadasys crocro*	Peripheral	Native
Perciformes	Sciaenidae	*Bairdiella ronchus*	Peripheral	Native
Perciformes	Sciaenidae	*Cynoscion praedatorius*	Peripheral	Native
Perciformes	Sciaenidae	*Menticirrhus americanus*	Peripheral	Native
Perciformes	Sciaenidae	*Paralonchurus dumerilii*	Peripheral	Native
Perciformes	Sciaenidae	*Umbrina broussonnetii*	Peripheral	Native
Perciformes	Polynemidae	*Polydactylus virginicus*	Peripheral	Native
Perciformes	Mugilidae	*Agonostomus monticola*	Peripheral	Native
Perciformes	Mugilidae	*Joturus pichardi*	Peripheral	Native
Perciformes	Mugilidae	*Mugil curema*	Peripheral	Native
Perciformes	Mugilidae	*Mugil liza*	Peripheral	Native
Perciformes	Cichlidae	*Amatitlania nigrofasciata*	Secondary	Native
Perciformes	Cichlidae	*Amphilophus hogaboomorum*	Secondary	Endemic

Table 7.6 (Continued)

Order (18)	Family (42)	Species (167)	Salinity	Status
Perciformes	Cichlidae	*Amphilophus longimanus*	Secondary	Native
Perciformes	Cichlidae	*Amphilophus trimaculatus*	Secondary	Native
Perciformes	Cichlidae	*Archocentrus centrarchus*	Secondary	Native
Perciformes	Cichlidae	*Chortiheros wesseli*	Secondary	Endemic
Perciformes	Cichlidae	*Chuco microphthalmus*	Secondary	Native
Perciformes	Cichlidae	*Cribroheros alfari*	Secondary	Native
Perciformes	Cichlidae	*Cribroheros robertsoni*	Secondary	Native
Perciformes	Cichlidae	*Criptoheros cutteri*	Secondary	Native
Perciformes	Cichlidae	*Criptoheros spilurus*	Secondary	Native
Perciformes	Cichlidae	*Herotilapia multispinosa*	Secondary	Native
Perciformes	Cichlidae	*Hypsophrys nicaraguensis*	Secondary	Native
Perciformes	Cichlidae	*Mayaheros urophthalmus*	Secondary	Native
Perciformes	Cichlidae	*Oreochromis mossambicus*	Secondary	Introduced
Perciformes	Cichlidae	*Oreochromis niloticu*	Secondary	Introduced
Perciformes	Cichlidae	*Parachromis dovii*	Secondary	Native
Perciformes	Cichlidae	*Parachromis friedrichsthalii*	Secondary	Native
Perciformes	Cichlidae	*Parachromis loisellei*	Secondary	Native
Perciformes	Cichlidae	*Parachromis managuensis*	Secondary	Native
Perciformes	Cichlidae	*Parachromis motaguensis*	Secondary	Native
Perciformes	Cichlidae	*Rocio octofasciata*	Secondary	Native
Perciformes	Cichlidae	*Thorichthys aureus*	Secondary	Native

Order (18)	Family (42)	Species (167)	Salinity	Status
Perciformes	Cichlidae	*Vieja maculicauda*	Secondary	Native
Perciformes	Labrisomidae	*Labrisomus nuchipinnis*	Peripheral	Native
Perciformes	Dactyloscopidae	*Dactyloscopus tridigitatus*	Peripheral	Native
Perciformes	Blenniidae	*Lupinoblennius vinctus*	Peripheral	Native
Perciformes	Eleotridae	*Dormitator latifrons*	Peripheral	Native
Perciformes	Eleotridae	*Dormitator maculatus*	Peripheral	Native
Perciformes	Eleotridae	*Eleotris amblyopsis*	Peripheral	Native
Perciformes	Eleotridae	*Eleotris perniger*	Peripheral	Native
Perciformes	Eleotridae	*Eleotris picta*	Peripheral	Native
Perciformes	Eleotridae	*Erotelis smaragdus*	Peripheral	Native
Perciformes	Eleotridae	*Gobiomorus dormitor*	Peripheral	Native
Perciformes	Eleotridae	*Gobiomorus maculatus*	Peripheral	Native
Perciformes	Eleotridae	*Leptophilypnus fluviatilis*	Peripheral	Native
Perciformes	Gobiidae	*Awous banana*	Peripheral	Native
Perciformes	Gobiidae	*Bathygobius soporator*	Peripheral	Native
Perciformes	Gobiidae	*Ctenogobius boleosoma*	Peripheral	Native
Perciformes	Gobiidae	*Ctenogobius fasciatus*	Peripheral	Native
Perciformes	Gobiidae	*Ctenogobius sagitulla*	Peripheral	Native
Perciformes	Gobiidae	*Ctenogobius stigmaticus*	Peripheral	Native
Perciformes	Gobiidae	*Evorthodus lyricus*	Peripheral	Native
Perciformes	Gobiidae	*Gobionellus oceanicus*	Peripheral	Native
Perciformes	Gobiidae	*Lophogobius cyprinoides*	Peripheral	Native

Table 7.6 (Continued)

Order (18)	Family (42)	Species (167)	Salinity	Status
Perciformes	Gobiidae	*Sicydium gymnogaster*	Peripheral	Native
Perciformes	Gobiidae	*Sicydium multipunctatum*	Peripheral	Native
Perciformes	Gobiidae	*Sicydium plumieri*	Peripheral	Native
Perciformes	Gobiidae	*Sicydium punctatum*	Peripheral	Native
Perciformes	Microdesmidae	*Microdesmus carri*	Peripheral	Native
Perciformes	Acanthuridae	*Acanthurus bahianus castelnau*	Peripheral	Native
Perciformes	Sphyraenidae	*Sphyraena barracuda*	Peripheral	Native
Perciformes	Sphyraenidae	*Sphyraena guachancho*	Peripheral	Native
Pleuronectiformes	Paralichthyidae	*Citharichthys abbotti*	Peripheral	Native
Pleuronectiformes	Paralichthyidae	*Citharichthys arenaceus*	Peripheral	Native
Pleuronectiformes	Paralichthyidae	*Citharichthys gilberti*	Peripheral	Native
Pleuronectiformes	Paralichthyidae	*Citharichthys macrops*	Peripheral	Native
Pleuronectiformes	Paralichthyidae	*Citharichthys spilopterus*	Peripheral	Native
Pleuronectiformes	Achiridae	*Achirus lineatus*	Peripheral	Native
Pleuronectiformes	Achiridae	*Trinectes fonseecensis*	Peripheral	Native
Pleuronectiformes	Achiridae	*Trinectes maculatus*	Peripheral	Native
Tetraodontiformes	Tetraodontidae	*Sphoeroides testudineus*	Peripheral	Native

The list is arranged systematically. Gray highlighted lines represent species belonging to the class Chondrichthyes, and the non-highlighted lines represent species belonging to the Class Actinopterygii. Salinity refers to the three commonly used salinity tolerance classes of Myers (1949). The classifications for status are based on geographical distributions. A native fish is a species that has naturally occurring populations in Honduras, and also occurs naturally in at least one other country. Endemic refers to native species that are restricted in distribution to Honduras. Exotic Refers to species that are introduced into natural systems in Honduras, but are native to some other area of the world. This table also includes one introduced order, one introduced family, and six introduced species.

The high dominance level of peripheral fishes in Honduran freshwater ecosystems is also common to all Central American countries (see Kihn Pineda et al., 2006; Matamoros et al., 2009; Angulo et al., 2013; McMahan et al., 2013). Although, this pattern remains largely unexplained, it has been hypothesized that several factors might be affecting the paucity of primary freshwater fishes in the country and all of Central America. The first factor might be the relatively younger age of native primary freshwater fish clades occurring in the country (Matamoros et al., 2012, 2015; Tagliacollo et al., 2015). Those primary fishes are limited to the families Characidae (four species), Heptapteridae (two species), and Gymnotidae (two species) (Matamoros et al., 2009) and are thought to have arrived in Central America after the final closure of the Panamanian Isthmus (Matamoros et al., 2015).

A second factor that might explain the depauperate native primary Honduran freshwater fish fauna could be because Honduras is completely located in the Chortis Block, a geological feature that encompasses all of Honduras and El Salvador, and extends to the Nicaraguan Graben to the south, and is bounded on the north by the Río Motagua Fault (Rogers et al., 2007). The Chortis block originated in the eastern Pacific Ocean during the Eocene and drifted in the Pacific Ocean until it reached its current position and collated with the Mayan Block (Keppie and Morán-Zenteno, 2005). During that time, the Chortis Block was an oceanic island; oceanic islands are normally depleted of primary freshwater fish faunas. Neither of the above-mentioned hypotheses have been empirically tested and the mechanisms that have prevented diversification of the Nuclear Middle American native freshwater fish fauna, to which Honduras is a portion of, remains unknown.

Whereas species richness, in general, is a working tool for a conservation measurement in a region, an important component of species richness is the number of endemic species found in a given region. This measurement is widely used to determine hot-spots and important conservation areas. The freshwater fishes endemic to Honduras are only four (Table 7.6), *Tlaloc portillorum, Poecilia hondurensis, Amphilophus hogaboomorum*, and *Chortiheros wesseli*. Besides *P. hondurensis,* which is widespread along the Honduran Caribbean Coast from reaches of the Río Patuca in eastern Honduras to the middle reaches of the Río Motagua in western Honduras, the remaining three species show very restricted distributions. *Chortiheros wesseli* occurs only in rivers in the area of Pico Bonito National Park near La Ceiba, Atlántida, in north-central Honduras. The known distribution of *A. hogaboomorum* is restricted to lowland areas of the Río Choluteca in southern Honduras, and *T. portillorum* is only known from a few localities in central to south-central Honduras near the cities of Siguatepeque and Comayagua, and a number of localities in the towns of Lepaterique and Ojojona near Tegucigalpa.

Although efforts have been made to document the native freshwater fish fauna of Honduras (e. g., Matamoros et al., 2009), these efforts have focused at the major river drainage scale and many tributaries, moderate-sized to tiny drainages, have been neglected. As a result, large sections of the country remain understudied and have relatively few records, especially those of the Río Lempa and the upper reaches of the ríos Ulúa and Chamelecón in northwestern Honduras as well as those of the ríos Coco and Patuca in northeastern Honduras. Also, understudied is the elevational distributions of the Honduran freshwater fish fauna.

7.4.5 Amphibians and Reptiles: An Updated Taxonomic List of Species

The herpetofauna of Honduras is known to contain 145 species of amphibians and 270 species of reptiles (Table 7.7). Of the seven Central American countries, only the two southernmost, Costa Rica (with about 420 species) and Panama (about 450 species) have been reported to have more species of amphibians and reptiles than does Honduras. However, in percentages of the total herpetofauna species of a given country being endemic to that country, Honduras is far and away the Central American country with the highest percentage of an endemic herpetofauna. The total known Honduran herpetofauna is 415 species, with 128 of those species endemic (30.8%) to the country (updated from McCranie, 2015, *In Press*). Of the total known amphibian species, 42.8% (62 of 145 species) are Honduran endemics, whereas of the total reptile species, 24.4% are Honduran endemics (66 of 270 known species). Of the two Central American countries known to contain more amphibian and reptilian species than does Honduras, Costa Rica has only 16.9% of its herpetofauna endemic to that country; Panama has only 14.5% endemic to that country. Those percentages and numbers of the Honduran species included in Table 7.7 also contain five species for which their descriptions have yet to be published, but the manuscripts containing their descriptions have been accepted and are in press. Therefore, those species are listed by an uppercase letter in Table 7.7, instead of in listing them by their new species name.

One mountain range in Honduras, the Cordillera Nombre de Dios, stands out among all other mountain ranges in all of Central America by having an unusually high number of amphibian and reptilian species known to be

Table 7.7 A List of the Amphibians and Reptiles Known to Occur in Honduras

Order	Family	Genus and Species	Status
Class Amphibia			
Anura	Bufonidae	*Atelophryniscus chrysophorus*	Extinct*
Anura	Bufonidae	*Incilius campbelli*	Peripheral
Anura	Bufonidae	*Incilius coccifer*	Widespread
Anura	Bufonidae	*Incilius ibarrai*	Peripheral
Anura	Bufonidae	*Incilius leucomyos*	Widespread*
Anura	Bufonidae	*Incilius luetkenii*	Widespread
Anura	Bufonidae	*Incilius porteri*	GD not defined*
Anura	Bufonidae	*Incilius valliceps*	Widespread
Anura	Bufonidae	*Rhaebo haematiticus*	Widespread
Anura	Bufonidae	*Rhinella horribilis*	Widespread
Anura	Centrolenidae	*Cochranella granulosa*	Peripheral
Anura	Centrolenidae	*Espadarana prosoblepon*	Widespread
Anura	Centrolenidae	*Hyalinobatrachium chirripoi*	Peripheral
Anura	Centrolenidae	*Hyalinobatrachium crybetes*[1]	Tiny GD*
Anura	Centrolenidae	*Hyalinobatrachium fleischmanni*	Widespread
Anura	Centrolenidae	*Sachatamia albomaculata*	Widespread
Anura	Centrolenidae	*Teratohyla pulverata*	Widespread
Anura	Centrolenidae	*Teratohyla spinosa*	Peripheral
Anura	Craugastoridae	*Craugastor anciano*	Likely Extinct*
Anura	Craugastoridae	*Craugastor aurilegulus*	Small GD*
Anura	Craugastoridae	*Craugastor chac*	Widespread
Anura	Craugastoridae	*Craugastor charadra*	Peripheral
Anura	Craugastoridae	*Craugastor chrysozetetes*	Extinct*
Anura	Craugastoridae	*Craugastor coffeus*	GD not documented*

Table 7.7 (Continued)

Order	Family	Genus and Species	Status
Anura	Craugastoridae	*Craugastor cruzi*	Extinct*
Anura	Craugastoridae	*Craugastor cyanochthebius*	Tiny GD*
Anura	Craugastoridae	*Craugastor emleni*	Small GD*
Anura	Craugastoridae	*Craugastor epochthidius*	Small GD*
Anura	Craugastoridae	*Craugastor fecundus*	Small GD*
Anura	Craugastoridae	*Craugastor fitzingeri*	Peripheral
Anura	Craugastoridae	*Craugastor laevissimus*	Widespread
Anura	Craugastoridae	*Craugastor laticeps*	Peripheral
Anura	Craugastoridae	*Craugastor lauraster*	Widespread
Anura	Craugastoridae	*Craugastor loki*	Extirpated
Anura	Craugastoridae	*Craugastor megacephalus*	Peripheral
Anura	Craugastoridae	*Craugastor merendonensis*	Likely Extinct*
Anura	Craugastoridae	*Craugastor milesi*	Small GD*
Anura	Craugastoridae	*Craugastor mimus*	Peripheral
Anura	Craugastoridae	*Craugastor noblei*	Widespread
Anura	Craugastoridae	*Craugastor olanchano*	Small GD*
Anura	Craugastoridae	*Craugastor omoaensis*	Likely Extinct*
Anura	Craugastoridae	*Craugastor pechorum*	Small GD*
Anura	Craugastoridae	*Craugastor rostralis*	Peripheral
Anura	Craugastoridae	*Craugastor rupinius*	Peripheral
Anura	Craugastoridae	*Craugastor saltuarius*	Small GD*
Anura	Craugastoridae	*Craugastor stadelmani*	Small GD*
Anura	Craugastoridae	*Craugastor* sp. A (*In Press*)	Tiny GD*

Table 7.7   (Continued)

Order	Family	Genus and Species	Status
Anura	Craugastoridae	*Craugastor* sp. B (*In Press*)	Small GD*
Anura	Craugastoridae	*Pristimantis cerasinus*	Peripheral
Anura	Craugastoridae	*Pristimantis ridens*	Widespread
Anura	Eleutherodactylidae	*Diasporus diastema*	Peripheral
Anura	Eleutherodactylidae	*Eleutherodactylus planirostris*	Introduced
Anura	Hylidae	*Anotheca spinosa*	GD not known
Anura	Hylidae	*Bromeliohyla bromeliacia*	Peripheral
Anura	Hylidae	*Dendropsophus ebraccatus*	Widespread
Anura	Hylidae	*Dendropsophus microcephalus*	Widespread
Anura	Hylidae	*Duellmanohyla salvavida*	Small GD*
Anura	Hylidae	*Duellmanohyla soralia*	Peripheral
Anura	Hylidae	*Ecnomiohyla miliaria*	Peripheral
Anura	Hylidae	*Ecnomiohyla salvaje*	Small GD*
Anura	Hylidae	*Exerodonta catracha*	Widespread*
Anura	Hylidae	*Isthmohyla insolita*	Tiny GD*
Anura	Hylidae	*Isthmohyla melacaena*	Tiny GD*
Anura	Hylidae	*Plectrohyla calvata*	Tiny GD*
Anura	Hylidae	*Plectrohyla chrysopleura*	Small GD*
Anura	Hylidae	*Plectrohyla dasypus*	Tiny GD*
Anura	Hylidae	*Plectrohyla exquisita*	Tiny GD*
Anura	Hylidae	*Plectrohyla guatemalensis*	Widespread
Anura	Hylidae	*Plectrohyla hartwegi*	Peripheral
Anura	Hylidae	*Plectrohyla matudai*	Peripheral

Table 7.7 (Continued)

Order	Family	Genus and Species	Status
Anura	Hylidae	*Plectrohyla psiloderma*	Small GD
Anura	Hylidae	*Ptychohyla euthysanota*	Peripheral
Anura	Hylidae	*Ptychohyla hypomykter*	Widespread
Anura	Hylidae	*Ptychohyla salvadorensis*	Widespread
Anura	Hylidae	*Ptychohyla spinipollex*	Small GD*
Anura	Hylidae	*Scinax boulengeri*	Peripheral
Anura	Hylidae	*Scinax staufferi*	Widespread
Anura	Hylidae	*Smilisca baudinii*	Widespread
Anura	Hylidae	*Smilisca manisorum*	Peripheral
Anura	Hylidae	*Smilisca phaeota*	Widespread
Anura	Hylidae	*Smilisca sordida*	Peripheral
Anura	Hylidae	*Tlalocohyla loquax*	Widespread
Anura	Hylidae	*Tlalocohyla picta*	Widespread
Anura	Hylidae	*Trachycephalus typhonius*	Widespread
Anura	Leptodactylidae	*Engystomops pustulosus*	Widespread
Anura	Leptodactylidae	*Leptodactylus fragilis*	Widespread
Anura	Leptodactylidae	*Leptodactylus melanonotus*	Widespread
Anura	Leptodactylidae	*Leptodactylus savagei*	Widespread
Anura	Leptodactylidae	*Leptodactylus silvanimbus*	Small GD*
Anura	Microhylidae	*Hypopachus barberi*	Peripheral
Anura	Microhylidae	*Hypopachus elegans*	Peripheral
Anura	Microhylidae	*Hypopachus variolosus*	Widespread
Anura	Phyllomedusidae	*Agalychnis callidryas*	Widespread
Anura	Phyllomedusidae	*Agalychnis moreletii*	Widespread

Table 7.7 (Continued)

Order	Family	Genus and Species	Status
Anura	Phyllomedusidae	*Agalychnis saltator*	Peripheral
Anura	Phyllomedusidae	*Agalychnis taylori*	Widespread
Anura	Phyllomedusidae	*Cruziohyla calcarifer*	Peripheral
Anura	Ranidae	*Lithobates brownorum*	Widespread
Anura	Ranidae	*Lithobates forreri*	Widespread
Anura	Ranidae	*Lithobates maculatus*	Widespread
Anura	Ranidae	*Lithobates vaillanti*	Widespread
Anura	Ranidae	*Lithobates warszewitschii*	Peripheral
Anura	Rhinophrynidae	*Rhinophrynus dorsalis*	Widespread
Total Anura	Species 103	Endemic Species 33	32.0%
Caudata	Plethodontidae	*Bolitoglossa carri*	Tiny GD*
Caudata	Plethodontidae	*Bolitoglossa cataguana*	Tiny GD*
Caudata	Plethodontidae	*Bolitoglossa celaque*	Small GD*
Caudata	Plethodontidae	*Bolitoglossa conanti*	Widespread*
Caudata	Plethodontidae	*Bolitoglossa decora*	Small GD*
Caudata	Plethodontidae	*Bolitoglossa diaphora*	Small GD*
Caudata	Plethodontidae	*Bolitoglossa dofleini*	Widespread
Caudata	Plethodontidae	*Bolitoglossa dunni*	Small GD
Caudata	Plethodontidae	*Bolitoglossa longissima*	Tiny GD*
Caudata	Plethodontidae	*Bolitoglossa mexicana*	Widespread
Caudata	Plethodontidae	*Bolitoglossa nympha*	Widespread
Caudata	Plethodontidae	*Bolitoglossa odonnelli*	Peripheral
Caudata	Plethodontidae	*Bolitoglossa oresbia*	Small GD*
Caudata	Plethodontidae	*Bolitoglossa porrasorum*	Widespread*

Table 7.7   (Continued)

Order	Family	Genus and Species	Status
Caudata	Plethodontidae	*Bolitoglossa striatula*	Peripheral
Caudata	Plethodontidae	*Bolitoglossa synoria*	Small GD*
Caudata	Plethodontidae	*Cryptotriton nasalis*	Small GD*
Caudata	Plethodontidae	*Cryptotriton necopinus*	Tiny GD*
Caudata	Plethodontidae	*Dendrotriton sanctibarbarus*	Tiny GD*
Caudata	Plethodontidae	*Nototriton barbori*	Tiny GD*
Caudata	Plethodontidae	*Nototriton brodiei*	Peripheral
Caudata	Plethodontidae	*Nototriton lignicola*	Small GD*
Caudata	Plethodontidae	*Nototriton limnospectator*	Small GD*
Caudata	Plethodontidae	*Nototriton mime*	Tiny GD*
Caudata	Plethodontidae	*Nototriton nelsoni*	Tiny GD*
Caudata	Plethodontidae	*Nototriton oreadorum*	Tiny GD*
Caudata	Plethodontidae	*Nototriton picucha*	Tiny GD*
Caudata	Plethodontidae	*Nototriton tomamorum*	Tiny GD*
Caudata	Plethodontidae	*Oedipina capitalina*	Status Unknown*
Caudata	Plethodontidae	*Oedipina chortiorum*	Peripheral
Caudata	Plethodontidae	*Oedipina elongata*	Peripheral
Caudata	Plethodontidae	*Oedipina gephyra*	Tiny GD*
Caudata	Plethodontidae	*Oedipina ignea*	Widespread
Caudata	Plethodontidae	*Oedipina kasios*	Small GD*
Caudata	Plethodontidae	*Oedipina leptopoda*	Small GD*
Caudata	Plethodontidae	*Oedipina petiola*	Tiny GD*
Caudata	Plethodontidae	*Oedipina quadra*	Widespread*
Caudata	Plethodontidae	*Oedipina stuarti*	GD Unknown*
Caudata	Plethodontidae	*Oedipina taylori*	Peripheral
Caudata	Plethodontidae	*Oedipina tomasi*	Tiny GD*
Total Caudata	Species 40	Endemic Species 29	72.5%
Gymnophiona	Dermophiidae	*Dermophis mexicanus*	Widespread

Table 7.7 (Continued)

Order	Family	Genus and Species	Status
Gymnophiona	Dermophiidae	*Gymnopis multiplicata*	Widespread
Total Gymnophiona	Species 2	Endemic Species 0	0%
Total Amphibia	Species 145	Endemic Species 62	42.8%
Class Reptilia			
Crocodylia	Alligatoridae	*Caiman crocodylus*	Widespread
Crocodylia	Crocodylidae	*Crocodylus acutus*	Widespread
Total Crocodylia	Species 2	Endemic Species 0	0%
Squamata (L)	Anguidae	*Abronia montecristoi*	Tiny GD
Squamata (L)	Anguidae	*Abronia salvadorensis*	Small GD
Squamata (L)	Anguidae	*Mesaspis moreletii*	Widespread
Squamata (L)	Corytophanidae	*Basiliscus plumifrons*	Peripheral
Squamata (L)	Corytophanidae	*Basiliscus vittatus*	Widespread
Squamata (L)	Corytophanidae	*Corytophanes cristatus*	Widespread
Squamata (L)	Corytophanidae	*Corytophanes hernandesii*	Peripheral
Squamata (L)	Corytophanidae	*Corytophanes percarinatus*	Peripheral
Squamata (L)	Corytophanidae	*Laemanctus longipes*	Widespread
Squamata (L)	Corytophanidae	*Laemanctus serratus*	Peripheral
Squamata (L)	Corytophanidae	*Laemanctus waltersi*	Small GD*
Squamata (L)	Corytophanidae	*Laemanctus* sp. A (*In Press*)	Small GD*
Squamata (L)	Dactyloidae	*Anolis allisoni*	Small GD
Squamata (L)	Dactyloidae	*Norops amplisquamosus*	Tiny GD*
Squamata (L)	Dactyloidae	*Norops beckeri*	Widespread
Squamata (L)	Dactyloidae	*Norops bicaorum*	Tiny GD*
Squamata (L)	Dactyloidae	*Norops biporcatus*	Widespread
Squamata (L)	Dactyloidae	*Norops capito*	Widespread
Squamata (L)	Dactyloidae	*Norops carpenteri*	Peripheral

Table 7.7 (Continued)

Order	Family	Genus and Species	Status
Squamata (L)	Dactyloidae	*Norops crassulus*	Widespread
Squamata (L)	Dactyloidae	*Norops cupreus*	Widespread
Squamata (L)	Dactyloidae	*Norops cusuco*	Small GD*
Squamata (L)	Dactyloidae	*Norops heteropholidotus*	Widespread
Squamata (L)	Dactyloidae	*Norops johnmeyeri*	Small GD*
Squamata (L)	Dactyloidae	*Norops kreutzi*	Small GD*
Squamata (L)	Dactyloidae	*Norops laeviventris*	Widespread
Squamata (L)	Dactyloidae	*Norops lemurinus*	Widespread
Squamata (L)	Dactyloidae	*Norops limifrons*	Widespread
Squamata (L)	Dactyloidae	*Norops loveridgei*	Small GD*
Squamata (L)	Dactyloidae	*Norops mccraniei*	Widespread
Squamata (L)	Dactyloidae	*Norops morazani*	Small GD*
Squamata (L)	Dactyloidae	*Norops muralla*	Tiny GD*
Squamata (L)	Dactyloidae	*Norops nelsoni*	Tiny GD*
Squamata (L)	Dactyloidae	*Norops ocelloscapularis*	Small GD*
Squamata (L)	Dactyloidae	*Norops oxylophus*	Peripheral
Squamata (L)	Dactyloidae	*Norops petersii*	Peripheral
Squamata (L)	Dactyloidae	*Norops pijolense*	Small GD*
Squamata (L)	Dactyloidae	*Norops purpurgularis*	Small GD*
Squamata (L)	Dactyloidae	*Norops quaggulus*	Widespread
Squamata (L)	Dactyloidae	*Norops roatanensis*	Tiny GD*
Squamata (L)	Dactyloidae	*Norops rodriguezii*	Peripheral
Squamata (L)	Dactyloidae	*Norops rubribarbaris*	Small GD*
Squamata (L)	Dactyloidae	*Norops sagrei*	Introduced
Squamata (L)	Dactyloidae	*Norops sminthus*	Small GD*
Squamata (L)	Dactyloidae	*Norops uniformis*	Widespread
Squamata (L)	Dactyloidae	*Norops unilobatus*	Widespread
Squamata (L)	Dactyloidae	*Norops utilensis*	Tiny GD*
Squamata (L)	Dactyloidae	*Norops wampuensis*	Small GD*
Squamata (L)	Dactyloidae	*Norops wellbornae*	Widespread
Squamata (L)	Dactyloidae	*Norops wermuthi*	Peripheral

Table 7.7   (Continued)

Order	Family	Genus and Species	Status
Squamata (L)	Dactyloidae	*Norops wilsoni*	Small GD*
Squamata (L)	Dactyloidae	*Norops yoroensis*	Widespread*
Squamata (L)	Dactyloidae	*Norops zeus*	Widespread*
Squamata (L)	Diploglossidae	*Diploglossus bivittatus*	Widespread
Squamata (L)	Diploglossidae	*Diploglossus montanus*	Small GD
Squamata (L)	Diploglossidae	*Diploglossus scansorius*	Small GD*
Squamata (L)	Eublepharidae	*Coleonyx mitratus*	Widespread
Squamata (L)	Gekkonidae	*Hemidactylus frenatus*	Introduced
Squamata (L)	Gekkonidae	*Hemidactylus haitianus*	Introduced
Squamata (L)	Gekkonidae	*Hemidactylus mabouia*	Introduced
Squamata (L)	Gymnophthalmidae	*Gymnophthalmus speciosus*	Widespread
Squamata (L)	Iguanidae	*Ctenosaura bakeri*	Tiny GD*
Squamata (L)	Iguanidae	*Ctenosaura flavidorsalis*	Small GD
Squamata (L)	Iguanidae	*Ctenosaura melanosterna*	Small GD*
Squamata (L)	Iguanidae	*Ctenosaura oedirhina*	Tiny GD*
Squamata (L)	Iguanidae	*Ctenosaura quinquecarinata*	Widespread
Squamata (L)	Iguanidae	*Ctenosaura similis*	Widespread
Squamata (L)	Iguanidae	*Iguana iguana*	Widespread
Squamata (L)	Leiocephalidae	*Leiocephalus varius*	Introduced
Squamata (L)	Mabuyidae	*Marisora brachypoda*	Widespread
Squamata (L)	Mabuyidae	*Marisora roatanae*	Widespread
Squamata (L)	Phrynosomatidae	*Sceloporus schmidti*	Small GD
Squamata (L)	Phrynosomatidae	*Sceloporus squamosus*	Widespread

Table 7.7 (Continued)

Order	Family	Genus and Species	Status
Squamata (L)	Phrynosomatidae	*Sceloporus variabilis*	Widespread
Squamata (L)	Phrynosomatidae	*Sceloporus* sp. A (*In Press*)	Small GD*
Squamata (L)	Phrynosomatidae	*Sceloporus* sp. B (*In Press*)	Widespread*
Squamata (L)	Phyllodactylidae	*Phyllodactylus palmeus*	Small GD*
Squamata (L)	Phyllodactylidae	*Phyllodactylus paralepis*	Tiny GD*
Squamata (L)	Phyllodactylidae	*Phyllodactylus tuberculosus*	Widespread
Squamata (L)	Phyllodactylidae	*Thecadactylus rapicauda*	Widespread
Squamata (L)	Polychrotidae	*Polychrus gutturosus*	Widespread
Squamata (L)	Scincidae	*Mesoscincus managuae*	Peripheral
Squamata (L)	Scincidae	*Plestiodon sumichrasti*	Widespread
Squamata (L)	Sphaerodactylidae	*Aristelliger georgeensis*	Tiny GD
Squamata (L)	Sphaerodactylidae	*Aristelliger nelsoni*	Tiny GD*
Squamata (L)	Sphaerodactylidae	*Gonatodes albogularis*	Widespread
Squamata (L)	Sphaerodactylidae	*Sphaerodactylus alphus*	Tiny GD*
Squamata (L)	Sphaerodactylidae	*Sphaerodactylus continentalis*	Widespread
Squamata (L)	Sphaerodactylidae	*Sphaerodactylus dunni*	Widespread*
Squamata (L)	Sphaerodactylidae	*Sphaerodactylus exsul*	Tiny GD*
Squamata (L)	Sphaerodactylidae	*Sphaerodactylus glaucus*	Peripheral
Squamata (L)	Sphaerodactylidae	*Sphaerodactylus guanajae*	Tiny GD*
Squamata (L)	Sphaerodactylidae	*Sphaerodac. leonardovaldesi*	Tiny GD*

Table 7.7 (Continued)

Order	Family	Genus and Species	Status
Squamata (L)	Sphaerodactylidae	*Sphaerodactylus millepunctatus*	Peripheral
Squamata (L)	Sphaerodactylidae	*Sphaerodactylus poindexteri*	Tiny GD*
Squamata (L)	Sphaerodactylidae	*Sphaerodactylus rosaurae*	Small GD*
Squamata (L)	Sphenomorphidae	*Scincella assata*	Peripheral
Squamata (L)	Sphenomorphidae	*Scincella cherriei*	Widespread
Squamata (L)	Sphenomorphidae	*Scincella incerta*	Widespread
Squamata (L)	Teiidae	*Ameiva fuliginosa*	Extirpated
Squamata (L)	Teiidae	*Aspidoscelis deppii*	Widespread
Squamata (L)	Teiidae	*Aspidoscelis motaguae*	Widespread
Squamata (L)	Teiidae	*Cnemidophorus ruatanus*	Widespread
Squamata (L)	Teiidae	*Holcosus festivus*	Widespread
Squamata (L)	Teiidae	*Holcosus undulatus*	Widespread
Squamata (L)	Xantusiidae	*Lepidophyma flavimaculatum*	Widespread
Squamata (L)	Xantusiidae	*Lepidophyma mayae*	Peripheral
Total lizard	Species 107	Endemic Species 38	35.5%
Squamata (S)	Anomalepididae	*Anomalepis mexicanus*	Widespread
Squamata (S)	Boidae	*Boa imperator*	Widespread
Squamata (S)	Boidae	*Corallus annulatus*	Widespread
Squamata (S)	Charinidae	*Ungaliophis continentalis*	Widespread
Squamata (S)	Colubridae	*Chironius grandisquamis*	Widespread
Squamata (S)	Colubridae	*Coluber mentovarius*	Widespread
Squamata (S)	Colubridae	*Dendrophidion apharocybe*	Peripheral
Squamata (S)	Colubridae	*Dendrophidion percarinatum*	Widespread
Squamata (S)	Colubridae	*Dendrophidion rufiterminorum*	Widespread

Table 7.7 (Continued)

Order	Family	Genus and Species	Status
Squamata (S)	Colubridae	*Drymarchon melanurus*	Widespread
Squamata (S)	Colubridae	*Drymobius chloroticus*	Widespread
Squamata (S)	Colubridae	*Drymobius margaritiferus*	Widespread
Squamata (S)	Colubridae	*Drymobius melanotropis*	Peripheral
Squamata (S)	Colubridae	*Ficimia publia*	Widespread
Squamata (S)	Colubridae	*Lampropeltis abnorma*	Widespread
Squamata (S)	Colubridae	*Leptodrymus pulcherrimus*	Widespread
Squamata (S)	Colubridae	*Leptophis ahaetulla*	Widespread
Squamata (S)	Colubridae	*Leptophis depressirostris*	Peripheral
Squamata (S)	Colubridae	*Leptophis mexicanus*	Widespread
Squamata (S)	Colubridae	*Leptophis modestus*	Widespread
Squamata (S)	Colubridae	*Leptophis nebulosus*	Peripheral
Squamata (S)	Colubridae	*Mastigodryas alternatus*	Peripheral
Squamata (S)	Colubridae	*Mastigodryas dorsalis*	Widespread
Squamata (S)	Colubridae	*Mastigodryas melanolomus*	Widespread
Squamata (S)	Colubridae	*Oxybelis aeneus*	Widespread
Squamata (S)	Colubridae	*Oxybelis brevirostris*	Peripheral
Squamata (S)	Colubridae	*Oxybelis fulgidus*	Widespread
Squamata (S)	Colubridae	*Oxybelis wilsoni*	Tiny GD*
Squamata (S)	Colubridae	*Phrynonax poecilonotus*	Widespread
Squamata (S)	Colubridae	*Pseudoelaphe flavirufa*	Widespread
Squamata (S)	Colubridae	*Rhinobothryum bovallii*	Peripheral
Squamata (S)	Colubridae	*Scolecophis atrocinctus*	Widespread

Table 7.7   (Continued)

Order	Family	Genus and Species	Status
Squamata (S)	Colubridae	*Senticolis triaspis*	Widespread
Squamata (S)	Colubridae	*Spilotes pullatus*	Widespread
Squamata (S)	Colubridae	*Stenorrhina degenhardtii*	Widespread
Squamata (S)	Colubridae	*Stenorrhina freminvillii*	Widespread
Squamata (S)	Colubridae	*Tantilla armillata*	Widespread
Squamata (S)	Colubridae	*Tantilla excelsa*	Tiny GD*
Squamata (S)	Colubridae	*Tantilla gottei*	Tiny GD*
Squamata (S)	Colubridae	*Tantilla impensa*	Peripheral
Squamata (S)	Colubridae	*Tantilla lempira*	Small GD*
Squamata (S)	Colubridae	*Tantilla olympia*	Tiny GD*
Squamata (S)	Colubridae	*Tantilla psittaca*	Small GD*
Squamata (S)	Colubridae	*Tantilla schistosa*	Widespread
Squamata (S)	Colubridae	*Tantilla stenigrammi*	Tiny GD*
Squamata (S)	Colubridae	*Tantilla tritaeniata*	Tiny GD*
Squamata (S)	Colubridae	*Tantilla vermiformis*	Small GD
Squamata (S)	Colubridae	*Tantillita lintoni*	Widespread
Squamata (S)	Colubridae	*Trimorphodon quadruplex*	Widespread
Squamata (S)	Dipsadidae	*Adelphicos quadrivirgatum*	Widespread
Squamata (S)	Dipsadidae	*Amastridium sapperi*	Peripheral
Squamata (S)	Dipsadidae	*Clelia clelia*	Widespread
Squamata (S)	Dipsadidae	*Coniophanes bipunctatus*	Widespread
Squamata (S)	Dipsadidae	*Coniophanes fissidens*	Widespread
Squamata (S)	Dipsadidae	*Coniophanes imperialis*	Widespread
Squamata (S)	Dipsadidae	*Coniophanes piceivittis*	Widespread
Squamata (S)	Dipsadidae	*Conophis lineatus*	Widespread
Squamata (S)	Dipsadidae	*Crisantophis nevermanni*	Widespread

Table 7.7   (Continued)

Order	Family	Genus and Species	Status
Squamata (S)	Dipsadidae	*Cubophis brooksi*	Tiny GD*
Squamata (S)	Dipsadidae	*Dipsas bicolor*	Widespread
Squamata (S)	Dipsadidae	*Enuliophis sclateri*	Peripheral
Squamata (S)	Dipsadidae	*Enulius bifoveatus*	Tiny GD*
Squamata (S)	Dipsadidae	*Enulius flavitorques*	Widespread
Squamata (S)	Dipsadidae	*Enulius roatanensis*	Tiny GD*
Squamata (S)	Dipsadidae	*Erythrolamprus mimus*	Widespread
Squamata (S)	Dipsadidae	*Geophis damiani*	Tiny GD*
Squamata (S)	Dipsadidae	*Geophis fulvoguttatus*	Peripheral
Squamata (S)	Dipsadidae	*Geophis hoffmanni*	Widespread
Squamata (S)	Dipsadidae	*Geophis nephodrymus*	Tiny GD*
Squamata (S)	Dipsadidae	*Geophis rhodogaster*	Peripheral
Squamata (S)	Dipsadidae	*Hydromorphus concolor*	Widespread
Squamata (S)	Dipsadidae	*Imantodes cenchoa*	Widespread
Squamata (S)	Dipsadidae	*Imantodes gemmistratus*	Widespread
Squamata (S)	Dipsadidae	*Imantodes inornatus*	Widespread
Squamata (S)	Dipsadidae	*Leptodeira nigrofasciata*	Widespread
Squamata (S)	Dipsadidae	*Leptodeira rhombifera*	Widespread
Squamata (S)	Dipsadidae	*Leptodeira septentrionalis*	Widespread
Squamata (S)	Dipsadidae	*Ninia diademata*	Widespread
Squamata (S)	Dipsadidae	*Ninia espinali*	Widespread
Squamata (S)	Dipsadidae	*Ninia maculata*	Widespread
Squamata (S)	Dipsadidae	*Ninia pavimentata*	Widespread
Squamata (S)	Dipsadidae	*Ninia sebae*	Widespread
Squamata (S)	Dipsadidae	*Nothopsis rugosus*	Peripheral
Squamata (S)	Dipsadidae	*Omoadiphas aurula*	Tiny GD*

Table 7.7   (Continued)

Order	Family	Genus and Species	Status
Squamata (S)	Dipsadidae	*Omoadiphas cannula*	Tiny GD*
Squamata (S)	Dipsadidae	*Omoadiphas texiguatensis*	Tiny GD*
Squamata (S)	Dipsadidae	*Oxyrhopus petolarius*	Widespread
Squamata (S)	Dipsadidae	*Pliocercus elapoides*	Widespread
Squamata (S)	Dipsadidae	*Pliocercus euryzonus*	Peripheral
Squamata (S)	Dipsadidae	*Rhadinaea decorata*	Widespread
Squamata (S)	Dipsadidae	*Rhadinella anachoreta*	Peripheral
Squamata (S)	Dipsadidae	*Rhadinella godmani*	Widespread
Squamata (S)	Dipsadidae	*Rhadinella kinkelini*	Widespread
Squamata (S)	Dipsadidae	*Rhadinella lisyae*	Tiny GD*
Squamata (S)	Dipsadidae	*Rhadinella montecristi*	Widespread
Squamata (S)	Dipsadidae	*Rhadinella pegosalyta*	Tiny GD*
Squamata (S)	Dipsadidae	*Rhadinella tolpanorum*	Tiny GD*
Squamata (S)	Dipsadidae	*Sibon annulatus*	Peripheral
Squamata (S)	Dipsadidae	*Sibon anthracops*	Widespread
Squamata (S)	Dipsadidae	*Sibon carri*	Widespread
Squamata (S)	Dipsadidae	*Sibon dimidiatus*	Widespread
Squamata (S)	Dipsadidae	*Sibon longifrenis*	Peripheral
Squamata (S)	Dipsadidae	*Sibon manzanaresi*	Small GD*
Squamata (S)	Dipsadidae	*Sibon miskitus*	Small GD*
Squamata (S)	Dipsadidae	*Sibon nebulatus*	Widespread
Squamata (S)	Dipsadidae	*Tretanorhinus nigroluteus*	Widespread
Squamata (S)	Dipsadidae	*Tropidodipsas fischeri*	Peripheral
Squamata (S)	Dipsadidae	*Tropidodipsas sartorii*	Widespread
Squamata (S)	Dipsadidae	*Urotheca decipiens*	Peripheral
Squamata (S)	Dipsadidae	*Urotheca guentheri*	Peripheral

Table 7.7   (Continued)

Order	Family	Genus and Species	Status
Squamata (S)	Dipsadidae	*Xenodon angustirostris*	Widespread
Squamata (S)	Elapidae	*Hydrophis platurus*	Marine
Squamata (S)	Elapidae	*Micrurus alleni*	Peripheral
Squamata (S)	Elapidae	*Micrurus browni*	Peripheral
Squamata (S)	Elapidae	*Micrurus diastema*	Widespread
Squamata (S)	Elapidae	*Micrurus nigrocinctus*	Widespread
Squamata (S)	Elapidae	*Micrurus ruatanus*	Tiny GD*
Squamata (S)	Leptotyphlopidae	*Epictia ater*	Widespread
Squamata (S)	Leptotyphlopidae	*Epictia martinezi*	Tiny GD*
Squamata (S)	Leptotyphlopidae	*Epictia magnamaculata*	Peripheral
Squamata (S)	Leptotyphlopidae	*Epictia phenops*	Peripheral
Squamata (S)	Loxocemidae	*Loxocemus bicolor*	Widespread
Squamata (S)	Natricidae	*Storeria dekayi*	Widespread
Squamata (S)	Natricidae	*Thamnophis fulvus*	Widespread
Squamata (S)	Natricidae	*Thamnophis marcianus*	Widespread
Squamata (S)	Natricidae	*Thamnophis proximus*	Widespread
Squamata (S)	Sibynophiidae	*Scaphiodontophis annulatus*	Widespread
Squamata (S)	Sibynophiidae	*Scaphiodontophis venustissimus*	Peripheral
Squamata (S)	Typhlopidae	*Amerotyphlops costaricensis*	Widespread
Squamata (S)	Typhlopidae	*Amerotyphlops stadelmani*	Widespread*
Squamata (S)	Typhlopidae	*Amerotyphlops tycherus*	Widespread*
Squamata (S)	Typhlopidae	*Indotyphlops braminus*	Introduced
Squamata (S)	Viperidae	*Agkistrodon howardgloydi*	Widespread
Squamata (S)	Viperidae	*Atropoides indomitus*	Widespread*

Table 7.7 (Continued)

Order	Family	Genus and Species	Status
Squamata (S)	Viperidae	*Atropoides mexicanus*	Widespread
Squamata (S)	Viperidae	*Bothriechis guifarroi*	Widespread*
Squamata (S)	Viperidae	*Bothriechis marchi*	Small GD*
Squamata (S)	Viperidae	*Bothriechis schlegelii*	Widespread
Squamata (S)	Viperidae	*Bothriechis thalassinus*	Peripheral
Squamata (S)	Viperidae	*Bothrops asper*	Widespread
Squamata (S)	Viperidae	*Cerrophidion wilsoni*	Widespread
Squamata (S)	Viperidae	*Crotalus simus*	Widespread
Squamata (S)	Viperidae	*Porthidium nasutum*	Widespread
Squamata (S)	Viperidae	*Porthidium ophryomegas*	Widespread
Total Snake	Species 144	Endemic Species 28	19.4%
Testudinata	Cheloniidae	*Caretta caretta*	Marine
Testudinata	Cheloniidae	*Chelonia mydas*	Marine
Testudinata	Cheloniidae	*Eretmochelys imbricata*	Marine
Testudinata	Cheloniidae	*Lepidochelys olivacea*	Marine
Testudinata	Chelydridae	*Chelydra acutirostris*	Widespread
Testudinata	Chelydridae	*Chelydra rossignonii*	Widespread
Testudinata	Dermochelyidae	*Dermochelys coriacea*	Marine
Testudinata	Kinosternidae	*Kinosternon albogulare*	Widespread
Testudinata	Kinosternidae	*Kinosternon leucostomum*	Widespread
Testudinata	Kinosternidae	*Staurotypus triporcatus*	Peripheral
Testudinata	Emydidae	*Trachemys emolli*	Peripheral
Testudinata	Emydidae	*Trachemys scripta*	Introduced

Table 7.7   (Continued)

Order	Family	Genus and Species	Status
Testudinata	Emydidae	*Trachemys venusta*	Widespread
Testudinata	Geoemydidae	*Rhinoclemmys annulata*	Widespread
Testudinata	Geoemydidae	*Rhinoclemmys areolata*	Peripheral
Testudinata	Geoemydidae	*Rhinoclemmys funerea*	Peripheral
Testudinata	Geoemydidae	*Rhinoclemmys pulcherrima*	Widespread
Total Testudinata	Species 17	Endemic Species 0	0%
Total Reptilia	Species 270	Endemic Species 66	24.4%
Total Herpetological	Species 415	Endemic Species 128	30.8%

[1] Considered by some to be a synonym of *Hyalinobatrachium colymbiphyllum*.

Abbreviations used are: GD (geographical distribution); L (lizards); and S (snakes). An asterisk (*) indicates that species is endemic to Honduras. The classifications for status are based on geographical distributions (historical and current) both within and outside of Honduras. Unpublished data of JRM are also included in this table (see text). All attempts at classifying the conservation status of any given species are fruitless given the uncontested accelerated deforestation and habitat destruction currently underway throughout Honduras and surrounding countries.

either endemic to that range, or otherwise Honduran endemics. The Cordillera Nombre de Dios is in north-central Honduras and is currently known to contain 117 species of amphibians and reptiles, of which 37 (31.6%) are either endemic to that mountain range, or are otherwise endemic to Honduras (JRM unpublished data). No comparative data have been offered for any mountain range in any other Central American country, but it is doubted that any other of those mountain ranges could match that a high number of endemic species.

7.5   Biodiversity Threats

As has been noted above, Honduras has a large diversity of the floral and animal groups as discussed herein. Honduras is notable among all of Central America in having a high level of endemic species of amphibians (42.8%) and reptiles (24.4%), whereas the vascular plants (2.8%) and other more mobile groups treated above have low levels of endemism (0.0% of bats, 0.0% of medium and large-sized mammals, 0.001% of birds, and 2.5% of freshwater fishes). It has also been noted (McCranie, 2011) that the country has in place a robust set of national parks, wildlife reserves, etc., and even a World Heritage Site. Unfortunately, all of those "protected parks" are only paper parks. The accelerated rate of the human devastation of all forests, even currently close to the center of that World Heritage Site, continues unchallenged by the governmental agencies in charge of protecting those paper parks, despite a number of laws in place to enforce stopping that forest destruction. That governmental nonaction, should it continue for another 20–30 years, will bring doom to the vast majority of the Honduran natural world, and eventually will severely impact the human population, and its already declining quality of life. Soon, if not already happening, "natural disasters" should no longer be called natural disasters because there will not be any natural world left in the country to help lessen the damages those disasters bring. Those disasters are surely going to continue, as is most likely the continuation of the human deterioration of the quality of life.

Keywords

- threats
- vertebrate taxonomic groups

References

Angiosperm Phylogeny Group (APG IV), (2016). An update of the angiosperm phylogeny group classification for the orders and families of flowering plants. *Botanical Journal of the Linnean Societcy*, *181*, 1–20.

Angulo, A., Garita-Alvarado, C. A., Bussing, W. A., & Lopez, M. I., (2013). Annotated checklist of the freshwater fishes of continental and insular Costa Rica: Additions and nomenclatural revisions. *Check List*, *9*, 987–1019.

Atlas Geográfico de Honduras, (2006). *Ediciones Ramsés*, Tegucigalpa.

Baird, A. B., Braun, J. K., Mares, M. A., et al., (2015). Molecular systematic revision of tree bats (*Lasiurini*): Doubling the native mammals of the Hawaiian Islands. *Journal of Mammalogy*, *96*, 1255–1275.

Baird, A. B., Marchán-Rivadeneira, M. R., Perez, S. G., & Baker, R. J., (2012). Morphological analysis and description of two new species of *Rhogeessa* (Chiroptera: Vespertilionidae) from the Neotropics. *Occasional Papers, Museum of Texas Tech University*, *307*, 1–25.

Baker, R. J., Solari, S., & Hoffmann, F., (2002). A new Central American species from the *Carollia brevicauda* Complex. *Occasional Papers, Museum of Texas Tech University*, *2017*, 1–12.

Bonta, M., & Anderson, D. L., (2002). *Birding Honduras, A Checklist and Guide*. EcoArtes, S. de R. L., Tegucigalpa.

Boyles, J. G., Cryan, P., McCracken, G., & Kunz, T., (2011). Economic importance of bats in agriculture. *Science*, *332,* 41–42.

Cajas, A., (2009). *Bats in Maya Art.* Asociación FLAAR Mesoamerica, Guatemala City.

Carter, D. C., Pine, R. H., & Davis, W. B., (1966). Notes on middle American bats. *The Southwestern Naturalist, 11*, 488–499.

Castañeda, F., Herrera, L., & Pereira, S., (2013). Behaviour of two male jaguars scavenging on a marine dolphin in Honduras. *CATNews, 58*, 11–12.

Chapman, A., (1978). *Master of the Animals: Oral Traditions of the Tolupan Indians, Honduras.* Gordon and Breach London.

Clare, E. L., Adams, A., Maya,-Simões, A., etal. (2013). Diversification and reproductive isolation: cryptic species in the only New World high-duty cycle bat, *Pteronotus parnelli. BMC Evolutionary Biology, 13*, 1–18.

Davis, W. B., & Carter, D. C., (1962). Notes on Central American bats with description of a new subspecies of *Mormoops. The Southwestern Naturalist, 7*, 64–74.

Davis, W. B., Carter, D. C., & Pine, R. H., (1964). Noteworthy records of Mexican and Central American bats. *Journal of Mammalogy, 45*, 375–387.

Dolan, P., & Carter, D., (1979). Distributional notes and records for Middle American Chiroptera. *Journal of Mammalogy, 60*, 644–649.

Dunn, M., Estrada, N., & Smith, D., (2016). The coexistence of Baird's tapir (*Tapirus bairdii*) and indigenous hunters in northeastern Honduras. *Integrative Zoology, 7*, 429–438.

Eger, J. L., (2008). Family Molossidae. In: Gardner, A. L., (ed.). *Mammals of South America.* vol. 1, Marsupials, xenarthrans, shrews and bats. University of Chicago Press, Chicago. pp. 399–440.

Espinal, M., & Mora, J. M., (2016). Noteworthy record of *Eptesicus brasiliensis* (Vespertilionidae) in Honduras. *Ceiba, 53*, 77–80.

Espinal, M., Mora, J. M., & O'Reilly, C. M., (2016). The occurrence of the Peale's free-tailed bat (*Nyctinomops curispinosus*, Molossidae) in Central America. *Caribbean Journal of Science, 49*, 79–82.

Estrada, N., (2004). Notes about the relative abundance and hunting of Baird's tapir in the Rus-Rus region of La Moskitia, Honduras: A proposed biological reserve. *Tapir Conservation, 13*, 28–29.

Estrada, N., (2006). Humans attacked by a Baird's tapir (*Tapirus bairdii*) in the Sierra de Agalta National Park, Olancho, Honduras. *Tapir Conservation, 15*, 13–14.

Fagan, J., & Komar, O., (2016). *Field Guide to the Birds of Northern Central America.* Peterson Field Guides, New York.

Fash, B., & Fash, W., (1989). Copán temple 20 and the house of bats. In: Greene, R. M., (ed.). *Seventh Palenque Round Table,* vol. IX. Northern Illinois University, Dekalb. pp. 61–67.

Gallardo, R., (2014). *Guide to the Birds of Honduras.* Mountain Gem Tours, Tegucigalpa.

Gamero, I., (1978). *Mamíferos de Mi Tierra.* Banco Central de Honduras. Primera Edición. Tegucigalpa.

Gonthier, D., & Castañeda, F., (2013). Large- and medium-sized mammal survey using camera traps in the Sikre River in the Río Plátano Biosphere Reserve, Honduras. *Tropical Conservation Science, 6*, 584–581.

Goodwin, G., (1942). Mammals of Honduras. *Bulletin of the American Museum of Natural History, 79*, 107–195.

Handley, C. O., (1959). A revision of American bats of the genera *Euderma* and *Plecotus. Proceedings of the United States National Museum, 110*, 95–246.

Handley, C. O., (1960). Descriptions of new bats from Panama. *Proceedings of the United States National Museum, 112*, 459–479.

Helgen, K. M., Pinto, C. M., Kays, R., et al., (2013). Taxonomic revision of the olingos (*Bassaricyon*), with description of a new species, the Olinguito. *ZooKeys, 324*, 1–83.

Hernández, D. J., (2015). Programa para la Conservación de los Murciélagos de Honduras (PCMH). In: Rodríquez-Herrera, B., & Sánchez, R., (eds.). *Estrategia Centroamericana Para la Conservación de los Murciélagos*. Universidad de Costa Rica, San José. pp. 41–55.

Holdridge, L. R., (1967). *Life Zone Ecology*. Revised edition. Tropical Science Center, San José, Costa Rica.

Hoofer, S. R., & Van den Bussche. R. A., (2003). Molecular phylogenetics of the chiropteran family Vespertilionidae. *Acta Chiropterologica, 5*, 1–63.

House, P. R., Lagos-Witte, S., Ochoa, L., Mejía, T., & Rivas, M., (1995). *Plantas Medicinales Comunes de Honduras*. Primera edición, Tegucigalpa.

House, P., Rodríguez, I., & Ferrufino, L., (1999). Rescatando el conocimiento popular en la base de datos de plantas medicinales de Honduras. *Revista Ciencia y Tecnología, 4*, 47–49.

Hurtado, N., & Pacheco, V., (2014). Análisis filogenético del género *Mimon* Gray, (1847). (Mammalia, Chiroptera, Phyllostomidae) con la descripción de un nuevo género. *Therya, 5*, 751–791.

IUCN, (2017). *The IUCN Red List of Threatened Species. Version 2017-2*, http://www.iucnredlist.org. Downloaded on 14 September 2017.

Jones, J. K., (1964a). Bats new to the fauna of Nicaragua. *Transactions of the Kansas Academy of Science, 67*, 506–508.

Jones, J. K., (1964b). Bats of western and southern Mexico. *Transactions of the Kansas Academy of Science, 67*, 509–516.

Jones, J. K., (1966). Bats from Guatemala. *University of Kansas Publications, Museum of Natural History, 16*, 439–472.

Keppie, J. D., & Morán-Zenteno, D. J., (2005). Tectonic implications of alternative Cenozoic reconstructions for southern Mexico and the Chortis Block. *International Geology Review, 47*, 473–491.

Kihn Pineda, P. H., Cano, E. B., & Morales, A., (2006). Peces de las aguas interiores de Guatemala. In: Cano, E. B., (ed.). *Biodiversidad de Guatemala*. vol. 1. Universidad del Valle de Guatemala, Guatemala City. pp. 457–486.

Knox Jones, J., Arroyo-Cabrales, J., & Owen, R., (1988). Revised checklist of bats (Chiroptera) of Mexico and Central America. *Occasional Papers, The Museum, Texas Tech University, 120*, 1–34.

Koeppe, C. E., & De Long, G. C., (1958). *Weather and Climate*. McGraw-Hill Book Company, New York.

Kreft, H., & Jetz, W., (2007). Global patterns and determinants of vascular plant diversity. *Proceedings of the National Academy of Sciences, 104*, 5925–5930.

Kunz, T. H., Braun de Torrez, E., Bauer, D., Lobova, T., & Fleming, T. H., (2011). Ecosystem services provided by bats. *Annals of the New York Academy of Science, 1223*, 1–38.

LaVal, R., (1969). Records of bats from Honduras and El Salvador. *Journal of Mammalogy, 50*, 819–822.

López-Wilchis, R., Guevara-Chumacero, L., Pérez, N., & Barriga-Sosa, L., (2012). Taxonomic status assessment of the Mexican populations of funnel-eared bats, genus *Natalus* (Chiroptera: Natalidae). *Acta Chiropterologica, 14*, 305–316.

Mantilla-Meluk, H., (2014). Defining species and species boundaries in *Uroderma* (Chiroptera: Phyllostomidae) with the description of a new species. *Occasional Papers, Texas Tech University, 325*, 1–29.

Marineros, L., & Martínez, F., (1998). *Guía de Campo de los Mamíferos de Honduras.* Instituto Nacional de Ambiente y Desarrollo/INADES. Tegucigalpa.

Matamoros, W. A., Kreiser, B. R., & Schaefer, J. F., (2012). A delineation of Nuclear Middle America biogeographical provinces based on river basin faunistic similarities. *Reviews in Fish Biology and Fisheries, 22*, 351–365.

Matamoros, W. A., McMahan, C. D., Chakrabarty, P., Albert, J. S., & Schaefer, J. F., (2015). Derivation of the freshwater fish fauna of Central America revisited: Myers' hypothesis in the 20-first century. *Cladistics, 31*, 177–188.

Matamoros, W. A., Schaefer, J. F., & Kreiser, B. R., (2009). Annotated checklist of the freshwater fishes of continental and insular Honduras. *Zootaxa, 38*, 1–38.

McCann, N., (2015). *The Conservation of Baird's Tapir* (Tapirus bairdii) *in Honduras.* Thesis, Cardiff University.

McCarthy, T., Davis, W. B., Hill, J., et al. (1993). Bat (Mammalia: Chiroptera) records, early collectors, and faunal lists from Northern Central America. *Annals of Carnegie Museum, 62*, 191–228.

McCranie, J. R., (2011). *The Snakes of Honduras: Systematics, Distribution, and Conservation.* Society for the Study of Amphibians and Reptiles, Contributions to Herpetology.

McCranie, J. R., (2015). A checklist of the amphibians and reptiles of Honduras, with additions, comments on taxonomy, some recent taxonomic decisions, and areas of further studies needed. *Zootaxa, 3931,* 352–386.

McCranie, J. R. (2018) The lizards, crocodiles, and turtles of Honduras. Systematics, distribution, and conservation. *Bulletin of the Museum of Comparative Zoology. Special Publication Series, 2,* 1—666..

McCranie, J. R., Harrison, A., & Valdés Orellana, L., (2017). Updated population and habitat comments about the reptiles of the Swan Islands, Honduras. *Bulletin of the Museum of Comparative Zoology at Harvard University, 161*, 265–284.

McCranie, J. R., Townsend, J. H., & Wilson, L. D., (2006). *The Amphibians and Reptiles of the Honduran Mosquitia.* Krieger Publishing Company, Malabar, Florida.

McCranie, J. R., Wilson, L. D., & Köhler, G., (2005). *Amphibians & Reptiles of the Bay Islands and Cayos Cochinos, Honduras.* Bibliomania, Salt Lake City, Utah.

McMahan, C. D., Matamoros, W. A., Álvarez Calderón, F. S., et al. (2013). Checklist of the inland fishes of El Salvador. *Zootaxa, 3608*, 440–456.

Medina-Fitoria, A., Saldaña, O., Martínez, J., & Pérez, J., (2015). Nuevos reportes sobre los murciélagos (Mammalia: Chiroptera) de Nicaragua, América Central, con la adición de siete nuevos registros de especies. *Mastozoología Neotropical, 22*, 43–54.

Mejía-Ordoñez, M. T., & House, P. R., (2008). *Especies de Preocupación Especial en Honduras.* Secretaria de Recursos Naturales y Ambiente. Dirección General de Biodiversidad (DiBio) Tegucigalpa.

Merida, J., & Cruz, G., (2014). Confirmación de la presencia del oso hormiguero gigante *Myrmecophaga tridactyla centralis* (Xenarthra: Myrmecophagidae) en la Reserva Biósfera Río Plátano, Departamento de Gracias a Dios, Honduras, con descripción y comentarios sobre su estatus taxonómico. *Edentata, 15*, 9–15.

Monroe, Jr., B. L., (1968). *A Distributional Survey of the Birds of Honduras.* Ornithological Monographs, *7*, 1–458.

Mora, J. M., (2012). Big Red Bat *Lasiurus egregius* (Vespertilionidae) in Honduras. *The Southwestern Naturalist, 57*, 104–105.

Mora, J. M., (2016). Clave para la identificación de las especies de murciélagos de Honduras. *Ceiba, 54*, 93–117.

Mora, J. M., & López, L. I., (2012). First record of the Hoary Bat (*Lasiurus cinereus*, Vespertilionidae) for Honduras. *Ceiba, 51*, 89–90.

Mora, J. M., Marineros, L., & López, L. I., (2014). First record of the striped yellow-eared bat, *Vampyriscus nymphaea*, (Stenodermatinae, Phyllostomidae) in Honduras. *Caribbean Journal of Science, 48*, 49–51.

Myers, G. S., (1949). Salt-tolerance of fresh-water fish groups in relation to zoogeographical problems. *Bijdragentot de Dierkunde, 28*, 315–322.

Nelson, C. H., (2008). *Catálogo de las Plantasvasculares de Honduras: Espermatofitas. Guaymuras y Secretaría de Recursos Naturales y Ambiente*, Tegucigalpa.

Petracca, L., Frair, J., Cohen, J., et al., (2017). Robust inference on large-scale species habitat use with interview data: The status of jaguars outside protected areas in Central America. *Journal of Applied Ecology.* doi: 10.1111/1365-2664.12972.

Portillo, H., & Elvir, F., (2013). Distribución de felinos silvestres en las áreas protegidas de Honduras. *Revista Mexicana de Mastozoología, 3*, 6–15.

Portillo, H., & Elvir, F., (2015). Registros y distribución potencial del jaguar (*Panthera onca*) en Honduras. *Revista Mexicana de Mastozoología, 5*, 55–65.

Portillo, H., & Elvir, F., (2016). Distribución potencial de la jagüilla (*Tayassu pecari*) en Honduras. *Revista Mexicana de Mastozoología, 6*, 15–23.

Rabinowitz, A., (2014). *An Indomitable Beast, the Remarkable Journey of the Jaguar.* Island Press, Washington.

Rogers, R. D., Mann, P., & Emmet, P. A., (2007). Tectonic terrains of the Chortis block based on integration of regional aeromagnetic and geologic data. *Geological Society of America, Special Papers, 428*, 65–88.

Schank, C., Mendoza, E., García Vettorazzi, M., et al., (2015). Integrating current range-wide occurrence data with species distribution models to map the potential distribution of Baird's Tapir. *Tapir Newsletter, 24*, 33.

Scheffler, T., Hirth, G., & Hasemann, G., (2012). The Gigante Rockshelter: Preliminary observations on an early to late Holocene occupation in southern Honduras. *Latin American Antiquity, 23*, 597–610.

Secretaria de Energía, Recursos Naturales, Ambiente y Minas (2014). V. *Informe Nacional de Biodiversidad.* CBD, GEF, PNUD, Tegucigalpa.

Solari, S., Hoofer, S., Larsen, P., Baker, R., et al., (2009). Operational criteria for genetically defined species: Analysis of the diversification of the small fruit-eating bats, *Dermanura* (Phyllostomidae: Stenodermatinae). *Acta Chiropterologica, 11*, 279–288.

Tagliacollo, V. A., Duke-Sylvester, S. M., Matamoros, W. A., Chakrabarty, P., & Albert, J. S., (2015). Coordinated dispersal and preisthmian assembly of the Central American ichthyofauna. *Systematic Biology, 66*, 183–196.

Valdez, R., & La Val, R., (1971). Records of bats from Honduras and Nicaragua. *Journal of Mammalogy, 52,* 247–250.

Velazco, P. M., & Patterson, B. D., (2013). Diversification of yellow-shouldered bats, genus *Sturnira* (Chiroptera: Phyllostomidae), in the New World tropics. *Molecular Phylogenetics and Evolution, 68,* 683–698.

Velazco, P. M., & Simmons, N. B., (2011). Systematics and taxonomy of great striped-faced bats of the genus *Vampyrodes* Thomas, (1900). (Chiroptera: Phyllostomidae). *American Museum Novitates, 3710,* 1–35.

Zea, C., & Cruz, J., (2006). *Evaluación Nacional Forestal: Resultados del Inventario de Bosques y Árboles, 2005–2006,* Report No. TCP/HON/3001(A), Tegucigalpa.

Zúniga, E., (1989). *La Estructura y el Comportamiento de la Atmósfera.* CETNA, Tegucigalpa.

CHAPTER 8

Biodiversity in Mexico: State of Knowledge

**PATRICIA KOLEFF, TANIA URQUIZA-HAAS,
SYLVIA P. RUIZ-GONZÁLEZ, DIANA R. HERNÁNDEZ-ROBLES,
ALICIA MASTRETTA-YANES, ESTHER QUINTERO, and
JOSÉ SARUKHÁN***

National Commission for the Knowledge and Use of Biodiversity (CONABIO),
Liga Periférico Insurgentes Sur 4903, Col. Parques del Pedregal, Tlalpan 14010,
Mexico City, Mexico, E-mail: pkoleff@conabio.gob.mx

8.1 Introduction

Mexico is one of the most diverse countries in the world (Figure 8.1), of the more than 190 nations, regarding its flora and vertebrate fauna (Figures 8.2 and 8.3). It is part of a select group of 17 countries recognized as megadiverse—these countries are home to 60–70% of the planet's known biological diversity (Mittermier and Goettsch-Mittermier, 1992). Moreover, Mexico hosts numerous different landscapes across its territory where biological diversity is intrinsically related to cultural diversity.

Its great diversity is attributed to many biogeographical and environmental factors. As has been known for centuries, biodiversity is unevenly distributed in the world; tropical zones have higher species diversity for most groups in both marine and terrestrial systems (Jablonski et al., 2013). Moreover, Mexico is located at the intersection of two major biogeographic regions, the Nearctic—of northern affinity—and the Neotropical—of southern affinity. Therefore, a large representation of the species from the northern temperate zones and many elements that originated in the Amazon Basin occur in the country along with adapted biota to dry and semi-dry conditions (Sarukhán et al., 2012). In addition, its position in the intertropical convergence zone, the shape and size of the territory (13th largest country worldwide with 2 million square kilometers) are determinants of its diversity.

---

*Coauthors: Francisca Acevedo, Angélica Cervantes, Sarita C. Frontana-Uribe, A. Margarita Hermoso-Salazar, Susana Ocegueda-Cruz, Dulce Parra-Toríz, Rafael Ramírez, M. Alicia Reséndiz-López, Alberto Romo-Galicia, and Adriana Valera.

Mexico's natural diversity is not only expressed in its number of species, but also in the vast life forms, evolutionary lineages, biotic communities, ecosystems, landscapes and seascapes. For example, almost all of the world's terrestrial ecosystems are represented in the country due to its complex topography, geology and diversity of climates; the later influenced by the funnel shape of the country, mountain ranges, prevailing winds and the oscillations of the high-pressure subtropical belt (Espinosa et al., 2008). Moreover, the different environments have resulted in numerous groups of organisms showing a high degree of endemism, which gives the country a remarkable differentiation of its biota between regions. For example, in the Cactaceae family—plants native to the Americas for which there are estimated to be 1750 species (Christenhusz & Byng, 2016)—at least 498 of the 677 recognized species (that belong to 62 genera) in Mexico are endemic (Guzmán et al., 2003; Villaseñor, 2016). In addition, in several genera of this family, the degree of endemism can reach a very high proportion, up to more than 90%. Those high levels of endemism, in turn, explain the high levels of beta diversity among biotic communities and the development of different local knowledge systems for its use, management, and conservation.

In particular, Mexico stands out as a center of origin of many species' groups, which linked to the enormous cultural diversity intrinsically related to its biodiversity resulted in a high genetic diversity and in a center of domestication and diversification of numerous cultivated species (at least 100 species; Perales and Aguirre, 2008; Bellon et al., 2009). Many of which are crops of global importance and inherent to our culture, such as maize (*Zea mays*), squashes (*Cucurbita* spp.), beans (*Phaseolus* spp.), chilies (*Capsicum* spp.), avocado (*Persea* spp.), cocoa (*Theobroma* spp.) and vanilla (*Vanilla* spp.), among others. This diversity modified by man is also known as humanized biodiversity (Perales and Aguirre, 2008), and it plays a fundamental role in our daily life. The conservation of genetic resources contained in humanized biodiversity and their wild relatives is strategic for food security, particularly in the face of global change (Bellon et al., 2011; Khoury et al., 2014) (Figures 8.1–8.3).

8.2 Background

The study of biodiversity in Mexico has a long history, dating back to prehispanic times; nevertheless, the knowledge of the usefulness and the management practices of different species were only partly incorporated in the new colonial society. The impulse of explorations in the New World, and the

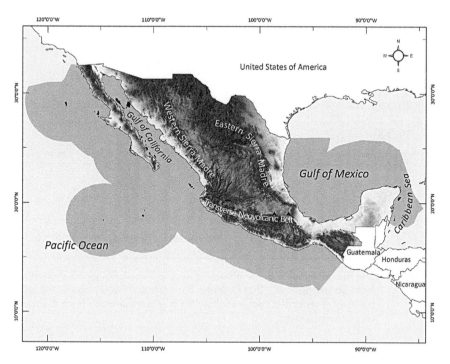

Figure 8.1 Map of Mexico and its exclusive economic zone showing the seas that surround the country, the main Sierras and the Neovolcanic Belt with two of its highest volcanoes (indicated with a black triangle, from left to right: Popocatépetl, 5,426 m; Citlaltépetl or Pico de Orizaba 5,636 m).

work of remarkable naturalists at different times during the XVI, XVII and XVIII centuries contributed importantly to increase the knowledge of our biodiversity (Sarukhán et al., 2016). As in other countries, over the past 100 years, the study of biological wealth has been related to the establishment and consolidation of scientific collections, tied to the development of public teaching and research centers (Escobar et al., 2009). Recognizing the importance of scientific collections, specimen databases were the main axes upon which the National Biodiversity Information System (SNIB, for its Spanish acronym) started to be developed. Created in 1992, the National Commission for the Knowledge and Use of Biodiversity (CONABIO) has the legal mandate to develop and keep the SNIB updated. CONABIO also has the mission to promote, coordinate, support and carry out activities aimed at increasing the knowledge of biological diversity, its conservation and sustainable use for the well-being of society (CONABIO, 2017).

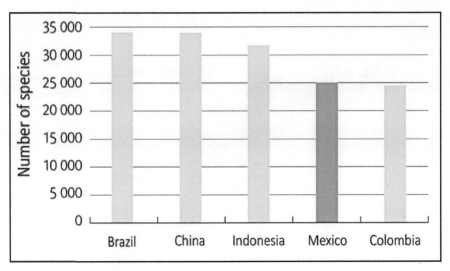

Figure 8.2   The five countries with the greatest diversity of vascular plants (Reprinted from Sarukhán et al., 2017). Note: For comparative purposes, the total number of recorded species follow the highest estimates from the most recent assessments.

In 1998, CONABIO coordinated the first national effort to synthesize biodiversity information from bibliographic sources, which resulted in a descriptive synopsis of the biodiversity of Mexico that enabled the country to fulfill the commitments made in signing and ratifying the Convention on Biological Diversity (CBD). Later, as an effort to update that study, an unprecedented scientific assessment was launched in 2005 to assess the state of knowledge, the status of the components and function of biodiversity, the impact that different policies have had on it, as well as the human and institutional capacities needed for its conservation and sustainable management. This work entitled '*Natural Capital of Mexico*' was inspired by the Millennium Ecosystem Assessment conceptual framework (MA, 2005), and aimed to organize and make this information widely available to large sectors of society to guide decision-making and public policies. Almost 800 scientists, government officials, and nongovernmental organization members have participated in this effort. Further, key priority issues for future attention, as well as new research areas and options for the conservation and sustainable management of Mexico's threatened biodiversity were highlighted in several publications (Sarukhán et al., 2010, 2012, 2015). We base most of the information contained in this chapter on the first four volumes published

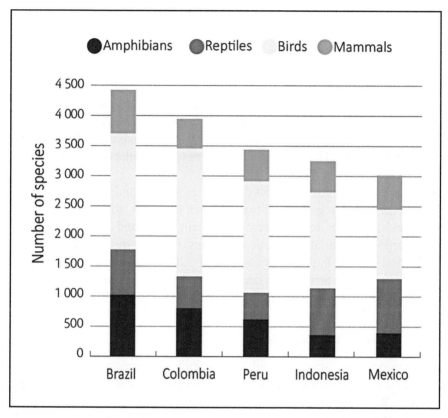

Figure 8.3 The five countries with the greatest diversity of vertebrates (Reprinted from Sarukhán et al., 2017). Note: For comparative purposes, the total number of recorded species follow the highest estimates from the most recent assessments.

so far and on an updated synopsis (Sarukhán et al., 2017) containing information of almost all the 56 chapters that confirm this body of work on the biodiversity and ecosystem services of Mexico.

8.3 Ecosystems

The wide variety of terrestrial, freshwater and marine ecosystems throughout the country is probably the most evident expression of Mexico's biodiversity. In the terrestrial realm ecosystems vary from the lush evergreen tropical humid forests in the southeast to the cactus-rich shrublands in the north and center of the country, passing through tropical dry and semi-dry forests with thick undergrowth, the alpine and sub-alpine grasslands on top of the high mountains, the epiphyte rich cloud forests on tropical mountains, and the temperate forests, across most of the length and breadth of the country on the mountain sierras.

Freshwater ecosystems are equally diverse as the hydrographic network spans over almost 633,000 km in length and estuarine and marine wetlands are distributed along the 11,122 km long Mexican coastline (Olmsted, 1993). The National Wetland Inventory identified 6,331 wetlands and wetland complexes, covering 10.03 million ha, equivalent to 5% of the country's surface (CONAGUA, 2012).

In the marine realm, the great diversity of species and ecosystems are a result of the different geological features such as slopes, abyssal plains, oceanic islands, trenches and submerged mountain ranges and due to its position within large-scale variable oceanic influences of the central-western Atlantic and central-eastern Pacific (Tittensor et al., 2010; Sarukhán et al., 2017). Mexico has also the privilege to home the world's second largest barrier reef shared with Belize, Guatemala, and Honduras.

8.3.1 Terrestrial Ecosystems

The physiognomy, structure and composition of terrestrial communities in Mexico is very varied as has been shown in different studies that classify vegetation types (for example, Miranda and Hernández-X., 1963; Rzedowski, 1978; INEGI, 1997), many of which also include classifications and descriptions of plant communities linked to aquatic environments, such as mangroves, coastal dunes, aquatic vegetation along the rivers, among other ecosystems.

One of the most recent and broadly used spatially explicit vegetation types classification, at 1:250,000 scale, recognizes more than 50 vegetation types in the country (INEGI, 2013) which are nested in broader groups.

In the last decades, ecoregions have been widely recognized as geographic areas with distinctive natural communities in similar topographic and environmental conditions. This regionalization allows a hierarchical system of classification of North American environments in different levels of aggregation. Due to the diversity of ecosystems in Mexico, a more detailed level of clustering was proposed which included 96 terrestrial ecoregions (INEGI et al., 2008; Challenger and Soberón, 2008).

To provide a general view of Mexican ecosystems, we present a brief description of level I ecoregions (CEC, 1997), based mainly on a review by Urquiza-Haas et al. (2011, and references therein) (Figure 8.4).

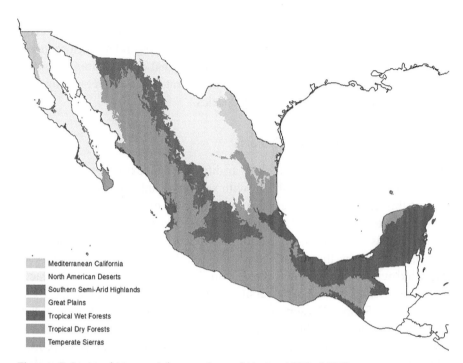

Figure 8.4   Level I terrestrial ecoregions of Mexico (CEC, 1997).

8.3.1.1   Great Plains

Located in northeastern Mexico, its extension in the country is 106,603 km²
(5.4% of the continental surface). The landscapes of the great Mexican plains
are composed mainly of xerophilous scrub, between desert conditions and the
warmer and wetter tropical thorn forest. Dominant vegetation types are the
Tamaulipecan thorn scrub (28.4%) and the submontane scrub (12.8%). In the
first, thorny bushes of compound leaves predominate; the vegetation is diverse
and dense, with between 60 and 80 species of arboreal and shrub plants from
1 to 15 m (e.g., *Acacia farnesiana, A. amentacea, Celtis pallida, Cordia bois-
sieri, Havardia pallens, Prosopis glandulosa, Ebenopsis ebano*), and a density
of between 14,000 and 30,000 individuals per hectare. The low stratum is
dominated by annual and perennial grasses (dominated by *Bouteloua* spp.).
Large populations of bisons (*Bison bison*), Mexican gray wolf (*Canis lupus
baileyi*), pronghorns (*Antilocapra americana*) and grizzly bears (*Ursus arctos
nelsoni*) used to roam in this ecoregion, their populations were greatly reduced
or extirpated before the mid of the twentieth century (e.g., Mexican gray wolf,

grizzly bear). Over 40% of the area of this ecoregion in the Mexican portion is being used for agriculture and cattle ranching; one of the most threatened ecosystems is the Tamaulipecan thorn scrub, whose extension has been reduced from 45% to 28% (i.e., 38% reduction of its original cover). Moreover, this vegetation type is highly fragmented and deteriorated due to overgrazing and overexploitation of some of its woody species.

8.3.1.2   North American Deserts

This ecoregion is characterized by its aridity, seasonal extreme temperatures and a vegetation of cacti and shrubs with xerophytic adaptations. In Mexico, 80% of this ecoregion is covered in by some type of scrub vegetation. The microphyll desert scrubland (dominated by *Larrea tridentata* and *Flourensia cernua*) is the type of vegetation that occupies the greatest extension (34%), followed by the rosette-like scrubland desert (18%) and the mixed scrubland with columnar cacti (8.6%). In the deserts of Mexico, the cacti family reaches its maximum diversity, particularly in the Chihuahuan Desert. While the Sonoran Desert presents the greatest structural variation of the desert vegetation, with abundant types of cacti such as the saguaro (*Carnegiea gigantea*) and opuntias, as well as numerous species of agaves (CEC, 1997). A variety of small globose cacti and tall semisucculent plants are also found in this ecoregion (i.e., several species of *Yucca*). It is estimated that the Sonoran Desert has 21 endemic genera of plants and about 2,500 species of plants, while the Chihuahuan Desert contains 16 endemic genera of woody plants and around 3,500 plant species (Challenger, 1998). Its fauna is also rich and unique, as an example, rattlesnakes (*Crotalus* spp.) reach their maximum diversity in this ecoregion, with 21 species, eight of which are endemic to Mexico. Nearly 10% of this region has been transformed as a result of human activities, in particular to irrigated agriculture in several areas near and around river currents and lagoons within the Chihuahuan Desert. Although cities and irrigated agriculture occupy a relatively small area, the excessive use of water, and the use of agricultural inputs have had far-reaching negative impacts on freshwater ecosystems of this region (Contreras and Lozano, 1994). Mining and cattle ranching have also had a great ecological impact in this ecoregion.

8.3.1.3   Mediterranean California

It extends along a strip of territory from Oregon to Baja California. In Mexico, it occupies a relatively small area (1.3% of the continental surface). It is the only part of the continent with a Mediterranean climate, characterized by hot and dry summers, temperate winters, and a very low annual rainfall with winter rains; it also presents a complex topography composed by mountains, hills, and plains (CEC, 1997). Most of this ecoregion is covered by chaparral (69%), a fire-prone scrubland adapted to low fertile soils and composed of evergreen shrubs of the genera *Adenostoma, Quercus, Rhus, Ceanothus, Arctostaphylos, Cercocarpus*, and *Fremontodendron*. Among the Californian Mediterranean vertebrate species are several species of amphibians (e.g., toads: *Anaxyrus boreas, A. californicus*; Californian Chorus frog: *Pseudacris cadaverina*), reptiles (e.g., Granite Night Lizard: *Xantusia henshawi*, Legless lizards: *Anniella geronimensis, A. pulchra*), birds (Nuttall's Woodpecker: *Dryobates nuttallii;* California Thrasher: *Toxostoma redivivum*) and mammals (California Ground Squirrel: *Spermophilus beecheyi*) (Mellink, 2002); of which, for example, *A. geronimensis* and the San Quintin kangaroo rat (*Dipodomys gravipes*) are endemic to the Baja California Mediterranean. The Mediterranean biome across the globe is particularly sensitive to land use and climate change (Sala et al., 2000); in the Mexican portion 15% of this ecoregion has been lost to agriculture, induced or cultivated grasslands and urbanization. Increasing wildland-urban interface represents not only habitat fragmentation but human habitation promotes the introduction of invasive alien species and intensifies the severity of wildfires (Underwood et al., 2009).

8.3.1.4 Southern Semi-Arid Highlands

This ecoregion is covered mainly by grasslands (31.5%) and in transition zones, by xerophytic scrub and oak forests (8%). Natural semidesert grasslands measure between 20 and 70 cm high and are dominated by species of the Poaceae family, in particular, the genus *Bouteloua*. In the rainy season, many species of other families, in particular, of Asteraceae are abundant. Among the mammal species characteristic of the grasslands we find the American badger (*Taxidea taxus*), the pronghorn (*Antilocapra americana*), the bison (*Bison bison*) and the black-tailed prairie dog (*Cynomys ludovicianus*); the latter is distributed almost exclusively in the ecoregion and it is considered a key species of grassland ecosystems. Among the species of grassland birds are the Savannah sparrow (*Passerculus sandwichensis*), the Horned lark (*Eremophila alpestris*), the Burrowing owl (*Athene cunicularia*) and the Prairie falcon (*Falco mexicanus*). Grassland ecosystems and their associated species, are greatly threatened by the impact of human activities; the structure and composition of most of the natural grasslands in the country has been altered due to overgrazing, human-induced exotic grasses have dominated native grasslands or they have been altogether replaced by cultivars.

8.3.1.5 Temperate Sierras

This ecoregion covers about 25% of the Mexican territory. It comprises the main sierras and mountains of the country: Eastern Sierra Madre, Western Sierra Madre, Southern Sierra Madre, Sierra Madre of Chiapas and Oaxaca and the Transvolcanic Belt, with the highest volcanoes (>5,000 m) in Mexico: Citlaltépetl, Popocatépetl, and Iztaccíhuatl (Figure 8.1). The average elevation of the ecoregion is 1,828 m. In mountainous areas, coniferous and oak forests are the dominant vegetation. This ecological region contains practically all the cloud forests (these forests cover around 0.96% of the continental surface). Mexico is the primary center of diversification of the genera *Pinus* and *Quercus* in the Western Hemisphere; both are among the most representative and economically important temperate climate trees. These genera dominate the temperate forests and although the diversity of canopy species is relatively low locally, the herbaceous and shrub layers are highly diverse; Oak and pine forests also present a great ecotype heterogeneity. The diversity of these genera in Mexico is extraordinary: between 32 and 40% of the *Quercus* species (161 species, of which 67.7% are endemic) and 50% of *Pinus* (43) species known in the world are present in the country (Farjon, 1996; Valencia, 2004). The following genera stand out among boreal affinity trees found in oak and pine forests: *Abies, Arbutus, Crataegus, Cupressus, Juglans, Juniperus,* and *Pseudotsuga,* and in humid ravines and riverbanks species of the genera *Alnus, Buddleja, Fraxinus, Platanus, Populus, Prunus, Salix,* and *Sambucus.* This ecoregion harbors a great diversity of vertebrates, many of them endangered because of habitat loss, in particular, cloud forests have a high endemism rate of plants, reptiles, amphibians and mammals (Williams-Linera et al., 2002). Among possibly extinct, critically endangered or endangered species exclusive to this ecoregion are amphibians such as *Lithobates omiltemanus, L. pueblae, L. tlaloci, Ambystoma leorae, Charadrahyla trux, Plectrohyla cembra, P. cyanomma, Pseudoeurycea anitae,* and mammals such as rabbits *Romerolagus diazi* and *Sylvilagus insonus),* shrews (*Sorex stizodon*), mice (*Megadontomys nelsoni, Microtus oaxacensis, M. umbrosus*) and squirrels (*Spermophilus perotensis*). This ecoregion has the highest human population density in the country, as it concentrates 40% of the Mexican population; land use change, expansion of urban centers, forest fires and unsustainable logging are among the greatest threats to the biodiversity of this ecoregion. Most of the original vegetation cover has disappeared or has been degraded, only 46% remains in primary condition and an additional 28% is secondary vegetation, i.e., vegetation that is in different processes of recovery or deterioration (INEGI, 2009); some regions, like Sierra de Juaréz, maintain a high proportion of forest cover through active and sustainable management of its resources by the local community (Pazos Almada, 2016).

8.3.1.6   Tropical Dry Forests

This discontinuous ecological region varies in altitude from sea level up to 3,000 m in zones with tropical climate characterized by a dry season that varies from 5 to 8 months and with intense precipitation during the summer months. Low-stature deciduous forests occupy most of the area of this ecoregion (63%) and are found in areas with annual rainfall of up to 1,200 mm, on stony hillsides, with shallow, sandy or clay soils with strong surface drainage. The vegetation is usually short (from 4 to 10 m, tallest trees are less than 15 m) and most of its species lose their leaves during the drought period. It has a reduced herbaceous cover, only evident during the rainy season, while; vines and succulent plants, mainly of the genera *Agave* and *Opuntia* are abundant. Columnar cacti are also frequent (*Stenocereus* and *Cephalocereus*). The deciduous forest possesses a great floristic diversity and a high number of endemism. Approximately 6,000 species inhabit this type of vegetation, of which 60% are endemic. Some of the representative species are of the genera *Bursera, Acacia, Lysiloma, Croton, Ceiba* and *Mimosa*. At a continental scale Mexican dry forests harbor the highest levels of endemicity of all neotropical dry forests; in terms of its vertebrate fauna, 31% of all endemic species in Mexico are found in dry forests and 11% are unique to this ecosystem (Ceballos and García, 1995), some examples are the Magdalena rat (*Xenomys nelsoni*), the grayish mouse opossum (*Tlacuatzin canescens*), the Allen's woodrat (*Hodomys alleni*), and the pygmy spotted skunk (*Spilogale pygmaea*), the Tehuantepec Striped snake (*Geagras redimitus*), and the Citreoline trogon (*Trogon citreolus*).

Tropical dry ecosystems have been transformed through human agricultural land use over millennials, but more intensively since the 1970 s; nearly over 39% of this ecoregion is used by humans (mostly agricultural and livestock activities), while 50% of the remaining vegetation in this ecoregion is in some state of succession.

8.3.1.7   Tropical Wet Forests

This ecoregion spans from the Gulf of Mexico's Coastal Plain in the north of Veracruz to the Yucatan Peninsula and the lower portions of the Sierra Madre in Chiapas. It is also located in the Pacific Coastal Plain. It covers about 14% of the national territory and presents a tropical climate with an annual average temperature between 20 and 26° C, while annual rainfall can reach more than 4,500 mm. The landscape is composed mainly by lower elevation mountains. The Gulf Coastal Plain is characterized for presenting an extensive network of rivers (e.g., Panuco, Papaloapan, Coatzacoalcos, Grijalva, Usumacinta). On the contrary, the karstic plains of the Yucatan Peninsula lack surface water, but have a unique intricate network of underground rivers and sinkholes.

Evergreen forests (35% of the ecoregion) develop in areas with annual average rainfall greater than 2,000 mm or even between 1,600 and 1,700 mm depending on its distribution throughout the year, they have a high arboreal stratum over 30 m although some trees can reach up to 75 m and up to 3 m in diameter (e.g., *Guatteria anomala, Swietenia macrophylla, Terminalia amazonia*), but usually the trunks of the trees measure between 30 and 60 cm. This forest is characterized for housing a great variety of vegetative forms; besides trees and bushes, palms and herbaceous plants are very abundant in the lower strata, while in the upper strata the foliage of vines can be found at more than 40 m from the ground; another characteristic way of life of this stratum are strangler species (*Ficus* and *Clusia*) and epiphytic plants, mainly belonging to the Bromeliaceae, Araceae and Orchidaceae families. Medium-stature semievergreen forests (27% of the ecoregion) share most of the physiognomic characteristics of evergreen forests but develop in areas with less rainfall (100 to 1,300 mm); trees reach up to 20 to 35 m in height, and in between 25 to 50% of the trees (e.g., *Bursera simaruba, Alseis yucatanensis, Zuelania guidonia, Coccoloba barbadensis, C. spicata, Vitex gaumeri*) lose their foliage in the 3–5 months dry season. The Mayan nut (*Brosimum alicastrum)*, is a characteristic species of these forests, along with *Manilkara zapota, Pimenta dioica* and *Aphananthe monoica* in most of its distribution.

The ecoregion harbors a highly diverse fauna, some of the vertebrates found in this ecoregion in Mexico (most with distributions extending south of the country) are the Yucatan Brown Brocket (*Mazama pandora*), the Northern-tailed Armadillo (*Cabassous centralis*) and the Four-eyed opossum (*Metachirus nudicaudatus*) and several species of bats (e.g., *Eumops hansae, Noctilio albiventris*) and rodents (e.g., *Orthogeomys lanius, Liomys salvini*); among bird species are the endangered Tuxtla Quail-Dove (*Geotrygon carrikeri*). There are also charismatic and key species like the Howler Monkeys (*Alouatta pigra, A. palliata*), the Spider Monkey (*Ateles geoffroyi*), the tapir (*Tapirus bairdii*), the Keel-billed toucan (*Ramphastos sulfuratus*), the Harpy Eagle (*Harpia harpyja*), the Ocellated Turkey (*Meleagris ocellata*) and crocodiles (*Crocodylus moreletii* and *C. acutus*). Tropical rainforests are among the most threatened ecosystems worldwide (Bradshaw et al., 2008) and in Mexico, conservative estimates show that only about 17% of its original cover remains in primary condition (INEGI, 2013), while a further 35% is in some successional state or has been degraded by extractive activities. Of the remaining cover, 42% of the area is distributed in patches of less than 80 km^2, which can show irreversible biodiversity loss (Sarukhán et al., 2017).

8.3.2 Inland Water and Coastal Ecosystems

The following section presents a general overview of aquatic ecosystems in Mexico, based mainly on descriptions by Lara-Lara et al. (2008a and references therein).

Mexico's complex orography has also resulted in a wide variety of freshwater ecosystems. The rivers and streams constitute a hydrographic network of almost 633,000 km long; most of the annual runoff (87%) goes through 51 main rivers (CONAGUA, 2016). Within the 320 watersheds in the Mexican territory, there are over 14,000 water bodies, of which 70 are natural lakes with surfaces of more than 10 ha, and most of the rest are artificial. These ecosystems harbor a great number of species, around 65% of those reported in North America (Miller et al., 2005). The oases in the Cuatro Ciénegas Valley stand out as a captivating landscape with unique biodiversity, another notable example are the lakes found across the Trans-Mexican Volcanic Belt which harbors some of the most extensive and important aquatic plant and algal communities in the country, and are a key part of the migratory route of North American birds.

Inland wetlands had not been well inventoried in the past (Mitsch and Hernandez, 2013). Nowadays, 142 wetlands (8.7 million ha) have been recognized for their international importance by the Ramsar Convention (Ramsar, 2017). According to their geomorphology, of the total inventoried registered wetlands (interior and coastal) by CONAGUA (2012), 2,406 are marshes, 536 lacustrine, 1,932 fluvial, 965 estuarine and 492 artificial. There is also a record of approximately 125 major sinkholes or cenotes located in the Yucatan Peninsula (southeast Mexico) and eight oases with a surface greater than 10 ha. Extensive inland and saltwater marshes can be found in the deltaic regions of rivers flowing to the Gulf of Mexico in the southeast of the country (e.g., Pantanos de Centla). Popal marshes are dominated by broadleaf plants such as *Pontederia, Sagittaria* and *Thalia,* while "tular" marshes are dominated by *Typha* or other grass-like plants (Mitsch and Hernandez, 2013). The Cuatro Ciénegas basin, one of the driest in the Chihuahuan desert, is considered a biodiversity oasis and a unique environment; it contains in its more than 200 springs and other aquatic habitats such as shallow swamps, streams and pools over 70 endemic species (e.g., fish, mollusks and crustaceans), abundant living stromatolites and other microbial communities (Souza et al., 2006; Mitsch and Hernandez, 2013).

Along the 11,122 km of Mexico's coastline, a great variety of landscapes can be found such as mangrove forests, dunes, coastal lagoons and coral reefs, among other formations. Coastal wetlands in Mexico include about 118 major complexes and at least another 538 smaller systems (ca. 1.6 million hectares) that are often composed of different wetland types. For instance, Laguna de Términos, one of the largest wetlands, contains mangrove swamps, coastal dunes vegetation, freshwater swamps, flooded vegetation, lowland forest, palms, spiny scrubs, forests, secondary forests, and seagrass beds (Mitsch and Hernandez, 2013).

Coastal ecosystems are a result of the interaction between the ocean, the land and the atmosphere and are usually highly dynamic and productive. Aside from harboring highly rich biotic communities, coastal ecosystems are critical for various biological processes such as the reproduction of many fish, birds and mammal species. Because Mexico has coastlines exposed to different ocean systems, there is a clear variability of these ecosystems.

8.3.2.1 Mangroves

Mangroves are composed by mangrove trees or shrubs (1 to up to 30 m in height) that occur along bays, coastal lagoons, estuaries, and beaches without surge along the Pacific, the Gulf of California, the Gulf of Mexico and the Caribbean coasts. Five mangrove species are distributed in the country: *Rhizophora mangle, Laguncularia racemosa, Conocarpus erectus, Avicennia germinans* and *Rhizophora harrisonnii,* the first four being the most widely distributed and abundant. Because these ecosystems have high primary productivity, they represent a key energy source for marine environments and are also habitat for a great number of marine and estuarine species (many of which are fishery species). In 2015 mangroves covered 775,555 hectares (or 5.1% of the world's mangroves) in contrast to the 856,405 hectares estimated for 1981; nearly 63% of mangroves are within protected areas (Troche-Souza et al., 2016).

8.3.2.2 Intertidal Strip and Coastal Dunes

These ecosystems are found along most of Mexico's coastline. Because of their highly diverse geological features, they house an important regional biodiversity, particularly of infauna species. Coastal dunes are highly dynamic systems mainly due to their sandy substrate and the fact that they are exposed to strong winds resulting in sand displacement that buries part of the vegetation at different times, they are also prone, in some areas, to floods and droughts. Coastal dunes harbor a vast flora diversity, a recent inventory registered 2,075 plant species, of which 95 are considered to be mainly associated to these ecosystems, the rest are often species from surrounding ecosystems such as xerophytic shrubs, pine forests, tropical dry forests and mangroves (Espejel et al., 2017).

8.3.2.3 Coastal Lagoons

Along Mexico's coast, there are a wide variety of coastal lagoons (e.g., estuaries, marshes, bays, coves). These ecosystems are highly productive due to photosynthetic activity by phytoplankton, macroalgae, aquatic vegetation, and mangroves. According to a study, there are around 400 fish, 50 mollusk and 90 crustacean species associated to the country's coast zones (Contreras, 2005). The coastal lagoons in the Baja California Peninsula are also key breeding places for the North American gray whales (*Eschrichtius robustus*).

8.3.2.4 Macroalgae-Dominated Ecosystems

The coasts of the Baja California Peninsula and the Gulf of California have high macroalgae species richness, which form ecosystems dominated mainly by *Macrocystis pyrifera* in the Pacific Coast and by *Sargassum* species in the Gulf of California where they can be found at great densities (up to 150,000 wet tons) along with two endemic species, *Eucheuma uncinatum* and *Chondracantus scuarrulosus,* which have high growth rates and productivity. These highly productive and diverse ecosystems house relevant fishing species such as abalones, sea urchins, sea cucumbers, lobsters and snails, and other species like sea turtles.

8.3.2.5 Coral Reefs

Coral reefs are the most biologically diverse marine ecosystems. Mexico possesses coral reefs along the Pacific, the Gulf of Mexico and in the Caribbean coasts. In the Pacific Ocean, coral reefs are relatively small and geographically isolated. In the Gulf of Mexico, along the coast of the states of Veracruz and Campeche, these reefs have high species diversity, with about 60 coral and 100 fish species. In the Caribbean Sea, Mexico hosts part of the Mesoamerican Reef system, the longest barrier in the Western hemisphere extending about 1,000 km along the coasts of Mexico, Belize, Guatemala, and Honduras (Garcia-Salgado et al., 2008).

Freshwater, estuarine and coastal ecosystems are of great social and economic importance because they provide crucial ecosystem services, such as water provision and maintenance of water quality, buffering of flood flows, erosion control, carbon sinks, provision of nursery habitats for fishery species, coastline stabilization and protection against weather events (e.g., tropical storms, hurricanes), climate regulation, recreation and cultural services, among others (Aylward et al., 2005; Barbier et al., 2011). Thus the loss of these ecosystem services directly impacts human well-being and represents significant economic costs. For example, a study in the Gulf of California estimated that the loss of a hectare of mangrove fringe would cost local economies approximately US$ 605,290 over 30 years in fisheries landings (Aburto-Oropeza et al., 2008).

Despite their importance, these ecosystems are among the most threatened natural systems globally (Barbier et al., 2011) and in Mexico. Many human civilizations have established in the coasts and near water bodies for millennia, thus they have been greatly impacted by human activities such as water extraction, habitat loss from land use change (e.g., agriculture, aquaculture, urban and tourist developments), hydrologic modifications to build water dams (over 5,000 in the country), pollution and eutrophication resulting from productive activities, introduction of exotic species, ocean warming and overfishing, among others (Contreras-Espinosa and Warner, 2004; Martínez et al., 2006; Garcia-Salgado et al., 2008). For instance, in the Comarca Lagunera region in Sonora at least 92 springs and 2,500 km of rivers have dried up, surface waters and water tables have decreased. Of the nearly 200 species of freshwater fish from this region, 15 became extinct, while 120 were considered in danger of extinction more than 20 years ago (Contreras and Lozano, 1994). Another relevant threat to biodiversity are changes to the hydrological flow; for example; the widening of a canal that connects a coastal lagoon (Laguna Cuyutlán) with the sea from 100 to 300m, provoked the failure of nesting efforts of several waterbirds (Mellink and López-Riojas, 2007).

8.3.3 Marine Ecosystems

The following section presents a general overview of marine ecosystems in Mexico, mostly based on descriptions by Lara-Lara et al. (2008b, and references therein).

The geographic location of our country between the oceanic influences of the central-western Atlantic, the central-eastern Pacific, the Caribbean Sea and the Gulf of California (or Sea of Cortez) explains to a large extent its enormous marine species and ecosystems diversity. Mexico is the country with the 12th largest marine extension (3,149,920 km^2 of exclusive economic zone and 231,813 km^2 of territorial sea), which implies an ample potential for the sustainable use of coastal-marine resources.

The great diversity of marine ecosystems in Mexico emerges from the variety of oceanic relief features, such as continental slopes, abyssal plains, oceanic islands, moats and submarine mountain ranges, and from the vertical heterogeneity that results from depth, water layers, currents and the upwelling systems. There are many ways in which marine environments have been characterized based on different features, for instance, life zones (i.e., pelagic or benthic), environmental attributes or biotopes (e.g., sandy or rocky zones) and biological communities. Coastal or marine ecosystems have also been differentiated according to their position in the continental platform, they can also be classified as euphotic and aphotic environments based on light availability. Additionally, there have been approaches to map marine zones based on the distribution of water masses and the biological communities that characterize them.

One of the more comprehensive maps of the North American marine regions was carried out by the Commission for Environmental Cooperation (CEC, 2009), the result has three nested levels. Level I reflects differences at ocean basin level. Of the 21 ecoregions at this level, eight are within Mexico's exclusive economic zone (Figure 8.5). Level II shows benthic and pelagic environments' distribution considering their depth and topography, 28 level II ecoregions were defined for Mexico. Level III ecoregions were determined at a finer scale for the neritic region to reflect locally significant variations, there are 24 ecoregions within the Mexican continental platform.

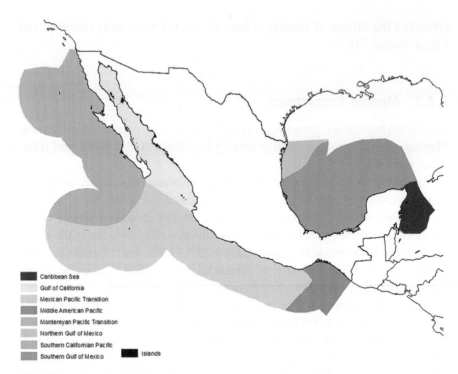

Caribbean Sea
Gulf of California
Mexican Pacific Transition
Middle American Pacific
Montereyan Pacific Transition
Northern Gulf of Mexico
Southern Californian Pacific
Southern Gulf of Mexico   Islands

Figure 8.5   Level I Marine ecoregions of Mexico (CEC, 2009).

8.3.3.1   Pelagic Ecosystems

Pelagic ecosystems are usually differentiated in the ones that have a stable low primary production and the ones that have high primary production pulses, mainly due to fertilizing inputs from rivers or during upwelling seasons. Upwelling systems constitute functional units that contribute with around 50% of fisheries catches, as well as influencing the distribution patterns of larvae of marine organisms (Escobar et al., 2008). In Mexico, the main upwelling systems are in the Western coast of Baja California, the continental and peninsular margin of the Gulf of California, Cabo Corrientes (in the shore where the Gulf of California and the Southern Californian Pacific meet), Golfo de Tehuantepec (in the coast of the Middle American Pacific), in the southern Gulf of Mexico (Yucatan and Cabo Catoche upwelling) and some areas in the Caribbean Sea (CONABIO et al., 2007).

8.3.3.2 Benthic Ecosystems

Benthic ecosystems are usually dominated by polychaetes, crustaceans, echinoderms and mollusks species. Mexico has a great number echinoderm's species, some with high commercial value (like the red and purple sea urchins, *Strongylocentrotus franciscanus* and *S. purpuratus*).

Deep sea ecosystems are found in depths greater than 200 m, within them a wide variety of geological structures and habitats can be found, such as submarine canyons, abyssal plains, seamounts, deep-water coral and sponge reefs and hydrothermal vents. Despite the fact that most of them have a very low biological productivity, these ecosystems are characterized by high species diversity, endemic to these environments (between 21 to 250 species have been found in a sample of 0.25 m² seamud) and it is estimated that worldwide they could be harboring up to 10 million species, including microbial mats, anemones, echinoderms, tube worms, mollusks, crustaceans and fishes (Lara-Lara et al., 2008b; UNEP, 2006; CONABIO et al., 2007).

Within these ecosystems submarine canyons, seamounts, oceanic trenches and oceanic dorsals stand out due to their structural and functional complexity. The interaction between their geological shapes and water currents results in a high biological productivity and thus are important habitats and feeding areas for many benthic species and some cetaceans (WWF et al., 2001; UNEP, 2006; CONABIO et al., 2007). Other key ecosystems, habitat for many fish species, are deep-water coral reefs that form at great oceanic depths (up to 2,000 m) in temperatures from 8 to 4°C, in the absence of light, and with sponges' aggregations that can grow up to 19 m and live over 100 years. Because of their slow growth rates these ecosystems are particularly vulnerable to any disturbance that may affect their structure. On the other hand, at hydrothermal vents—aphotic environments with temperatures up to 400°C, highly toxic chemicals and hydrostatic pressure over 200 atm—only some organisms can exist, such as hyperthermophilic microorganisms and tube worms. In Mexico, two hydrothermal vents with depths over 2,000 m have been identified in the Pacific, one near the coast of Colima, and the other in the Gulf of California.

Even though there are no quantitative studies on the extension of marine ecosystems that has been lost or degraded, there is evidence that they have been severely impacted as a result of human activities such as pollution, mass tourism and overfishing; trawling and toxic waste dumping represent additional major threats to deep sea environments (CONABIO et al., 2007; Sarukhán et al., 2012). There is an urgent need to increase our knowledge on these ecosystems, by promoting research studies and the implementation of monitoring programs. Some important steps have been taken in this direction, in 2008 an expedition was made in the deep sea of the Gulf of California (Aburto-Oropeza et al., 2010). Also, CONABIO has established a monitoring system for Mexico and neighboring countries that allows the identification of important marine processes and threats at a high spatial and temporal resolution (e.g., of upwelling events, coral reef bleaching, algal blooms) (Cerdeira-Estrada and López-Saldaña, 2011; Cerdeira-Estrada, 2016).

8.3.4 Insular Ecosystems

The Mexican insular territory (MIT) is a strategic resource for the country, comprising a set of more than 4,111 island elements (islands, reefs and cays) mainly of federal jurisdiction located in coastal and marine areas with a total surface of more than 5,100 km^2 (INEGI, 2015). Because of its insular territory Mexico's exclusive economic zone is 1.6 times greater than the continental surface, where Mexico exercises sovereign rights for the exploration, exploitation and conservation of natural resources, for example, oil, mining, fisheries, among others (CANTIM, 2012; Figure 8.5).

Islands are natural fragments of habitats where unique biological communities have evolved. The Mexican islands account for at least 8% of the country's known plants and animals and endemism are higher than in any other ecosystem, therefore MIT harbors an important part of our unique biodiversity in a very small area. MIT also offers a key territory for reproduction, nesting, resting and feeding of many migratory species, mainly birds and mammals. For example, 2 million birds of 12 species nest in three islands of San Benito Archipielago, which represents the highest concentration of marine birds in the Nororiental Pacific (CANTIM, 2012).

Because insular species evolved with few or no predators or competitors they are particularly vulnerable to the introduction of non-native invasive species (Lara-Lara et al., 2008a). It is estimated that invasive species can increase the probability of extinction up to 40% more for insular species than for continental species. Worldwide, most of the extinctions documented during the last 400 years (about 50–75% of the total) have occurred on islands, and 67% have been caused by invasive species. In Mexico, introduced mammal species, such as cats, rats, goats, sheeps, dogs, and pigs have caused severe impacts on insular ecosystems. Fortunately, joint actions by governmental institutions (mainly the Navy and Environmental ministries) carried out for more than 20 years with the outstanding participation of civil society (in particular, Grupo de Ecología y Conservación de Islas, GECI), and academic institutions have led to the eradication of exotic species from many Mexican islands, which is an important step for their ecological restoration (Aguirre-Muñoz et al., 2008; Samaniego et al., 2009; Latofski-Robles et al., 2017).

8.4 Species Diversity

The biota of the world is not yet fully known. It is estimated that the nearly 2 million described species (Wheeler et al., 2012; Hinchliff et al., 2015) represent approximately 18% of the total estimated to inhabit the planet (Mora et al., 2011). During the last decades, about 18,000 new plants and animals have been discovered each year (IISE, 2011) and thus every effort to account for the described species in the world becomes outdated as soon as it is published, in addition to the uncertainty brought by the "synonym problem" and the "taxon concept problem" (Lucarini et al., 2015; Willis and Bachman, 2016). Estimates of the potential number of species on earth vary even more greatly from 10 to 50 million (May, 1998; Wilson, 2000; Mora et al., 2011). Thus, reliable estimates of the total number of species, both potential and known, especially in less well-known taxonomic groups are unlikely to be made for many decades (Chapman, 2009), although several initiatives and researchers are attempting to fill these knowledge gaps (Blackwell, 2011; Zhang, 2011; Guiry, 2012; Christenhusz and Byng, 2016). In Mexico, according to recent data by Martínez-Meyer et al. (2014), about 100,000 species of 56 taxonomic groups have been described in more than 200 years of systematic collection and documentation (or 8.59% of the world's cataloged species in those groups). Several estimates point out that Mexico could hold in between 10 and 12% of the world's species (Llorente-Bousquets and Ocegueda, 2008; Sarukhán et al., 2017). However, these estimates are conservative because there are taxonomic groups practically unknown in the country, such as microscopic fungi, viruses and other microorganisms, and even some families of insects and other invertebrates and families of plants that have been poorly collected and studied due to their inconspicuousness, apparent low economic importance, little public interest, lack of taxonomists, or because they are difficult to collect or inhabit environments difficult to access.

The number of species recorded in a given area is the most widely used indicator of diversity and, in a certain way, it is a first step in the assessment of progress of the state of knowledge of the biota. Considering the actual and estimated number of species, Wheeler et al. (2012) concluded that an ambitious goal to describe 10 million species in less than 50 years is attainable based on the progress made in the last 250 years by worldwide collections, existing experts, technological innovation and collaborative teamwork.

In Mexico, albeit its long taxonomic tradition and the vast number of taxonomic revisions that have been conducted in the country, it is still difficult to get a reliable and precise number of species for every biological group, mainly due to different taxonomic opinions regarding scientific names validity and the concept of species. In general, it has been considered practically impossible to verify all the species identifications and records; furthermore, the number of reviews or monographs available is relatively low (Villaseñor, 2016).

Here, we present a general synthesis of the number of species recorded within the main taxonomic groups, except bacteria, virus, and microfungi. We provide a general framework of the species described in the world, and those registered in Mexico and recorded in the taxonomic catalogs of the SNIB (CAT-CONABIO, 2017), as well as the number of endemic taxa reported in the literature. SNIB nomenclatural catalogs are supported by recent publications (descriptions and monographs) and experts to keep the information updated. The continued work of taxonomists is crucial for a complete inventory of our biota. Also, the spatial patterns of species diversity have been studied at different scales, particularly to describe species richness and beta diversity patterns of better-known groups, such as plants and vertebrates (Koleff et al., 2008 and references therein). Most groups present a latitudinal diversity gradient, with a larger number of species at lower latitudes, in the tropical southern part of Mexico, but this pattern does not usually coincide if only endemic species are considered. Also, altitudinal diversity gradients are evident along the country, which vary accordingly to taxa, scale, and slope. These patterns are not discussed here; however, there is a vast literature on the subject since there are a great number or research groups working on these issues in the country.

8.4.1 Flora

Plants are one of the best-known biological groups in the country (e.g., Rzedowski, 1991; Delgadillo-Moya, 2014; Delgadillo-Moya and Juárez-Martínez, 2014; Gernandt and Pérez de la Rosa, 2014; Villaseñor, 2016) and worldwide (Troudet et al., 2017). One of the latest estimations by Christenhusz and Byng (2016) counted the currently known, described and accepted number of plant species in the world in 330,237, of which approximately 308,312 are vascular plants. They determined that nearly

2,000 species are described every year; despite a reduction of financial and scientific support for basic natural history studies. The International Plant Names Index, the most comprehensive and regularly updated listing of scientific names of vascular plants documented in 2016—1,065,235 species names—with an average of 2.7 names per plant (Willis and Bachman, 2016). In Mexico, the long tradition in botany and the work of several remarkable botanists have contributed to the knowledge of the flora through intensive inventories in different regions housed in scientific collections in the country and abroad. Nevertheless, the knowledge of species is still considered unsatisfactory because of the scattered publications as well as a lack of a well-curated database that synthesizes the information and reconciles the multiple taxonomic criteria among specialists. About 28,132 species of plants have been recorded in Mexico in the CAT-CONABIO (see Table 8.1). Forzza et al. (2012) reported 25,036 vascular plants for Mexico (Figure 8.2); however, a recent taxonomic inventory recorded about 23,300 vascular plants—with an average of 1.3 synonyms per accepted species—, of a total that has been calculated in up to 30,000 (Rzedowski, 1991; Villaseñor, 2016; Table 8.1). Noteworthy is the high level of endemism of the Mexican flora, Villaseñor (2004) points out that 7.8% of the genera of the vascular flora are endemic; in terms of species the percentage of endemism is 49.8%, this means that half of our flora is not found anywhere else in the world, a figure surpassed only by South Africa and Brazil among mainland countries. Moreover, 27.3% of the endemic species are only found in one state (Villaseñor, 2016). The families with the highest proportion of endemism are shown in Figure 8.6. Other notable examples of species richness and endemism refer to several taxa: pines (*Pinus*: 49 species of 120 species in the world, 22 endemics), nopales and xoconostles (*Opuntia*: 93 species of 191–215 in the world, 62 endemic); and orchids (Orchidaceae: 1,263 species of 27,800 in the world, 585 endemic), which includes the vanilla, one of the world's most appreciated culinary item. Another case to be highlighted is *Lacandonia schismatica* (Triuridaceae), a small saprophyte plant (i.e., lacks chlorophyll and feeds on decomposing organic matter) found in the Selva Lacandona in 1987 (Martínez and Ramos, 1989), it surprised the scientific community for being the only plant of more than 280,000 known species with a different arrangement of reproductive structures. *L. schismatica* has only been described in Mexico, although in 2012 a very similar plant was discovered in a fragment of the Atlantic Forest in northeastern Brazil, the first record of a new species of *Lacandonia* in South America (Melo and Alves, 2012).

Table 8.1 Valid Species of Plants Described in the World and in Mexico (Only Selected Groups are Shown as Examples)

Group / Selected group Selected family	World Total number of species reported in the literature	Mexico Cataloged in Conabio*	Mexico Total number of species reported in the literature	Mexico Endemic reported in the literature	Mexico Endemic Catalogued in Conabio*
Bryophyta	13,000[a]	1,326	1,000[b]	77[b]	NA
Anthocerotophyta	150[c]	4	9[d]	3[d]	NA
Marchantiophyta	5,000[e]	282	592[d]	105[d]	NA
Vascular plants	308,312[f]	26,520	23,296[g,h,i,j,k]	11,551[g]	9,420
Pteridophyta	11,850[f]	1,143	1014[h]	188[h]	147
Gymnosperms	1,079[f]	176	156[i,j,k]	92[g,i,l]	74
Cycadaceae and Zamiaceae	337[f]	58	54[j,l]	48[l]	39
Angiosperms	295,383[f]	25,201	22,126[g]	11,694[g,m-s]	9,302
Asteraceae	24,700[f]	3,461	3,057[g]	1,995[g]	1,561
Cactaceae	1,750[f]	715[t]	677[g]	498[g]	490[t]
Crassulaceae	1,400[f]	397	372[g]	327[g]	262
Euphorbiaceae	8,525[f]	708	826[m]	458[m]	431
Fabaceae	19,500[f]	2,138	1,805[n]	850[n]	588
Lamiaceae	7,530[f]	702	591[o]	388[o]	409
Rubiaceae	13,620[f]	684	660[u]	364[g]	216
Solanaceae	2,600[f]	430	387[p]	133[p]	109
Orchidaceae	28,000[f]	1,263[v]	1,263[q]	585[q]	562[v]
Asparagaceae**	2,900[f]	473	445[g]	348[g]	348

Group	World	Mexico			
Selected group Selected family	Total number of species reported in the literature	Total number of species reported in the literature	Catalogued in Conabio*	Endemic reported in the literature	Endemic Catalogued in Conabio*
Arecaceae	2,600[f]	95[q]	139	40[q]	64
Bromeliaceae	3,475[f]	400[q]	441	290[q]	321
Poaceae	12,000[f]	1,213[s]	1,246	292[s]	403

*Complete classification, updated names and corresponding references are available at Taxonomic Catalogs (http://www.biodiversidad.gob.mx/especies/CAT.html). Species names for Mexico can be searched at Enciclovida (http://www.enciclovida.mx/).

** Includes Agavoidea sensu APG IV.

NA: not available.

Sources: [a]Goffinet & Shaw, 2009; [b]Delgadillo-Moya, 2014; [c]Renzaglia, Villarreal, & Duff, 2009; [d]Delgadillo-Moya & Juárez-Martínez, 2014; [e]Crandall-Stotler, et al., 2009; [f]Christenhusz & Byng, 2016; [g]Villaseñor, 2016; [h]Martínez-Salas & Ramos, 2014; [i]Gernandt & Pérez-de la Rosa, 2014; [j]Osborne et al., 2012; [k]Villanueva-Almanza & Fonseca, 2011; [l]Nicolalde-Morejón et al., 2014; [m]Martínez-Gordillo et al., 2002; [n]Delgado-Salinas & Torres-Colín, 2016; [o]Martínez-Gordillo et al., 2013; [p]Rodríguez et al., 2016; [q]Espejo-Serna, 2012; [r]García-Mendoza, 2011; [s]Dávila et al., 2016; [t]Guzmán (ined.); [u]Borhidi, 2012; [v]Solano-Gómez (ined.).

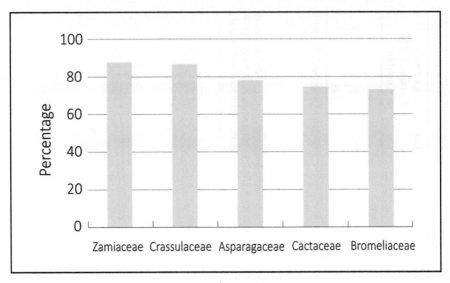

Figure 8.6 Vascular plants with highest proportion of endemism (Reprinted from Sarukhán et al., 2017).

8.4.2 *Fauna*

Animals represent the vast majority of currently described species worldwide. This kingdom is estimated to have a total of 1,525,728 described living species in 40 phyla, of which Arthropoda represents around 78.5% of the total (Zhang, 2013). However, knowledge varies greatly among groups, with vertebrates being the best-known organisms in the world.

8.4.2.1 Invertebrates

Considering the classification proposed by Zhang (2011, 2013), this biological group comprises 37 phyla, some of them have a long taxonomic tradition in the country such as the Arthropods, Mollusks, Echinoderms, Annelida, Cnidarians, and Porifera. Almost all phyla are represented by at least one species in Mexico. However, 17 phyla[1] (31,229 species worldwide, of which 2% are extant) are not included in the catalogs due to lack of knowledge, expert taxonomists or accessibility to publications of records in Mexico. The remaining 20 phyla sum 88,902 of recorded species out of 1,429,695 species recorded worldwide (i.e., 6.2%) (Table 8.2).

---

[1]Myxozoa, Xenoturbellida, Acoelomorpha, Orthonectida, Rhombozoa, Nematoda, Nematomorpha, Priapulida, Loricifera, Kinorhyncha, Entoprocta, Cycliophora, Echiura, Nemertea, Gastrotricha, Gnasthostomulida, and Micrognathozoa.

It is important to emphasize that the groups with the greatest number of species have not been fully studied, such as insects, one of the most numerous groups of animals. Although estimates are still uncertain, 65,275 species of insects have now been documented, of the nearly 100,000 species that are estimated to occur in Mexico (or, 6.3% of all species known worldwide). The most speciose insect orders are the Diptera, Coleoptera, and Lepidoptera, which accounts for 77% of all insects (Table 8.3), this agrees with the most speciose orders reported worldwide, although Coleoptera ranks first worldwide. For various groups of invertebrates, particularly marine, information gaps are more evident (Jiménez et al., 2016).

8.4.2.2 Vertebrates

Mexico hosts 5,757 vertebrate species, out of the 68,724 that have been described in the world, which represent 8.37% of the world diversity of vertebrates (see Table 8.4). It stands out as the fourth most diverse country in terms of total species richness and number of endemism (Figures 8.3 and 8.7), in particular, for mammals and reptiles. With 564 mammal species,—of which 514 are terrestrial and 50 are marine (Ceballos, 2014, Sánchez-Cordero et al., 2014)—the country ranks second only to Brazil with 720 species, and first in the number of marine mammals. The group of mammals in which Mexico stands out due to its diversity are bats, with 144 species (Sánchez-Cordero et al., 2014). A recent study evaluated the role of these species in the environmental services of pollination and pest control in agroecosystems and estimated that the economic benefit is more than 1 billion dollars in the world (Maine and Boyles, 2015). On the other hand, the marine vaquita (*Phocoena sinus*), endemic to the Upper Gulf of California, is perhaps the most charismatic species, but it faces a high risk of extinction due to incidental capture in fishing activities.

In our territory 908 reptile species have been described so far (56% are endemic); second only to Australia with 917 species, which holds a total land area 3.5 times greater. It is worth noting that around 65% of the 399 recorded amphibian species in Mexico are endemic and that seven of the 16 amphibian families contain more than 50% of the endemic species (Parra-Olea et al., 2014). Another relevant example, given the wider distribution of the avifauna, is the high proportion of endemic species (ca. 10%), for which Mexico ranks 11th in the world (see Figure 8.7).

The level of diversity and endemism in the ichthyofauna in marine and freshwater ecosystems of our country is also remarkable. Around 13.3% (or 2,224 species) of all known marine fish species have been recorded in

Table 8.2 Valid Species of Invertebrates (Not Arthropods) Described in the World and in Mexico (Only Selected Groups are Shown as Examples)

Group	World	Mexico			
	Total number of species reported in the literature	Total number of species reported in the literature	Catalogued in Conabio*	Endemic reported for Mexico	Endemic Catalogued in Conabio*
Porifera	8,659[w]	517[x]	367	174[x]	NA
Placozoa	1[w]	1[y]	1	NA	NA
Cnidaria	10,203[w]	823[z, aa – aw]	866	NA	NA
Ctenophora	187[w]	11[ax – az]	4	NA	NA
Chaetognata	170[w]	31[ba – bc]	23	NA	NA
Platyhelminthes	29,448[w]	1,015[bd]	1020	NA	NA
Mollusca	84,977[w]	6,074[be – bi]	3910	898[be, bg, bh, bi]	34
Annelida	17,388[w]	1,633[bj – bl]	1095	At least 215[bj, bl]	NA
Sipuncula	147[w]	23[bm, bn]	3	NA	NA
Rotifera	2,049[w]	303[b,o]	301	NA	NA
Acanthocephala	1,298[bp]	60[bq]	67	NA	NA
Phoronida	16[w]	1[br]	1	NA	NA
Bryozoa	6,008[w]	183[bs – bv]	41	NA	NA
Brachiopoda	392[w]	20[bw – bz]	4	NA	NA
Tardigrada	1,167[w]	40[ca]	40	NA	NA
Onychophora	183[w]	4[cb, d]	4	NA	NA

Group	World	Mexico			
	Total number of species reported in the literature	Total number of species reported in the literature	Cataloged in Conabio*	Endemic reported for Mexico	Endemic Catalogued in Conabio*
Echinodermata	7,550[w]	643[cc]	776	at least 7 [cc, cd]	7
Tunicata (Urochordata)	2,804[w]	91 [ba, ce – ck]	32	NA	NA

* Complete classification, updated names and corresponding references are available at Taxonomic Catalogs (http://www.biodiversidad.gob.mx/especies/CAT.html). Species names for Mexico can be searched at Enciclovida (http://www.enciclovida.mx/).

NA: not available.

Sources: [w] Zhang, 2013; [x] Carballo et al., 2014; [y] Mayén-Estrada, 2007; [z] Abeytia et al., 2013; [aa] Breedy&Guzmán, 2002; [ab] Breedy&Guzmán, 2007; [ac] Breedy et al., 2009; [ad] Breedy&Guzmán, 2011; [ae] Breedy&Guzmán, 2016; [af] Brusca&Trautwein, 2005; [ag] Cairns, 2007; [ai] Cairns et al., 2009; [aj] Cutress&Pequegnat, 1960; [ak] Horta-Puga&Carricart-Ganivet, 1993; [al] Dahlgren, 1989; [am] Reyes-Bonilla et al., 2005; [an] Viada& Cairns, 2007; [ao] ZarcoPerelló et al., 2013; [ap] Carlgren, 1951; [aq] González-Muñóz et al., 2013; [ar] González-Muñóz et al., 2012; [as] González-Muñóz et al., 2015; [at] Potts, 2016; [au] Calder & Cairns, 2009; [av] Calder et al., 2009; [aw-ax] Moss, 2009; [ay] Gamero-Mora, et al., 2015; [az] Ruiz-Escobar et al., 2015; [ba] Hendrickx&Brusca, 2005; [bb] Hernández-Flores et al., 2009; Gasca& Loman-Ramos, 2014; [bc] Tovar & Suárez-Morales, 2007; [bd] García-Prieto et al., 2014; [be] Castillo-Rodríguez, 2014; [bf] Contreras-Arquieta, 2000; [bg] Hershler et al., 2011; [bh] Thompson, 2011; [bi] Naranjo-García, 2014; [bj] Tovar-Hernández et al., 2014; [bk] Fragoso& Rojas, 2014; [bl] Oceguera-Figueroa & León-Règagnon, 2014; [bm] Hermoso-Salazar, et al., 2013; [bn] Frontana-Uribe et al., *In Press,* [bo] Sarma, 2008; [bp] Amin, 2013; [bq] García-Prieto et al., 2014; [bs] Santagata, 2009; [bs] Brusca&Hendrickx, 2005; [bt] Winston &Maturo, 2009; [bu] Sosa-Yáñez et al., 2015; [bv] Bertsch& Aguilar Rosas, 2016; [bw] Brusca, 2005; [bx] Cooper, 1977; [by] Santagata&Tunnell, 2009; [bz] Simon &Hendrickx, 2015; [ca] Kaczmarek et al., 2011; [cb] Fernández-Álamo& Rojas, 2007; [cc] SolisMarín et al., 2014; [cd] SolisMarín&Laguarda-Figueras, 2010; [ce] Bertsch& Aguilar-Rosas, 2016; [cf] Castellanos & Suárez-Morales, 2009; [cg] Hereu et al., 2006; [ch] Hereu et al., 2010; [ci] Hereu& Suárez-Morales, 2014; [cj] Hereu et al., 2014; [ck] Rocha et al., 2012 [cl] Peck, 1975.

Table 8.3 Valid Species of Arthropods in the World and in Mexico (Only Selected Groups are Shown as Examples)

Group / Selected group	World — Total number of species reported in the literature	Mexico — Total number of species reported in the literature	Mexico — Catalogued in Conabio*	Mexico — Endemic reported in the literature	Endemic Catalogued in Conabio*
Arthropoda	**1,257,047**[cl]	**77,460**	**42,578**	**13,007**[cm–dk]	**1,745**
Crustacea	67,735[cl]	4,956[cm,cn,dl-dr]	4,595	At least 227[cm–cp]	48
Branchiopoda	-	348[cm,dl]	214	1[cm]	1
Remipedia	-	2[cn,dm]	2	2[cm,cn]	2
Cephalocarida	-	1[dn]	1	NA	NA
Maxillopoda	-	617[dl,do,dp]	933	At least 15[cm]	15
Ostracoda	-	883[dl]	325	NA	NA
Malacostraca	-	3,105[cm,dl,dq,dr]	3,120	At least 209[cm–co]	30
Arachnida	112,442[cl]	5,751[cq,ds]	5,040	2,269[cq,cr]	3
Amblypygi	163[cl]	27[cq]	22	19[cq]	NA
Araneae	43,678[cl]	2,295[cq]	2383	1,759[cr]	NA
Opiliones	6,534[cl]	238[cq]	242	218[cq]	1
Pseudoscorpiones	3,533[cl]	159[cq]	144	110[cq]	NA
Scorpiones	1,988[cl]	258[cq]	232	162[cq]	2
Solifugae	1,113[cl]	79[cq]	57	ND	NA
Thelyphonida	110[cl]	4[cq]	2	0[cr]	NA
Mesostigmata	11,419[cl]	507[ds]	306	NA	NA
Ixodida	892[cl]	100[ds]	357	NA	NA
Trombidiformes	25,766[cl]	1,208[ds]	1134	NA	NA
Sarcoptiformes	16,173[cl]	801[ds]	357	NA	NA

| Group | World | Mexico | | | |
Selected group	Total number of species reported in the literature	Total number of species reported in the literature	Cataloged in Conabio*	Endemic reported in the literature	Endemic Cataloged in Conabio*
Pycnogonida	1,335 [cl]	42 [dt]	41	NA	NA
Xiphosura	4 [dt]	1 [du]	1	0 [eb]	0
Myriapoda	11,999 [cl]	661 [cs, ct, dv]	588	281 [cs, ct]	176
Hexapoda	1,063,533 [cl]	66,049 [cr, cu – dk, dw-dy]	35,617	10,230	1,517
Collembola, Diplura, Protura	9,954 [cl]	774 [dtw-dy]	627	ND	NA
Insecta	1,053,578 [cl]	65,275	34,990	10,230 [cr, cu-de]	1,517
Coleoptera	389,487 [cl]	15,692 [cr, cu-db]	13,483	6,140 [cr, cu-db]	896
Lepidoptera	158,423 [cl]	14,507 [dc]	3,315	656 [dc]	261
Hymenoptera	153,088 [cl]	7,516 [cr, dd, de, dz, ea]	6,410	1,263 [cr, dd, dz, ea]	NA
Diptera	156,774 [cl]	20,000 [ec]	6,154	715 [cr]	NA

* Complete classification, updated names and corresponding references are available at Taxonomic Catalogs (http://www.biodiversidad.gob.mx/especies/CAT.html). Species names for Mexico can be searched at Enciclovida (http://www.enciclovida.mx/).

NA: not available.

Source: [cl] Zhang, 2013; [cm] Mercado-Salas et al., 2013; [cn] Álvarez et al., 2013; [co] Villalobos &Alvarez, 2008; [cp] Álvarez et al., 2014; [cq]Francke, 2014; [cr]Morrone& Márquez, 2008; [cs]Cupul-Magaña, 2013; [ct] Bueno-Villegas et al., 2004; [cu] Cifuentes-Ruiz & Zaragoza-Caballero, 2014; [cv] Morón et al., 2014; [cw]Morrone, 2014; [cx] Navarrete-Heredia & Newton, 2014; [cy] Noguera, 2014; [cz] Ordóñez-Reséndiz et al., 2014; [da] Zaragoza-Caballero & Pérez-Hernández, 2014; [db] Zurita-García et al., 2014; [dc] Llorente-Bousquets et al., 2014; [dd] Coronado-Blanco & Zaldívar-Riverón, 2014; [de] Ruiz-Cancino et al., 2014; [df] Acosta-Gutiérrez, 2014; [dg] Contreras-Ramos et al., 2014; [dh] Contreras-Ramos & Rosas, 2014a, [di] Contreras-Ramos & Rosas, 2014b, [dj] García-Aldrete, 2014; [dk] González-Soriano & Novelo-Gutiérrez, 2014; [dl] García-Madrigal et al., 2012; [dm] Neiber et al., 2012; [dn] De Troch et al., 2000; [do] Poore, 2012; [dp]Christoffersen& De Assis, 2013; [dq] Escobar-Briones, 2004; [dr] Escobar-Briones, 2002; [ds] Pérez et al., 2014; [dt]Munilla, 2002; [du] Gómez-Aguirre, 2002; [dv] Llorente-Bousquets & Ocegueda, 2008; [dw] Palacios-Vargas 2014; [dx] Palacios-Vargas & García-Gómez, 2014; [dy] Palacios-Vargas & Figueroa, 2014; [dz]Nikolaevna et al., 2014; [ea]Rios-Casanova, 2014; [eb] Smith et al., 2017; [ec]Morón & Valenzuela, 1993.

Table 8.4 Valid Species of Vertebrates Described in the World and in Mexico

Group	World	Mexico			
	Total number of species reported in the literature	Total number of species reported in the literature	Catalogued in Conabio*	Endemic reported in the literature	Endemic cataloged in Conabio*
Vertebrates	**68,724**	**5,757**	**5,709**	**At least 1,322**	**1,026**
Pisces	34,274 [ed]	2,763 [ee]	2,781	At least 289 [ee]	NA
Amphibia	7,621 [ef]	399 [ef]	389	258 [eg-ek]	261
Reptilia	10,450 [el]	908 [el]	887	509 [em-ev]	501
Aves	10,672 [ew]	1,123–1,150 [ex]	1,106	104 [ey]	104
Mammalia	5,707 [ez]	564 [fa]	546	162 [fa]	160

* Complete classification, updated names and corresponding references are available at Taxonomic Catalogs (http://www.biodiversidad.gob.mx/especies/CAT.html). Species names for Mexico can be searched at Enciclovida (http://www.enciclovida.mx/).

NA: not available.

Sources: [ed]Eschmeyer et al., 2017; [ee] Espinosa-Pérez, 2014; [ef]Frost, 2017; [eg] Parra-Olea et al., 2014; [eh] Campbell et al., 2014; [ei] Campbell et al., 2014; [ej] Reyes-Velasco et al., 2015; [ek]Rovito& Parra-Olea, 2015; [el]Uetz et al., 2017; [em]Flores-Villela & García-Vázquez, 2014; [en]Bryson et al., 2014; [eo]Campbell, 2015; [ep] Campbell et al., 2016; [eq] Campillo et al., 2016; [er] Edwards et al., 2016; [es]Grummer & Bryson, 2014; [et]Grünwald et al., 2015; [eu]Köhler et al., 2014a, [ev]Köhler et al., 2014b, [ew] IOC WorldBirdList, 2017; [ex] Navarro-Sigüenza et al., 2014; [ey] Berlanga et al., 2015; [ez] Wilson &Reeder, 2011; [fa] Sánchez-Cordero et al., 2014.

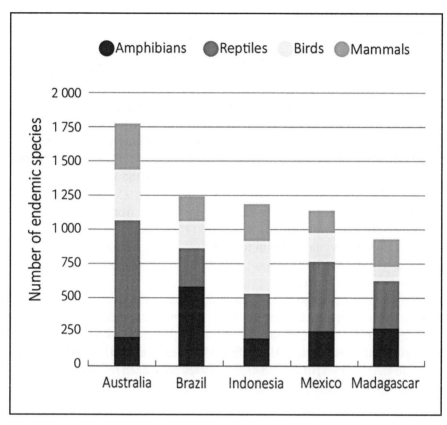

Figure 8.7 The five countries with the greatest diversity of endemic vertebrates (Reprinted from Sarukhán et al., 2017).

Mexico. In freshwater environments, 505 fish species have been recorded in the country—of which 289 are endemic (Espinosa-Pérez, 2014). Within the endemic species are the Goodeids, a group of about 41 species of fish that inhabit exclusively epicontinental waters in the center of the country and that present a peculiar form of reproduction and embryonic nutrition (Domínguez, 2010).

8.4.3 Other Groups

8.4.3.1 Algae

Algae are a complex heterogeneous group which ranges in size from a single cell to giant kelps of over 60 m long (Brodie and Lewis, 2007), as a consequence there are numerous uncertainties regarding the organisms that "fit" into the many species' concepts of algae. In general, algae are considered to be aquatic, oxygen-evolving photosynthetic autotrophs that are unicellular, colonial or constructed of filaments or composed of simple tissues (Guiry, 2012). Algae sometimes are considered within plants (Christenhusz and Byng, 2016); but they are not a monophyletic group and indeed can be referred to three or four kingdoms (depending on the system of classification): Eubacteria, Chromista, Protista, and Plantae. Guiry (2012) estimated conservatively that there are 72,500 algal species in the world, based on species' numbers in phyla and classes included on the online taxonomic database AlgaeBase. Only 43,918 species grouped in 15 phyla have been described. For diatoms, worldwide estimates range from a conservative number of 20,000 (Guiry, 2012) to 200,000 (Mann and Droop, 1996). But this latter estimate would result in four or five diatom species for every algal species as Guiry (2012) points out; also these numbers show the taxonomic uncertainties that still persist.

In Mexico, none of the algal taxonomic studies give an estimation of the total number of algae species (*sensu lato*). Nonetheless, species richness has been reported for specific groups in various regions (Pedroche and Sentíes, 2003; Hernández-Becerril, 2014; Oliva-Martínez et al., 2014; León-Tejera, 2017). In CAT-CONABIO 4,728 algae species have been recorded, which is by no means an exhaustive checklist (Table 8.5).

8.4.3.2 Fungi

With 900 million years of evolutionary history, the Fungi kingdom has an enormous biological diversity; fungi play a key role in the environment as decomposers and through symbiotic relationships with many organisms. They also provide enzymes and active pharmaceutical components important for human health, while some can also have devastating impacts on human health and on agricultural production (Galagan et al., 2005). Up until recently, advances in the taxonomic study of fungi was hindered by problems, such as too few morphological characters, noncorresponding characters among taxa, and convergent morphologies (Blackwell, 2011). Best estimates of fungal diversity came from extrapolations from sites where fungi were well known; based on a 6 to 1 ratio between fungi and plant species, Hawksworth (1991) estimated 1.5 million species of fungi worldwide. Recent advancements using DNA sequencing identified about 10 phyla as members of the monophyletic kingdom, and estimates based on these methods suggest there are between 3.5 and 5.1 million species of fungi

Table 8.5 Valid Species of Algae and Fungi Described in the World and in Mexico (Only Selected Groups are Shown as Examples)

Group Selected groups	World Total number of species reported in the literature	Mexico Total number of species reported in the literature	Catalogued in Conabio*	Endemic reported in the literature	Endemic cataloged in Conabio*
Algae Sensu lato	43,918[fb]	At least 4,728	4,728	At least 95	At least 46
Bacillariophyta	5,000[fc]	NA	1,693	NA	NA
Cyanobacteria	2,698[fd]	159[fe]	348	NA	NA
Chlorophyta	1,300[fc]	342[ff, fg]	459	7th	16
Dinophyta	2,000[fl]	NA	661	NA	ND
Haptophyta	287[fj]	NA	43	NA	ND
Rhodophyta	6,150[fc]	331[ff]	1,112	71[fk]	ND
Ochrophyta	2,094[fc]	251[fl]	274	At least 17th	30
Fungi	**99,426[fm]**	**7,222[fn, fo]**	**4,474**	**NA**	**NA**
Ascomycota	64,163[fm]	2,722[fo]	1,789	NA	NA
Basidiomycota	31,515[fm]	At least 4,500[fn]	2,566	NA	NA
Chytridiomycota, Glomeromycota & Zygomycota	1,946[fm]	At least 104[fp]	120	NA	NA

* Complete classification, updated names and corresponding references are available at Taxonomic Catalogs (http://www.biodiversidad.gob.mx/especies/CAT.html). Species names for Mexico can be searched at Enciclovida (http://www.enciclovida.mx/).

NA: not available.

Source: [fb]Guiry, 2012; [fc]Appeltans et al., 2012; [fd]Nabout et al., 2013; [fe] León-Tejera, *En prep*, [ff] Ortega et al., 2001; [fg]Pedroche et al., 2005; [th] Norris, 2010; [fi]Taylor et al., 2008; [fj] Jordan, 2004; [fk] Norris, 2014; [fl]Pedroche et al., 2008; [fm] Blackwell, 2011; [fn] Aguirre-Acosta et al., 2014; [fo] Herrera-Campos et al., 2014; [fp]Chimal-Sánchez et al., 2016.

(O'Brien et al., 2005; Blackwell, 2011). Only about 99,000 species have been described thus far (Kirk et al., 2008). This accounts for an increase of more than 60,000 described species over 65 years. In Mexico, the fungal diversity estimations are still based on a known ratio between the number of species of different groups (plants or insects) and the fungi. Based on Hawksworth (2001) methods, Guzmán (1998a, b; 2008) estimated more than 200,000 species. Aguirre-Acosta et al. (2014) reported 6,500 known species, and estimated in between 90,000 and 110,000 of macrofungi based on a relationship of macrofungi:plants of 1:2 and 1:5 in temperate and tropical areas, respectively. These authors also considered the number of species calculated by Guzmán (1998a, b) and the general knowledge of fungal species in Mexico and concluded that the study of this group is still incipient and that an urgent review of the recorded species is needed because of the uncertain identifications, mainly with microfungi. García-Sandoval (2015) listed 4,477 species of mainly macrofungi based on a deeper analysis of literature (Table 8.5).

8.5 Genetic Diversity

Genetic diversity is the basis of evolution. It governs species ability to adapt to environmental changes, including those caused by human disturbances. Population genetics studies may reveal important aspects about species needed to implement conservation practices successfully, such as demographic history, population genetic structure, patterns of local adaptation and isolation, by distance or environment. Knowledge of the genetic structure and diversity of populations has also applications for public health, agriculture, livestock production, fisheries, and forestry. For instance, epidemiologic studies can use genetic data to track the spread of a disease; breeding programs can be assisted by molecular markers; and reforestation programs can select trees to grow in a target environment-based on previously known patterns of genetic structure and local adaptation. However, for this to be possible it is necessary to generate considerable genetic data for each species.

In Mexico, genetic diversity studies have focused mostly on the phylogeography of wild species, on describing the genetic diversity levels of crop species and more recently on uncovering their domestication history. Up until 2008, practically all of these studies were performed with few neutral molecular markers, such as chloroplast and mitochondrial DNA sequences, microsatellites and AFLPs (reviewed in Piñero et al., 2008), but since a few years ago it became feasible to obtain genomic data cost-efficiently, even for nonmodel species. As a result, studies with thousands of SNPs are starting to appear for both wild species (e.g., Leaché et al., 2013; Mastretta-Yanes et al., 2014; Zarza et al., 2016) and crops and their wild relatives (e.g., Guerra-García et al., 2017; Romero Navarro et al., 2017; Aguirre-Liguori et al., 2017). Whole genomes and transcriptomes descriptions of Mexican species are almost restricted to crops, of which maize has more studies (e.g., Schnable et al., 2009; Ganal et al., 2011), but recently other species like chili (Kim et al., 2014), common bean (Vlasova et al., 2016; Rendón-Anaya et al., 2017) and avocado (Ibarra-Laclette et al., 2015) are also starting to be analyzed.

Phylogeographic studies have been performed mostly on highlands (>1,800 m), particularly with conifer and cloud forests species (reviewed in Mastretta-Yanes et al., 2015), to a lesser extent in lower elevation rainforests, dry forests and deserts (e.g., Bryson et al., 2011; Rovito et al., 2012; Suárez-Atilano et al., 2014) and considerably less for marine ecosystems (for the Gulf of Mexico reviewed in Neigel, 2009), where studies have focused mostly on wide distributed taxa along America's coasts. On the aggregate, some of the main conclusions of these studies are that: (1) the genetic diversity of Mexican species is highly structured because the main mountain ranges and the Tehuantepec Isthmus act as barriers or isolated habitats; (2) the Pleistocene climate fluctuations have caused species range fragmentation, but have allowed population persistence and thus accumulation of genetic diversity and in many cases considerably deep divergence times among populations. There is an important link between the high species richness of Mexico and the evolutionary processes behind it, this points towards the need of considering—particularly in conservation planning—the geographic areas that in the long-term promote genetic diversity accumulation and population differentiation.

As for crop species, the genetic diversity of agrobiodiversity is the result not only of environmental and biological factors, but of human management and artificial selection, or in other words evolution under domestication (Casas and Parra, 2016). As mentioned above, Mexico is an important center of origin, domestication and diversification of numerous crops, with an even larger number of wild relatives distributed in the country. These crop wild relatives are particularly important for extending the genetic pool of crop species, because they did not suffer the domestication bottleneck and thus harbor larger genetic diversity than their domesticated counterparts. Also, they have adapted to different environmental conditions, coevolved with pests to which they can have resistance, and can show traits that could have a potential for adaptation to new conditions (Maxted et al., 2013).

Research on crop species domesticated in Mexico first focused on describing the levels of genetic variation within the crop or among its agronomic varieties (reviewed in Piñero et al., 2008; Bellon et al., 2009). Results show that the large genetic diversity within Mexican cultivars tends to show a variation gradient, instead of being structured according to agro-morphological groups (e.g., Arteaga et al., 2016). Also, for several species it has been found that the domesticated populations do not necessarily have less genetic diversity than their wild counterpart (e.g., Otero-Arnaiz et al., 2005; Guerra-García et al., 2017), as opposed to what is expected following a domestication bottleneck. This points to more complex domestication processes than what is normally depicted and suggest that traditional management has an important role on shaping crops genetic diversity, even currently. With the onset of new sequencing technologies, uncovering this domestication history has gained resolution and it is allowing to detect the particular geographic region within Mexico where domestication occurred, and patterns of gene flow and hybridization involved in domestication (e.g., Hufford et al., 2012; Guerra-García et al., 2017).

Thanks to the above-mentioned research on wild, crop and crop wild relative species, Mexico is likely the Latin American country with more genetic studies on its native taxa. However, the number of taxa studied is relatively small compared to the magnitude of Mexico's diversity, and efforts have been concentrated on only few groups (around 1% of the known species). The upcoming challenge is not only to generate data for the rest of the species, but also to systematically store and connect genetic data with the rest of biodiversity information. To achieve this, Mexico has started to develop the necessary human and technical capacities, particularly with an emphasis on crop species and their wild relatives (CONABIO, 2016a).

8.6 Knowledge for Conservation Practices

Biodiversity conservation practice requires a solid scientific basis; for instance, to prioritize areas for effective protection of biodiversity, specially endangered species, as mandated by the CBD's Aichi Biodiversity Targets 11 and 12. Nevertheless, huge knowledge gaps still persist, even in the more basic aspects, due to on one hand, the relatively low investment on the scientific apparatus (Martínez et al., 2006), and on the other, to the immense biological diversity present in the country.

The vast majority of known species have not been comprehensively documented, assessed, nor protected. While there is a good level of knowledge for some groups such as mammals and birds, others such as fishes, invertebrates and even many groups of plants are poorly known, and most of the species have not been assessed in order to determine their conservation status and threats. In Mexico, there are two main legal instruments to identify and prioritize conservation needs: The National Red List (SEMARNAT, 2010) and the Priority Species List (SEMARNAT, 2014). The main objectives of these lists, as with any other assessment system, is to develop tailored plans for species conservation—including their habitats—in the most accurate way possible.

The National Red List also provides legal protection status; but so far only about 2.4% (i.e., 2,606) of all described species in Mexico are currently listed, of which only 1,371 (about 1.26% of those described) hold a risk category (i.e., endangered or threatened); there is a significant bias towards vertebrates, and within plants the bias is towards cacti, orchids, and cycads (Figure 8.8). Global data follows similar trends, as only 2.5% of described species have been assessed under the IUCN system, while nearly 43% of all chordate species were assessed by 2009 (Stuart et al., 2010). These figures are extremely low, considering that as much as 16 to 33% of all species of

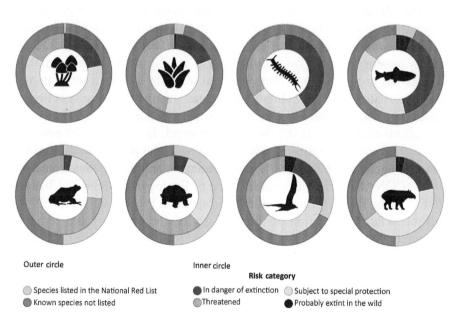

Outer circle

◯ Species listed in the National Red List
⬤ Known species not listed

Inner circle

Risk category

⬤ In danger of extinction ◯ Subject to special protection
◯ Threatened ⬤ Probably extint in the wild

Figure 8.8 Proportion of species included in the Mexican Red List (Reprinted from Sarukhán et al., 2017; icons made by Freepik from www.flaticon.com).

vertebrates are estimated to be globally threatened or endangered, and that threats are greater for tropical regions (Dirzo et al., 2014).

Moreover, only 314 species in the National Red List (SEMARNAT, 2010) have been assessed using a systematic risk assessment method (MER, for its Spanish acronym), representing 22.9% of those that have a risk category and 0.3% of described species. The rest of species in the Red List were listed based on expert criteria but without following a formal or replicable method. Some studies (e.g., Burgman, 2002; Brito et al., 2010; Ramírez and Quintero, 2016) have found that those systems are subjective and can be biased, especially for species with scant or a lack of information. Therefore, national assessments need to be modified to include quantitate thresholds, and use independent criteria to avoid problems with gaps of knowledge. Completing the best possible national assessment is a pressing issue, as species and populations are disappearing faster than we can assess them (as in South Africa, Raimondo et al., 2013). Having a faster and better system to assess species and ensuring a more accurate National Red List will benefit our country's conservation strategies.

Focusing on species remains an important conservation strategy in the country, as reflected by actions that target critically endangered or extirpated species, as well as charismatic, keystone or endemic to the country, like the Mexican wolf, bison, howler and spider monkeys, gray whale, vaquita (*Phocoena sinus*), jaguar (*Panthera onca*), manatee (*Trichechus manatus manatus*), marine turtles (*Lepidochelys kempii, L. olivacea*), California condor (*Gymnogyps californianus*), among others (see examples in Carabias et al., 2010). Efforts directed towards species usually imply the conservation of their habitats, which can eventually bring benefits to the ecosystem or other species maximizing the outcomes of conservation efforts (c.f. Caro, 2010). With this in mind, in 2011, a joint process was carried out to identify priority species with the participation of government agencies—CONABIO, the National Commission on Protected Areas (CONANP), the National Institute of Ecology and Climate Change (INECC), the Federal Attorney for Environmental Protection (PROFEPA), and the Wildlife Direction of the Ministry of Environment and Natural Resources (SEMARNAT)—and more than 100 experts from 37 different academic institutions and nongovernmental organizations across the country. According to Mexico's wildlife law, species or populations can be included in this list if they meet at least one of the following criteria: (a) are strategically important for other species and habitats conservation; (b) are important for maintaining ecosystem structure and function; (c) are endemic and threatened with extinction; (d) are of highly social, cultural, scientific or economic interest. The first National Priority Species List for Conservation was published in March 2014, and includes 372 species and subspecies of plants and animals (SEMARNAT, 2014). This is a first step towards maximizing limited human and financial resources for conservation actions in a megadiverse country.

Another relevant effort towards maximizing resources for conservation and linking primary biodiversity information with the decision-making process began in 2005, as part of the CBD Program of Work on Protected Areas to advance the establishment and maintenance of comprehensive, effectively managed and ecologically representative national and regional systems of PAs. Mexico's conservation assessments coordinated by CONABIO and CONANP compiled more than 4,500 biodiversity information layers (species to ecosystems), from published and assembled maps (e.g., vegetation types, species richness and endemism) to primary biodiversity data (i.e., species occurrence records) from which more than 2,400 vertebrate and plant species distribution models were generated. This information served as a basis to conduct systematic conservation planning (SCP) assessments for terrestrial and freshwater environments (Urquiza-Haas et al., 2008; Koleff et al., 2009; Lira-Noriega et al., 2015) and spatial multicriteria analysis based on the results of the SCP assessments were conducted to pinpoint on one hand, wilderness areas with greatest biological diversity that harbor species that are threatened with extinction, and on the other areas that have lost their conservation value, but are in great need of restoration due their former biological value and to increase ecosystem connectivity and habitat recovery for the most vulnerable species (Urquiza Haas et al., 2012; CONABIO, 2016b; Tobón et al., 2017).

An analysis to determine priority marine areas for conservation was published in 2007 (CONABIO et al., 2007). For this study, major oceanographic processes were determined, and analyzed considering other criteria; geomorphological, geological, biogeographical, physiographic characteristics, as well as conservation objects (biological communities or geographical formations). As a result, 85 priority marine, coastal and island sites were identified.

8.7 Concluding Remarks

Mexico harbors an extraordinary biological and ecological diversity as documented in a significant body of scientific literature and biodiversity assessments generated in the country and abroad, and from the knowledge accumulated by indigenous groups. The magnitude of Mexican biodiversity also means that enormous knowledge gaps still remain. Some of the biggest gaps in biodiversity knowledge relate to large and taxonomically complex groups with few (or no) specialists, including many invertebrates, fungi, algae, microorganisms, and marine life in general; many of which are relevant for human, animal and plant health. There is a global recognition that countries cannot make decisions regarding major environmental problems without the essential support from the best scientific knowledge available (Sarukhan et al., 2012). Biodiversity governance will require the attention to different aspects in the science-policy interface across scales (local, subnational, national, regional, and global), for which knowledge will need to be generated based on questions and themes of different stakeholder interests (Sarukhán et al., 2015; Soberón and Peterson, 2015).

A key aspect in the science domain is the worldwide taxonomic impediment, which represents a big obstacle to increase our biodiversity knowledge. Tackling this issue successfully will need different strategies, including the promotion of the systematic taxonomic research, genetic analysis, the implementation of novel and innovative tools and continued investment in human and institutional capacities (Samper, 2004).

While many gaps still need to be filled, Mexico has developed human capacities, and infrastructure to compile and generate knowledge that should provide the basis for making informed decisions concerning the use of our natural resources (Sarukhán et al., 2010). Mexico's Ecosystem Assessment refers to the need to promote applied research on different aspects of genetic diversity to inform species' conservation and recovery programs, as well as sustainable use and restoration projects and the management of genetic resources in the case of the cultivated plants that originated in Mexico (Sarukán et al., 2015). For instance, conserving the genetic diversity of crop species and their wild relatives, will need a deeper understanding on the effect of management practices, gene flow between wild and domesticated forms, cultural aspects regarding their use, and other biocultural processes, in order to apply different conservation strategies than those normally applied to other wild taxa. For this, both *in situ* (on farm and natural ecosystems) and *ex situ* (gene banks) conservation policies need to be implemented in a complementary way, but historically *ex situ* conservation has been more supported; even though the "campesino" smallholders play an important role at conserving large and diverse crop populations. Also, increasing applied research is needed for those organisms that pose a threat to native species and human health, such as invasive species, plagues, GMOs, and microorganisms.

The generation of knowledge for the sustainable management (uses, conservation and restoration) of the natural capital should be a national priority considering the great opportunity that it represents for the development and generation of benefits for the entire population. Exemplary cases exist, that show it is possible to harmonize the conservation and sustainable management of Mexico's biological diversity, with tangible benefits for the population (Sarukhán et al., 2015; Carabias et al., 2010); like the community forest management strategies that result in the maintenance of its biodiversity by managing the landscape sustainably (Merino Pérez and Martínez Romero, 2014). Other examples refer to the sustainable use of highly valued fisheries, such as the Pacific red lobster (*Panilurus interruptus*), certified since 2004 by the Marine Stewardship Council in Baja California. Mexico, as most megadiverse countries, but especially considering the vast knowledge accumulated in the many cultures that inhabit the country, has a major challenge in documenting and applying scientific and traditional knowledge for maintaining its biodiversity and cultural heritage. Research on community resource management has often proved that traditional and communal management practices could significantly contribute toward a model of sustainable development.

Acknowledgments

To all participants of the Natural capital of Mexico and those who have contributed towards the development of the National Biodiversity Information System.

Keywords

- biological diversity
- conservation
- knowledge
- Mexico

References

Aburto-Oropeza, O., Caso, M., Erisman, B., & Ezcurra, E., (2010). *Bitácora del Mar Profundo: Una Expedición por el Golfo de California,* Instituto Nacional de Ecología, Uc Mexus, Scripps Institution of Oceanography, Mexico.

Aburto-Oropeza, O., Ezcurra, E., Danemann, G., Valdez, V., Murray, J., & Sala, E., (2008). Mangroves in the Gulf of California increase fishery yields. *Proc. Natl. Acad. Sci., 105,* 10456–10459.

Aguirre-Acosta, E., Ulloa, M., Aguilar, S., Cifuentes, J., & Valenzuela, R., (2014). Biodiversidad de hongos en México. *Rev. Mex. Biodivers., 85,* 76–81.

Aguirre-Liguori, J. A., Tenaillon, M. I., Vázquez-Lobo, A., Gaut, B. S., Jaramillo-Correa, J. P., Montes-Hernandez, S., et al., (2017). Connecting genomic patterns of local adaptation and niche suitability in teosintes. *Mol. Ecol.*, *26*, 4226–4240.

Aguirre-Muñoz, A., Croll, D. A., Donlan, C. J., Henry III, R. W., Hermosillo, M. A., Howald, G. R., & Samaniego, -H. A., (2008). High-impact conservation: Invasive mammal eradications from the islands of western Mexico. *AMBIO*, *37*, 101–107.

Arteaga, M. C., Moreno-Letelier, A., Mastretta-Yanes, A., Vázquez-Lobo, A., Breña-Ochoa, A., Moreno-Estrada, A., et al., (2016). Genomic variation in recently collected maize landraces from Mexico. *Genomics Data*, *7*, 38–45.

Aylward, B., Bandyopadhyay, J., & Belausteguigotia, J. C., (2005). Freshwater Ecosystem Services. In: *Millenium Ecosystem Services*, pp. 213–255.

Barbier, E. V., Hacker, S. D., Kennedy, C., Knoch, E. W., Stier, A. C., & Silliman, V. R., (2011). The value of estuarine and costal services. *Ecol. Monogr.*, *81*, 169–193.

Bellon, M. R., Barrientos, -P. A. F., Colunga, -G. P., Perales, H., Agüero, J. A. R., Serna, R. R., & Zizumbo, -V. D., (2009). Diversidad *y* conservación de recursos genéticos en plantas cultivadas. In: *Capital Natural de México*, vol. II, *Estado de Conservación y Tendencias de Cambio*. CONABIO: Mexico. pp. 355–382.

Bellon, M. R., Hodson, D., & Hellin, J., (2011). Assessing the vulnerability of traditional maize seed systems in Mexico to climate change. *Proc. Nat. Acad. Sci. USA.*, *108*, 13432–13437.

Blackwell, M., (2011). The Fungi: 1, 2, 3…5.1 million species? *Amer. J. Bot.*, *98*, 426–438.

Bradshaw, C. J. A., Sodhi, N. S., & Brook, B. W., (2008). Tropical turmoil: A biodiversity tragedy in progress. *Frontiers in Ecology and the Environment*, *7*, 79–87.

Brito, D., Ambal, R. G., Brooks, T., De Silva, N., Foster, M., Hao, W., et al., (2010). How similar are National Red Lists and the IUCN Red List? *Biological Conservation*, *143*, 1154–1158.

Brodie, J., & Lewis, J., (2007). *Unraveling the Algae: The Past, Present, and Future of Algal Systematics*. CRC Press, Boca Raton.

Bryson, R. W., García, -V. U. O., & Riddle, B. R., (2011). Phylogeography of middle American gophersnakes: Mixed responses to biogeographical barriers across the Mexican transition zone. *J. Biogeography*, *38*, 1570–1584.

Burgman, M. A., (2002). Are listed threatened plant species actually at risk? *Austral. J. Bot.*, *50*, 1–13.

CANTIM [Comité Asesor Nacional del Territorio Insular Mexicano], (2012). *Estrategia Nacional Para la Conservación y el Desarrollo Sustentable del Territorio Insular Mexicano*. Secretaría de Medio Ambiente *y* Recursos Naturales, Instituto Nacional de Ecología, Comisión Nacional de Áreas Naturales Protegidas, Secretaría de Gobernación, Secretaría de Marina-Armada de México *y* Grupo de Ecología *y* Conservación de Islas, A. C.: Mexico and Ensenada.

Carabias, J., Sarukhán, J., De la Maza, J., & Galindo, C., (2010). *Patrimonio Natural de México*. Cien casos de éxito. CONABIO, Mexico.

Caro, T., (2010). *Conservation by Proxy: Indicator, Umbrella, Keystone, Flagship and Other Surrogate Species*. Island Press, Washington, D.C.

Casas, A., & Parra, F., (2016). La domesticación como proceso evolutivo. Manejo de biodiversidad *y* evolución dirigida por las culturas del nuevo mundo. In: Casas, A., Torres-Guevara, J., & Parra, F., (eds.). *Domesticación en el Continente Americano.* MéxicoPerú, pp. 133–150.

CAT-CONABIO, (2017). *Catálogos de Autoridades Taxonómicas.* Comisión Nacional para el Conocimiento *y* Uso de la Biodiversidad. www.biodiversidad.gob.mx/especies/CAT. html.

Ceballos, G., & Garcia, A., (1995). Conserving neotropical biodiversity: The role of dry forests in western Mexico. *Conservation Biology, 9,* 1349–1353.

Ceballos, G., (2014). *Mammals of Mexico.* JHU Press, Baltimore, Maryland.

CEC, (1997). *North American Terrestrial Ecoregions.* Commission for Environmental Cooperation: Montréal.

CEC, (2009). *Marine Ecoregions of North America.* Commission for Environmental Cooperation: Montréal.

Cerdeira, E. S., & López, S. G., (2011). A novel Satellite-based ocean monitoring system for Mexico. *Ciencias Marinas, 37,* 237–247.

Cerdeira, E. S., (2016). Construcción de capacidades para el monitoreo de ecosistemas marinos de México, In: *Capital Natural de México,* vol. IV, *Capacidades Humanas e Institucionales.* CONABIO: Mexico, pp. 179–182.

Challenger, A., & Soberón, J., (2008). Los ecosistemas terrestres, In: *Capital Natural de México,* vol. I: *Conocimiento Actual de la Biodiversidad.* CONABIO: Mexico, pp. 87–108.

Challenger, A., (1998). *Utilización y Conservación de los Ecosistemas Terrestres de México: Pasado, Presente y Futuro.* CONABIO-Instituto de Biología, UNAM-Agrupación Sierra Madre: Mexico.

Chapman, A. D., (2009). *Numbers of Living Species in Australia and the World.* Heritage 2nd edn., 84.

Christenhusz, M. J. M., & Byng, J. W., (2016). The number of known plants species in the World and its annual increase. *Phytotaxa., 261,* 201–217.

CONABIO (coord.), (2016b). *Prioridades Para la Conservación y Restauración de la Biodiversidad de México: Resiliencia, Refugios y Conectividad.* Comisión Nacional para el Conocimiento *y* Uso de la Biodiversidad, Mexico.

CONABIO-CONANP-TNC-Pronatura, (2007). *Análisis de Vacíos y Omisiones en Conservación de la Biodiversidad Marina de México: Océanos, Costas e Islas.* Comisión Nacional para el Conocimiento *y* Uso de la Biodiversidad, Comisión Nacional de Áreas Naturales Protegidas, The Nature Conservancy- Programa México, Pronatura, A. C. Mexico.

CONABIO, (1998). *La Diversidad Biológica de México: Estudio de País.* Comisión Nacional para el Conocimiento *y* Uso de la Biodiversidad, Mexico.

CONABIO, (2016a). *Estrategia Nacional sobre Biodiversidad de México y Plan de Acción 2016–2030,* Comisión Nacional para el Conocimiento *y* Uso de la Biodiversidad: México.

CONABIO, (2017). *25 años de evolución.* Comisión Nacional para el Conocimiento *y* Uso de la Biodiversidad, México.

CONAGUA, (2012). *Inventario Nacional de Humedales.* http://www.conagua.gob.mx/Contenido.aspx?n1=4&n2=180&n3=180 (Accessed Aug 28, 2017).

CONAGUA, (2016). *Estadísticas del Agua en México.* Comisión Nacional del Agua, Secretaría de Medio Ambiente *y* Recursos Naturales, Mexico.

Contreras, F., (2005). Humedales costeros Mexicanos. In: Abarca, F. J., & Herzig, M., (eds.). *Manual Para el Manejo y la Conservación de los Humedales en México*. Secretaría de Medio Ambiente, Recursos Naturales y Pesca-U. S. Fish & Wildlife Service-Arizona Game and Fish Department-North American Wetlands Conservation Council: Mexico, pp. 1–25.

Contreras, S., & Lozano, M. L., (1994). Water, endangered fishes, and development perspectives in arid lands of Mexico. *Conservation Biology*, *8*, 379–387.

Contreras-Espinosa, F., & Warner, B. G., (2004). Ecosystem characteristics and management considerations for coastal wetlands in Mexico. *Hydrobiologia*, *511*, 233–245.

Delgadillo-Moya, C., & Juárez-Martínez, C., (2014). Biodiversidad de Anthocerotophyta y Marchantiophyta en México. *Rev. Mex. Biodivers.*, *85*, 106–109.

Delgadillo-Moya, C., (2014). Biodiversidad de Bryophyta (musgos) en México. *Rev. Mex. Biodivers.*, *85*, 100–105.

Dirzo, R., Young, H. S., Galetti, M., Ceballos, G., Isaac, N. J. B., & Collen, B., (2014). Defaunation in the Anthropocene. *Science*, *345*, 401–406.

Domínguez, O., (2010). Conservación de goodeidos, familia en alto riesgo. In: Carabias, J., et al., (eds.). *Patrimonio Natural de México. Cien Casos de Exito*. CONABIO, México, pp. 48–49.

Escobar, E., Maass, M., et al., (2008). Diversidad de procesos funcionales en los ecosistemas. In: *Capital Natural de México,* vol. I. Conocimiento actual de la biodiversidad. CONABIO, México, pp. 161–189.

Escobar, F., Koleff, P., & Rös, M., (2009). Evaluación de capacidades para el conocimiento: el Sistema Nacional de Información sobre Biodiversidad como un estudio de caso, In: CONABIO-PNUD, (ed.). *México: Capacidades Para la Conservación y el uso Sustentable de la Biodiversidad*. Comisión Nacional para el Conocimiento y Uso de la Biodiversidad – Programa de las Naciones Unidas para el Desarrollo: México, pp. 23–49.

Espejel, I., Jiménez-Orocio, O., Castillo-Campos, G., Garcillán, P. P., Álvarez, L., Castillo-Argüero, S, et al., (2017). Flora en playas y dunas costeras de México. *Acta Botánica Mexicana*, *121*, 39–81.

Espinosa, D., Ocegueda, S, Aguila, Z. C., Flores, V. O., & Llorente-Bousquets, J., (2008). El conocimiento biogeográfico de las especies y su regionalización natural. In: *Capital Natural de México,* vol. I. Conocimiento actual de la biodiversidad. Conabio, México, pp. 33–65.

Espinosa-Pérez, H., (2014). Biodiversidad de peces en México. *Rev. Mex. Biodivers.*, *85*, 450–459.

Farjon, A., (1996). Biodiversity of *Pinus* (Pinaceae) in Mexico: Speciation and palaeo-endemism. *Bot. J. Linn. Soc.*, *121*, 365–384.

Forzza, R. C., Baumgratz, J. F. A., Bicudo, C. E. M., Canhos, D. A., Carvalho, Jr, A. A., Coelho, M. A. N., et al., (2012). New Brazilian floristic list highlights conservation challenges. *BioScience*, *62*, 39–45.

Galagan, J. E., Henn, M. R., Ma, L. J., Cuomo, C. A., & Birren, B., (2005). Genomics of the fungal kingdom: Insights into eukaryotic biology. *Genome Research*, *15*, 1620–1631.

Ganal, M. W., Durstewitz, G., Polley, A., Bérard, A., Buckler, E. S., Charcosset, A., et al., (2011). A large maize (*Zea mays* L.) SNP genotyping array: Development and germplasm genotyping, and genetic mapping to compare with the B73 reference genome. *PLoS ONE*, *6*, e28334.

García-Salgado, M., Nava-Martínez, G., Bood, N., Mcfield, M., Molina-Ramírez, A., Yañez-Rivera, B., et al., (2008). Status of coral reefs in the Mesoamerican region. In: Wilkinson, C., (ed.). *Status of Coral Reefs of the World.* Global Coral Reef Monitoring Network and Reef and Rainforest Research Centre, Townsville, Australia, pp. 253–264.

García-Sandoval, R., (2015). *Catálogo de los Hongos de México.* Informe final de estancia Postdoctoral. CONABIO, Mexico.

Gernandt, D. S., & Pérez-de la Rosa, J. A., (2014). Biodiversidad de Pinophyta (coníferas) en México. *Rev. Mex. Biodivers., 85*, 126–133.

Guerra-García, A., Suárez-Atilano, M., Mastretta-Yanes, A., Delgado-Salinas, A., & Piñero, D., (2017). Domestication genomics of the open-pollinated scarlet runner bean (*Phaseolus coccineus* L.). *Frontiers in Plant Science, 8*, 1891.

Guiry, M. D., (2012). How many species of Algae are there? *J. Phycol., 48*, 1057–1063.

Guzmán, G., (1998a). Análisis cualitativo *y* cuantitativo de la diversidad de los hongos en México (Ensayo sobre el inventario fúngico del país). In: Halffter, G., (ed.). *La Diversidad Biológica de Iberoamérica II*, Acta zoológica Mexicana, nueva serie vol. Especial, CYTED e Instituto de Ecología, Xalapa., pp. 111–175.

Guzmán, G., (1998b). Inventorying the fungi of Mexico. *Biodiversity and Conservation, 7*, 369–384.

Guzmán, G., (2008). Diversity and use of traditional Mexican medicinal fungi. A review. *Intern. J. Med. Mushrooms, 10*, 209–2017.

Guzmán, U., Arias S., & Dávila, P., (2003). *Catálogo de Cactáceas Mexicanas.* UNAM-CONABIO: Mexico.

Hawksworth, D. L., (1991). The fungal dimension of biodiversity: Magnitude, significance, and conservation. *Mycological Research, 95*, 641–655.

Hawksworth, D. L., (2001). The magnitude of fungal diversity: The 1.5 million species estimate revisited. Paper presented at the Asian mycologlal congress, (2000). (AMC 2000), incorporating the 2nd Asia-pacific mycological congress on biodiversity and biotechnology, and held at the University of Hong Kong on 9–13 July 2000. *Mycological Research, 105*, 1422–1432.

Hernández-Becerril, D. U., (2014). Biodiversidad de algas planctónicas marinas cyanobacteria, prasinophyceae, euglenophyta, chrysophyceae, dyctyophyceae, eustigmatophyceae, parmophyceae, raphidophyceae, bacillariophyta, cryptophyta, haptophyta, dinoflagellata en México. *Rev. Mex. Biodivers., 85*, 44–53.

Hinchliff, C. E., Smith, S. A., Allman, J. F., Burleigh, J. G., Chaudhary, R., Coghill, L. M., & Gude, K., (2015). Synthesis of phylogeny and taxonomy into a comprehensive tree of life. *Proc. Nat. Acad. Sci., 112*, 12764–12769.

Hufford, M. B., Xu, X., Van Heerwaarden, J., et al., (2012). Comparative population genomics of maize domestication and improvement. *Nature Genetics, 44*, 808–811.

Ibarra-Laclette, E., Méndez-Bravo, A., Pérez-Torres, C. A., Albert, V. A., Mockaitis, K., Kilaru, A., et al., (2015). Deep sequencing of the Mexican avocado transcriptome, an ancient angiosperm with a high content of fatty acids. *BMC Genomics, 16*, 599.

IISE, (2011). *Retro SOS 2000–2009: A Decade of Species Discovery in Review. Tempe, AZ.* International Institute for Species Exploration. Retrieved from http://www.esf.edu/species/documents/sosretro.pdf.

INEGI, (1997). *Diccionario de Datos de Uso de Suelo y Vegetación: Escalas 1:250, 000 y 1: 1,000,000 (alfanumérico).* Dirección General de Geografía, Instituto Nacional de Estadística, Geografía e Informática: Aguascalientes.

INEGI, (2009). *Conjunto de Datos Vectoriales de la Carta de Uso del Suelo y Vegetación, Serie IV (Continuo Nacional), Escala 1: 250000.* Instituto Nacional de Estadística y Geografía, Aguascalientes.

INEGI, (2013). *Conjunto de Datos Vectoriales de la Carta de uso Del Suelo y Vegetación, Serie V (Continuo Nacional), Escala 1: 250000.* Instituto Nacional de Estadística y Geografía, Aguascalientes.

INEGI, (2015). *Catálogo del Territorio Insular Mexicano.* Subgrupo del Catálogo de Islas Nacionales del Grupo Técnico para la Delimitación de las Zonas Marítimas Mexicanas. INEGI, SEGOB, SEMAR, SEMARNAT, SRE, SCT, INECC, CONANP, UNAM: Aguascalientes.

INEGI-CONABIO-INE, (2008). *Ecorregiones Terrestres de México, Escala 1: 1,000,000.* Mexico.

Jablonski, D., Belanger, C. L., Berke, S. K., Huang, S., Krug, A. Z., Roy, K., et al., (2013). Out of the tropics, but how? Fossils, bridge species, and thermal ranges in the dynamics of the marine latitudinal diversity gradient. *Proc. Nat. Acad. Sci., 110,* 10487–10494.

Jiménez, R., Koleff, P., et al., (2016). La informática de la biodiversidad: una herramienta para la toma de decisiones, In: *Capital Natural de México,* vol. IV, *Capacidades humanas e institucionales.* CONABIO: Mexico, pp. 143–195.

Khoury, C. K., Bjorkman, A. D., Dempewolf, H., Ramirez-Villegas, J., Guarino, L., Jarvis, A., et al., (2014). Increasing homogeneity in global food supplies and the implications for food security. *Proc. Nat. Acad. Sci., 111,* 4001–4006.

Kim, S., Park, M., Yeom, S.-I., et al., (2014). Genome sequence of the hot pepper provides insights into the evolution of pungency in *Capsicum* species. *Nat. Genet., 46,* 270–278.

Kirk, P. M., Cannon, P. F., Minter, D. W., & Stalpers, J. A., (2008). Dictionary of the Fungi. *Mycological Research, 106,* 507–508.

Koleff, P., Soberón, J., Arita, H. T., Dávila, P., Flores-Villela, O., et al., (2008). Patrones de diversidad espacial en grupos selectos de especies. In: *Capital Natural de México,* vol. I. Conocimiento actual de la biodiversidad. CONABIO: Mexico, pp. 323–364.

Koleff, P., Tambutti, M., March, I. J., Esquivel, R., Cantú, C., Lira-Noriega, A., et al., (2009). Identificación de prioridades y análisis de vacíos y omisiones en la conservación de la biodiversidad de México. In: *Capital Natural de México,* vol. II. Estado de conservación y tendencias de cambio. CONABIO: México, pp. 651–718.

Lara-Lara, J. R., Arreola Lizárraga, J. A., Calderón Aguilera, L. E., Camacho Ibar, V. F., de la Lanza Espino, G. et al., (2008a). Los ecosistemas costeros, insulares y epicontinentales, In: *Capital natural de México, vol. I: Conocimiento actual de la biodiversidad.* CONABIO: México, pp. 109–134.

Lara-Lara, J. R., Arenas, F. V. A., Bazán, G. C., Díaz, C. V., Escobar, -B. E., et al., (2008b). Los ecosistemas marinos. In: *Capital Natural de México,* vol. I. Conocimiento actual de la biodiversidad. CONABIO: México, pp. 135–159.

Latofski-Robles, M., Méndez-Sánchez, F., Aguirre-Muñoz, A., Jáuregui-García, C., Salizzoni-Chávez, K., & Fernández-Ham, G., (2017). Diagnóstico de especies exóticas invasoras en seis Áreas Naturales Protegidas Insulares, a fin de establecer actividades para su manejo. Reporte de actividades del año 2 (2016) dentro del proyecto GEF 00089333 "Aumentar las capacidades de México para manejar especies exóticas invasoras a través de la implementación de la Estrategia Nacional de Especies Invasoras." *Grupo de Ecología y Conservación de Islas,* A. C., Ensenada, B. C., México.

Leaché, A. D., Harris, R. B., Maliska, M. E., & Linkem, C. W., (2013). Comparative species divergence across eight triplets of spiny lizards (*Sceloporus*) using genomic sequence data. *Genome Biol. Evol.*, *5*, 2410–2419.

León-Tejera, H., (2017). *Catálogo de Autoridades Taxonómicas de Cyanoprocariota Marinos de México.* Informe final Proyecto KT016. CONABIO. Facultad de Ciencias, UNAM.

Lira-Noriega, A., Aguilar, V., Alarcón, J., Kolb, M., Urquiza-Haas, T., González-Ramírez, L., et al., (2015). Conservation planning for freshwater ecosystems in Mexico. *Biol. Conserv.*, *191*, 357–366.

Llorente-Bousquets, J., & Ocegueda, S., (2008). Estado del conocimiento de la biota. In: *Capital Natural de México,* vol. I. Conocimiento actual de la biodiversidad. CONABIO: Mexico. pp. 283–322.

Lucarini, D., Gigante, D., Landucci, F., Panfili, E., & Venanzoni, R., (2015). The an *Archive* taxonomic Checklist for Italian botanical data banking and vegetation analysis: Theoretical basis and advantages. *Plant Biosyst.*, *149*, 958–965.

MA [Millennium Ecosystem Assessment], (2005). *Ecosystems and Human Well-Being: Biodiversity Synthesis.* World Resources Institute, Washington, D.C.

Maine, J. J., & Boyles, J. G., (2015). Bats initiate vital agroecological interactions in corn. *Proc. Nat. Acad. Sci.*, *112*, 12438–12443.

Mann, D. G., & Droop, S. J. M., (1996). Biodiversity, biogeography and conservation of diatoms. In: *Biogeography of Freshwater Algae.* Springer, Netherlands, pp. 19–32.

Martínez, E., & Ramos, C. H., (1989). Lacandoniaceae (Triuridales): Una nueva familia de México. *Ann. Missouri Bot. Gard.*, *76*, 128–135.

Martinez, M. L., Gallego-Fernandez, J. B., Garcia-Franco, J. G., Moctezuma, C., & Jimenez, C. D., (2006). Assessment of coastal dune vulnerability to natural and anthropogenic disturbances along the Gulf of Mexico. *Environmental Conservation*, *33*, 109–117.

Martínez-Meyer, E., Sosa-Escalante, J. E., & Álvarez, F., (2014). El estudio de la biodiversidad en México ¿una ruta con dirección? *Rev. Mex. Biodivers.*, *85*, 1–9.

Mastretta-Yanes, A., Moreno-Letelier, A., Piñero, D., Jorgensen, T. H., & Emerson, B. C., (2015). Biodiversity in the Mexican highlands and the interaction of geology, geography and climate within the Trans-Mexican Volcanic Belt. *J. Biogeogr.*, *42*, 1586–1600.

Mastretta-Yanes, A., Zamudio, S., Jorgensen, T. H., Arrigo, N., Alvarez, N., Piñero, D., & Emerson, B. C., (2014). Gene duplication, population genomics, and species-level differentiation within a tropical mountain shrub. *Genome Biol. Evol.*, *6*, 2611–2624.

Maxted, N., Kell, S., Brehm, J. M., Jackson, M., Ford-Lloyd, B., Parry, M., et al., (2013). *Crop Wild Relatives and Climate Change, Plant Genetic Resources and Climate Change.* CABI Wallingford, UK.

May, R. M., (1988). How many species are there on earth? *Science (Washington)*, *241*, 1441–1449.

Mellink, E., & Riojas-López, M. E., (2017). The demise of a tropical coastal lagoon as breeding habitat for ground-nesting waterbirds: Unintended, but anticipated consequences of development. *Coastal Management*, *45*, 253–269.

Mellink, E., (2002). El límite sur de la región mediterránea de Baja California, con base en sus tetrápodos endémicos. *Acta Zool. Mex.*, *85*, 11–23.

Melo, A., & Alves, M., (2012). The discovery of *Lacandonia* (Triuridaceae) in Brazil. *Phytotaxa.*, *40*, 21–25.

Merino-Pérez, L., & Martínez-Romero, A. E., (2014). *A Vuelo de Pájaro: Las Condiciones de las Comunidades con Bosques Templados en México*. Comisión Nacional para el Conocimiento y Uso de la Biodiversidad: Mexico.

Miller, R. R. M., Norris, W. L., & Miller, S. M. R. R., (2005). *Freshwater Fishes of Mexico*. Museum of Zoology, University of Michigan, Chicago.

Miranda, F., & Hernández-X., E., (1963). Los tipos de vegetación de México y su clasificación. *Bol. Soc. Bot. México.*, *28*, 29–179.

Mitsch, W. J., & Hernandez, M. E., (2013). Landscape and climate change threats to wetlands of North and Central America. *Aquatic Sciences*, *75*, 133–149.

Mittermier, R. A., & Goettsch-Mittermier, C., (1992). La importancia de la diversidad biológica de México. In: Sarukhan, J., & Dirzo, R., (eds.). *Mexico ante los Retos de la Biodiversidad*. Comisión Nacional para el Conocimiento y Uso de la Biodiversidad, Mexico.

Mora, C., Tittensor, D. P., Adl, S., Simpson, A. G. B., & Worm, B., (2011). How many species are there on earth and in the ocean? *PLOS Biology*, *9*, 1–8.

Neigel, J. E., (2009). Population genetic and biogeography of the gulf of Mexico. *Gulf of Mexico: Origins, Waters and Biota*, *1*, 1353–1369.

O'Brien, H. E., Parrent, J. L., Jackson, J. A., Moncalvo, J. M., & Vilgalys, R., (2005). Fungal community analysis by large-scale sequencing of environmental samples. *Applied and Environmental Microbiology*, *71*, 5544–5550.

Oliva-Martínez, M. G., Godínez-Ortega, J. L., & Zuñiga-Ramos, C. A., (2014). Biodiversidad del fitoplancton de aguas continentales en México. *Rev. Mex. Biodivers.*, *85*, 54–61.

Olmsted, I., (1993). Wetlands of Mexico. In: Whigham, D., Dykyjová, D., & Hejny, S., (eds.). *Wetlands of the World: Inventory, Ecology and Management*, vol. 1. Kluwer Academic Publishers: London, UK. pp. 637–677.

Otero-Arnaiz, A., Casas, A., Hamrick, J. L., & Cruse-Sanders, J., (2005). Genetic variation and evolution of *Polaskia chichipe* (Cactaceae) under domestication in the Tehuacán Valley, central Mexico. *Mol. Ecol.*, *14*, 1603–1611.

Parra-Olea, G., Flores-Villela, O., & Mendoza-Almeralla, C., (2014). Biodiversidad de anfibios en México. *Rev. Mex. Biodivers.*, *85*, 460–466.

Pazos, A. B., (2016). Community-Governed multifunctional landscapes and forest conservation in the Sierra Norte of Oaxaca, Mexico. Master of Science in Environmental Studies Dissertation. Florida International University.

Pedroche, F. F., & Sentíes, G. A., (2003). Ficología marina Mexicana. Diversidad y problemática actual. *Hidrobiológica*, *13*, 23–32.

Perales, H. R., & Aguirre, J. R., (2008). Biodiversidad humanizada, In: *Capital Natural de México*, vol. I. Conocimiento actual de la biodiversidad. CONABIO: México, pp. 565–603.

Piñero, D., Caballero-Mellado, J., Cabrera-Toledo, D., Canteros, C. E., Casas, A., et al., (2008). La diversidad genética como instrumento para la conservación y el aprovechamiento de la biodiversidad: estudios en especies Mexicanas. In: *Capital Natural de México*, vol. I. Conocimiento actual de la biodiversidad. CONABIO: Mexico, pp. 437–494.

Raimondo, D. C., Von Staden, L., & Donaldson, J. S., (2013). Lessons from the conservation assessment of the South African Megaflora. *Ann. Missouri Bot. Gard.*, *99*, 221–230.

Ramírez, R. S., & Quintero, E., (2016). Assessing extinction risk for Mexican dry and cloud forest rodents: A case study. *J. Biodiversity & Endangered Species, 4*, 1–7.

Ramsar, Ramsar Sites Information Services. http://archive.ramsar.org/cda/en/ramsar-activities-rsis/main/ramsar/1-63-97_4000_0.

Rendón-Anaya, M., Montero-Vargas, J. M., Saburido-Álvarez, S., Vlasova, A., Capella-Gutierrez, S., Ordaz-Ortiz, J. J., et al., (2017). Genomic history of the origin and domestication of common bean unveils its closest sister species. *Genome Biology, 18*, p. 60.

Romero Navarro. J. A., Willcox, M., Burgueño, J., Romay, C., Swarts, K., Trachsel, S., et al., (2017). A study of allelic diversity underlying flowering-time adaptation in maize landraces. *Nat. Genet., 49*, 476–480.

Rovito, S. M., Parra-Olea, G., Vásquez-Almazán, C. R., Luna-Reyes, R., & Wake, D. B., (2012). Deep divergences and extensive phylogeographic structure in a clade of lowland tropical salamanders. *BMC Evol. Biol., 12*, 255.

Rzedowski, J., (1978). *Vegetación de México*. Limusa: México.

Rzedowski, J., (1991). Diversidad *y* orígenes de la flora fanerogámica de México. *Acta. Bot. Mex., 14*, 3–21.

Sala, O. E., Chapin III, F. S., Armeso, J. J., Berlow, E., Bloomfield, J., et al., (2000). Global biodiversity scenarios for the year 2100. *Science, 287*, 1770–1774.

Samaniego, -H. A., Aguirre-Muñoz, G. R., Howald, M., Félix-Lizárraga, J., Valdez-Villavicencio, R., Tershy, B. R., et al., (2009). Eradication of black rats from Farallón de san Ignacio and San Pedro Mártir Islands, Gulf of California, México, Arcata, *Proceedings of the 7th California Islands Symposium,* Institute for Wildlife Studies: Arcata, CA.

Samper, C., (2004). Taxonomy and environmental policy. *Proc. Royal Soc., 359*, 721–728.

Sánchez-Cordero, V., Botello, F., Flores-Martínez, J. J., Gómez-Rodríguez, R. A., Guevara, L., et al., (2014). Biodiversidad de Chordata (Mammalia) en México. *Rev. Mex. Biodivers., 85*, 496–504.

Sarukhán, J., Carabias, J., Koleff, P., & Uriquiza-Haas, T., (2012). *Capital Natural de México: Acciones Estratégicas Para su Valoración, Preservación y Recuperación.* CONABIO: Mexico.

Sarukhán, J., García Méndez, G., et al., (2016). La formación de recursos humanos vinculada al manejo de la biodiversidad: aciertos *y* limitaciones, In: *Capital Natural de México,* vol. iv. Capacidades humanas e institucionales. CONABIO: Mexico, pp. 371–418.

Sarukhán, J., Koleff, P., Carabias, J., Soberón, J., Dirzo, R., et al., (2017). *Capital Natural de México.* Síntesis: Evaluación del conocimiento *y* tendencias de cambio, perspectivas de sustentabilidad, capacidades humanas e institucionales. Comisión Nacional para el Conocimiento *y* Uso de la Biodiversidad: Mexico.

Sarukhán, J., Koleff, P., Carabias, J., Soberón, J., Dirzo, R., Llorente-Bousquets, J., et al., (2010). *Natural Capital of Mexico. Synopsis: Current Knowledge, Evaluation, and Prospects for Sustainability.* CONABIO: Mexico.

Sarukhán, J., Urquiza-Haas, T., Koleff, P., Carabias, J., Dirzo, R., et al., (2015). Strategic actions to value, conserve, and restore the natural capital of megadiversity countries: The case of Mexico. *BioScience, 65*, 164–173.

Schnable, P. S., Ware, D., Fulton, R. S., et al., (2009). The B73 maize genome: Complexity, diversity, and dynamics. *Science, 326*, 1112–1115.

SEMARNAT, (2010). Norma Oficial Mexicana NOM-059-Semarnat-2010, Protección ambiental – Especies nativas de México de flora *y* fauna silvestre, categorías de riesgo *y* especificaciones para su inclusión, exclusión o cambio, lista de especies en riesgo. *Diario Oficial de la Federación*, 30 de diciembre de 2010, Mexico.

SEMARNAT, (2014). Acuerdo por el que se da a conocer la lista de especies y poblaciones prioritarias para la conservación. *Diario Oficial de la Federación*, 5 de marzo de 2014.

Soberón, J., & Peterson, A. T., (2015). Biodiversity governance: A tower of Babel of scales and cultures. *PLoS Biol.*, *13*, e1002108.

Souza, V., Espinosa-Asuar, L., Escalante, A. E., Eguiarte, L. E., Farmer, J., Forney, L., et al., (2006). An endangered oasis of aquatic microbial biodiversity in the Chihuahuan desert. *Proc. Nat. Acad. Sci.*, *103*, 6565–6570.

Stuart, S. N., Wilson, E. O., McNeely, J. A., Mittermeier, R. A., & Rodríguez, J. P., (2010). The Barometer of Life. *Science*, *328*, 177.

Suárez-Atilano, M., Burbrink, F., & Vázquez-Domínguez, E., (2014). Phylogeographical structure within Boa constrictor imperator across the lowlands and mountains of Central America and Mexico. *J. Biogeography*, *41*, 2371–2384.

Tittensor, D. P., Mora, C., Jetz, W., Lotze, H. K., Ricard, D., Berghe, E. V., & Worm, B., (2010). Global patterns and predictors of marine biodiversity across taxa. *Nature*, *466*, 1098–1101.

Tobón, W., Urquiza-Haas, T., Koleff, P., Schröter, M., Ortega-Álvarez, R., Campo, J., et al., (2017). Restoration planning to guide Aichi targets in a megadiverse country. *Conserv. Biol.*, *31*, 1086–1097.

Troche-Souza, C., Rodríguez-Zúñiga, M. T., Velázquez-Salazar, S., Valderrama, L. L., Villeda-Chávez, E., Alcántara-Maya, A., et al., (2016). *Manglares de México: Extensión, Distribución y Monitoreo (1970/1980–2015)*. Comisión Nacional para el Conocimiento y Uso de la Biodiversidad: Mexico.

Troudet, J., Grandcolas, P., Blin, A., Vignes-Lebbe, R., & Legendre, F., (2017). Taxonomic bias in biodiversity data and societal preferences. *Sci. Rep.*, *7*, 9132.

Underwood, E. C., Viers, J. H., Klausmeyer, K. R., Cox, R. L., & Shaw, M. R., (2009). Threats and biodiversity in the Mediterranean biome. *Divers. Distrib.*, *15*, 188–197.

UNEP (United Nations Environment Program), (2006). *Ecosystems and Biodiversity in Deep Waters and High Seas*. UNEP Regional Seas Reports and Studies No. 178. UNEP/IUCN, Switzerland.

Urquiza-Haas, T., Cantú, C., Koleff, P., & Tobón, W., (2011). Caracterizacion de las ecorregiones terrestres: diversidad biológica, amenazas y conservación. In: Koleff, P., & Urquiza-Haas, T., (coords.). *Planeación para la Conservación de la Biodiversidad Terrestre en México: Retos en un País Megadiverso*. Comisión Nacional para el Conocimiento y Uso de la Biodiversidad–Comisión Nacional de Áreas Naturales Protegidas: Mexico.

Urquiza-Haas, T., Kolb, M., Koleff, P., Lira-Noriega, A., & Alarcón, J., (2008). Methodological approach to identify Mexico's terrestrial priority sites for conservation. *Gap Bulletin*, *16*, http://pubs.usgs.gov/gap/gap16/pdf/gap16. pdf.

Urquiza-Haas, T., Tobón, W., Kolb, M., Cuevas, M. L., & Koleff, P., (2012). *Anexo 2. Zonas de Atención Prioritaria, en Conabio*. Desarrollo territorial sustentable: Programa especial de gestión en zonas de alta biodiversidad. Comisión Nacional para el Conocimiento y Uso de la Biodiversidad, Mexico, pp. 35–39.

Valencia-Ávalos, S., (2004). Diversidad del género *Quercus* (*Fagaceae*) en México. *Bol. Soc. Bot. Méx.*, *75*, 33–53.

Villaseñor, J.L. (2004). Los géneros de plantas vasculares de la flora de México. *Bol. Soc. Bot. Méx.*, 75, 105-135.

Villaseñor, J. L., (2016). Checklist of the native vascular plants of Mexico. *Rev. Mex. Biodivers.*, *87*, 559–902.

Vlasova, A., Capella-Gutiérrez, S., Rendón-Anaya, M., et al., (2016). Genome and transcriptome analysis of the Mesoamerican common bean and the role of gene duplications in establishing tissue and temporal specialization of genes. *Genome Biology*, *17*, 32.

Wheeler, Q. D., Knapp, S., Stevenson, D. W., Stevenson, J., Blum, S. D., Boom, B. M., Borisy, G. G., et al., (2012). Mapping the Biosphere: Exploring species to understand the origin, organization and sustainability of Biodiversity. *Syst. Biodivers.*, *10*, 1–20.

Williams-Linera, G., Manson, R. H., & Isunza, V. E., (2002). La fragmentación del bosque mesófilo de montaña *y* patrones de uso del suelo en la región oeste de Xalapa, Veracruz, México. *Madera y Bosques*, *8*, 73–89.

Willis, K. J., & Bachman, S., (2016). *State of the World's Plants*, *84*. https://stateoftheworldsplants.com/report/sotwp_2016. pdf.

Wilson, E. O., (2000). A global biodiversity map. *Science*, *29*, 2279.

WWF-IUCN-WCPA (World Wildlife Fund, International Union for Conservation of Nature and Natural Resources), (2001). *The Status of Natural Resources on the High-Seas*. WWF/IUCN, Gland, Switzerland.

Zarza, E., Faircloth, B. C., Tsai, W. L. E., Bryson, R. W., Klicka, J., & McCormack, J. E., (2016). Hidden histories of gene flow in highland birds revealed with genomic markers. *Mol. Ecol.*, *25*, 5144–5157.

Zhang, Z. Q., (2011). Animal biodiversity: An introduction to higher-level classification and taxonomic richness. *Zootaxa*, *12*, 7–12.

Zhang, Z. Q., (2013). Animal biodiversity: An outline of higher-level classification and survey of taxonomic richness (Addenda 2013). *Zootaxa*, *3703*, 1–82.

CHAPTER 9

Biodiversity in Paraguay

PAMELA MARCHI,[1] FREDERICK BAUER,[2] PIER CACCIALI,[3] ALBERTO YANOSKY,[4] MARCELO DUJAK,[5] CHRISTIAN DUJAK,[6] MICHELLE CAMPI,[7] SIGRID DRECHSEL,[8] and BERNARDO CAÑIZA[9]

[1]Asociación Etnobotánica Paraguaya (AEPY), Ecuador 450 c/Bruno Guggiari. 2420, Lambaré-Paraguay, E-mail: pamepy@gmail.com

[2] Museo Nacional de Historia Natural del Paraguay, km 10.5, Ruta Mcal. José F. Estigarribia, 2160, San Lorenzo-Paraguay/Facultad de Ciencias Exactas y Naturales (FACEN), Departamento de Biología, Área Zoología, Campus UNA, San Lorenzo, Paraguay

[3]Instituto de Investigación Biológica del Paraguay, Del Escudo 1607, 1425, Asunción, Paraguay

[4]Director of Guyra Paraguay/Researcher CONACYT-PRONII, Biocentro Guyra Paraguay, Asunción, Paraguay.

[5]Categorized researcher CONACYT-PRONII, Asunción, Paraguay. E-mail: marcelodujak@gmail.com

[6]Centre for Research in Agricultural Genomics (CRAG) CSIC-IRTA-UAB-UB, Campus UAB, Bellaterra, Barcelona, Spain

[7]Universidad Nacional de Asunción, Facultad de Ciencias Exactas y Naturales. Laboratorio de Análisis de Recursos Vegetales, área Micología

[8]Asociación Etnobotánica Paraguaya (AEPY), Ecuador 450 c/Bruno Guggiari. 2420, Lambaré-Paraguay

[9]Laboratorio de Análisis de Recursos Vegetales FaCEN-UNA, 1039, San Lorenzo-Paraguay

9.1 Introduction

Paraguay is a country located in a south-central position within South America, and it limits with Bolivia at the north, Brazil at the east, and Argentina at the western and southern borders. Paraguay is divided into two halves by the homonymous river, leaving two regions: Occidental Region (also known as "Chaco") with a surface of 246,925 km²; and the Oriental Region, a little smaller, has 159.827 km² (Figure 9.1). Each region has a specific climate, geology, soil, and vegetation traits, and the differences between both sides increase with the distance from west to east (Mereles, 2007).

The average annual temperature varies between 19°C and 26°C, with the coolest area located in the south of the country. The annual precipitation ranges from an average of 400 mm in the westernmost part of the country, to 1900 mm in the east where the moistest forests are located. The rainfall has a marked seasonality, and the humid season corresponds to the beginning of the warm season, starting in October or November, and in summer the temperature can reach over 40°C in many areas of the country (Gauto and Stauffer, 2017).

Figure 9.1   Geography of Paraguay showing the inner political divisions. The capital (Asunción) is highlighted in red.

Ecologically, most of the authors follow Dinerstein et al. (1995) recognizing five ecoregions: Dry Chaco, Humid Chaco, Pantanal, Cerrado, and Alto Paraná Atlantic Forest (known in Paraguay as "BAAPA" by its acronym in Spanish) (Figure 9.2). Gauto and Stauffer (2017) state that there is an additional ecoregion located in the southernmost part of Paraguay referred as Southern Mesopotamian Grasslands, which is a natural grassland environment. The limits for this ecoregion are not well defined and it is considered a transition zone with ecological characteristics of different ecoregions (Batrina, 2007).

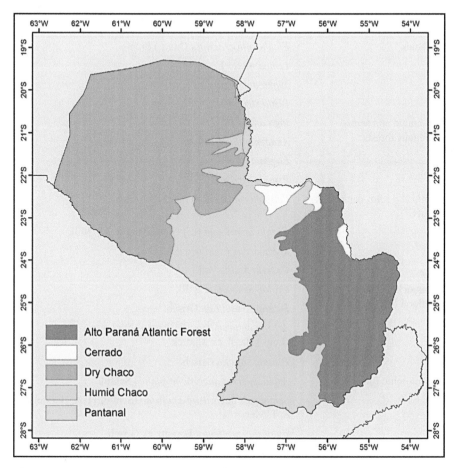

Figure 9.2 Ecoregional arrangement of Paraguay, according to Dinerstein et al. (1995).

9.2 Vegetation

The main types of natural vegetation currently observable in the country: wetlands, cerrado, savannas and forests. Mereles (2007) classifies the forests in four types: (i) subhumid and semi-deciduous forest; (ii) humid and semi-deciduous forest; (iii) xeromorphic forest; and (iv) floodable and hydrophillic riparian forest (Pin et al., 2009). Natural vegetation has been affected in recent years, being destroyed due to intense anthropic activities related to monocultures and farming. Some species that can be found in the different plant formations are mentioned in Table 9.1.

Table 9.1   Species Found in the Different Plant Formations

Wetlands	*Eichhornia crassipes* (Mart.) Solms
	Lemna minor L.
	Typha domingensis Pers.
	Pistia stratiotes L.
Sub-humid and semi-deciduous forests	*Inga affinis* DC.
	Anadenanthera columbrina (Vell.) Brenan
	Eugenia uniflora L.
	Pisonia aculeata L.
Humid and semi-deciduous forests	*Peltophorum dubium* (Spreng.) Taub.
	Cordia americana (L.) Gottschling & J.S. Mill.
	Myrocarpus frondosus Allemão.
	Cedrela fissilis Vell.
Riparian hydrophilic forests and flood forests	*Croton urucurana* Baill.
	Prosopis ruscifolia Griseb.
	Astronium urundeuva (Allemão) Engl. var. *candollei* (Engl.) Hassl. ex Mattick
	Pisonia zapallo Griseb.
Xeromorphic forests	*Aspidosperma quebracho-blanco* Schltdl.
	Schinopsis quebracho-colorado (Schltdl.) F.A. Barkley & T. Mey.
	Bulnesia sarmientoi Lorentz ex Griseb.
	Pereskia nemorosa Rojas Acosta
Cerrados	*Annona dioica* A. St.-Hil.
	Butia paraguayensis (Barb. Rodr.) L.H. Bailey
	Croton solanaceus (Müll. Arg.) G.L. Webster
	Stevia entreriensis Hieron.
Savannas	*Stevia rebaudiana* (Bertoni) Bertoni
	Cyclolepis genistoides Gillies ex D. Don
	Acrocomia aculeata (Jacq.) Lodd. ex Mart.
	Jatropha isabelliae Müll. Arg.

Climate regulates the types of natural plant formations in the two regions at each side of the Paraguay river, and the soil characteristics determine what species are dominant, frequent, and abundant for each plant formation (Mereles, 2007).

The country is known worldwide for the great number of medicinal plants used by its people, as 90% of the population ingests medicinal and aromatic plants daily in typical drinks such as *mate* (hot) or *tereré* (cold) (Pin et al., 2009).

9.2.1 Vascular Plants

Paraguay does not have an exact number of flora species because vegetation studies are still ongoing; therefore, quantitative records are approximations. Mereles (2007) presents around 6,500 to 7,000 species in vascular flora; this includes Dicotyledons, Monocotyledons, and Pteridophytes. In a list of plants collected by Vogt in the Western or Chaco Region (2011, 2012, 2013), 1651 species have been mentioned for this region to date.

9.2.2 Lichens

The records for lichens and lichenized fungi in the country start towards the end of the nineteenth century, recorded by different botanists, mycologists, and lichenologists, such as A. Muller, G. Malme, C. Spegazzini, L. Ferraro, A. Schinini, and E. Bordas. Paraguay has published records of 91 species distributed in 20 families, most of which were collected in the Eastern Region of Paraguay (Ramella and Perret, 1990). The highest number of species identified to date is found in the Parmeliaceae, Lecideaceae, Collemaceae, Physciaceae, Pertusariaceae and Usneaceae families (http://www.pybio.org/).

9.2.3 Bryophytes

Until now a total of 257 moss species, 80 liverwort species, and 3 hornwort species are recorded for Paraguay. The largest moss families recorded for Paraguay so far, are Fissidentaceae (25 species), Pilothricaceae (23), Pottiaceae (20), Dicranaceae (18), Sematophyllaceae (16), Bryaceae (15), and Hypnaceae (14). The Pottiaceae is the most diverse family with 16 genera and 20 species. 43 taxa are apparently endemic to Paraguay. Regarding the leafy liverworts, the Lejeuneaceae with 13 genera and 31 species is the largest and most diverse family recorded for Paraguay, whereas the most representative thalloid liverwort family is the Ricciaceae (1 genus and 10 species). Only tree hornwort species are known from Paraguay: *Phaeoceros laevis*, *Phaeoceros tenuis*, and *Phytomatoceros bulbiculosus*. Due to the few bryological studies in Paraguay, the bryophyte flora from this country is still underestimated. As long as more sampling across the country is carried out, these numbers will increase (O'Shea and Price, 2008; Söderström et al., 2013; Cañiza et al., 2017).

9.2.4 Fungi

The first records of fungi from Paraguay were made by the mycologist Carlos Spegazzini in the year 1883 when he first published "Fungi Guaranitici. Pugillus I" in the Anales de la Sociedad Científica Argentina, where he described 15 species from the genus *Agaricus* L. (Spegazzini, 1883). During his life, he published another 16 catalogs of Fungi from Paraguay and he was the first mycologist to promote the study of fungi in the country. After a considerable span, mycology in Paraguay was abandoned until the year 1998 when two mycologists from Argentina, Orlando Popoff, and Jorge Wright described a preliminary check-list of wood-inhabiting polypores from the country, recording 88 species (including 12 new records) (Popoff and Wright, 1998) and later Popoff (2000) presented an unpublished thesis work that included some specimens of corticioid and polyporoid fungi from Paraguay.

Recently the national university (Universidad Nacional de Asunción) established a working group focused on fungi research. Since 2013 the group is working to improve the previously scattered knowledge of the Paraguayan fungi, adding new records for the country: *Leucoagaricus lilaceus, Marasmius haematocephalus, Marasmius ferrugineus*, and *Geastrum violaceum* (Flecha et al., 2013; De Madrignac et al., 2013; Campi et al., 2013). This group also was in charge of the publication of the first book of Paraguayan fungi, from "Laguna Blanca," an area located in San Pedro Department (Estern Paraguay).

Regarding the diversity of the taxa studied, most (if not all of them) belong only to the Basidiomycota phylum and include gasteroid species and polyporoid to agaric taxa. The gasteroid fungi recorded include *Cyathus* species (Campi and Maubet, 2015a; Maubet et al., 2017), *Geastrum, Pisolithus, Scleroderma* (Campi et al., 2015; Campi and Maubet, 2015b); *Vascellum, Calvatia* (Campi and Maubet, 2016), *Tulostoma* (Campi et al., 2016a), *Battarrea* (Campi et al., 2017a), *Clathrus, Phallus* and *Itajahya* (Campi et al., 2017b). Regarding the agaricoid species, in the last four years, only *Suillus granulatus* (Campi et al., 2015), *Coprinus comatus* (Campi et al., 2016b), *Xeromphalina tenuipes, Filoboletus gracilis* (Campi et al., 2017c) and *Laccaria fraterna* (Campi et al., 2017d) were recorded for the first time for the country. The polyporoid species recorded for the country are *Donkia pulcherrima* (Campi et al., 2017e), a rare hydnoid species; along with a new species: *Amylosporus guaraniticus*, found on a sewer in 2015 in the *Universidad Nacional de Asunción*. After two years, finally, the new species was named after the Guarani folk who inhabited the center and south of South America being the first new species with molecular data for the country.

The diversity of species from Paraguay recorded still remains low considering that only four years of study were carried out in the last years and that most species are recorded from the Central department which is a very populated area. According to a current work carried out by Hawksworth and Lücking (2017) there has to be at least 3.2 to 5 million species of fungi and considering the remarkable low number of species recorded from Paraguay and the advances in molecular data in the last decades this number should increase dramatically in the next few years.

9.2.5 Plants Cultivated in Paraguay

In Paraguay, plant species of different families are cultivated. In the extensive production, five species are the most important nationally and globally. Soybean, *Glycine max* (L.) Merr is one of them, and it is the most important due to the economic income it generates in the country. In the harvest 2016/17, a national sowing area of 3,388,790 ha was estimated, and the harvest reached the 4th position worldwide in export and 6th in greater production (CAPECO, 2017). The varieties most used in agricultural production derive from improved seeds from neighboring countries that were adapted to our agricultural area. Currently, there also exist own and improved varieties "La Milagrosa," "SOJAPAR R24 and R19" (IPTA, 2017). Maize, *Zea mays* L., with a sowing area of 1,009,226 ha, has a diverse germplasm pool. and 10 native maize races with a diversity of phenotypes are known, the most popular are: "*Avati moroti*," "*Avati ti*," "*Avati guapy*," "*Avati mita*," "*Sape moroti*," "*Sape pyta*," "*Tupi moroti*," "*Tupi pyta*," "*Pichinga aristado*" and "*Pichinga Redondo*" (Salhuana and Machado, 1999). The bread wheat, *Triticum aestivum* subsp. *aestivum* L. is a species sown on a surface of 493,924 ha. In the Eastern Region of the country, the harvest is mostly exported to neighboring countries and other regions. Among the most used varieties there are "*Itapúa 70*" and "*Itapúa 75*." Other species cultivated in a smaller and extensive production are sunflower, *Helianthus annuus* L., with 30,000 ha. and canola, *Brassica napus* L. with 25,000 ha, both sowed in the southern part of the country (CAPECO, 2017; MAG, 2008).

Yerba mate, *Ilex paraguayensis* A.St.-Hil., an autochthonous species of economical, cultural and social interest, is cultivated in the Eastern Region in the departments of Guairá, Itapúa, Alto Paraná and Caazapá, in an area of 18,750 ha (MAG, 2013). For centuries, indigenous natives Guaranies have traditionally used this specie and this tradition was transmitted to the Paraguayans. The leaves of this specie have antioxidant and medicinal properties among others (Bracesco et al., 2011; Heck and De Mejia, 2007), and they are consumed in the country as a cold or hot drink known as "Tereré and Mate." The "*Ka'a he'ẽ*," *Stevia rebaudiana* Bertoni, is another autochthonous species such as yerba mate. In the Guarani language "*Ka'a he'ẽ*" means "sweet leaf or sweet grass," this species has sweetening properties stevioside and rebaudioside-like. It is cultivated in Itapúa, Alto Paraná, Amambay, San Pedro and Caaguazú departaments, and it has 3 varieties "*Eirete*," "*La Criolla*" and "AKH-L1" (Jimenez et al., 2010).

In the small rural production sector, there are cultivated species with local landraces derived from species introduced to the country for centuries. Among them there are fruit species such as: orange, *Citrus x sinesis* L., Grapefruit, *Citrus x paradisi* Macfad., Mandarin, *Citrus reticulata* Blanco and lemon, *Citrus x limon* L. There are also native species such as peanuts "*manduvi*" in guarani, *Arachis hypogaea* L., and their varieties are known as "*Manduví guaicurú*," "*Manduví pyta*," "*Manduví huí*" and "*Manduví pyta'ĩ*." The cassava or "*Mandió*," *Manihot esculenta* Crantz. is sown by stakes (vegetative reproduction) and it presents the varieties: "*Sa'y jú y*," "*Pomberí*," "*Mandió hovy*." Sweet potato or "*Jety*," *Ipomoea batatas* Lam. presents the varieties, "*Jety morotî*" or "*Jety avá*" of white pulp, "*Jety sa y jú*" with pale yellow skin and "*Jety pytá guazú*" with pink skin. And finally there are the "*Anda'ĩ*" or pumpkin, *Cucurbita* sp., and the bean or "*Kumanda*," *Phaseolus* sp. (MAG, 2008).

9.3 Biocultural Diversity

The diversity of life is not limited to the diversity of species and ecosystems of the planet, but also includes the cultural and linguistic diversity of human societies (Maffi, 2010). Biocultural diversity is a three-dimensional network constituted by biological, cultural and linguistic diversity, in which a society is related to it (Maffi, 2010).

The Guaraní are ancestral inhabitants of the current Paraguayan territory and part of the territories of bordering countries such as Bolivia, Argentina, Brazil (Azevedo *et al.*, 2009). Paraguay, in addition to containing the

Guaraní peoples belonging to the Guaraní linguistic family, includes four others: Zamuco, Mataco Mataguayo, Maskoy and Guaicurú (Zanardini and Biedermann, 2006). In total, these five linguistic families add up to a total of 19 to 20 indigenous peoples, according to the authors (Zanardini and Bierdermann, 2006; Melià, 2010; DGEEC 2014).

Our country notoriously expresses its indigenous cultural influence, especially the influence of the Guaraní culture. The Paraguayan is mostly the result of the mixture between Guaraní and European Indians who arrived from the colonization period in the sixteenth century. Paraguay is officially a bilingual country in which the native language of Guaraní and Castilian or Spanish is spoken and a third "hybrid," the Guaraní-Paraguayan dialect (Melià, 2012). There are other expressions or cultural manifestations that occur in material aspects such as: (i) the use of yerba mate, *Ilex paraguariensis* A. St.-Hil., native species present in the subtropical forests of Paraguay, and whose processed leaves are used in traditional drinks such as mate and tereré; and (ii) the extensive knowledge about autochthonous medicinal plants, both are examples that show a clear connection between biological diversity, language and culture.

The Guaraní, in particular, left an important legacy about knowledge and knowledge about nature, plants, animals, fungi and other elements of the environment, these elements only fulfill their material functions, but in turn are woven in the cosmos of sociocultural significations and representations of the Guaraní universe (Cadogan, 1959, 1968, 1973).

Numerous explorers already at the time of the colony and later other researchers, mainly focused on material aspects such as medicine, botany, zoology, geography, linguistics, etc., worked on documenting indigenous traditional knowledge about the natural sciences. To cite only some of the numerous records, is the work "Enciclopedia Guaraní-Castellano de Ciencias Naturales *y* Conocimientos Paraguayos" (posthumous work published in 1985); Dr. Carlos Gatti compiled approximately 10,000 Guaraní and guaranized terms related to the natural sciences and Paraguayan knowledge. This is probably the most extensive work that the Guaraní lexicon contains about nature. For this great research Gatti was based on historical writings of explorers, naturalists and missionaries such as: Father Antonio Ruiz de Montoya (Tesoro de la Lengua Guaraní, 1639), Father Pedro de Montenegro (Materia Médica Misionera, 1710), Félix de Azara (Apuntamientos para la Historia Natural de los cuadrúpedos *y* aves del Paraguay *y* Rio de la Plata, 1802–1805), among others. It is also worth mentioning other reference authors such as José Sánchez

Labrador, Arnoldo de Winkelried Bertoni, León Cadogan, Dionisio M. González Torres, and undoubtedly many others who contributed to the art in question.

No doubt this diversity in the lexicon is a reflection of the knowledge, knowledge and management of the elements (material and symbolic) that make up the diversity present in the ecosystems they inhabit. The naturalist scientist and scholar of the Guaraní culture, the Swiss Moisés Bertoni, affirmed that the Guaraní, as a result of their knowledge and manipulation of nature, handled the taxonomic principles of gender and family, before western science applied this system, in the 1700'. In his works, "La civilización Guaraní, parte III: Conocimientos, capitulo XVI" (published first in 1927) and his posthumous work the botanical dictionary "Introducción a las plantas usuales y útiles del Paraguay" (posthumous work published in the beginning in 1940), he argues his hypothesis because he had recognized that the Guaraní grouped the species into 40 families and 1,100 Guaraní botanical genera (Bertoni, 2004, 2012).

The Guaraní are very knowledgeable about the medicinal properties of native plants, thanks to this legacy Paraguay has developed a strong tradition in the practice of folk medicine, which was also enriched with knowledge about other species brought by Europeans from the time of the colony, sixteenth century (Bertoni, 2004). One of the most complete and current contemporary publications corresponds to the book of "Plantas Medicinales del Jardín Botánico de Asunción" (Pin et al., 2009), this presents a sample of 309 medicinal species of traditional use in Paraguay, of which 209 species are native and 100 exotic species, all of them present in the ethnobotanical nursery of the Botanical and Zoological Garden of Asunción (JBZA). 90% of the Paraguayan population consumes medicinal and aromatic plants in the tereré or mate (Pin et al., 2009), in the markets and streets of Paraguay it is very common to find stands selling medicinal plants, "*the yuyera*" (selling herbs), is a fundamental part of this social representation, it is she who knows the properties of the plants and recommends the uses of the species according to the affection that afflicts the client. It also provides the "*pohâ ro'ysâ,*" plants that have the property of refreshing the body in the intense and usual hot days, and that are applied to the water for the tereré, Paraguay's traditional drink made from dried, parched and crushed leaves of *Ilex paraguariensis* A. St.-Hil. According to Didier Roguet in Pin et al. (2009) registered 510 species of medicinal plants that are commercialized in the markets of Asunción, capital of Paraguay. These data position Paraguay first because it is the site with the greatest diversity of medicinal plants found in a single location.

9.4 Animals

9.4.1 Mammals

Despite the relatively small area and lack of important topographic assemblages, Paraguay holds a high diversity due to the confluence of some major ecoregions (Spichiger et al., 1995). The most updated list of mammals from Paraguay, recorded a diversity of 181 native species (de la Sancha et al., 2017).

The most abundant orders of Paraguayan mammals are Chiroptera and Rodentia respectively (de la Sancha et al., 2017). Bats in Paraguay are feared (and then often killed) because people rarely distinguish between a non-blood-eater and a vampire (only *Desmodus rotundus* feeds on mammal's blood in Paraguay), which is vector of rabies (Schneider et al., 2009). The largest Paraguayan bat is the Greater Spear-Nosed Bat (*Phyllostomus hastatus*) that can weigh around 100 g (Smith, 2009; Knörnschild and Tschapka, 2012). The diversity of rodents in Paraguay could be underestimated due to the lack of adequate sampling according to de la Sancha et al. (2017) because usually this order has the highest species richness in countries of Central and South America. Most of Paraguayan rodents are mice, with some exceptions like the squirrels (Family Sciuridae) recorded just in 2008 for the first time in Paraguay (D'Elía et al., 2008). Another noteworthy rodent from Paraguay is the Capybara, which has a strong social behavior and it is the world's largest rodent, whose herds are frequently seen in rivers or lagoons (Maldonado-Chaparro and Blumstein, 2008; Campos-Krauer et al., 2014).

A major problem in Paraguay related to rodents is the Hantavirus. Chu et al. (2003) proved that there are different strains of Hantavirus in Paraguay, and at least 11 species of mice are hosts for Hantaviruses. Nevertheless, deceases are not common and many native people of Paraguay have antibodies against Hantaviruses.

Several large-sized species constitute the most emblematic mammals in Paraguay, and for instance, the jaguar (*Panthera onca*), which is also the largest predator of South America, is an iconic species in Paraguay. This species, together with the puma (*Puma concolor*) are the biggest representatives of Carnivora in Paraguay, and are endangered species, which are threatened by habitat loss (as they require large areas to gait) and by illegal hunting. Farmers usually give economic incentives to all those who kill these two felines because very often jaguars and pumas predate on cattle, which represents an economic loss for the farmers. There are also five species of smaller cats (*Leopardus* spp.) three of which are threatened by extinction

(SEAM [Environment Secretary] Resolution 524/06). In addition, to felids, there are also some canids in Paraguay and the largest is the Manned Wolf (*Chrysocyon brachyurus*), a tall (90 cm at the shoulder) long-legged fox characterized by a thick mane, distributed in most parts of Paraguay with the exception of Dry Chaco (Cartes et al., 2013). The other Paraguayan canids are small foxes that are abundant, with one exception: the Bush Dog (*Speothos venaticus*). These animals are very rare due to their lifestyle living most of the time underground, and restricted to the easternmost part of Paraguay (Zuercher and Villalba, 2002).

Armored mammals are also present in Paraguay. The Order Cingulata has 11 species of armadillos in the country (de la Sancha et al., 2017). The world's largest armadillo (*Priodontes maximus*) was quite common once, but now it was almost led to the extinction, and records of this species are very rare (Smith, 2007). On the contrary, there are also smaller armadillos including some of the world's smallest species such as Southern Three-banded Armadillo (*Tolypeutes matacus*) and the Chaco Fairy Armadillo (*Calyptophractus retusus*). The first of these in Paraguay is known as "Tatú Bolita" which means, "little ball armadillo" because when is threatened this armadillo closes himself very tightly taking the shape of a ball. The Chaco Fairy Armadillo is a beautiful animal, extremely rare that only occasionally leaves its burrow system. It has small eyes (almost blind) and huge claws, and a shield of plates at the back of the body.

Other important components of the fauna on mammals from Paraguay are the peccaries (Family Tayassuidae). There are three different species: *Pecari tajacu* (Collared Peccary), *Tayassu pecari* (White-lipped Peccary), and *Parachoerus wagneri* (Tagua or Chaco Peccary) (de la Sancha et al., 2017). The last of them was known from fossils and described upon preHispanic archeological remains and believed extinct (Rusconi, 1930). However, it was rediscovered in the driest part of the Paraguayan Chaco (Wetzel et al., 1975; Wetzel, 1977) thus, in Paraguay it is considered a living fossil. This species is less abundant than the other two and became an important target for hunters, and in order to preserve this intriguing linage, some projects were erected to breed animals in captivity and reintroduce them in the wild (Benirschke and Heuschele, 1993).

Paraguay also holds some deer species: *Mazama* spp. (single tine antlers), *Blastocerus dichotomus* (8 to 10 multitined antlers), and *Ozotoceros bezoarticus* (three tines antlers) (de la Sancha et al., 2017). Peccaries and deer (Order Cetartriodactyla) are an important source of hunting, along with some armadillos and other medium-sized mammals (Hill et al., 1997). Traditionally, deer of the genus *Mazama* are used to treat snakebites (Cartes, 2007).

Regarding monkeys, Paraguay has five species and each of them belongs to a different Family (de la Sancha et al., 2017). *Alouatta caraya* is the most hunted species by native communities in the eastern Paraguay (Hill et al., 1997; Cartes, 2007). This is also the largest Paraguayan monkey called Howler, because males howl loudly and they can be heard from far. All Paraguayan monkeys are diurnal but one: *Aotus azarae.*

Another Order in Paraguay is Pilosa, which contains two anteaters: the Giant Anteater (*Myrmecophaga tridactyla*) and the Southern Tamandua (*Tamandua tetradactyla*) (de la Sancha et al., 2017). These animals are common, and slow, and then are frequently killed by cars on roads. Finally, two orders (Lagomorpha and Perisodactyla) have only one representative each of them: *Sylvilagus brasiliensis* (Brazilian Cottontail) and *Tapirus terrestris* (Lowland Tapir) respectively.

Additionally, there is one order of marsupials (Didelphimorpha) in Paraguay, which contains 18 species, most of them known in the last decade (de la Sancha et al., 2007, 2012; de la Sancha and D'Elía 2015; Smith and Owen 2015). The most common marsupial in Paraguay is *Didelphis albiventris,* which can be found even in human dwellings not far from big cities.

9.4.2 Birds

Being probably the most conspicuous and visible animal wildlife, birds have been reported and talked about in some early works about Paraguayan nature. Since the Jesuitical missions in Sánchez Labrador writings from 1767 (1968), through Felix de Azara's notes on Paraguayan birds (1805). In the nineteenth century almost nothing else was published, but since the early twentieth century A. W. Bertoni published some papers on Paraguayan avifauna (1901, 1907), and Laubmann also did so on Paraguayan birds in two volumes (1939, 1940). In the 1980's with the creation of the Inventario Biológico Nacional (National Biological Inventory), which over the years turned into the National Museum of Natural History of Paraguay (Museo Nacional de Historia Natural del Paraguay, MNHNP) more has been worked on different taxa, and quite a lot on birds (Lopez, 1985; Hayes, 1988). Ornithologists that work or have worked with this Museum are Nancy López and Luis Amarilla.

Most work has been done on avifaunal assemblages, species' checklists, and biogeographic information with researchers from different foreign institutions like national History museum from Geneva and others. Since the 1990 lot has been researched and published on ornithological records, mainly through the work of Floyd Hayes at the beginning of that decade and afterwards, with the creation of Guyra Paraguay (NGO dedicated to the conservation of birds and their ecosystems) the works of Alberto Madroño, Rob Clay, Arne Lesterhuis, Alberto Esquivel (who gently provided information for this section), Hugo del Castillo and others.

The most comprehensive and complete works with lists that are considered compile almost all of the species of Paraguay are those of Guyra Paraguay (2004) and Narosky and Yzurieta (2006), which list more than 680 species of birds. In addition, the record of the Yellow-faced Parrot, *Alipiopsitta xanthops*, must be added (Álvarez and Amarilla, 2012).

According to these sources, mainly native Neotropical species occur in Paraguay, but also several Nearctic migratory species, mainly Scolopacidae, among them the near threatened (NT) Buff-breasted Sandpiper, *Callidris subruficollis* (IUCN in BirdLife, 2017a); There are also a few birds that are considered cosmopolitan, such as the barn owl, *Tyto alba*, and the Cattle Egret, *Bubulcus ibis*, and two exotic species that are naturalized, especially in urban environments (the House Sparrow, *Passer domesticus*, and the Domestic Pigeon, *Columba livia*).

9.4.3 Amphibians and Reptiles

The herpetology in Paraguay had made some advances, with the publication of a checklist of amphibians (Brusquetti and Lavilla, 2006) and a posterior identification guide where some new records were incorporated (Weiler et al., 2013) summarizing a total of 85 species. The last published account of Paraguayan reptiles, recorded a diversity of 175 species (Cacciali et al., 2016a), with some recent additions (Cacciali et al., 2016b; Carvalho, 2016; Cacciali et al., 2017a, b) adding a total of 180 native species.

The two groups of amphibians in Paraguay are Gymnophiona (three species) and Anura (82 species) (Brusquetti and Lavilla, 2006; Weiler et al., 2013). The very rare caecilians are known for a few records: *Chtonerpeton indisctinctum* (one record), *Luetkenotyphlus brasiliensis*, and *Siphonops paulensis* (two records each one) (Brusquetti and Lavilla, 2006; Weiler et al., 2013). Anurans (frogs and toads) are more commonly seen, or heard during the reproductive season. Two families are also a little rare, and are known based on a single species from a single locality as is the case of *Limnomedusa macroglossa* (Alsodidae) and *Crossodactylus schmidti* (Hylodidae). The rarest species of amphibians, are distributed in the eastern side of the country.

One of the commonest families of amphibian anurans of Paraguay is Bufonidae, which contains the widely distributed *Rhinella schneideri* commonly known as Kururu toad, being "kururu" the guarani name for this species in reference to his mating call: "kururururururururu…" (Weiler et al., 2013). The genus *Rhinella* in Paraguay is composed by eight species. In the same family, there are some small blackish toads, called *Melanophryniscus* (Brusquetti and Lavilla, 2006; Weiler et al., 2013). These are species with commercial value, because they are requested as pets (Weiler et al., 2013). These little toads have usually striking yellow and/or red spots contrasting the black background color.

Peculiar frogs in Paraguay are the members of the Family Ceratophryidae. These are known as horned toads. These anurans have large mouths with teeth and their bite can be painful. Due to its large size, these animals can feed on small vertebrates, and especially other frogs (Scott and Aquino, 2005). Moreover, people in the countryside are actually afraid of these frogs with the belief that they are poisonous. Similar in aspect, but smaller in size are the little escuerzos of the family Odontophrynidae (genera *Odontophrynus* and *Proceratophrys*), which also are believed to be poisonous. Sometimes these frogs can be really aggressive when somebody attempts to catch them.

The two more diverse families of frogs in Paraguay are Leptodactylidae (true South American frogs) and Hylidae (tree frogs) (Brusquetti and Lavilla, 2006; Weiler et al., 2013). The body shape of the Leptodactylidae resembles the European Ranidae: long and strong hind legs, elongate body, and pointed muzzle. There are 27 species in this family. Probably the most peculiar frogs in this group are the *Physalaemus*, whose members have two black inguinal spots (one on each side) that are dermic glands used for defense since they segregate a toxic substance, but they also resemble a pair of big eyes that can persuade predation. Another peculiar group of frogs in this family, is the genus *Pseudopaludicola*, which contains some of the tiniest anurans in Paraguay. Most of the males of Paraguayan frogs call at night, but these small frogs call either during the day or at night. And the genus that gives the name to the family is *Leptodactylus*. These are usually small to medium size frogs with a reticulated color pattern and strong hind legs. In this genus is the largest Paraguayan frog: *Leptodactylus labyrinthicus* (Labyrinth frog). This species can reach almost 20 cm of snout-vent length, and males have massive forearms. The coloration of this frog is basically olive brown dorsally with reddish flanks, and segregates a toxic substance when is manipulated. Another quite toxic species is *Leptodactylus laticeps*, a beautiful frog called "coral frog" because its coloration: white background with black irregular spots, and red marks or dots in each black spot.

Tree frogs in Paraguay is a rather diverse group. Some species of the genus *Dendropsophus* reach not much more than 2 cm, and a species of Gladiator frog (*Boana faber*) easily reach 10 cm. Males of this last species have a claw in the pollex, used for harsh fights. Another big frog in Paraguay is *Trachycephalus typhonius* (White-lipped tree frog), which is common in the whole country. In Spanish is called "*Rana lechosa*" [Milky frog] because when grabbed it segregates a very sticky white substance that is difficult to wash even with soap. In this family, there is also the swimming frog *Pseudis platensis*. The tadpole of this frog can reach 16 cm while adults only 6 cm (Weiler et al., 2013). More than once, the press gave an alarm about fishers that caught "a radioactive fish with legs" in the river, without knowing that it was a tadpole of *P. platensis*. Finishing with the amphibians, another group of hilid frogs are worth to mention: the genus *Phyllomedusa*. This is a genus of green tree frogs extended from North America to Argentina, and they climb more than jump. Frogs of the genus *Phyllomedusa* have a nice green with some contrasting colors that make them very attractive for the pet market worldwide.

With respect to the reptiles in Paraguay, there are 3 species of Crocodyilia, 10 chelonians, and 163 Squamata. In the country, there are two tortoises (genus *Chelonoidis*) and eight turtles (Families Kinosternidae and Chelidae). The tortoises in the country are more frequently found in some dry areas of the Paraguayan Chaco. The bigger species *Chelonoidis carbonaria* is a beautiful animal with a blackish background color and yellow or orange spots, very used as pet. Water turtles are rarer, and some species of the genus *Phrynops* are known from a few specimens and in some cases, the taxonomic status is not clear. A common turtle in the central part of the Eastern Region of Paraguay is *Hydromedusa tectifera*, commonly known as Snake-Necked Turtle, because its neck can have the same length of the carapax. Another peculiar species is the *Kinosternon scorpioides*, known as Scorpion Mud Turtle, which has a movable plastron that is closed tightly when the animal is hidden in the shell (Norman, 1994).

The most common species of Crocodylia in Paraguay is *Caiman yacare* and the epithet refers to the colloquial name given to the animal in Guarani. This species, together with *C. latirostris* are widely used in the leather industries and are regulated by CITES (Brazaitis et al., 1996; CITES, 2017). The third species of Crocodylia in Paraguay is *Paleosuchus palpebrosus*, which is known based upon a single record in the northern part of the Oriental Region (Scott et al., 1991; Cacciali et al., 2016). In Paraguay this species is very enigmatic, and further attempts to find more species have failed.

Regarding Squamata, Paraguay holds an important diversity, described below.

- **Iguania:** One of the larger lizards in Paraguay is the *Iguana iguana* frequently found in bushes or tree branches along the Paraguay River in the Pantanal (Cacciali, 2007). The False Chameleon *Polychrus acutirostris* is also in this group. Similar to true chameleons, this animal can change its color and move each eye independently allowing a 360° vision. The most diverse group of Iguania in Paraguay is *Tropidurus*, which contains seven species, and three of them (*T. lagunablanca, T. tarara*, and *T. teyumirim*) recently described and endemic to Paraguay (Carvalho, 2016). Paradoxically, the two more diverse lizard genera in the Neotropics are *Anolis* and *Liolaemus*, but in Paraguay these are represented by one and two species respectively, and the country is in the edge of their ranges.
- **Gekkota:** Formerly, all geckos in Paraguay were considered Gekkonidae, but some recent molecular analyzes showed that the origins of some members of the family are different (Gamble et al., 2008), and then the geckos in Paraguay are mainly Phyllodactylidae (arrived to the Neotropics in the Paleogene, whereas the genus *Lygodactylus* (Gekkonidae) migrated to America during the Late Paleogene/Neogene (Gamble et al., 2011). The only introduced reptile in Paraguay (*Hemidactylus mabouia*) belongs also to this family, and is colonizing several important cities in the country (Baldo et al., 2008; Cacciali and Motte, 2009).
- **Scincomorpha:** This is a heterogeneous group, containing skinks (Mabuyidae), spectacled lizards (Gymnophthalmidae) and green runners and large lizards (Teiidae). Some Gymnophthalmidae are almost unknown, as the case of the rare *Bachia bresslaui* known by a single record. These are animals of small size, inhabiting grasslands, and living in burrows; what makes them very difficult to see. The Family Teiidae are mostly runners (genera *Ameiva, Ameivula, Kentropyx*, and *Teius*), but also some big lizards such as the tegu lizards (*Salvator* spp.) and *Dracaena paraguayensis*, largely used in the leather market (Norman, 1994).
- **Diploglossa:** This group is composed by a single family in Paraguay (Anguidae) and a single genus (*Ophiodes*) with five species. These all are legless lizards, often mistaken with snakes.
- **Amphisbaenia:** The worm lizards are a strange group of animals with a scale arrangement around the body unique among Squamata,

eyes under the skin and a morphology completely adapted to an underground lifestyle. Twelve species are present in Paraguay, and all need to be carefully examined for an accurate identification. The only easily recognizable species is *Amphisbaena microcephalum* that has a shovel-shaped nose.

- **Scolecophidia:** This is a group of small snakes, adapted to fossorial habits, with some characteristics such as eyes reduction covered by cephalic shields, bright and smooth scales, no differentiation between dorsal and ventral scalation, and reduction of teeth. The larger species is *Amerotyphlops brongersmianus*, which is present from Venezuela to Argentina (McDiarmid et al., 1999).

- **Boidea:** Completely opposite to the previous group, Boidea comprise large snakes including one of the world's largest snake *Eunectes murinus*, known as Anaconda. Some smaller species are the *Boa constrictor* and *Epicrates cenchria*. These snakes kill by constriction (suffocating their prey), but none of them (including big anacondas) are strong enough to break bones as commonly believed. Given the big size of these reptiles, their skins are valuable for leather market.

- **Colubroidea:** This is a diverse and heterogeneous group of snakes. Here are included coral snakes, pitvipers, rattlesnakes, racers, water snakes, and more; with sizes from 20 cm (genus *Sibynomorphus*) to 2 m (genus *Hydrodynastes*). Elapidae (coral snakes of the genus *Micrurus*) and Viperidae (pitvipers of the genus *Bothrops* and the rattlesnake *Crotalus durissus*) are the two families of venomous snakes. Nevertheless, Viperidae is the only with a real medical importance, since there is no record of bites by coral snakes.

Traditionally, the Family Colubridae was the most diverse in the country; but after a systematic revision this family was divided, and members of Colubridae have a Nearctic origin, whereas Dipsadidae is a family originated in the Neotropics (Grazziotin et al., 2012). Thus, most of the diversity in Paraguay is currently in the Family Dipsadidae. Paraguayan ophidians with Nearctic origin are some Sipo snakes (*Chironius* spp.), Indigo Snake (*Drymarchon corais*), Rio Tropical Racer (*Mastigodryas biffosatus*), and the Tiger Rat Snake (*Spilotes pullatus*), among others. Snakes with Neotropical origin have more diversity, with genera containing many species such are: *Oxyrhopus* (three species), *Apostolepis* (four species), *Sibynomorphus* (four species), *Phalotris* (five species), *Thamnodynastes* (five species), *Lygorphis* (five species), *Philodryas* (ten species), and *Erythrolamprus* (11 species).

9.4.4 Fishes

As for the studies on the fish fauna, it is worth mentioning two great contributions. On the one hand, the expeditions of the *Muséum d'histoire naturelle de la Ville de Genève*, Switzerland, which generated numerous publications on both the fish and their parasites (Weber, 1986; Gery et al., 1987; Petter and Dlouhy, 1985) but little worked with local experts. On the other hand, the PROVEPA project (Vertebrates of Paraguay Project, by its initials in Spanish), of the Naturhistoriska Riksmuseet in Stockholm, Sweden, which together with the National Museum of Natural History of Paraguay (MNHNP by its initials in Spanish) formed a large part of the latter's ichthyic collections in the 90 s (Mandelburger et al., 1996). The first collections of the MNHNP were the product of expeditions of the Peace Corps of the United States (Ramlow, 1989).

From the PROVEPA project, a lot of new information about the fish fauna was generated and with the collections formed and active in loans of specimens, work done by both foreign experts, for example, describing new species (Kullander, 2009; Kullander and Lucena, 2013), and national experts such as Hector Vera (Vera et al., 2008, 2012, 2017) stand out. Also, the organization Natura Vita made three important contributions in the form of books that compiled many of the species of Paraguay (Neris et al., 2008, 2010; Villalba, 2012).

However, the most complete work that lists all the species registered in the country so far is the checklist by Koerber et al. (2017). In this work 307 species of fish are compiled.

The fish richness of Paraguay corresponds to the river systems of the Paraguay and Paraná rivers, for which almost all species are shared with Argentina, Brazil and/or Bolivia, however, there could be endemic species of Paraguay in endorreic systems in the northern Chaco. The high diversity despite not having marine or coastal systems, is due to the warm waters of the tropical to subtropical system of the Paraguay and Paraná rivers. This diversity ranges from a few species of Chondrichthyes (rays of the genus Potamotrygon), through a large majority of Actinopterygii Teleosteii, to a single species of Sarcopterygii, the South American Lungfish *Lepidosiren paradoxa*. As for the Actinopterygii, the orders that make up most of the Paraguayan ichthyofauna are Characiformes and Siluriformes; then, with fewer species, Cichliformes, Cyprinodontiformes and Gymnotiformes, and finally with two species each, Clupeiformes, Beloniformes, Perciformes, and Pleuronectiformes and with one species, Synbranchiformes.

Fish and fishing have a deep social and cultural importance in the country as an alternative protein source to beef and as a widespread recreational activity.

9.4.5 Invertebrates

Invertebrates recorded in Paraguay belong to at least 10 classes that, in turn, comprise numerous species, some of which are exclusive of the region. Although there are few systemized information sources on invertebrate species, and despite the scarcity of studies on the subject, at least 3,000 invertebrate species have been recorded, distributed in 236 families (Table 9.2). Furthermore, in addition to the lack of studies in the area, 336,000 hectares are deforested every year due to the fast expansion of mechanized agriculture and extensive livestock farming in the name of progress, with disastrous consequences for biodiversity.

The most studied subject is entomofauna, comprising a high diversity of both aquatic and terrestrial species belonging to around 20 orders, 180 families, 1,246 genera, and 2,878 species recorded and identified nationwide (Table 9.3). The *Coleoptera* and *Lepidoptera* orders are the most numerous, represented by 1,442 and 843 species respectively.

Table 9.2 Number of Species, Genera and Families of the Different Classes of Invertebrates Recorded in Paraguay

	Number of families	Number of genera	Number of species
Arachnida	31	63	122
Bivalvia	4	8	13
Chilopoda	1	1	1
Decapoda	3	6	7
Diplopoda	2	2	2
Gastropoda	10	13	22
Insecta	180	1246	2878
Isopoda	2	2	2
Oligochaeta	1	1	1
Turbelaria	1	1	1
Total (10):	**235**	**1,343**	**3,049**

Table 9.3 Number of Species, Genera and Families of the Different Orders of Insects Recorded in Paraguay

	Number of families	Number of genera	Number of species
Blattaria	1	11	26
Coleoptera	40	349	1442
Dermaptera	1	1	16
Dipetera	27	49	74
Ephemeroptera	1	1	1
Heteroptera	18	103	169
Homoptera	14	55	68
Hymenoptera	16	58	88
Isoptera	1	1	1
Lepidoptera	38	521	843
Mantodea	2	12	25
Megaloptera	1	1	1
Odonata	3	19	21
Phasmida	1	3	6
Planipennia	4	11	22
Psocoptera	1	1	2
Saltatoria	8	47	70
Thysanoptera	1	1	1
Thysanura	1	1	1
Trichoptera	1	1	1
Total (20):	**180**	**1,246**	**2,878**

9.5 Protected and Conservation Areas

According to the Environment Secretary (SEAM), the national territory currently has 103 protected wild areas (ASP), 10 of which are national parks. Among these natural parks, the *Parque Nacional Defensores del Chaco* is the largest with 721,000 ha. The remaining ASPs are distributed under the following categories: natural reserve, national reserve, ecological reserve, biological reserve, scientific reserve, managed resources reserve, private reserve, protected landscape, natural monument, scientific monument, biological sanctuary, wildlife sanctuary, national reserve zone, and national park reserve.

Under public domain there are 42 protected wild areas (ASP); 47 are under private domain, 8 are under self-governed entities such as Itaipu and Yacyreta Hydroelectrics, 3 reserves for the biosphere, and 3 RAMSAR sites.

Only some of the ASPs have an approved and current management plan, the rest do not have the same, are in the process of obtaining, or must be updated.

9.6 Threatened Species

Paraguay currently has official lists of protected species recognized by resolutions issued by the Secretariat of the Environment (SEAM by its initials in Spanish), these are Resolutions No. 2242/06 and 2243/06 that list the protected species threatened with extinction and the protected species in danger of extinction. These lists are valid since more than ten years, which is why they will be updated based on workshops that have been carried out since mid–2017. It is expected that by the mid of 2018 the new listings will be valid. The methodology applied to these workshops is that of the IUCN red lists, both for the 2006 and 2018 lists.

The categories of the current resolutions do not correspond exactly to the IUCN categories because they have been adapted to the national legislation (Law No. 96/92 "On Wildlife"). Thus, the category of protected species threatened with extinction (Res. No. 2242/06), corresponds to the category vulnerable (VU) of the IUCN, and the category of protected species threatened by extinction (Res. No. 2243/06), corresponds to the categories of endangered (EN) and critically endangered (CR) of the IUCN.

These lists are the result of workshops with several specialists, however, none of them was published in scientific journals, except the work that corresponds to the herpetofauna (Motte et al., 2009). In this paper, the lists do not correspond exactly to the official ones of the resolutions of the SEAM, since in the following years after 2006, they continued being reviewed by the specialists.

According to these resolutions, still in force in 2017, 40 species of plants are considered protected species threatened with extinction (mainly Orchidaceae, Cactaceae and Bromeliaceae, beside some medicinal species of the genus *Baccharis* [Asteraceae], and a few timber species), also 10 species of mammals. (mainly Didelphimorphia and Chiroptera), 64 of birds, 35 of reptiles, 8 of amphibians and 17 of invertebrates. In addition, protected species in danger of extinction are 81 species of plants (mainly timber species in addition to the families mentioned above), 35 of mammals, 56 of birds, 12 of reptiles, 9 of amphibians, and 16 of invertebrates.

The fish have separate lists and categories different from the rest of the wildlife, since they are governed by their own law of protection and use, Law No. 3556/08 "On Fishing and Aquaculture," and therefore they are listed in another resolution (Res. SEAM No. 1563/09). It categorizes 15 species as VU (vulnerable), and 5 as EN (endangered), mostly commercial species and a few species of which there are few records in the country.

There are few plant species that in Paraguay are considered extinct in the wild, among them the endemic Sweet Leaf, *Stevia rebaudiana* (Bertoni) Bertoni, which however has many strains or cultivars. Also, two species of birds are considered extinct in Paraguay, the Brazilian Merganser, *Mergus octosetaceus*, which however could still be present (Clay et al., 2003), and the Glaucous Macaw, *Anodorrhynchus glaucus*, possibly extinct in the wild worldwide, since the last records date back to the 1960 s, according to Bird-Life International (2017b).

In Paraguay, it is quite evident that the main threat faced by wildlife species refers to habitat loss, caused by the conversion of natural ecosystems to productive soils (pastures and agricultural crops), however, in the case of timber species there is a direct pressure due to the use of these species as well as on species that are used as medicinal, especially those that come from ecosystems that are in rapid degradation.

9.7 The Paraguayan Ecoregional Scope

Paraguay has been in the middle of several debates among the ecological division of the country with conflictual situations for ecosystems, ecoregions and biomes. During the last years, it appears to be a consensus given a contribution based on expert consultations to understand ecoregions in the western part of the country (Mereles et al., 2013). Technical classifications for Paraguay were included in pioneer studies which identified landscapes and their relation with vegetation (Koppen, 1884; Holdridge, 1947; Peel et al., 2007), and life zones (1947) with the identification of the temperate forest of the country, dry forests in the western part and humid forests in the eastern part. The country is divided into two distinct regions by the Paraguay River which runs south to flow waters into the Paraná River. The River and the associated wetlands on both sides, gives rise to flooded grasslands, marshes and other aquatic habitats which are very well differentiated from the forests. In the myriad of habitat types, we find permanently flooded marshes and floodable wax-palm savannas. In the 70 s, the region received different classification systems and the most holistic was Cabrera and Willink (1973), based on different research. Sanjurjo (1989) proposed forest ecoregions for Eastern Paraguay. Mereles et al. (2013) refer to the historical approach developed by Udvardy, who for the first time proposed for Paraguay (a) a rainy forest, (b) gran Chaco, (c) cerrado, and (d) pampas. This approach was later used for a thorough analysis of ecoregions developed by Dinerstein et al. (2005). Hueck (1978) studied forests and nonforest vegetal communities in South America and mentioned for the first time, the Pantanal, in the northernmost corner of the country. A very thorough analysis of habitats and natural communities was compiled and published by the *Centro de Datos para la Conservación* (CDC, 1990) for priorities areas for the conservation of eastern Paraguay. This is the most widely cited and basis for the ecoregional approach of eastern Paraguay. After this, the focus was the western part of the country, the Chaco, in which the identification and classification of ecoregions was developed by Fundación Chaco and Geosurvey (1992) where 12 biomes are identified in this part of the country.

One of the most widely recognized ecoregions of the country is the presently known as Alto Parana Atlantic Forest (APAF), part of an ecoregion shared with Argentina and Brazil. At the national level and according to the (subecoregions) CDC (1990), this APAF embraces Amambay, Selva Central and Alto Paraná. The APAF is also recognized in the widely known Latin-American Ecoregions from Dinerstein et al. (2005) (Figure 9.1). This latter identifies five ecoregions for the country: Dry Chaco, Humid Chaco, Pantanal, Cerrado and Atlantic Forest (APAF), with similarities to those proposed by Cabrera and Willink (1973). This approach is the most widely cited and referenced, which defines ecoregions with a greater precision such as the case of the Pantanal, the Atlantic Forest and the southern boundary of the Cerrado Ecoregion. Guyra Paraguay published Important Bird Areas (Clay et al., 2005) and therein and based on the ornithogeographic regions proposed a new ecoregion called Southern Grasslands or Missions' Grasslands in the southern part of the country. After Mereles et al. (2013) and ratified officially by a Resolution from the Maximum Environmental Authority in Paraguay (SEAM Resolution 614/13), Paraguay's western region holds five ecoregions while six are defined for eastern Paraguay (Table 9.4, Figure 9.3).

From the landscape perspective three of them, Ñeembucú, Litoral Central and Humid Chaco are very similar habitat and ecotypes, marshes and grasslands, partially flooded during the different seasons with some patchy and sparse low forests and shrubby areas. The Humid Chaco is differentiated

Table 9.4 Ecoregions Officially Recognized Based on SEAM Resolution 614/13

Officially recognized ecoregions	Area (km²)
1. Médanos	7,576.8
2. Cerrado	12,279.2
3. Pantanal	42.023.1
4. Humid Chaco	51,927.6
5. Dry Chaco	127,211.6
6. Aquidaban	10,700
7. Amambay	9,207
8. Alto Paraná	33,510
9. Selva Central	38,400
10. Litoral Central	26,310
11. Ñeembucú	35,700

by the presence of the wax palm *Copernicia,* with extensive palm lands. The three ecoregions are regulated and influenced by rivers, Paraguay, Paraná and Pilcomayo Rivers with other important tributaries. Four other ecoregions though differentiated by ecological features form the (Alto Paraná) Atlantic Forest, a high, humid forest that originally covered around 9 million hectares, of which remain little more of 10%. Though now after this resolution, the Cerrado only exists in the northern part of western Paraguay, sufficient evidence exists of the southern boundaries of Cerrado in eastern Paraguay with areas of expansion and isolated patches within the ranges of the Atlantic Forest. The Pantanal is a very unique and distinctive landscape, in some way similar to the Humid Chaco with dense tall grasslands mostly flooded year around and with extensive palm lands of *Copernicia.* The ecoregion known as Médanos is a very unique landscape for Paraguay, originally included into the Dry Chaco Forest, the area is clearly different by sand dunes with a particular feature and a highly increasing vulnerability due to deforestation and erosion. The Dry Chaco is the largest ecoregion in Paraguay and is subject to high levels of deforestation. With more than 12,000 km², the ecoregion is being cleared for livestock production at a rate of 500–1,800 hectares/day.

Paraguay is located at the confluence of different ecoregions, all of them are transboundary and shared with the neighboring countries. No ecoregions are unique or endemic to Paraguay. Perhaps the most unknown ecoregion is the Médanos together with the Pantanal, though the southern grasslands have not been yet officially recognized. The Atlantic Forest is shared with Argen-

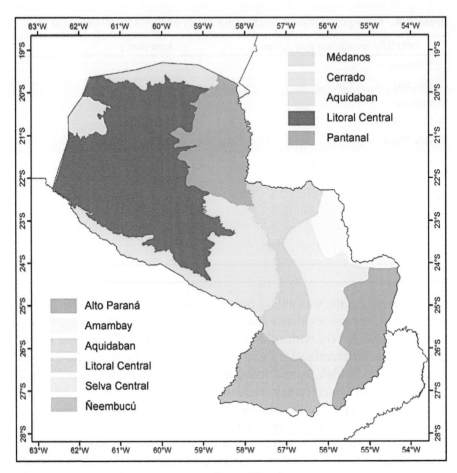

Figure 9.3   Map of ecoregions officially used by SEAM.

tina and Brazil, the Pantanal with Bolivia and Brazil, the Chaco (Argentina, Bolivia, and a small part in Brazil), the Cerrado with Brazil (and Bolivia). The Southern Grasslands or Missions' Grasslands (Pampas) is shared with Argentina in close connectivity and widely with Rio Grande do Sul's State in Brazil and Uruguay.

When depicted in a map all watercourses and bodies, Paraguay is a huge wetland. Except for limited areas in the Dry Chaco and the Médanos Ecoregions, the entire country is rich in freshwater features. The entire country was assigned to four global biogeographic regionalizations of Earth's freshwater systems by Abell et al. (2008). Freshwater habitats, grasslands and different woody landscapes (forests, shrubby savannas) characterize Paraguay, with two distinct regions, the eastern part with little less than 50% of the country area with more than 95% of the population has provided the basis for the economic growth and development of the country, giving room for agriculture (livestock, grains especially soya) and provided water for three hydroelectric dams. Ecoregions in Eastern Paraguay are better seen and understood in conservation sites, especially protected areas. Extensive areas of Atlantic Forest have been converted, originally into areas of cattle grazing, and later displaced by soya, or artificial lakes such as Itaipú and Yacyreta lakes, because of the hydroelectric binational companies.

At a finer analysis, Paraguay has advanced in the identification of ecosystems. Based on the ecoregional planning developed by TNC (2005) and with the support given by the USGS and Guyra Paraguay1, 101 ecosystems were identified and clearly mapped, and incorporated into the Inter American Biodiversity Information Network (IABIN). These ecosystems were the basis to present the gap analysis of ecosystems done by Guyra Paraguay and SEAM to CBD COP8 in 2006 (Curitiba, Brazil) (Rodas et al., 2006). This study concluded that 55 (54%) of the ecosystem are not protected nationally and 14 (14%) are poorly protected in national conservation units. Only 24 (23.7%) ecosystems are well represented with more than 10% of their area within the National System of Protected Areas (Table 9.5).

Table 9.5 Representativeness of Ecosystems in Paraguay's Protected Areas

Ecoregion	Area	Country (%)	No. of ecosystems
Dry Chaco	17,484,326	42	41
Humid Chaco	12,858,489	32	33
Atlantic Forest	8,591,121	21	11
Cerrado	819,101	2	11
Pantanal	198,494	1	5
Other areas	723,669	2	-
Total	**40,675,200**	**100**	**101**

[1] This digitalization was based on an adaptation and revision of the geospatial database of the ecosystems of South America, produced by Roger Sayre, Jacquie Bow and Carmen Josse, of the United States Geological Survey, Conservation Geospatial and The Nature Conservancy, respectively. The mentioned work, was able to identify 659 ecosystems for South America, developing a system of criteria and names to describe ecological systems derived from expert knowledge.

Keywords

• Paraguayan ecoregional scope • threatened species

References

Abell, R., Thieme, M. L., Revenga, C., Bryer, M., Kottelat, M., Bogutskaya, N., Coad, B., Mandradk, N., et al., (2008). Freshwater ecoregions of the world: A new map of biogeographic units for freshwater biodiversity conservation. *BioScience*, *58*(5), 403–414.

Álvarez, A., Amarilla, L., & Fernández, R., (2012). Primer registro reportado de loro keréu (*Amazona xanthops*) en la ecorregión cerrado del Paraguay. *Compendio de Ciencias Veterinarias*, *2*(2), 7–11.

Azara, F. M., (1945). *Descripción e Historia del Paraguay y del Río de La Plata*. Apuntamientos para la Historia Natural de los cuadrúpedos *y* aves del Paraguay *y* Rio de la Plata. Editorial Babel, *Buenos Aires*, 1802.

Azevedo, M., Antonio Brand, Ana Maria Gorosito, Egon Heck, Bartomeu Melià, Jorge Servín, (2008). Los pueblos en las fronteras Argentina, Brasil y Paraguay. Bartomeu Melià, Asunción. p. 24.

Baldo, D., Borteiro, C., Brusquetti, F., García, J. E., & Prigioni, C., (2008). Reptilia, Gekkonidae, *Hemidactylus mabouia, Tarentola mauritanica:* Distribution extension and anthropogenic dispersal. *Check List*, *4*(4), 434–438. http://dx.doi.org/10.15560/4.4.434.

Batrina, L., (2007). Contexto geográfico general. In: Salas-Dueñas, D., & Facetti, J., (eds.). *Biodiversidad del Paraguay, una Aproximación a Sus Realidades*. Fundación Moises Bertoni, USAID, GEF/BM, Asunción. pp. 26–32.

Benirschke, K., & Heuschele, W. P., (1993). Proyecto Taguá: The Giant Chaco peccary *Catagonus wagneri* conservation project. *Int. Zoo Yb.*, *32*(1), 28–31. http://dx.doi.org/10.1111/j.1748-1090.1993.tb03511.

Bertoni, A. W., (1901). *Aves Nuevas del Paraguay*. Catálogo de las Aves del Paraguay – Annal Cient., Parag, *1*, 1–216.

Bertoni, A. W., (1907). Segunda Contribución á la Ornitología Paraguaya. *Nuevas Especies Paraguayas – Rev. Inst. Paraguay*, pp. 298–309.

Bertoni, M., (2004). La civilización Guaraní. Parte III: Conocimientos. La higiene guaraní *y* su importancia científica *y* práctica. La medicina guaraní. Conocimientos Científicos. Ex Sylvis, *Alto Paraná*. p. 1927.

Bertoni, M., (2012). Diccionario botánico "Introducción a las plantas usuales *y* útiles del Paraguay." *Obra Póstuma, MAG*. p. 1940.

BirdLife International, (2017a). *Calidris subruficollis*. (amended version published in 2016) The IUCN red list of threatened species 2017: e. T22693447A111804064. http://dx.doi.org/10.2305/IUCN. UK. 2017–1. RLTS. T22693447A111804064.

BirdLife International, (2017b). *Species Factsheet: Anodorhynchus glaucus*. Downloaded from http://www.birdlife.org on 07/11/2017. Recommended citation for factsheets for more than one species: BirdLife International (2017) IUCN Red List for birds. Downloaded from http://www.birdlife.org on 06/11/2017.

Bracesco, N., Sanchez, A. G., Contreras, V., Menini, T., & Gugliucci, A., (2011). Recent advances on *Ilex paraguariensis* research: Minireview. *J. Ethnopharmacol., 136*(3), 378–384.

Brazaitis, P., Rebêlo, G. H., Yamashita, C., Odierna, E. A., & Watanabe, M. E., (1996). Threats to Brazilian crocodilian populations. *Oryx*, *30*(4), 275–284. https://doi.org/10.1017/S0030605300021773.

Brusquetti, F., & Lavilla, E. O., (2006). Lista comentada de los anfibios de Paraguay. *Cuad. Herpetol.*, *20*, 3–79.

Cabrera, A., & Willink, A., (1973). *Biogeografía de América Latina*. Programa Regional de Desarrollo Científico *y* Tecnológico, Serie de Biología, OEA, Washington.

Cacciali, P., & Motte, M., (2009). Nuevos registros de *Hemidactylus mabouia* (Sauria: Gekkonidae) en Paraguay. *Cuad. Herpetol.*, *23*(1), 41–44.

Cacciali, P., (2007). Diversidad de anfibios *y* reptiles en Paraguay. In: *Biodiversidad del Paraguay, Una Aproximación a sus Realidades*. Fundación Moisés Bertoni, Asunción, pp. 109–117.

Cacciali, P., Cabral, H., Ferreira, V. L., & Köhler, G., (2016b). Revision of *Philodryas mattogrossensis* with the revalidation of *P. erlandi* (Reptilia: Squamata: Dipsadidae). *Salamandra*, *52*(4), 293–205.

Cacciali, P., Martínez, N., & Köhler, G., (2017a). Revision of the phylogeny and chorology of the tribe Iphisini with the revalidation of *Colobosaura kraepelini* Werner, (1910). (Reptilia, Squamata, Gymnophthalmidae). *ZooKeys*, *669*, 89–105. http://dx.doi.org/10.3897/zookeys.669.12245.

Cacciali, P., Morando, M., Medina, C. D., Köhler, G., Motte, M., & Avila, L. J., (2017b). Taxonomic analysis of Paraguayan samples of *Homonota fasciata* Duméril & Bibron (1836) with the revalidation of *Homonota horrida* Burmeister (1861) (Reptilia: Squamata: Phyllodactylidae) and the description of a new species. *Peer. J.*, *5*, e3523. http://dx.doi.org/10.7717/peerj.3523.

Cacciali, P., Scott, N., Aquino, A. L., Fitzgerald, L. A., & Smith, P., (2016a). The reptiles of Paraguay: Literature, distribution, and an annotated taxonomic checklist. *Special Publications of the Museum of Southwestern Biology*, *11*, 1–373.

Cadogan, L., (1959). Ayvu rapytã. Textos míticos de los Mbyá-Guaraní del Guairá. Universidad de Sao Paulo, Facultad de Filosofia, *Ciencias e Letras. Bol.*, *227*(5).

Cadogan, L., (1968). Chonó Kybwvrá: Aporte al conocimiento de la Mitología Guaraní. *Suplemento Antropológico de la Revista del Ateneo Paraguayo*, *3*(1–2), 55–158.

Cadogan, L., (1973). Ta-ngy Puku: Aportes a la etnobotánica de algunas especies arbóreas del paraguay Oriental. *Suplemento Antropológico de la Universidad Católica de Asunción*, *7*, 7–59.

Campi, M. G., & Maubet, Y., (2015b). Especies de *Geastrum* (Geastraceae, Basidiomycota) nuevos registros para Paraguay. *Steviana*, *7*, 79–88.

Campi, M. G., Maubet, Y. E., & Britos, L., (2015). Mycorrhizal fungi associated with plantations of *Pinus taeda* L. from the National University of Asunción, Paraguay. *Mycosphere*, *6*(4), 486–492.

Campi, M. G., Trierveiler-Pereira, L., & Maubet, Y. E., (2017b). New records of Phallales from Paraguay. *Mycotaxon*, *132*(2), 361–372.

Campi, M., & Maubet, Y., (2015a). *Cyathus poeppigii* (Agaricales, Basidiomycetes): Nuevo registro para Paraguay. *Steviana*, *7*, 74–78.

Campi, M., & Maubet, Y., (2016). Contribución a la micobiota gasteroide de Paraguay: Nuevos registros de *Calvatia rugosa* y *Vascellum pampeanum*. *Steviana*, *8*(1), 43–49.

Campi, M., Madrignac, B., Rivas, A. F., & Gullón, M. A., (2013). *Geastrum violaceum* Rick (Geastraceae, Basidiomycota), nuevo registro para Paraguay. *Reportes Científicos de la Facen, 4*(2), 15–18.

Campi, M., Mancuello, C., Maubet, Y., & Niveiro, N., (2017d). *Laccaria fraterna* (Cooke & Mass.: Sacc.) Pegler, (1965). (Agaricales, Basidiomycota) associated with exotic *Eucalyptus* sp. in northern Argentina and Paraguay. *Check List, 13*(4), 87–90.

Campi, M., Maubet, Y., Armoa, J., & Sandoval, P., (2017e). *Donkia pulcherrima* (Polyporales, Phanerochaetaceae*)* una especie hidnoide poco conocida, nueva cita para el Paraguay. *Steviana, 9*(1), 25–30.

Campi, M., Maubet, Y., Miranda, B., Armoa, J., & Cristaldo, E., (2017c). Dos nuevas citas de Mycenaceaes para el Paraguay: *Xeromphalina tenuipes* & *Filoboletus gracilis*, un interesante agarical poroide. *Steviana, 9*(1), 16–24.

Campi, M., Maubet, Y., Morerno, G., & Altés, A., (2016a). *Tulostoma cyclophorum* (Basidiomycota, Agaricomycetes) nuevo registro para el Paraguay. *Boletín de la Sociedad Micologica de Madrid. 40*, 129–132.

Campi, M., Miranda, B., & Maubet, Y., (2016). *Coprinus comatus* (OF Müll.) Pers. (Agaricaceae-Basidiomycota), hongo de interés medicinal *y* gastronómico, nueva cita para el Paraguay. *Steviana, 8*(2), 68–74.

Campi, M., Moreno, G., Maubet, Y., & Weiler, A., (2017a). *Battarrea phalloides* (Basidiomycota, Agaricomycetes) nuevo para el Chaco paraguayo. *Boletín de la Sociedad Micologica de Madrid.*

Campos-Krauer, J. M., Wisely, S. M., Benitez, I. K., Robles, V., & Golightly, R. T., (2014). Rango de hogar *y* uso de hábitat de carpinchos en pastizales recién invadido en el Chaco Seco de Paraguay. *Therya, 5*(1), 61–79.

Cañiza, B. D., Peralta, D. F., & Suárez, G. M., (2017). New records and extension of bryophytes for Paraguay. *Cryptogamie, Bryologie, 38*(4), 393–410.

CAPECO, (2017). Cámara Paraguaya de Exportadores *y* Comercializadores de Cereales *y* Oleaginosas, Estadísticas, Área de Siembra, Producción *y* Rendimiento. Asunción- Paraguay. Available in: http://capeco.org.py.

Cartes, J. L., (2007). Patrones de Uso de los Mamíferos del Paraguay: Importancia sociocultural *y* económica. In: *Biodiversidad del Paraguay: Una Aproximación a sus Realidades*. Fundación Moisés Bertoni: Asunción, pp. 167–186.

Cartes, J. L., Giordano, A. J., & Mujica, C. M. N., (2013). The Maned Wolf (*Chrysocyon brachyurus*) in Paraguay. In: *Ecology and Conservation of the Maned Wolf: Multidisciplinary Perspectives*. CRC Press, Boca Raton, pp. 235–247.

Carvalho, A. L. G., (2016). Three new species of the *Tropidurus spinulosus* group (Squamata: Tropiduridae) from Eastern Paraguay. *Am. Mus. Novit., 3853*, 1–44.

CDC, (1990). Áreas prioritarias para la conservación en la Región Oriental del Paraguay. CDC/DPNVS/SSRNMA/MAG. Asunción.

Chu, Y. K., Owen, R. D., González, L. M., & Jonsson, C. B., (2003). The complex ecology of Hantavirus in Paraguay. *Am. J. Trop. Med. Hyg., 69*(3), 263–268.

CITES, (2017). Appendices I, II and III of the Convention on International Trade in Endangered Species of Wild Fauna and Flora. Available at https://cites.org/sites/default/files/eng/app/2017/E-Appendices-2017–10–04. pdf.

Clay, R. P., Egea, J., & Castillo, H., (2005). Ecorregiones de Paraguay. In: *Guyra Paraguay*. Atlas de las Aves de Paraguay. Asunción, Paraguay, Guyra Paraguay, pp. 6–9.

Clay, R., Villanueva, S., Fernandez, D., & Fernandez, E., (2003). Brazilian merganser may survive in Paraguay. *TWSG News, 14,* 52.

D'Elía, G., Mora, I., Myers, P., & Owen, R. D., (2008). New and noteworthy records of Rodentia (Erethizontidae, Sciuridae, and Cricetidae) from Paraguay. *Zootaxa, 1784,* 39–57.

De Azara, F., (1805). Apuntamientos para la Historia Natural de los Páxaros del Paraguay *y* Rio de la Plata – Madrid Imprenta de la Viuda Ibarra.

De la Sancha, N. U., & D'Elía, G., (2015). Additions to the Paraguayan mammal fauna: The first records of two marsupials (Didelphimorphia, Didelphidae) with comments on the alpha taxonomy of *Cryptonanus* and *Philander*. Mammalia, *79,* 343–356. https://doi. org/10.1515/mammalia-2013–0176.

De la Sancha, N. U., D'Elía, G., & Teta, P., (2012). Systematics of the subgenus of mouse opossums *Marmosa* (*Micoureus*) (Didelphimorphia, Didelphidae) with noteworthy records from Paraguay. *Mamm. Biol., 77,* 229–236.

De la Sancha, N. U., López-González, C., D'Elía, G., Myers, P., Valdez, L., & Ortiz, M. L., (2017). An annotated checklist of the mammals of Paraguay. *Therya, 8*(3), 241–260. https://doi.org/10.12933/therya-17–473 ISSN 2007–3364.

De la Sancha, N. U., Solari, S., & Owen, R. D., (2007). First records of *Monodelphis kunsi* Pine (Didelphimorphia, Didelphidae) from Paraguay, with an evaluation of its distribution. *Mastozool. Neotrop., 14,* 241–247.

De Madrignac, B., Campi, M., & Flecha, A., (2013). Nuevos registros del género *Marasmius* (Basidiomycota-Marasmiaceae) para la región de la reserva natural Laguna Blanca, San Pedro-Paraguay. *Reportes Científicos De La Facen, 4*(2), 5–10.

Dinerstein, E., Olson, D., Graham, D., Webster, A., Primm, S., Bookbinder, M., & Ledec, G., (1995). *Una evaluación del Estado de Conservación de las Ecorregiones Terrestres de América Latina y el Caribe*. Washington, WWF, Banco Mundial.

Dirección General de Estadística, Encuestas *y* Censos (DGEEC), (2014). Pueblos Indígenas en el Paraguay, Resultados Finales de Población *y* Viviendas 2012. III Censo Nacional de Población *y* Viviendas para Pueblos Indígenas. Fernando de la Mora.

Flecha, A. M., De Madrignac, B., & Campi, M., (2013). Nuevo registro de Leucoagaricus lilaceus Singer (Agaricomycetes-Agaricaceae) para Paraguay. *Reportes Científicos De La Facen, 4*(2), 11–14.

Fundación Chaco Paraguayo *y* Geosurvey SRL, (1992). *Áreas Prioritarias para la Conservación en la Región Occidental del Paraguay, Tercera Aproximación*. Convenio Subsecretaría de Estado de Recursos Naturales *y* Medio Ambiente *y* Fundación Chaco Paraguayo. Asunción, Paraguay.

Gamble, T., Bauer, A. M., Colli, G. R., Greenbaum, E., Jackman, T. R., Vitt, L. J., & Simons, A. M., (2011). Coming to America: Multiple origins of new world geckos. *J. Evol. Biol., 24,* 231–244. https://doi.org/10.1111/j.1420–9101.2010.02184.

Gamble, T., Bauer, A. M., Greenbaum, E., & Jackman, T. R., (2008). Out of the blue: a novel, transAtlantic clade of geckos (Gekkota, Squamata). *Zool. Scr., 37*(4), 355–366.

Gatti, C., (1985). *Enciclopedia Guaraní-Castellano de Ciencias Naturales y Conocimientos Paraguayos*. Obra póstuma. Arte Nuevo Editores, Asunción.

Gauto, I., & Stauffer, F., (2017). *Palmeras del Paraguay*. Guía de identificación. Proyecto Paraguay Biodiversidad, Itaipú Binacional. Asunción, Paraguay.

Géry, J., Mahnert, V., & Dlouhy, C., (1987). Poissons characoïdes non Characidae du Paraguay. *Revue suisse de Zoologie, 94*, 357–464.

Grazziotin, F., Zaher, H., Murphy, R. W., Scrocchi, G., Benavides, M. A., Zhang, Y. P., & Bonatto, S. L., (2012). Molecular phylogeny of the New World Dipsadidae (Serpentes: Colubroidea): A reappraisal. *Cladistics, 28*, 437–459. https://doi.org/10.111 1/j.1096–0031.2012.00393.

Guyra Paraguay, (2004). *Lista Comentada de las Aves de Paraguay, Annotated checklist of the Birds of Paraguay*. Asunción.

Hawksworth, D. L., & Lücking, R., (2017). Fungal diversity revisited: 2.2 to 3.8 million species. *Microbiology Spectrum, 5*(4), 1–17.

Hayes. F. E., & F. E. Areco De Medina., (1988). *Notes on the ecology of the avifauna of Chore, Department of San Pedro, Paraguay*. El Hornero 13, 59–70.

Heck, C. I., & De Mejia, E. G., (2007). Yerba mate tea (*Ilex paraguariensis*): A comprehensive review on chemistry, health implications, and technological considerations. *J. Food Science, 72*, 138–151.

Hill, K., Padwe, J., Bejyvagi, C., Bepurangi, A., Jakugi, F., Tykuarangi, R., & Tykuarangi, T., (1997). Impact of hunting on large vertebrates in the Mbaracayu reserve, Paraguay. *Conserv. Biol., 11*(6), 1339–1353. https://doi.org/10.1046/j.1523–1739.1997.96048.

Holdridge, L. R., (1947). Determination of world plant formations from simple climatic data. *Science, 105*(2727), 367–368.

Hueck, K., (1978). Mapa de la vegetación de América del Sur. In: *Los Bosques de Sudamérica" Eachborn – Sociedad Alemana de Cooperación Técnica*.

Instituto LIFE, (2016). Ecorregiones del Paraguay – Definición de prioridades em conservación.

IPTA, (2017). *Reunión de investigación en Soja, Resúmenes expandidos Cap. Miranda: Editores Dujak Christian and Rodriguez Patricia*. Instituto Paraguayo de Tecnología Agraria. Capitán Miranda-Itapúa. Paraguay.

Jiménez, T., Cabrera, G., Álvarez, E., & Gómez, F., (2010). Evaluation of the content of Stevioside and Rebaudioside A in a population of *Stevia rebaudiana* Bertoni (kaâ heê) commercially cultivated. A preliminary study. *Memorias del Instituto de Investigaciones en Cienciasde la Salud, 8*(1), 47–53.

Knörnschild, K., & Tschapka, M., (2012). Predator mobbing behavior in the Greater Spear-Nosed Bat, *Phyllostomus hastatus. Chiropt. Neotrop., 18*(2), 1132–1135.

Koerber, S., Vera-Alcaraz, H. S., & Reis, R. E., (2017). Checklist of the Fishes of Paraguay (CLOFPY). *Ichthyological Contributions of Peces Criollos, 53*, 1–99.

Köppen, W., (1884). Die wärmezonen der erde, nach dauer der heissen, gemässigten und kalten zeit, und nach der wirkung der wärme auf die organische welt betarchtet. *Meterologische Zeitschrift, 1*, 215–226.

Kullander, S. O., & Lucena, C. A. S., (2013). *Crenicichla gillmorlisi*, a new species of cichlid fish (Teleostei: Cichlidae) from the Paraná river drainage in Paraguay. *Zootaxa, 3641*(2), 149–164.

Kullander, S. O., (2009). *Crenicichla mandelburgeri*, a new species of cichlid fish (Teleostei: Cichlidae) from the Paraná river drainage in Paraguay. *Zootaxa, 2006*, 41–50.

Laubmann, A., (1939). *Wissenschaftliche Ergebnisse der Deutschen Gran ChacoExpedition*, Die Vogel von Paraguay 1 – Stuttgart Strecker und Schroder.

Laubmann, A., (1940). *Wissenschaftliche Ergebnisse der Deutschen Gran Chaco-Expedition*, Die Vogel von Paraguay 2 – Stuttgart Strecker und Schroder.

Lopez, N., (1985). Avifauna del Departamento de Alto Paraguay, *El Volante Migratorio, 4*, pp. 9–13.

Maffi, L., (2010). Part I. Biocultural diversity: Conceptual framework. In: Maffi, L., & Ellen Woodley, (eds.). *Biocultural Diversity Conservation, A Global Sourcebook*. Earthscan, London, pp. 3–22.

MAG, (2008). Segundo informe nacional sobre el estado de los recursos fitogeneticos de importancia para la alimentación *y* la agricultura. Dirección de investigacion agricola. *Proyecto GCP/GLO/190/SP*. Asunción, Paraguay.

MAG, (2013). Plan Nacional de Yerba Mate. Propuesta Tecnica por Ejes Tematicos. *MAG-DIRECCIÓN DE INVESTIGACION AGRICOLA*. Asunción, Paraguay.

Maldonado-Chaparro, A., & Blumstein, D. T., (2008). Management implications of capybara (*Hydrochoerus hydrochaeris*) social behavior. *Biol. Cons., 141*, 1945–1952. https://doi.org/10.1016/j.biocon.2008.05.005.

Mandelburger, D., Medina, M., & Romero, M. O., (1996). Los peces del inventario biológico nacional. In: Romero Martínez, O., (ed.). *Colecciones de Flora y Fauna del Museo Nacional de Historia Natural del Paraguay*. MNHNP, Asunción, pp. 285–330.

Maubet, Y., Campi, M., Armoa, J., & Cristaldo, E., (2017). Nuevas citas de Cyathus Haller (Agaricaceae, Basidiomycetes) pa-ra Paraguay *y* ampliación de la distribución del género. *Steviana, 9*(1), 31–39.

McDiarmid, R. W., Campbell, J. A., & Touré, T. A., (1999). *Snake Species of the World*. vol. 1, Herpetologist's League, p. 511.

Melià, B., (2010). Lenguas Indígenas en el Paraguay *y* Políticas Lingüísticas. *Curriculo sem Fronteiras, 10*(1), 12–32.

Melià, B., (2012). El guaraní desde que el Paraguay es independiente. *Cuadernos Hispano-americanos, 744*, 39–54.

Mereles, F., (2007). La diversidad vegetal en Paraguay. In: Salas-Dueñas, D., & Facetti, J., (eds.). *Biodiversidad del Paraguay, una Aproximación a sus Realidades*. Fundación Moises Bertoni, USAID, GEF/BM, Asunción, pp. 89–104.

Mereles, F., Cartes, J. L., Clay, R., Cacciali, P., Paradeda, C., Rodas, O., & Yanosky, A. A., (2013). Análisis cualitativo para la definición de las ecorregiones de Paraguay occidental. *Paraquaria Nat., 1*(2), 12–20.

Montenegro, P., (2007). *Materia Médica Misionera*. 1ª ed. EdUNam, Posadas. p. 1710.

Montoya, A., (1639). *Tesoro de la Lengua Guaraní*. Madrid, , Iuan Sanchez, p. 830.

Motte, M., Nuñez, K., Cacciali, P., Brusquetti, F., Scott, N., & Aquino, A. L., (2009). Categorización del estado de conservación de los Anfibios *y* Reptiles de Paraguay. Cuadernos de Herpetología, *23*(1), 5–18.

Narosky, T., & Yzurieta, D., (2006). *Guía Para la Identificatión de las Aves del Paraguay*. Vasquez Mazzini Editores, Buenos Aires.

Neris, N., Koehn, C., Villalba, F., Ruiz Díaz, G., & Franco, G., (2008). *Guía Ilustrada de los Peces más Comunes del Paraguay*. Zamphiropolos, Asunción.

Neris, N., Villalba, F., Kamada, D., & Viré, S., (2010). *Guía de Peces del Paraguay*. Itaipú Binacional/Natura Vita, Asunción.

Norman, D., (1994). *Anfibios y Reptiles del Chaco Paraguayo, Tomo I*. Private printing, San José, Costa Rica.

O'shea, B. J., & Price, M. J., (2008). An updated checklist of the mosses of Paraguay. *Tropical Bryology, 29*, 6–37.

Peel, M. C., Finlayson, B. L., & McMahon, T. A., (2007). Updated world map of the Köpepen-Geiger climate classification. *Hydrol. Earth Syst. Sci., 11*, 2633–1644.

Petter, A., & Dlouhy, C., (1985). Nematodes de Poissons du Paraguay. III. Camallanina. Description d'une espece et d'une sous-espece nouvelles de la famille des Guyanemidae. *Revue Suisse de Zoologie, 92*, 165–175.

Pin, A., González, G., Marín, G., Céspedes, G., Cretton, S., Christen, P., & Roguet, D., (2009). *Plantas Medicinales del Jardín Botánico de Asunción. Asociación Etnobotánica Paraguaya*. Asunción.

Popoff, O. F., & Wright, J. E., (1998). Fungi of Paraquay. I. Preliminary check-list of wood-inhabiting polypores (Aphyllophorales, Basidiomycota). *Mycotaxon, 67*(2), 323–340.

Popoff, O. F., (2000). *Novedades sobre Corticioides y Políporos (Aphyllophorales, Basidiomycota) del nordeste argentino y Paraguay.* Tesis Doctoral. Facultad de Ciencias Exactas, Físicas y Naturales, Universidad Nacional de Córdoba, Córdoba.

Ramella, L., & Perret, P., (1990). Notulae ad Floram paraquaiensem, 113–115. *Candollea, 45*, 655–670.

Ramlow, J., (1989). Lista de peces *y* sitios de colección de la sección de ictiologia del inventario biológico nacional/MNHNP (Junio, (1980). – Diciembre 1988). *Boletín del Museo Nacional de Historia Natural del Paraguay, 9*, 2–38.

Rodas, O., Sayre, R., Grosse, A., & Mosseso, J., (2006). Análisis de vacíos de conservación de ecosistemas del Paraguay. *GAP Analysis Bulletin Nro., 14*, 15–17.

Rusconi, C., (1930). Las especies fósiles Argentinas de pecaríes (Tayassuidae). *Anales del Museo Nacional de Historia Natural Bernardino Rivadavia, 36*, 122–241.

Salhuana, W., & Machado, V., (1999). *Races of maize in Paraguay*, considerations in organization and utilization of maize genetic resources United States Department of Agriculture, Agricultural Research Service and The Maize Program of the Paraguayan Ministry of Agriculture and Livestock. Publication No. 25.

Sánchez, L. J. S., (1968). "Peces *y* aves del Paraguay natural ilustrado, 1767." Compañía General Fabril Editora, (Buenos Aires), Argentina, 1–511.

Sanjurjo, M., (1989). Regiones forestales del Paraguay. *La Revista Crítica, 3*(7), 53–64.

Schneider, M. C., Romijn, P. C., Uieda, W., Tomayo, H., Fernandes, S. D., Belotto, A., et al., (2009). Rabies transmitted by vampire bats to humans: An emerging zoonotic disease in Latin America? *Rev. Panam. Salud. Pub, 25*(3), 260–269.

Scott, N. J., & Aquino, A. L., (2005). It's a frog-eat-frog world in the Paraguayan Chaco: Food habits, anatomy, and behavior of the frog-eating anurans. In: *Ecology & Evolution in the Tropics: A Herpetological Perspective*. University of Chicago Press, Chicago, pp. 243–259.

Scott, N. J., Aquino, A. L., & Fitzgerald, L. A., (1991). Distribution, habitats, and conservation of the caimans (Alligatoridae) of Paraguay. *Vida Silvestre Neotropical, 2*, 43–51.

Secretaría del Ambiente (SEAM), (2016). Subsector: Conservación en Áreas Silvestres Protegidas. In: *Estrategia Nacional y Plan de Acción para la Conservación de la Biodiversidad del Paraguay 2015–2020*. Programa de las Naciones Unidas para el Desarrollo (PNUD) *y* Fondo para el Medio Ambiente Mundial (FMAM). Asunción, 88–95.

Smith, P., & Owen, R. D., (2015). The subgenus *Micoureus* (Didelphidae: Marmosa) in Paraguay: Morphometrics, distributions, and habitat associations. *Mammalia, 79*, 463–471. https://doi.org/10.1515/mammalia-2014-0050.

Smith, P., (2007). Giant Armadillo *Priodontes maximus* (Kerr, 1792). *Handbook of the Mammals of Paraguay, No. 6*. Fauna Paraguay, Encarnación, p. 11.

Smith, P., (2009). Greater Spear-Nosed Bat *Phyllostomus hastatus* (Pallas, 1767). *Handbook of the Mammals of Paraguay, No. 37*. Fauna Paraguay, Encarnación, p. 17.

Söderström, L., Hagborg, A., Von Konrat, M., & Séneca, A., (2013). The Guaraní Land. Checklist of hornworts (Anthocerotophyta) and liverworts (Marchantiophyta) of Paraguay. *Polish Botanical Journal, 58*, 267–277.

Spegazzini, C. L., (1883). Fungi Guaranitici. Pugillus I. *Anales de la Sociedad Científica Argentina, 16*(5), 242–248.

Spichiger, R., Palese, R., Chautems, A., & Ramella, L., (1995). Origins, affinities and diversity hotspots of the Paraguayan dendrofloras. *Candollea, 50*, 515–537.

TNC, (2005). Evaluación Ecorregional del Gran Chaco Americano/Gran Chaco Americano Ecoregional Assessment. The Nature Conservancy (TNC), Fundación Vida Silvestre Argentina (FVSA), Fundación para el Desarrollo Sustentable del Chaco (DeSdel Chaco) *y* Wildife Conservation Society Bolivia (WCS). Edit. Fundación Vida Silvestre Argentina. Buenos Aires. http://www.nature.org/media/aboutus/dossier_eval_ecorregional_del_gran_chaco_americano.pdf.

Vera-Alcaraz, H. S., Graça, W. J., & Shibatta, O. A., (2008). *Microglanis carlae*, a new species of bumblebee catfish (Siluriformes: Pseudopimelodidae) from the río Paraguay basin Paraguay. *Neotrop. Ichthyol, 6*(3), 425–432.

Vera-Alcaraz, H. S., Pavanelli II, C. S., & Zawadzki III, C. H., (2012). Taxonomic revision of the *Rineloricaria* species (Siluriformes: Loricariidae) from the Paraguay River basin. *Neotropical Ichthyology, 10*(2), 285–311.

Vera-Alcaraz, H. S., Rios, G., & Emhart, V. J. D., (2017). Confirmation of the occurrence for seven fish species in Paraguay. *Ichthyological Contributions of Peces Criollos, 49*, 1–9.

Villalba, F., (2012). Peces del Paraguay: Guía de identificación de setenta especies. *Natura Vita*. Asunción.

Vogt, C., (2011). Composición de la Flora Vascular del Chaco Boreal, Paraguay I. Pteridophyta *y* monocotiledoneae. *Steviana, 3*, 13–47.

Vogt, C., (2012). Composición de la Flora Vascular del Chaco Boreal, Paraguay II. Dicotyledoneae: Acanthaceae-Fabaceae. *Steviana, 4*, 65–116.

Vogt, C., (2013). Composición de la Flora Vascular del Chaco Boreal, Paraguay III. Dicotyledoneae: gesneriaceae-zygophyllaceae. *Steviana, 5*, 5–40.

Weber, C., (1986). Les poissons-chats cuirasses de la Sous-famille des Hypostominae du Paraguay. Disertación (Maestría en Estudios Superiores). Nancy, Université de Nancy.

Weiler, A., Núñez, K., Airaldi, K., Lavilla, E., Peris, S., & Baldo, D., (2013). *Anfibios del Paraguay*. Facultad de Ciencias Exactas *y* Naturales, Universidad Nacional de Asunción, San Lorenzo.

Wetzel, R. M., (1977). The extinction of peccaries and a new case of survival. *Ann. N. Y. Acad. Sci., 288*, 538–544.

Wetzel, R. M., Dubois, R. E., Martin, R. L., & Myers, P., (1975). *Catagonus wagneri*, an "extinct" peccary, alive in Paraguay. *Science, 189*, 379–381. https://doi.org/10.1126/science.189.4200.379.

Zanardini, J., & Bierdermann, W., (2006). *Los Indígenas del Paraguay. 2ᵃ edn*. Itaipú Binacional. Asunción.

Zuercher, G. L., & Villalba, R. D., (2002). Records of *Speothos venaticus* Lund, (1842). (Carnivora, Canidae) in eastern Paraguay. *Mamm. Biol.*, *67*(3), 185–187. http://dx.doi.org/10.1590/1519–6984.02514BM.

Appendix

Lichens   *Candelaria fibrosa* (Fr.) Mull. Arg. (Photo: Ulf Drechsel. Source: www.pybio. org)., *Cladonia farinacea* (Vainio) A. Evans (Photo: Ulf Drechsel. Source: www.pybio.org).

Bryophytes   *Tortella humilis* (Hedw.) Jenn. (Photo: Bernardo Cañiza), *Sematophyllum subpinnatum* (Brid.) (Photo: Bernardo Cañizza).

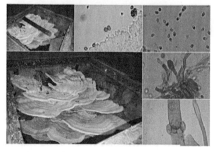

Fungi   *Vascellum pampeanum* (Photo: Michelle Campi), *Amylosporus guaraniticus* sp nov (Photo: Michelle Campi).

Vascular plants species   *Butia marmorii* Noblick (Author: Irene Gauto & Pamela Marchi), Specimens of *Syagrus romanzoffiana* (Cham.) Glassman (Author: Irene Gauto), *Miltonia flavescens* (Author: María Vera), *Cordia trichotoma* (Author: María Vera), *Tradescantia fluminensis* (Author: María Vera).

Mammals *Alouatta caraya* (Humboldt, 1812) (Author: Ulf Drechsel. Source: www.pybio.org), *Ozotoceros bezoarticus* (Author: Ulf Drechsel. Source: www.pybio.org).

Birds *Ara ararauna* (Author: Ulf Drechsel. Source: www.pybio.org), *Ramphastos toco* (Author: Ulf Drechsel. Source: www.pybio.org).

Amphibians *Hypsiboas raniceps* Cope, 1862 (Author: Sigrid Drechsel. Source: www.pybio.org), *Phyllomedusa sauvagei* Boulenger, 1882 (Author: Ulf Drechsel. Source: www.pybio.org).

Reptiles   *Geochelone carbonaria* (Author: Ulf Drechsel. Source: www.pybio.org), *Crotalus durissus* Linnaeus, 1758 (Author: Ulf Drechsel. Source: www. pybio.org).

Fishes   *Austrolebias cf. nigripinnis* (Author: Ulf Drechsel. Source: www.pybio.org), *Pterygoplichthys ambrosettii* (Author: Ulf Drechsel. Source: www.pybio.org

Invertebrates   *Sylviocarcinus australis* (Author: Ulf Drechsel. Source: www.pybio. org), *Automeris amoena* (Boisduval, 1875) (Author: Ulf Drechsel. Source: www.pybio.org).

CHAPTER 10

Biodiversity in North America (North of Mexico)

KANCHI N. GANDHI

Senior Nomenclatural Registrar, Harvard University Herbaria, Harvard University, Cambridge, MA 02138, USA; Nomenclature Editor & Member of Board of Directors, Flora of North America Association

10.1 Introduction

Biodiversity is as important as climate to a healthy, sustainable and prosperous planet. For Erik Solheim, the head of the United Nations Environment Program (UNEP), biodiversity has to be protected therefore for spiritual, ecosystem, and economic reasons. If we want to pass on a world that inspires the generations to come, the protection of every species is necessary since every species is indispensable to the maintenance of the stable ecosystems that provide the products we need for our well being. By 2020, we should aim to be at the same stage politically with biodiversity as we are with climate change.[1]

Almost 2,000 years ago, Pliny the Elder (Natural History XXXVI) voiced his concern on environment degradation; he remarked, "We quarry these mountains and haul them (marble) away for a mere whim … nature is flattened." A Native American saying is "We do not inherit the earth from our ancestors, we borrow it from our children." It has been well known that the present biodiversity reflects the vast evolution that has occurred on earth. "Living wild species are like a library of books still unread (for the most part). Our heedless destruction of them is akin to burning the library without ever having read its books" (Dingell, 1989). "Destroying rainforest for economic gain is like burning a Renaissance painting to cook a meal" (Wilson1a). Nevertheless, the ever-increasing global population, especially in Asia and Africa, poses challenges to preserve the existing biodiversity. It is also well known that for nourishing the population, the priority of the agricultural industry has been to produce a commodity and not sustenance.

Taxonomy and ecology hold the key to understand bio-diversity of any region. It is a challenge to estimate biodiversity because life on earth

---

[1] Manila, October 2017: Keynote speech to the 12th Conference of the Parties (COP 12) to the Convention on the Conservation of Migratory Speeches (CMS) of Wild Animals.

[1a] Edward O. Wilson in his interview with Jim Al-Khalili on BBC's radio show "The Life Scientific."

is mind-bogglingly diverse. Based on morphological characters, Linnaeus established a strategy to identify and quantify the species richness. Although the Linnaean system is still practiced today in various ways, it is complemented with other tools, especially with the DNA-based phylogeny.

The territory covered in this chapter encompasses the area north of Mexico plus Greenland. It measures about 21.5 million square kilometers, roughly equivalent to the land areas of Brazil, China, and India combined. It ranges between latitudes 26° and 85° N, and longitudes 15° and 173° E, and stretches from the Florida Keys northward to Ellesmere Island, and from Greenland westward to Attu Island in the Aleutian Archipelago. Widest in the north, where it is enclosed by the Arctic, Pacific and Atlantic oceans, the continent tapers sharply as it approaches the Gulf of Mexico.[2]

10.2 Climate

Climate, geology, and physiography play major roles in determining the distributions of soil classes, vegetation types, floras, and faunas in North America today (Jenny, 1941; Major, 1951; Brouillet and Whetstone, 1993). It is possibly the only continent that has every kind of climate and supports almost all major types of biomes (McDaniel et al., 2012).[3]

Climate may be understood simply as the composite weather of a region. Three major air masses affect the climate in North America: first, the cold, inherently dry, continental polar air mass of the frozen Arctic, strengthened in winter by snow-covered northwestern Canada and Alaska; second, the mild, moist maritime Pacific air mass that originates in the North Pacific and is carried eastward by the mid-latitude westerly winds; and third, the warm, moist tropical air mass originating in the Gulf of Mexico, Caribbean Sea, and subtropical western Atlantic. The climate in northeastern North America is also influenced, especially in winter, by a fourth air mass: the cold, moist oceanic North Atlantic air mass generated by the waters off Newfoundland, Labrador, and Greenland. Its influence is sporadic along the coast, not penetrating beyond the Appalachians and Cape Hatteras (cf. Figure 10.1).

10.3 Geology

In the past, the geology of North America was well illustrated and expressed by the classical maps of landforms produced by the cartographer Erwin J.

---

[2] The following information concerning North American biota was compiled from personal study and published data, the latter drawn especially from the *Flora of North America North of Mexico*, volume one, published in 1993.

[3] See: http://www.raiszmaps.com.

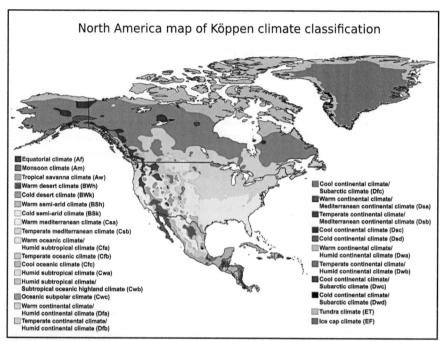

Figure 10.1 Köppen climate classification Alizifon (derived from World Köppen Classification [with authors]. svg; CC BY-SA 4.0).

Raisz in the 1960's and the data collected in the 1980's by the Decade of North American Geology Project.[4,5] These maps sought to reproduce realistically the continental landscape from information gathered on the ground. Today, data gathered by satellites with remote-sensing devices are transformed into images and maps that reveal the diversity and complexity of North America's geology.

10.3.1 Tectonic Geology

Most of North America rests on a single major plate, the North American Plate, while Baja California and California west of the San Andreas Fault lie on part of the Pacific Plate. These two plates are colliding along the deep Aleutian Trench. Strike-slip motions, that is, lateral motions along a fault line, dominate along the west coast of the United States around the San Andreas Fault and its associated faults. As a result, the Atlantic Ocean is widening, causing North America to drift slowly away from Europe. Mount

---

[4] See: http://www.raiszmaps.com.
[5] See National Centers for Environmental Information website: https://www.ngdc.noaa.gov/geomag/fliers/se–2004.shtml.

Denali in Alaska, previously known as Mount McKinley, is the highest point in North America at 6,190 m a.s.l. with Death Valley in California the lowest at 86 m below sea level (see Figure 10.2).

10.3.2 Glaciation

The geological history of the earth, known as the Quaternary, stretches back some 2.6 Ma. During this period, major glaciations that strongly affected North American biomes have occurred repeatedly. Glaciations result from both climatic and geological phenomena and modify the landscape, scouring the surface and producing surficial deposits over vast areas. These deposits might result from glacial erosion or, less directly, from proglacial lacustrine deposition or the transport of silt and sand by modified wind circulation patterns.

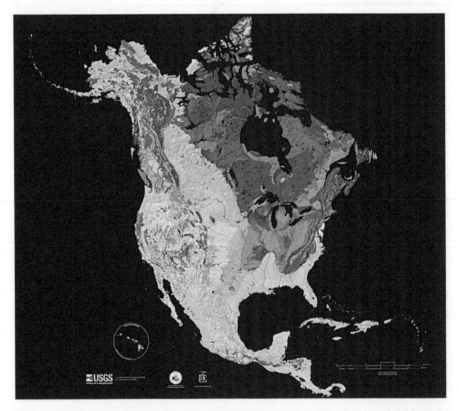

Figure 10.2 Age of the bedrock underlying North America, from red (oldest) to blue, green, yellow (newest).

The last major glaciation, the Wisconsin, began some 115,000 B.P. and ended about 7,000 years ago. It significantly defined the distribution of modern North American biota. At the point of its maximum extension about 18,000 years ago, ice covered nearly the whole of present-day Canada, the Arctic lowlands, and Greenland. Two major ice masses, confluent at that time, constituted the ice cover: The Cordilleran Ice Sheet in the Canadian Cordillera, and the Laurentide Ice Sheet covering half of the continent east of the Rocky Mountains. Sea levels lowered as much as 200 m and broad expanses of the continental shelf were exposed and thus available for plant colonization and migration. By 14,000 years ago, the ice sheets had receded perceptibly, particularly in the east, and an ice-free corridor was gradually developing between the Cordilleran and Laurentide ice masses. About 12,000 years ago, ice was receding everywhere and, by 10,000 years ago, the rate of recession was accelerating due to an increase in global temperatures. By 8,400 years ago, ice in the Canadian Cordillera and Arctic was only found in areas where glaciers persist today, and the Atlantic Coastal Plain was nearly as it is now.

Soon after this period, the ice receded permitting the sea to flow into the Hudson Bay area. This led to the separation of the Keewatin and Labrador domes that together had formed the Laurentide Ice Sheet. With this, the Wisconsin glaciation came to an end. The postglacial rebound in the earth's surface that followed led to the disappearance of the Champlain Sea, while the Great Lakes slowly assumed their present familiar contours. By 7,000 years B.P., the ice had disappeared from Keewatin, with a dome remaining on Baffin Island. The Labrador dome, for its part, slowly melted in central New Quebec and adjacent Labrador. During this same period, ice receded in Greenland, reaching its minimum range about 5,000 years ago only to readvance since then to the position it occupies today. During that same period, there was all but no ice left anywhere on continental North America, except in the glacier areas of the Cordillera.

10.4 Physiography – Geographical Features

River systems, surficial geology, bedrock types, and permafrost are features of North American geography that encroach upon the physiographical regions described further in the following subsections.

10.4.1 River Systems

North America has several major river basins including the St. Lawrence River, the Mississippi River system, the Rio Grande River system, the

Colorado River Basin, the Columbia River and its tributaries, the Fraser River, the Yukon River, the Mackenzie River Basin. Besides these, there are several regional river systems, for instance, the Saskatchewan North and Saskatchewan South, the Souris, and the Red rivers.

10.4.2 Surficial Geology

The northern half of the continent is covered with deposits such as tills and moraines that reflect the action of the ice sheets on the terrain. Whilst glacial action exposed many areas, the Central Valley of California was inundated by the sea. The fluvial and Aeolian action led to plain and loess deposits in the Great Plains and the Central Lowlands. In the mountainous, arid western United States, water and wind erosion played a major role in shaping the land, leaving bare rock and alluvial fans and plains. The Atlantic and Gulf coastal plains are today covered by coarse, stratified marine deposits of Cretaceous to Quaternary age, with marine limestone prominent in southern Florida, and fine-grained deposits present in the alluvial plains and deltas of the Yukon, Mackenzie, and Mississippi rivers.

10.4.3 Bedrock Geology

Unconsolidated deposits predominate in the Atlantic and Arctic coastal plains, in the western Great Plains, in intermontane basins and valleys, in the Central Valley of California, and in parts of Alaska. Intrusive igneous and metamorphic rocks, granites and gneiss dominate the Canadian and Greenland shields, Newfoundland, and parts of the Appalachian Mountains and adjacent Piedmont. Several batholiths are found in the western Cordillera, covering a large part of British Columbia and the Yukon. Sedimentary rocks dominate the stable platform surrounding the craton and cover the Great Plains and Central Lowlands, the St. Lawrence Lowlands, the Hudson Bay Basin, the Colorado Plateau, and the Wyoming Basin, and parts of Alaska, the Arctic Archipelago, and northern Greenland. Metamorphic formations have been folded notably in the Queen Elizabeth Archipelago, along the Appalachian Mountains and in the Ouachita Mountains, in the Rocky Mountains and Brooks Range, and along the Pacific Coast Ranges. Volcanic rocks are encountered mainly in the Aleutian Archipelago, in the interior of British Columbia, in the Columbia-Snake plateaus, and in various areas of the Basins and Ranges physiographic province.

10.4.4 Permafrost

The Arctic Ocean is almost completely covered by a permanent sea pack of ice. In winter, this continuous pack of ice extends southward to cover the Bering Sea as far south as the Alaska Peninsula in the west, reaching south of Greenland and along the Labrador Coast in the east.

Permafrost is ground that remains permanently frozen to a certain depth with the surface thawing only in summer. Continuous permafrost is found in the northern two-thirds of Greenland, the Canadian Arctic Archipelago, the Ungava Peninsula, Torngat Mountains, the south coast of Hudson Bay, northwestwardly to Great Bear Lake, the northern Yukon, and the Brooks Range, with discontinuous permafrost occurring elsewhere.

10.5 Physiography – Regions

The physiographic regions of North America have similar geological histories and maintain geographic continuity and distinctness. The *Arctic system* comprises the extreme north of the continent and borders the Arctic Ocean. The system encompasses about 1.3 million km^2, including intervening waters. The *Canadian and Greenland Shields* form the core of the continent. The Canadian Shield is a gently rolling peneplain, about 3,400 km long and 3,000 km wide, covering over 5 million km^2. The Greenland Shield drifted from the Canadian Shield during the Tertiary and is now separated from the mainland by Davis Strait. The north-south axis of the Greenland Shield is about 2,700 km long, the east-west axis, about 1,200 km. The *Appalachian system* consists of a series of roughly parallel plateaus, valleys, and ridges that extend for more than 3,000 km, from Newfoundland southwestward to central Alabama. The *Atlantic and Gulf Coastal Plains Province* forms a belt along the Atlantic Oceans and the Gulf of Mexico, extending from Cape Cod in Massachusetts south to the Mexican Border. The *Interior Highlands system* comprises two nearly disjunct areas, the Interior Low Plateau Province east of the Mississippi River, and the Ozark Plateau Province and the Ouachita Mountains Province to its west. The Interior Low Plateau Province is 470 km long and 400 km wide. The Ozark Plateau Province runs about 470 km north-to-south and 270 km east-to-west. The *Interior Plains system* stretches from the Appalachian Mountains, Interior Highlands, and Canadian Shield on the east to the Rocky Mountains on the west and from southern Texas northward to the Arctic Lowlands. It comprises three provinces: St. Lawrence Lowlands, Central Lowlands, and the Great Plains. The Interior Plains constitute the stable platform bordering the Precambrian Shield.

They cover one-third of the continent and are the most extensive system. The *Intermountain system* derives its name from being located between the Rocky Mountains and the Sierra Nevada–Cascades axis. It extends south of the Mexican border. The system comprises four distinct and geologically diverse provinces: Basin and Range, Colorado Plateau, Wyoming Basin, and Columbia River plateaus. It arose during the Laramide orogeny (Cretaceous and Tertiary) along with flanking mountain ranges. The intermountain region consists mainly of a series of plateaus that are the highest on the continent. The *Rocky Mountains system* occurs along the western edge of the stable craton of the continent and is part of the folded belt uplifted during the Laramide orogeny of the early Tertiary. The system runs northward some 5,000 km from the vicinity of Santa Fe, New Mexico, to Cape Hope, Alaska, with a bend westward near the Beaufort Sea. The *Pacific Interior system* extends 4,900 km from the southernmost Sierra Nevada in California to the mouth of the Yukon River in Alaska and is characterized by belts of rugged mountains and dissected uplands. The *Pacific Border system* fronts the Pacific Ocean. It is the longest North American system, extending for more than 7,000 km from western Alaska to Baja California.

10.6   Soils in North America

There are many kinds of soil in North America, and the geographic distribution of these soil orders is correlated to regional vegetation types and climate (Jenny, 1941; Steila, 1993). Soil classification provides a framework whereby the relationships among the kinds of soils may be compared and patterns among them studied. The soil classification scheme here follows *Soil Taxonomy*, prepared by the U.S. Soil Survey Staff (1975) (Table 10.1) (cf. Figure 10.3).

10.7   Vegetation (after Barbor and Christensen, 1993)

Historically, Joakim F. Schouw (1823), a Danish botanist, divided global plant distribution into 22 phytogeographic regions and assigned the "Kingdom of Asterias and Solidagines" and "Kingdom of Magnolias" to North America. According to Thorne (1993), due to the basically temperate nature of the flora of the United States of America and Canada, its floristic richness is considerably less than that of Mexico and several of the other nations to its south. Nonetheless, the North American continent supports a rich array of vegetation types, with most of the world's major plant formations

Table 10.1 Soil Orders and Derivation of Root Word

U.S. Soil Orders	Derivation of Root Word	Canadian Soil Orders
Entisols	(Artificial syllable)	Regosolic, some gleysolic
Vertisols	Latin: *verto*, turn upward	Some chernozemic
Inceptisols	Latin: *inceptum*, inceptive	Brunisolic, some gleysolic and podzolic
Ariisols	Latin: *aridus*, dry	Some solonetzic
Mollisols	Latin: *mollis*, soft	Chernozemic, some brunizolic, gleysolic, and solonetzic
Spodosols	Greek: *spodes*, wood ash	Podzolic
Alfisols	(Artificial syllable)	Luvisolic, some gleysolic and solonetzic
Ultisols	Latin: *ultimus*, ultimate	No equivalents
Oxisols	Greek: *oxy*, sharp, acid	No equivalents
Histosols	Greek: *histos*, tissue	Organic soils

represented (Barbor and Christensen, 1993). Tundra, boreal forests, montane conifer forest, temperate deciduous forest, prairie, steppe, woodlands of deciduous and evergreen trees, semi-arid desert scrub, Mediterranean scrub, and tidal marshes occupy significant portions of the continent. Most notably absent are extremely arid deserts and several tropical grassland, savanna, and forest types. An outline is shown in Figure 10.4.

10.7.1 Arctic Ecosystems

The Arctic, the region north of the climatic limit of boreal forest, is vegetated by a diverse assemblage of plant communities. Bliss distinguished the Low Arctic region, dominated by tundras, from the High Arctic region in which polar deserts are prevalent (Bliss, 1981). The term "tundra" refers to a continuum of treeless plant communities varying in stature from shrub heaths (1–5 m high) to low sedges, grasses, and cryptogams. Plant cover is virtually 100% in tundras, but generally less than 50% (often less than 5%) in polar ecosystems. Bliss also distinguished three arctic semidesert communities: (1) Cushion plant-cryptogam communities with *Draba corymbosa*, *Dryas integrifolia*, and *Saxifraga* spp., and lichens and mosses; (2) Cryptogam-herb vegetation, in which lichens and mosses account for 50–80%; and, (3)

II - North America

Legend

Af- Ferric Acrisols	E- RENDZINAS	Je- Eutric Fluvisols	Pg- Gleyic Podzols	We- Eutric Planosols
Ag- Gleyic Acrisols	Fh- Humic Ferralsols	Kh- Haplic Kastanozems	Pl- Leptic Podzols	Xh- Haplic Xerosols
Ah- Humic Acrisols	Fr- Rhodic Ferralsols	Kk- Calcic Kastanozems	Po- Orthic Podzols	Xk- Calcic Xerosols
Ao- Orthic Acrisols	Gc- Calcaric Gleysols	Kl- Luvic Kastanozems	Rc- Calcaric Regosols	Xl- Luvic Xerosols
Ap- Plinthic Acrisols	Gd- Dystric Gleysols	La- Albic Luvisols	Rd- Dystric Regosols	Yh- Haplic Yermosols
Bc- Chromic Cambisols	Ge- Eutric Gleysols	Lc- Chromic Luvisols	Re- Eutric Regosols	Yk- Calcic Yermosols
Bd- Dystric Cambisols	Gh- Humic Gleysols	Lf- Ferric Luvisols	Rx- Gelic Regosols	Yl- Luvic Yermosols
Be- Eutric Cambisols	Gm- Mollic Gleysols	Lg- Gleyic Luvisols	Sm- Mollic Solonetz	Zo- Orthic Solonchaks
Bh- Humic Cambisols	Gx- Gelic Gleysols	Lk- Calcic Luvisols	So- Orthic Solonetz	Water bodies (WA)
Bk- Calcic Cambisols	Hg- Gleyic Phaeozems	Lo- Orthic Luvisols	Th- Humic Andosols	Glaciers (GL)
Bx- Gelic Cambisols	Hh- Haplic Phaeozems	Mo- Orthic Greyzems	Tm- Mollic Andosols	Salt flats (ST)
Ch- Haplic Chernozems	Hl- Luvic Phaeozems	O- HISTOSOLS	To- Ochric Andosols	No Data (ND)
Ck- Calcic Chernozems	I- LITHOSOLS	Od- Dystric Histosols	Tv- Vitric Andosols	Land boundaries
Cl- Luvic Chernozems	Jc- Calcaric Fluvisols	Oe- Eutric Histosols	Vc- Chromic Vertisols	
De- Eutric Podzoluvisols	Jd- Dystric Fluvisols	Ox- Gelic Histosols	Vp- Pellic Vertisols	

Figure 10.3 North American soil.

Graminoid steppe, dominated by *Alopecurus alpinus* and *Luzula confusa* (Bliss, 1988).

Compared to temperate and boreal ecosystems, arctic ecosystems are quite young. During Pliocene times (6–3 million years B.P.), nearly all of the present Arctic was covered by coniferous forest. As might be expected, arctic plants are well adapted to grow at low temperatures. Optimal temperatures for shoot growth are generally between 10° and 20°C, and many species can maintain net photosynthesis at temperatures near 0°C.

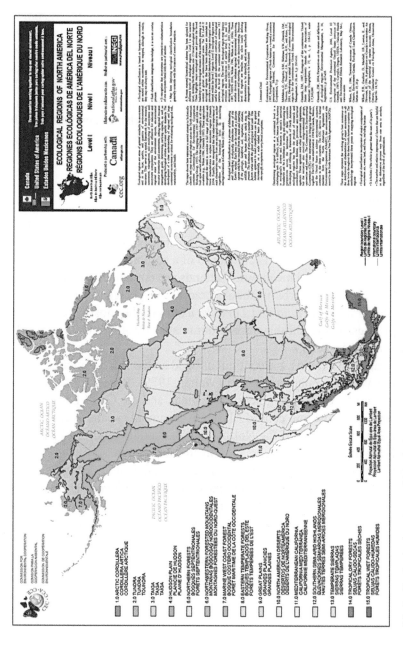

Figure 10.4 Ecological regions of North America From United States Environmental Protection Agency.

Arctic vegetation includes: tall shrub genera *Alnus*, *Betula*, and *Salix* (stature of 1 to 5 m); low shrub diminutive individuals (stature 40–60 cm) of the same *Alnus*, *Betula*, and *Salix* spp., *Arctostaphylos alpina*, *A. rubra*, *Empetrum nigrum*, *Ledum paluster*, *Rubus chamaemorus*, *Vaccinium uliginosum*, and *V. vitis-idaea*; low stature (10–20 cm) shrubs belonging to the Ericaceae, Empetraceae, and Diapensiaceae families; prominent tussocks of *Eriophorum vaginatum* (tussock grass); graminoid-moss tundras such as *Carex* and *Eriophorum* spp., and grasses, including *Alopecurus*, *Arctagrostis*, *Arctophila*, and *Dupontia*; plant cover in herb barrens varies with *Draba corymbosa*, *D. subcapitata*, *Minuartia rubella*, *Papaver radicatum*, *Puccinellia angustata*, and *Saxifraga oppositifolia*.

10.7.2 Boreal Forests

Boreal forest, or taiga, is typified by relatively dense stands of evergreen coniferous trees of modest stature and comparatively low species diversity. In addition, to forest stands, it is composed of a complex of plant communities that include bogs and meadows. Wetlands are dominated by *Sphagnum*, Ericaceous shrub, and herbs in the Asteraceae, Cyperaceae, Orchidaceae, and Poaceae. *Picea mariana* and *Larix laricina* are the most common tree species in bogs, while shrublands are dominated by *Alnus crispa*, *Betula glandulosa*, and *Salix* spp.

This formation extends as a continuous band across North America, coinciding with Walter's Zonobiome VIII (Walter, 1979). Boreal forest reaches its southernmost extent in southeastern Canada and the Great Lakes states and extends northwest into central Alaska. It commonly spans over 10° latitude and dominates 28% of the North America landmass north of Mexico, making it the most extensive formation in North America.

Larsen (1980) divided the North American taiga into seven regional zones: (1) the Alaskan region dominated by *Picea glauca*, *P. mariana*, *Alnus crispa*, *Populus balsamifera*, *Betula papyrifera*, and *Salix* spp.; (2) the Cordilleran region influenced by *Picea mariana* dominated forests; (3) the interior region with *Picea* spp., *Pinus contorta* var. *latifolia* and *Pinus banksiana*; (4) the Canadian Shield with *Picea mariana* and *Abies balsamea*; (5) Gaspé-Maritime region dominated by *Abies balsamea*; (6) the Labrador-Ungava region characterized by species of *Abies balsamea*, *Betula papyrifera*, *Larix laricina*, *Picea* spp., *Pinus banksiana*, *Populus balsamifera*, *P. tremuloides*, and *Thuja occidentalis*; and, (7) the northern boreal boundary populated by *Picea mariana*, *P. glauca*, and *Larix laricina*.

Elliott-Fisk (1988) classified upland ecosystems into closed forest, lichen woodland, and tundra-forest ecotone. Spruce-feathermoss forests are the most widespread closed forest type. Either *Picea glauca* or *P. mariana* may dominate the canopy of such forests, along with a nearly continuous bryophyte stratum of *Hylocomium splendens* and *Pleurozium schreberi*. The flora of understory vascular plants is most diverse beneath *Picea glauca–Abies balsamea*-dominated canopies. Lichen woodlands are characterized by scattered individuals of *Picea glauca* or *P. mariana* and an understory covered by the lichens *Stereocaulon paschale* or *Cladonia stellaris*.

Human impact on boreal forest landscapes includes timber and mining activities, fire events, and hydrologic modification. In more northern regions, destruction of permafrost accompanying human land use has resulted in landscape degradation. Nevertheless, extensive vigorous tracts of boreal forest remain and provide a major renewable timber resource. Since boreal wetlands are of limited economic value, many of these landscapes have remained virtually unaffected by human activity.

10.7.3 Eastern Deciduous Forests

The greater part of the eastern deciduous forest area has been subjected to significant human impact. The most arable lands was cleared for agriculture, and woodland was selectively cut and used for livestock grazing. The diverse array of forests is dominated by winter-deciduous trees (Braun, 1950). The closed canopies of these forests reach a height of over 25 m, with a complex understory. The broad northern transition of deciduous forest to taiga begins at about 45° N latitude. The western grassland–deciduous forest ecotone is often abrupt, though it meanders between 90° W and 100° W longitude. The eastern deciduous forest occupies approximately 11% of the continent and straddles Walter's (1979) Zonobiomes V (warm temperate maritime) and VI (temperate with a cold winter–nemoral). In the southeastern coastal plain, coniferous and broad-leaved evergreen trees become more common.

The mixed mesophytic forest canopy includes a diverse array of taxa including *Fagus grandifolia*, *Liriodendron tulipifera*, *Acer saccharum*, and various species of *Tilia*, *Aesculus*, *Quercus*, *Fraxinus*, and *Carya*. *Castanea dentata* was a prominent member of this community prior to the chestnut blight epidemic. These forests also possess very diverse herb and shrub floras.

The oak-hickory association extends westward to the grassland ecotone. In northern areas, these forests may be savanna-like, dominated by *Quer-*

cus macrocarpa. Farther south, *Q. stellata*, *Q. macrocarpa*, and *Carya* spp. dominate.

The *Fagus* and *Acer* (beech-maple) association predominates on glaciated terrain in Ohio, western Indiana, and southern Michigan. *Fagus grandifolia* and *Acer saccharum* are canopy dominants, with lesser numbers of *Fraxinus* and *Ulmus* spp. *Fagus grandifolia* is replaced by *Tilia americana* west of Lake Michigan in the maple-basswood association.

In the northern hardwoods association, the mixed conifer-hardwood forest is dominated by *Acer saccharum*, *Fagus grandifolia*, *Tilia americana*, *Tsuga canadensis*, *Pinus strobus*, and *P. resinosa*. Toward the northern portion of this transition, boreal species such as *Picea glauca* (in the west) and *Picea rubens* (in the east) become more important. *Picea mariana* and *Larix laricina* dominate boggy areas.

10.7.4 High Elevation Appalachian Ecosystems

Higher elevations of the Appalachian Mountains support a unique assemblage of sub-Alpine and alpine vegetation types distributed from New England to southwestern North Carolina. The upper elevational limit for deciduous forest is approximately 760 m in the Green Mountains of Vermont and White Mountains of New Hampshire. The transition to sub-Alpine vegetation rises to 1,280 m in the Catskill Mountains of New York. To the south, the highest elevations of the Appalachian complex are below the sub-Alpine transition until one reaches the Allegheny Mountains of West Virginia. Farther south in Tennessee and North Carolina, the Blue Ridge and Unaka mountains (including the Great Smoky Mountains) rise well above the 1,400–1,500 m elevational limit of deciduous forest.

As one nears the elevational limit of deciduous forest, *Picea rubens*, *Abies* spp., *Betula allegheniensis*, *B. lenta*, *B. papyrifera* (north), *Sorbus americana*, *Vaccinium* spp., and *Viburnum* spp. preponder. In the south, the endemic *Abies fraseri* is found, whereas *Abies balsamea* predominates to the north. In the snow-free areas, *Diapensia lapponica* is most common along with *Juncus trifidis*. Sedge meadows are populated mainly by *Carex bigelowii*. Other important species include *Arenaria groenlandica*, *Ledum groenlandicum*, *Vaccinium vitis-idaea*, and *V. uliginosum*.

On mountaintops and ridges between 1,700 and 1,800 m in the southern Appalachians, the most common plants include *Danthonia compressa*, *Potentilla canadensis*, *Rumex acetosella*, with *Polytrichum commune* and *P. juniperinum* in some locations. Closed-canopy shrub communities are dominated by *Kalmia latifolia*, *Leiophyllum lyonii*, *Rhododendron catawbiense*, and *R. maximum*.

10.7.5 Southeastern Coastal Plain Ecosystems

The southeastern coastal plain is composed of alluvial and marine sediments deposited since Mesozoic times and considerably reworked during the Pleistocene by fluvial and coastal geomorphic processes. It covers 3% of the North American landscape. This province is often classified as the southern mixed hardwood region of the deciduous forest, based on the assumption that successional change in the absence of disturbance such as fire would result in a climax forest dominated by a mixture of deciduous and evergreen angiosperms and evergreen gymno-ovulates. Christensen argued that the unique soils, topography, climate, and disturbance regime of the coastal plain make succession to such vegetation unlikely in many locations. Instead, upland sites are normally covered by pine-dominated forests, and low sites support a diverse array of paludal and alluvial wetlands (Christensen, 1988).

Barbor and Christensen divided the upland pine forests into three general types: (1) The Northern pine barrens that are confined to the coastal plain north of Delaware Bay with pine-oak forests dominated by *Pinus rigida*, *Quercus stellata*, and *Q. marilandica*; (2) The Xeric Sandhill pine forest that occurs in much of the southern coastal plain and is dominated by a broken overstory of *Pinus palustris*; Understory trees include *Diospyros virginiana*, *Nyssa sylvatica* var. *sylvatica*, and *Quercus laevis*; and, (3) The Mesic pine communities include so-called "flatwoods" and "savannas"(Barbor and Christensen, 1993). Flatwoods are generally dominated by an even-aged closed canopy of *Pinus palustris*, *P. taeda*, *P. serotina*, and/or *P. elliottii*. Abundant insectivorous plants (species of *Sarracenia*, *Pinguicula*, *Drosera*, and *Dionaea muscipula*) testify to the low availability of nutrients in savanna soils. In the absence of fire, broadleaf forests with *Quercus virginiana* dominate some coastal plain sites. In Florida and along the Gulf, *Magnolia grandiflora* and *Fagus grandifolia* are prominent overstory trees, along with *Quercus* spp., *Liquidambar styraciflua*, and *Pinus taeda*. In the Carolinas, such forests are dominated by *Carya* spp., *Fagus grandifolia*, *Quercus alba*, and *Q. laurifolia*.

For the vegetation of Alluvial Wetlands, Huffman and Forsythe proposed a 5-zoned classification: Zone I plant communities occupy permanent watercourses and impounded areas, e.g., *Alternanthera philoxeroides*, *Pontederia* and *Sagittaria* spp., *Azolla*, *Lemna*, and *Spirodela*; Zone II, river swamp forests, occupy river floodplains that are inundated for much of the year, with *Taxodium distichum*, *Chamaecyparis thyoides*, *Nyssa* spp. dominating; Zone III, or lower hardwood swamp forest, where soils are saturated for 40–50% of the year, but suffer water deficits in the late summer; here, dominant trees include *Quercus lyrata* and *Carya aquatica*; Zone IV forests occur in backwater and flat areas that are inundated only 20–30% of the year; *Quercus laurifolia*, *Q. phellos*, *Fraxinus pennsylvanica*, and *Ulmus americana* typify canopy trees; and, Zone V, transition to upland, comprises the highest locations of the active floodplain and includes levees, higher terraces and flats, sand ridges and dunes. Among the dominant trees here are *Quercus michauxii*, and *Q. falcata*, with *Q. nigra* on fine-textured soils, and *Quercus virginiana* on sandy sites.

In paludal (nonalluvial) wetlands, plant production is often nutrient limited. Dominant genera in these habitats include *Carex*, *Cladium*, *Juncus*, *Muhlenbergia*, *Panicum*, *Rhynchospora*, and *Scirpus*. Poor drainage in flat interstream areas results in peat accumulation and bog development over thousands of hectares. Pocosins, or shrub bog vegetation, are characteristic of the centers of such bogs. Bay forests are dominated by broad-leaved evergreen trees such as *Gordonia lasianthus*, *Magnolia virginiana*, and *Persea borbonia*. *Taxodium distichum* var. *imbricarium* often forms a monotypic, relatively closed canopy.

Within the contiguous United States, only at the very southern tip of Florida does the vegetation have a tropical aspect. The vast majority of this area is vegetated by wetlands of various kinds, or upland pine forest. Tropical hardwood hammocks do, however, occur. Broad-leaved evergreen taxa with definite tropical affinities include *Bursera simaruba*, *Ficus* spp., *Metopium toxiferum*, and *Swietenia mahogani*. Lianas and vines, as well as epiphytes in the Bromeliaceae and Orchidaceae, give these forests a particularly tropical appearance.

10.7.6 Grasslands

Grassland vegetation is dominated by herbaceous plants that form one or two canopy layers. Woody plants are restricted to local areas of distinctive topography, soil, or protection from fire. "Prairie" denotes relatively humid grasslands of nearly continuous cover, dominated by tallgrass species. "Plains," "high plains," or "steppe" denote more arid, open vegetation dominated by short grasses. The grasses are largely perennial, and they may be either rhizomatous sod-formers or nonrhizomatous bunch grasses.

Poaceae, along with Cyperaceae, dominate the biomass, but forbs account for most of the species, in particular, members of Asteraceae and Fabaceae.

As per the volumes 24 and 25 of the *Flora of North America* (Flora of North America Association, 1993), the floral area contains ca. 1,375 grass species. Ecotype differentiation among wide-ranging taxa is a common theme. Species richness increases with topographic diversity, precipitation, and distance from human disturbance in the past century. Few taxa are endemic to grasslands, however, Axelrod used this fact to support his contention that grasslands are geologically recent, dating back only 7–5 million years (Axelrod, 1985). In addition, large areas of grassland were displaced during Pleistocene glacial advances. Young & al. (1976) have suggested that the intermountain grassland is especially fragile and easily modified by human disturbance because of its evolutionary youth.

Although each regional grassland is floristically rich, a given local patch of grassland is dominated by only four or five grass species. Important growth forms include: perennial sod-forming (rhizomatous or stoloniferous) species; nonrhizomatous or short-rhizomatous perennial bunchgrass species; cool-season C_3 species active in fall, winter, and spring; warm-season C_4 species active in summer; and annuals. These growth forms are not mutually exclusive. Annual and perennials can be either cool-season or warm-season types, and certain species can be rhizomatous in home habitats, but nonrhizomatous in others.

Bunchgrasses tend to dominate the desert, intermountain, and California regions. Cool-season grasses, differently, tend to dominate north of 45° N in regions that have a July minimum temperature cooler than 8°C. They have also historically dominated in California. The several growth forms appear to exhibit metabolic, phenologic, and morphologic behaviors that adapt them to stresses such as drought, high temperatures, nutrient limitations, grazing, and fire. Bunchgrasses exhibit less resistance to intense grazing than do sod-forming grasses, leading Mack and Thompson (1982) to hypothesize that bunchgrasses evolved with lower grazing pressure than sod-formers.

The central grasslands consist of three major grassland types: (1) The tall-grass prairie extends from southern Manitoba to Texas and is dominated by *Andropogon gerardi*, *Schizachyrium scoparium*, and *Sorghastrum nutans*; (2) The shortgrass prairie extends southward into Mexico mainly at 1,100–2,500 m elevation along the flank of the Sierra Madre from the state of Chihuahua at the United States/Mexico border to 22° N in the state of Jalisco; The grasses are 20–70 cm tall. Generally, less than 50% of the ground is covered by vegetation; dominants include *Bouteloua curtipendula*, *B. gracilis*, and *B. hirsuta*; and, (3) the mixed-grass prairie is floristically the most complex of the central highlands, as it lies in an ecotone/tension zone between the tallgrass and shortgrass prairies, and thus contains taxa characteristic of both. As the annual climate varies between relatively humid and arid, species of tallgrass and shortgrass prairies are alternatively favored. Dominants of the shortgrass prairie include *Agropyron smithii, Bouteloua gracilis, Buchloë dactyloides, Hilaria jamesii, Koeleria cristata, Muhlenbergia torreyi, Sporobolus cryptandus*, and *Stipa comata*. Warm-season grasses predominate.

The warm desert grassland covers plateaus greater than 1000 m elevation at the edge of the Sonoran and Chihuahuan deserts in western Texas, southern New Mexico, southeastern Arizona, and northeastern Sonora. Within the past century, desert shrubs (mainly *Larrea divaricata* subsp. *tridentata, Flourensia cernua, Prosopis glandulosa*, and *P. velutina*) have expanded in cover, sometimes 20-fold, to become dominants.

The intermountain grassland extends from western Wyoming through northwestern Utah, southern Idaho, northern Nevada, northeastern California, and into the Columbia Basin of eastern Oregon and the Palouse area of southeastern Washington. *Artemisia* shrubs (*A. arbuscula, A. rigida, A. tridentata*) are typically associated with the cool-season bunchgrasses *Aristida longiseta, Elymus lanceolatus* (= *Agropyron dasystachyum*), *E. cinereus, Festuca idahoensis, Koeleria cristata, Pascopyrum smithii* (= *Agropyron smithii*), *Poa fendleri, P. nevadensis, P. sandbergii, Pseudoroegneria spicata* (= *Agropyron spicatum*), *Sitanion hystrix, Sporobolus airoides*, and *Stipa comata* (Risser, 1985).

The central California grassland covers 13–14% of California's area and evolved uniquely in a Mediterranean-type climate. Consequently, many of its pristine dominants were endemic. The original dominants at elevations below 100 m were the cool-season bunchgrasses *Stipa pulchra* and *S. cernua*. On gentle slopes just above valley floors, *Stipa* was joined by *Elymus glaucus, Festuca californica, Melica californica, M. torreyana, Muhlenbergia rigens*, and *Sitanion hystrix*. Bunchgrasses contributed 50% cover, with native annual grasses and annual and perennial forbs seasonally filling in the spaces between bunch.

10.7.7 *Desert Scrub*

Deserts occupy 5% of the North American landmass north of Mexico. They are found in the southwestern part of North America at elevations below 1,700 m where annual precipitation is less than 400 mm and highly variable from year to year. On a global scale, "true deserts" are areas receiving less precipitation than 120 mm/yr (Shmida, 1985). North American deserts receive more precipitation and qualify only as "semi-arid." They are dominated by desert scrub vegetation consisting of evergreen or drought-deciduous shrubs and subshrubs that provide interrupted cover, typically 5–25%, and possess low annual net productivity of less than 100 g/m². Associated growth forms are diverse and include phreatophytic trees, the roots of which penetrate to groundwater, stem succulents, bunchgrasses, ephemerals, and the ecophysiological $C_3/C_4/CAM$ photosynthetic syndromes.

10.7.8 Regional Deserts

Chihuahuan Desert vegetation can be classified into half a dozen major communities. The most common one is dominated by *Larrea divaricata* subsp. *tridentata*, associated with *Acacia neovernicosa, Agave lechuguilla, Dalea formosa, Ephedra* spp*., Flourensia cernua, Fouquieria splendens, Jatropha dioica, Koeberlinia spinosa, Krameria* spp., *Opuntia phaeacantha*, and a rich collection of *Yucca* species.

Sonoran Desert vegetation is characterized by an arboreal element. The diversity of growth forms reaches its peak in Arizona uplands and central Baja California. Overstorey elements include columnar cacti (e.g., *Carnegiea gigantea* and *Pachycereus pringlei*), or trees such as *Cercidium floridum, C. microphyllum, C. praecox, Olneya tesota, Pachycormus discolor*, and *Yucca valida*, the inflorescences of *Agave deserti* and *A. shawii*, and the bizarre boojum, *Fouquieria columnaris*. Tall shrubs include *Acacia greggii, Fouquieria splendens, Krameria grayi, K. parvifolia, Larrea divaricata* subsp. *tridentata, Lycium andersonii*, and *Simmondsia chinensis*. Subshrubs and small cacti include *Encelia farinosa* and several species of *Ambrosia, Jatropha*, and *Opuntia*. The composition of the herbaceous element depends on the season. Winter annuals flower in April and May, summer annuals flower in August and September. Few species are capable of doing both. Summer and winter annuals differ in morphology, photosynthetic pathway, phenology, and length of life. Total plant cover for all four-canopy layers can reach 25–50%.

The Sonoran Desert is rich in Cactaceae. Ecological literature is extensive for saguaro, *Carnegiea gigantea*, which visually characterizes the Sonoran landscape. It exhibits the classic CAM syndrome of traits, that is, shallow roots, slow growth rate, high water-use efficiency, and capability of tolerating relatively high tissue temperature, greater than 60°C in some cases. In addition, the saguaro has a life-span of more than 200 years.

In contrast to the Chihuahuan and Sonoran deserts, the Mojave has a low, 1 or 2 canopy layer and a rather monotonous aspect. Perennial plant density is typically less than 2000 individuals/ha, and cover is less than 10%. The bajadas are thoroughly dominated by *Larrea divaricata* subsp. *tridentata*, associated with *Ambrosia dumosa, Grayia spinosa, Lycium* spp.,

Krameria parvifolia, and *Yucca schidigera*. The Joshua-tree, *Y. brevifolia*, is more characteristic of the upper elevation limits of the desert than it is of the regional bajadas. Columnar succulents are absent.

Intermountain desert scrub has low species diversity and growth-form richness. In general, *Artemisia* spp. constitute more than 70% of the cover and more than 90% of the biomass. Species and subspecies of *Artemisia arbuscula, A. longiloba, A. nova,* and *A. tridentata* are segregated on gradients of moisture and elevation. Associated woody perennial include *Chrysothamnus nauseosus, C. viscidiflorus, Ephedra viridis, Grayia spinosa, Purshia tridentata,* and *Tetradymia glabrata*. Clearing, overgrazing, burning, and introductions of alien plants have led to significant modifications in the vegetation.

10.7.9 Mediterranean and Madrean Scrublands and Woodlands

In North America, Mediterranean and Madrean areas replace each other along an axis from central California to central Mexico. The northwest position of the axis is Mediterranean, the southeast portion Madrean. They share common flora and vegetation, but experience different climates. Mediterranean vegetation includes, from the most mesic to the most xeric, mixed evergreen forest, oak woodland and savanna, grassland, and several types of scrub, for instance, northern coastal scrub, chaparral, and southern coastal scrub.

Closely related Madrean vegetation extends from central Arizona to southern New Mexico, southwestern Texas, and south along the Sierra Madre Occidental to southern Durango, at about 22° N. Mexico is a center of oak speciation (ca. 150 species), and oak woodlands account for 5–6% of Mexico's land area. In North America, north of Mexico, however, Mediterranean and Madrean vegetations dominate only 1% of the land area.

The mixed evergreen forest has a rather dense, species-rich overstory of hardwood and needle-leaf evergreen trees. In Oregon and California, *Quercus chrysolepis* is a unifying taxon for this forest. Common associates include *Acer macrophyllum, Arbutus menziesii, Calocedrus decurrens, Chrysolepis chrysophylla, Lithocarpus densiflorus, Pinus coulteri, P. ponderosa, Pseudotsuga menziesii, P. macrocarpa, Quercus kelloggii,* and *Umbellularia californica*. Pines are an important element there, in particular, *Pinus cooperi, P. durangensis, P. engelmannii, P. leiophylla, P. lumholtzii,* and *Pinus ponderosa* var. *arizonica*.

Oak woodland generally lies below mixed evergreen forest, but floristic elements of the two may interdigitate, co-occur, or occupy different slope aspects. In Oregon and California, Oak woodland is characteristically two-layered, with an overstory canopy 5–15 m tall and 30–80% closed, which is composed of evergreen and deciduous trees; the most characteristic taxa are *Quercus agrifolia*, *Q. douglasii*, *Q. engelmannii*, *Q. garryana*, *Q. kelloggii*, *Q. lobata*, and *Q. wislizeni*, associated with *Aesculus californica*, *Juglans californica*, and *Pinus sabiniana*. In Arizona, New Mexico, and Texas, the Madrean oak woodland is characterized by *Quercus arizonica*, *Q. emoryi*, *Q. gambelii*, *Q. grisea*, *Q. hypoleucoides*, *Q. oblongifolia*, and *Q. rugosa*, and they are associated with *Pinus cembroides*, *P. edulis*, *P. leiophylla*, *P. ponderosa*, and *P. strobiformis*.

Chaparral is a dense, one-layered scrub, 1–3 m tall, composed of rigidly branched C_3 shrubs with small, sclerophyllous, leaves and extensive root systems. This scrub vegetation may be subdivided into three different types: (1) Californian chaparral occurring from Baja California to southern Oregon is characterized by *Adenostoma fasciculatum*, *Arctostaphylos* spp., *Ceanothus* spp., *Heteromeles arbutifolia*, *Garrya* spp., *Quercus dumosa*, *Rhamnus californica*, *R. crocea*, and *Rhus* spp.; (2) Madrean (or interior) chaparral extends from northern Mexico to northwestern Arizona, north central New Mexico, and western Texas, and is dominated by *Quercus turbinella*, associated with *Arctostaphylos pringlei*, *Ceanothus greggii*, *Cercocarpus betuloides*, *Garrya* spp., *Rhamnus* spp., *Rhus* spp., and several other scrub oaks; and, (3) southern coastal scrub, a variation of chaparral, dominates slopes below chaparral in southern California and in Baja California; dominant taxa include *Artemisia californica*, *Encelia californica*, *Lotus scoparius*, *Eriogonum fasciculatum*, *Malosma laurina*, *Salvia* spp., *Rhamnus ilicifolia*, *Rhus* spp., *Toxicodendron* spp., and *Yucca whipplei*.

10.7.10 Pacific Coast Coniferous Forest

This forest has been called the most productive vegetation in the world. It is dominated by a rich diversity of massive, long-lived tree species, underlain by equally rich canopies of shrubs, herbs, and cryptogams. It occupies a coastal, low elevation strip (0–1,000 m) that extends from 36° N in Monterey County, California, to 61° N at Cook Inlet, Alaska, an area of about 3% of the North American landmass.

The climate is maritime, with narrow diurnal 6°–10°C (daytime maximum minus night-time minimum) and seasonal 7°–17°C (mean of warmest month minus mean of coldest) thermoperiods. Hard frosts and persistent snow are uncommon. Studies in growth chambers with seedling conifer elements of this forest, such as *Sequoia sempervirens*, show that optimum growth occurs at zero thermoperiod. Annual precipitation is 800–3000 mm, 80% or more of which falls between 1 October–1 April, during the cool season. Summers are relatively dry, but they are moderated by cloud cover and fog drip.

Net annual productivity is 1,500–2,500 g/m², and standing above-ground biomass is 80,000–90,000 (230,000)g/m². Both exceed the values for any other forest. Furthermore, the productivity is high despite an average age of overstorey dominants at 400—1,200 years. Hardwoods are minor elements in this forest, generally restricted to specialized or seral habitats, and they have been declining in diversity and abundance for at least the past 1.5 million years. Conifers account for more volume over hardwoods by 1,000:1 for several reasons: Moderate temperatures permit net photosynthesis outside the normal growing season with possibly as little as 30% of annual net photosynthesis occurring during the growing season itself; South of 51° N, conifer needles are tolerant of the moderately high water stresses that develop during summer; and, conifers have high nutrient-use efficiency, with time periods between disturbances—on the order of several centuries—suiting the life cycles of populations with massive, long-lived individual specimens.

The northernmost forest, from Cook Inlet to the southern tip of Alaska, has a relatively simple overstory with *Tsuga heterophylla* the dominant species. A central section, much richer in tree species, extends from the southern Alaska–British Columbia border to the Oregon-California border. *Tsuga heterophylla* is dominant, and important associates include *Picea sitchensis* (along the most coastward strip), *Pseudotsuga menziesii*, *Abies amabilis*, and *Thuja plicata*. *Taxus brevifolia* is a common sub-canopy tree, and in the ground layer may be found many understory shrubs, ferns, herbs, bryophytes, and lichens. The southern part of this Pacific Coast forest, from the California-Oregon border to the Big Sur coast at 36° N, is dominated by *Sequoia sempervirens*. *Sequoia* occupies both riparian floodplains and slopes.

10.7.11 Western Montane Coniferous Forests

Coniferous forests cover the slopes of the Rocky Mountains and their extensions into Mexico (Sierra Madre Oriental and Sierra Madre Occidental), the Sierra Nevada—Cascade axis, the Coast Range of British Columbia, Washington, Oregon, and northern California, the Transverse and Peninsular ranges of southern California and Baja California, and scattered ranges and northern plateaus of the intermountain region. They extend from 65° N to 19° N latitude and from 100° W to 140° W longitude. North of Mexico, these forests dominate 7% of the landmass.

Rocky Mountain Forests are divided into four areas: (1) far north (approximately 53°–65° N latitude) forests are relatively simple, and they broadly integrate with surrounding lowland boreal forest. The common montane tree species, *Abies bifolia, Picea engelmannii*, and *Pinus contorta* are found here; (2) northern (45°–53°) forests have a Cascadian aspect, due to penetration by westerly storm tracks. Low elevations, 800–1,500 m, support a *Juniperus scopulorum–Pinus ponderosa* woodland intermixed with *Cercocarpus ledifolius* and *Purshia tridentata* shrubs; (3) south (35°–45°) zonation proceeds from a pinyon-juniper woodland through a *Pinus ponderosa* parkland, a *Pseudotsuga menziesii* forest (with *Abies concolor* and *Picea pungens*), and a *Pinus contorta* forest to a dense *Abies bifolia–Picea engelmannii* sub-Alpine forest (with an open *Pinus aristata* woodland on exposed sites); and, (4) Madrean (19°–35°) zonation typically proceeds from a pinyon-juniper woodland (north) or an oak-pine mixed evergreen forest (south) through a *Pinus ponderosa* parkland, and a dense *Pseudotsuga menziesii* forest (associated with *Abies concolor* and *Pinus flexilis*, among several other taxa with more Madrean affinities) to a sub-Alpine forest of *Picea* and *Abies*.

In the Pacific Northwest Montane Forests, montane vegetation above the temperate rainforest proceeds through an *Abies amabilis* zone to a *Tsuga mertensiana* zone and an *Abies lasiocarpa* sub-Alpine zone. Eastern slopes of the Cascade Range, and mountains to the east, exhibit a complex zonation that proceeds from a *Juniperus occidentalis–Artemisia* savanna through a *Pinus ponderosa* forest (sometimes with *Pseudotsuga menziesii*) to a mesic *Abies lasiocarpa* or *Tsuga mertensiana* forest above 1,500 m elevation.

Intermountain Region Montane Forests lie between the Cascade-Sierra axis and the Rocky Mountains. Elevations between 1,500 and 2,500 m have a pinyon-juniper woodland and coversan approximate area of 170,000 km². The woodland trees have rounded crowns that generally do not come into contact with each other. They overlie an understory of cold desert shrubs, C_3 bunchgrasses, and forbs with varying cover. Above the pinyon-juniper woodland lies an *Abies concolor* montane forest, succeeded by sub-Alpine woodland at ca. 3000 m elevation that is dominated by *Pinus flexilis* and/or *P. longaeva*.

10.7.12 Tidal Wetlands

Salt marshes are coastal meadowlands subject to periodic flooding by the sea. They are typically restricted to shorelines with low-energy waves, such as estuaries or areas behind barrier islands or spits, but otherwise they can occur on stable, subsiding, or rising coasts. The vegetation is usually a single, nearly closed layer of perennial herbs. The flora is rather simple. For example, tidal marshes along the Atlantic and Gulf coasts of North America combined support only 347 taxa of vascular plants; those along the Pacific Coast from the tip of Baja California to Point Barrow, 78 species; and, the marshes of Hudson Bay, 28.

Arctic salt marshes are scattered, and their short plants (3–5 cm tall) exhibit only 15–25% cover. Low marsh is dominated by *Puccinellia phryganodes*, and high marsh contains a mixture of *Atriplex patula, Calamagrostis deschampsioides, Carex bipartita, C. maritima, C. neglecta, C. ramenskii, C. salina, C. subspathacea, C. ursina, Chrysanthemum arcticum, Cochlearia officinalis, Dupontia fisheri, Festuca rubra, Glaux maritima, Plantago maritima, Potentilla egedii, Ranunculus cymbalaria, Salicornia europaea, Scirpus maritima, Senecio congestus, Stellaria crassifolia, S. humifusa, Suaeda maritima*, and *Triglochin maritima*.

The Atlantic Coast low marsh, from the Gulf of St. Lawrence to southern Florida, is dominated by cordgrass, *Spartina alterniflora*. Its upper marsh is dominated by *Spartina patens*, associated most commonly with *Distichlis spicata, Juncus gerardi, Limonium carolinianum, Salicornia virginica, Suaeda linearis*, and *Triglochin maritima*.

Along the United States portion of the Gulf of Mexico, cordgrass continues to dominate the low marsh. The high marsh is thoroughly dominated by *Juncus roemerianus*, but associates include *Batis maritima, Borrichia frutescens, Carex* spp., *Distichlis spicata, Fimbristylis castanea, Limonium carolinianum, Salicornia virginica, Scirpus* spp., *Sesuvium portulacastrum, Spartina patens*, and *Suaeda linearis*.

South of 29° N latitude, on both sides of the Florida peninsula, mangroves assume increasing dominance. *Rhizophora mangle* is the most seaward species, followed by a zone of *Avicennia germinans* (= *A. nitida*), *Laguncularia racemosa*, and *Conocarpus erecta*. The *Avicennia-Laguncularia* zone includes the salt marsh species *Batis maritima, Salicornia virginica, Sesuvium portulacastrum*, and *Suaeda linearis*. Sand, marl, or limestone substrates also support *Monanthochloë littoralis* and *Sporobolus virginicus* here.

In northern California, the temperate low marsh is dominated by *Spartina foliosa*, while the high marsh is dominated by *Salicornia virginica* (associated with its parasite *Cuscuta salina*), *Distichlis spicata, Frankenia grandifolia, Grindelia stricta, Jaumea carnosa, Limonium californicum, Suaeda californica,* and *Triglochin maritima.* From Oregon through western Alaska, *Puccinellia* spp. replaces cordgrass in an open low marsh, the high marsh being characterized by *Carex lyngbyei,* which also extends into brackish areas.

10.7.13 Beach and Frontal Dune Vegetation

"Beach" is that strip of the sandy substrate adjacent to open coast that extends from mean tide line to the top of the frontal dune or, in the absence of a frontal dune, to the farthest inland reach of storm waves. It is characterized by a maritime climate, high exposure to salt spray (oceanic coasts) and sandblast, and a shifting, sandy substrate with low water-holding capacity and low organic matter content. It is also subject to episodic overwash. Reviews of entire coastlines indicate that many beach taxa have wide latitudinal and longitudinal distributions, the limits of which relate more to substrate texture and chemistry, wave energy, and frequency of disturbance than to zonal climate. The dominant growth forms are rhizomatous perennial grass or prostrate, succulent, perennial forb. Woodiness, sclerophylly, and annuality are uncommon features. Plant cover tends to be low at 5—20%, and any given beach is characterized by only a dozen or fewer taxa.

In beaches facing the Arctic Ocean, the dominant taxa are: *Angelica lucida, Cochlearia officinalis* subsp. *groenlandica, Elymus mollis* (= *E. arenarius* subsp. *mollis* and *Lymus mollis*), *Festuca rubra, Honkenya peploides, Lathyrus japonicus, Ligusticum scoticum, Matricaria ambigua, Mertensia maritima,* and *Senecio pseudoarnica. Elymus* and *Festuca* are dominant grasses. On Pacific Coast beaches, common taxa include *Abronia latifolia, Ambrosia chamissonis, Atriplex leucophylla, Calystegia soldanella, Lathyrus littoralis,* and *Poa douglasii.* Introduced plants are *Ammophila arenaria* (which has largely displaced *Elymus mollis*), *Cakile maritima, C. edentula,* and *Mesembryanthemum chilense.* Along the United States shore of the Gulf of Mexico, 73 common taxa form three vegetation regions. *Uniola paniculata* dominates everywhere except Louisiana, where it is replaced by a dune ecotype of *Spartina patens.* Other important local or regionwide Gulf strand plants include *Cenchrus incertus, Croton punctatus, Ipomoea stolonifera, Iva imbricata, Oenothera humifusa, Panicum amarum, Schizachyrium maritimum,* and *Sesuvium portulacastrum.*

Along the Atlantic Coast, *Uniola paniculata* is the beach dominant, with *Andropogon virginicus, Cenchrus tribuloides, Croton punctatus, Diodia teres, Erigeron canadensis, Euphorbia* (= *Chamaesyce*) *polygonifolia, Heterotheca subaxillaris, Hydrocotyle bonariensis, Iva imbricata, Oenothera laciniata, Panicum amarum, Physalis maritima,* and *Triplasis purpurea* among the important associates. The mid-Atlantic zone is dominated by *Ammophila breviligulata,* most commonly associated with *Andropogon virginicus, Arenaria peploides, Artemisia stelleriana, Cakile edentula, Euphorbia polygonifolia, Lathyrus japonicus, Polygonum glaucum,* and *Solidago sempervirens.* Along the St. Lawrence River, beach dominants include the grasses *Ammophila breviligulata* and *Calamovilfa longifolia,* and the forbs *Cakile edentula, Corispermum hyssopifolium, Euphorbia polygonifolia,* and *Salsola kali.* Along the Newfoundland Coast (approximately 50°–52° N latitude), the dominance shifts from *Ammophila* to *Elymus.*

10.8   Flora

The Flora of North America has been built upon the cumulative wealth of information acquired since botanical studies began in the United States and Canada more than two centuries ago (Nuttall, 1818; Torrey and Gray, 1838; Flora of North America Association, 1993).

10.8.1   Lichens

Approximately 4,888 total true lichen species in 312 genera, highly diverse in size, form, and color, have been reported for North America.[6] These fascinating organisms have been found growing almost everywhere on the continent, except in the centers of large cities. Since lichens prefer unpolluted habitats, their abundant occurrence anywhere is an indication of the relevant area's atmospheric purity. Lichens with cyanobacteria as their algal partner contribute significant amounts of nitrogen to the ecosystem, as is witnessed in the forests of northwestern North America.

10.8.2   *Etracheophyta (Bryophytes, s.l.)*

Etracheophytes, the hepatics, anthocerotes, and mosses, comprise about 9.6% of the embryophyte species (etracheophytes and tracheophytes). It has

---

[6]Douglas Goldman (USDA; pers. comm.) The USDA's species no. excludes lichenicolous fungi. In contrast, Esslinger (2018) listed 5,561 total species in 755 genera (includes 615 lichenicolous fungi, 107 saprophytic fungi related to lichens or to lichenicolous fungi, and another 53 species of varying and/or uncertain biological status; https://www.ndsu.edu/pubweb/.

been estimated that 546 species in 125 genera of the hepatics and anthocerotes occur in the flora area, and there are 1,404 species in 333 genera of mosses reported for the floral area. According to Schofield (2004), there are 23 endemic genera, 20 mosses and three liverworts and, of these, 15 genera are essentially woodland or forest-dwelling. *Cryptothallus mirabilis*, a mycoheterotrophic liverwort, is entirely white and was once reported in Greenland.

A general lack of commercial value, small size, and inconspicuous place in the ecosystem has made the bryophytes appear to be of no use to most people. Now, however, contemporary plant scientists are considering bryophytes as sources of genes for modifying crop plants to withstand the physiological stresses of the modern world.

Both liverworts and mosses are often good indicators of environmental conditions. A correlation has been shown between metal distribution in mosses and that of stream sediments. Bryophytes have also been used as indicators of past climate and as indicators of soil quality in steppe forests. Bryophyte crusts, endowed with nitrogen-fixing Cyanobacteria, can contribute considerably to levels of soil nitrogen, particularly to dry rangeland soils. By intercepting sulfate ions, they prevent the formation of sulfuric acid that results in the leaching of valuable nutrients from the soil. During atmospheric precipitation episodes, bryophytes serve as filters before water reaches the soil, trapping dissolved pollutants washed from trees.

Sphagnum becomes saturated due to its tremendous water-holding capacity, and when the layer above the permafrost melts, the mosses release a vast volume of water all at once, creating problems for road-building engineers. In the state of Iowa, the genera *Barbula*, *Bryum*, and *Weissia* give important support to new road banks, helping to control erosion before larger plants became established.

10.8.3 Tracheophytes [Pteridophytes, Gymno-Ovulates (Gymnosperms and Angiosperms)]

The Flora of North America project includes ca. 1,917 genera of vascular plants, consisting of approximately 18,400 species.

10.8.3.1 Pteridophytes

Pteridophytes, the ferns and so-called fern allies, comprise about 3% of vascular plants; a total 451 species belonging to 77 genera and 25 families have been reported (Wagner and Smith, 1993). This assemblage of organisms includes very diverse elements such as aquatic quillworts (*Isoëtes*;

Lycopodiophyta); desert cliff brakes (*Pellaea*; Polypodiophyta), nearly leaf-less whisk-ferns (*Psilotum*; Psilotophyta), and weedy horsetails (*Equisetum*; Equisetophyta). Some pteridophytes, such as *Psilotum*, seem to have nearly lost their ability to speciate; many others, such as *Thelypteris*, are apparently actively evolving today.

Even in the most tropical areas of the United States, for example, in southern Florida, there are no native tree ferns. The tallest pteridophytes are the climbing ferns (*Lygodium*). They do not achieve their height by stems, but rather by extremely long leaves that clamber up other plants. Species in the flora have twining leaves that may reach 3–7 m!

With the exception of certain grape ferns (*Botrychium*) and *Isoëtes*, no modern pteridophyte displays true secondary growth: the tissues are all pri-mary, derived from a terminal meristem. Pteridophytes have no cambium, and they lack cork and secondary vascular tissues.

As in the case of the Angiosperm Phylogeny Group (APG) classifica-tion scheme, a community-derived classification based on the phylogeny at the ranks of family and below, the classification of pteridophytes follow the scheme of the Pteridophytes Phylogeny Group (PPG).

10.8.3.2   Gymno-Ovulates (Gymnosperms)

A total of 115 species belonging to 20 genera in six families have been reported for the floral area. These taxa belong to three of the four extant taxa, namely, Cycadophyta, Coniferophyta, and Gnetophyta, and grow naturally in North America (Eckenwalder, 1993). The fourth division, Ginkgophyta, was distrib-uted worldwide in the Mesozoic and Tertiary, but now has a single extant spe-cies, *Ginkgo biloba*, the maidenhair tree. It is cultivated in North America and other temperate areas, though it is apparently no longer found in natural stands.

The indigenous North American orders all possess pollen cones separate from seed cones, but differ in whether the cones of one or both sexes are simple or compound. In the Gnetophyta, both the pollen and the seed cones are compound. The seed cones of the Coniferophyta are also compound. However, their fertile dwarf shoots have been transformed into seed-bearing cone scales that deceptively resemble spore leaves.

The naturalized Australian genus *Callitris* stands apart from the native North American cupressoids. Species delimitation in Pinaceae has been rela-tively stable, perhaps because the economic importance of these plants led to early exploitation and taxonomic study throughout the continent. Taxaceae are so poorly represented in North America and the species so distantly related to one another that no controversy has arisen regarding their circumscriptions.

Some members attain dominance over immense geographic spans: some species of the Pinaceae in boreal forests; redwood (*Sequoia sempervirens*) along the coast of northern California; several species of *Juniperus* (together with pinyons) at moderate elevations in the southwestern United States and Mexico; and bald cypress (*Taxodium distichum*) in deep swamps of the southeastern United States. Their ranges and regions of dominance were considerably greater during the early Tertiary. Redwood, including *Sequoia sempervirens* and *Sequoiadendron giganteum* is the state tree of California. Furthermore, redwood is the only naturally occurring hexaploid conifer. It is one of only a few vegetatively reproducing conifers (from stump sprouts) and possibly the tallest tree species known. *Pinus longaeva*, a bristle-cone, is among the longest-lived life forms on earth.

Members of the Pinaceae are of major economic importance as producers of most of the world's softwood timber. Additionally, they are sources of pulpwood, naval stores, (for instance, tar, pitch, and turpentine), essential oils, and other forest products. All members of the family present in the flora, especially pines, are of varying importance to wildlife for food and cover. Many species, including most of the genera, are grown as ornamentals and shelter-belt trees, and for revegetation. Most commonly seen in cultivation in the flora area are species of *Abies, Cedrus, Larix, Picea, Pinus, Pseudotsuga*, and *Tsuga*, each of these genera being represented by numerous cultivars. *Keteleeria* and *Pseudolarix* are mainly botanical garden subjects. *Cathaya*, the most recently described genus, is in cultivation in North America (Callaghan, 2009).

10.8.3.3 Angiosperms

In the eighteenth century and early the nineteenth century, American floristic works followed the Linnaean sexual system of the classification (Walter, 1788; Michaux, 1803; Nuttall, 1818). Subsequently, a system based either on form relationship, that is, natural system or on phylogeny was practiced (Torrey and Gray, 1838; Gray, 1878; Britton and Brown, 1896–1898; Cronquist, 1968). Presently, the Angiosperm Phylogeny Group classification–4 is in usage.

Approximately 17,833 species belonging to 1,820 genera in 268 families of native or naturalized trees, shrubs, annual, biennial, or perennial herbs, with diverse habitats have been reported for the area. The vegetation includes C3, C4, short-day, long-day, and neutral-day plants. The Asteraceae are the most dominant family (ca. 2,415 species), followed by Fabaceae s.l. (ca. 1,800 species), and Poaceae (ca. 1,375 species). Several genera are large, possessing >100 species, e.g., *Carex* (ca. 480 species), *Astragalus* (ca. 354

species), *Penstemon* (ca. 240 species), *Eriogonum* (224 species), *Erigeron* (ca. 175 species), *Crataegus* (ca. 170 species), *Euphorbia* (ca. 140 species), and *Salix* (113 species). *Poa* (ca. 75 species) is the largest grass genus.

Yucca brevifolia, commonly known as the Joshua tree, yucca palm, tree yucca, and palm tree yucca, is an arborescent monocot. Some of the unique plants with striking morphology are: Aztec gilia (*Aliciella formosa* or *Gilia formosa*), Canelo Hills ladies' tresses (*Spiranthes delitescens*), century plant (*Agave americana*), coast live oak (*Quercus agrifolia*), corkwood (*Leitneria floridana*), dogwood (*Cornus* species), flame azalea (*Rhododendron calendulaceum*), hawthorn (*Crataegus* spp*.*), Indian paintbrush or prairie-fire (*Castilleja* spp.), mycoheterotrophites (*Monotropa uniflora*, *M. hypopitys* and *Voyria parasitica* (Florida), pallid Manzanita (*Arctostaphylos pallida*), pitcher plants, also known as cobra lilies, (*Darlingtonia* spp. and *Sarracenia* spp.), pyramid magnolia (*Magnolia fraseri*), Pando or quaking aspen (*Populus tremuloides*), scrub blazing star (*Liatris ohlingerae*), sagebrush (*Artemisia tridentata*), snow flower (*Sarcodes sanguinea*), Texas snowbells (*Styrax platanifolius* subsp. *texanus*), Tumbleweed (*Salsola* spp.), venus flytrap (*Dionaea muscipula*), wild buckwheat (*Eriogonum* spp.), Winkler pincushion cactus (*Pediocactus winkleri*), yellow mock aster (*Eastwoodia elegans*). To these, vegetation in the White Sands National Monument in the northern Chihuahuan Desert of New Mexico is added.

Tall trees include: *Liriodendron tulipifera* (ca. 49.38 m), *Fagus grandifolia* (ca. 37.34 m), *Acer rubrum* (ca. 36.7 m), *Juglans x royalis* (ca. 34.14 m), *Populus deltoides* (ca. 32.52 m), and *Quercus virginiana* (ca. 30.5 m). In contrast, the free floating *Wolffia* species are the smallest (0.4–1.6 mm). *Magnolia macrophylla* has the largest simple leaves (30–60 cm long), and *M. grandiflora* has the largest flowers (15–30 cm across). Pando's clonal colony (*Populus tremuloides*) that occupies 43 hectares in Utah has 47,000 trees, which have sprung forth from a single root system, making the pando the most massive single organism known. Salicylic acid, the active ingredient in aspirin, is found in the barks of willow (*Salix* sp.) and was used by Native Americans. *Salix* species have been found to be capable of absorbing uranium from soil (Table 10.2).

10.9   Endemism

Differently from other continents, North America has a much higher proportion of endemics. Of the known 20,349 species of Embryophytes (Bryophytes, Pteridophytes, Gymno-ovulates and Angiosperms) belonging to about 2,375

Table 10.2 Genera and Species—Estimates

Group	Total Genera	Total Species	Endemic Genera	Endemic Species
Lichens	312	4,888	N/A	N/A
Etracheophytes (Bryophytes s.l.)	457	1,950	23	ca. 400
Pteridophytes	77	451	0	ca. 220
Gymno-ovulates (Gymnosperms)	20	115	2	7
Angiosperms	ca. 2,799	17,833	ca. 700	9,377

genera, about 17,439 species are native, while 2,910 species are naturalized. Of the natives, about 10,000 are endemic species, thus making for more than 57% endemism. Floristically, California has the highest of number of species (ca. 6,382 (native: 5,287; naturalized: 1,095)), followed by Texas (ca. 5,812 (native: 4,481; naturalized: 1,331)), Florida (4,255 (native: 2,857; naturalized: 1,398)), and New York (ca. 4,000). The highest endemism among states have been recorded in California with ca. 2,281 species, whereas Texas has ca. 760 species, and Florida ca. 229 species. Of the ca. 9,377-angiosperm endemics, most of them are eudicots. In contrast, rare species are found in monocots and pteridophytes. More than 60% of the vascular genera of the floral area also occur in other areas of the world, especially in Europe and northern Asia indicating their common place in the then existing supercontinent, Laurasia.

Acknowledgments

For the most part, information on plant diversity has been compiled from the chapters included in Volume One of the *Flora of North America, North of Mexico*. In this regard, contributions from the following are acknowledged: Peter H. Raven (MO), Dr. Alan Graham (MO), Dr. James L. Zarucchi (MO), Dr. Richard H. Zander (MO), Dr. Luc Brouillet (MT), Dr. Michael G. Barbor, Norman L. Christensen, Dr. Peter Stevens (MO), Dr. Douglas Goldman (USDA), Dr. Alan R. Smith (UC), and Dr. George Yatskievych (TEX). Additionally, I acknowledge useful comments from Dr. Anthony Brach (A, GH), Dr. Subir Bandyopadhyay (CAL), Mr. Kevin J M Keane (visiting scholar to Harvard University), Dr. Bruce G. Baldwin (UC), Dr. John L. Strother (UC), Dr. B.K. Sadashiva Singh (ex National College, Bangalore), and

Mrs. Rajalakshmi Prasad (a biographer). Special thanks are extended to Prof. T. Pullaiah for the encouragement.

Keywords

- angiosperms
- endemism

- flora
- soils

References

Axelrod, D. I., (1985). Rise of the grassland biome, central North America. *Bot. Rev. (Lancaster), 51*, 63–201.

Barbor, M. G., & Christensen, N. L., (1993). Vegetation of North America North of Mexico. In: Amer, Fl. N. N., (ed.). *Mexico FNA Ed. Comm.,* vol. 1, pp. 97–131.

Bliss, L. C., (1981). North American and Scandinavian tundras and polar deserts. In: Bliss, L. C., et al., (eds.). *Tundra Ecosystems: A Comparative Analysis.* New York: pp. 8–24.

Bliss, L. C., (1988). Arctic tundra and polar desert biome. In: Barbor, M. G., & Billings, W. D., (eds.). *North American Terrestrial Vegetation.* New York: pp. 1–32.

Braun, E. L., (1950). *Deciduous Forests of Eastern North America.* Philadelphia: The Blakiston Company. p. xiv, 596 pages, illustrations, and maps.

Britton, N. L., & Brown, A., (1896–1898). *An Illustrated Flora of the Northern United States, Canada and the British Possessions From Newfoundland to the Parallel of the Southern Boundary of Virginia, and From the Atlantic Ocean Westward to the 102d Meridian.* New York.

Brouillet, L., & Whetstone, R. D., (1993). Climate and physiography of North America. In: Amer, Fl. N. N., (ed.). *Mexico FNA Ed. Comm.,* vol. 1, pp. 15–46.

Callaghan, C. B., (2009). The Cathay silver fir: Its Discovery and journey out of China. *Arnoldia, 66*(3), 15–25.

Christensen, N. L., (1988). Vegetation of the southeastern coastal plain. In: M. G. Barbour and W. D. Billings, North American Terrestrial Vegetation. Cambridge University Press; 448 pages. https://pubs.er.usgs.gov/publication/1015629

Cronquist, A., (1968). *The Evolution and Classification of Flowering Plants.* Boston.

Dingell, J. D., (1989). Foreword. In: Rohlf, D. J., (ed.). *The Endangered Species Act, Stanford Environmental Law Society.* Stanford, Calif. pp. 1–6.

Eckenwalder, J. E., (1993). Gymnosperm Classification. In: Amer, Fl. N. N., (ed.). *Mexico FNA Ed. Comm.,* vol. 1, pp. 267–271.

Elliott-Fisk, D. L., (1988). The Boreal Forest. In: Barbor, M. G., & Billings, W. D., (eds.). *North American Terrestrial Vegetation.* New York. pp. 33–62.

Esslinger, T. L., (2018). *A Cumulative Checklist for the Lichen-forming, Lichenicolous and allied Fungi of the Continental United States and Canada,* version 22, https://www.ndsu.edu/pubweb/.

Flora of North America Association. (1993+). In: Amer, Fl. N. N., (ed.). *Mexico FNA Ed. Comm.,* vols. 1–30. New York.

Gray, A., (1878–1895). *Synoptical Flora of North America,* vols. 1–2. New York.

Jenny, H., (1941). *Factors of Soil Formation: A System of Quantitative Pedology*. New York. McGraw-Hill Book Company, Inc.: New York; p. xii, 281 pages.

Larsen, J. A., (1980). *The Boreal Ecosystem*. New York.

Mack, R. N., & Thompson, J. N., (1982). Evolution in steppe with few large hooved animals. *Amer. Naturalist, 119,* 757–773.

Major, J., (1951). A functional, factorial approach to plant ecology. *Ecology, 32*(3), 392–412.

McDaniel, M., Sprout, E., Boudreau, D., & Turgeon, A., (2012). *North America: Physical Geography*. National Geographic Society, https://www.nationalgeographic.org/encyclopedia/north-america-physical-geography/, accessed on 11 Apr 2018.

Michaux, A., (1803). *Flora boreali-Americana,* vols. 1–2. Paris.

Nuttall, T., (1818). *The Genera of North American Plants, and a Catalogue of the Species, to the Year 1817*. Philadelphia.

Raisz, E. C., (1960). *Land-Use Map of the World: Land Utilization Map of the United States*. Cambridge, Mass.

Risser, P. G., (1985). Grasslands. In: Chabot, B. F., & Mooney, H. A., (eds.). *Physiological Ecology of North American Plant Communities*. New York. pp. 232–256.

Schofield, W. B., (2004). Endemic genera of bryophytes of North America (north of Mexico). *Preslia, 76,* 255–277.

Schouw, J. F., (1823). *Grundzüge Einer Allgemeinen Pflanzengeographie*. Gedruckt und verlegt bei G. Reimer, Berlin.

Shmida, A., (1985). Biogeography of the desert flora. In: Evenari, M., et al., (eds.). *Hot Deserts and Arid Shrublands*. vol. 1. Amsterdam. pp. 23–77.

Steila, D., (1993). The Soils of North America. In: Amer, Fl. N. N., (ed.). *Mexico FNA Ed. Comm.,* vol. 1. pp. 47–54.

Thorne, R. F., (1993). Phytogeography of North America North of Mexico. In: Amer, Fl. N. N., (ed.). *Mexico FNA Ed. Comm.,* vol. 1. pp. 132–153.

Torrey, J., & Gray, A., (1838–1843). *A Flora of North America: Containing Abridged Descriptions of All the Known Indigenous and Naturalized Plants Growing North of Mexico/Arranged According to the Natural System,* vols. 1–2. New York, London.

Wagner, W. H., & Smith, A. R., (1993). Pteridophytes of North America. In: Amer, Fl. N. N., (ed.). *Mexico, FNA Ed. Comm.,* vol. 1. pp. 247–266.

Walter, H., (1985). *Vegetation of the Earth and Ecological Systems of the Geo-biosphere, 3rd edn.* Springer-Verlag Berlin Heidelberg. p. XVI, 318.

Walter, T., (1788). *Flora Caroliniana Cambridge*. Mass, London.

Young, J. A., Evans, R. A., & Tueller, P. T., (1976). Great Basin plant communities – Pristine and grazed. In: Elston, R., (ed.). *Holocene Environmental Change in the Great Basin*. Reno. University of Nevada, pp. 186–215.

CHAPTER 11

Biodiversity in Venezuela

JOSÉ GRANDE

Professor of Botany and Ecology, MERF Herbarium, Faculty of Pharmacy and Bioanalysis, University of the Andes, Postal Code 5101, Mérida, Venezuela, Email: jose.r.grande@gmail.com

11.1 Introduction

Venezuela is a country completely located within the tropical belt, with a very rich and diversified biota. According to selected groups of organisms, mainly vertebrates and vascular plants, it is between the 10th and 15th place in the ranking of megadiverse countries of the world. It also occupies the 10th place in the list of Earth's most biodiverse countries, tied with the USA, according to a recently published survey, based on the BioD Index (https://news.mongabay.com/2016/05/top-10-biodiverse-countries/). Recent studies have stressed this fact, as well as the vulnerability of several of its most important ecosystems (Rodríguez et al., 2010; Villamizar and Cervigón, 2017), and most emblematic species of vascular plants (Llamozas et al., 2003) and animals (Rodríguez et al., 2015).

In the present revision, a synoptical view of Venezuelan biodiversity, including major monographic and revisionary works, is offered. Many specific studies, online checklists, and journalistic reports were considered, but they are cited only when not included in previous revisions or used in the species countings.

11.1.1 Geography, Geology, and Biogeography

Venezuela, whose official name is *República Bolivariana de Venezuela* (Bolivarian Republic of Venezuela) borders by the north with territorial seas of Dominican Republic, Netherland Antilles, Puerto Rico, Virgin Islands, Martinique, Guadeloupe, Trinidad & Tobago, and Dominica; by the south, with Federative Republic of Brazil and Republic of Colombia; by the east, with Atlantic Ocean and Cooperative Republic of Guyana; and, by the west, with Republic of Colombia. It has 23 states, a DC, and 12 island clusters included as federal dependencies. It is located at the northern tip of South America, between 00° 43'–16° 44' N, and 59° 48'–73° 11' W, with 916,445 km² of land surface, including 1,170 km² of insular territory. Isla de Aves is the northernmost point of emerged land, located at 15° 40' N, 63° 37' W, in the middle of

Caribbean Sea, extending Venezuelan sovereignty to 860,000 km² of marine areas (INE, 2014a; Marrero, 1964; http://www.isladeaves.org.ve/). Altitudinal ranges span from –8 m in Lagunillas (a sinking produced by oil exploitation in eastern coast of Lake Maracaibo), to 4,978 m in Pico Bolívar, (previously estimated between 5,007 and 5,010 m; Pérez et al., 2005). Human (resident) population is 27,227,930 according to last official census, made in 2011 (INE, 2014); estimated population for 2016, however, is 31,028,637 (INE, 2018).

Venezuelan geography encompasses a wide variety of landscapes with very interesting and old geologic history. The main physiographic regions here accepted follow Marrero (1964) and include, from west to east, Andean range, Lake Maracaibo basin, transitional zone of Lara and Falcón states, the Venezuelan Llanos or Orinoco plains (north of the Orinoco river), Venezuelan Coastal Cordillera, Guayana Shield, the Orinoco river delta and Caribbean islands. Venezuelan geography was summarized by Aguilera et al. (2007), geology by González de Juana et al. (1980), the stratigraphic lexicon (CIEN s/d; published from 1956), Urbani (2011), and the geologic map of Hackley et al. (2006). A synoptical review of Venezuelan geomorphology and soils is available in Schargel (2011). Hierarchic classification of orographic systems includes, in an increasing order of complexity, the terms Cerro (a single mountain), Sierra (mountain range), Serranía (major range), and Cordillera (several Serranías, more or less elongated); the term Macizo (massiff) is somewhat ambiguous, but usually refers to a more or less isodiametric mountain range. Following paragraphs explain main physiographic regions accepted in the present revision.

Andean region (*Los Andes*) includes three main ranges: Tamá massif (*Macizo del Tamá*), Cordillera de Perijá, and Cordillera de Mérida. Tamá massif, between Apure and Táchira states, reaches 3,613 m at Cerro El Cobre, whereas Cordillera de Perijá, in northern Zulia state, includes three main ranges: Sierra de Los Motilones, Sierra de Perijá (including Serranía de Valledupar), and Montes de Oca, reaching the highest points at Cerro Pintado (3,650 m) and Pico Tétari (3,750 m), in the central area. Cordillera de Mérida includes all the mountainous area between Táchira and Barquisimeto-Carora depressions, usually grouped into nine main subdivisions. Sierra Nevada de Mérida (maximum altitude [max. alt.]: Pico Bolívar, 4,978 m), Sierra (Nevada) de Santo Domingo, Sierra del Norte or Sierra de La Culata, Sierra del Sur or Cordillera de Tovar, Cordillera de Trujillo, Sierra de Calderas, Sierra del Rosario, Sierra de Barbacoas, and Sierra de Portuguesa. Andean region includes, besides several types of forests and azonal xerophilous vegetation with clear Caribbean affinities (Morrone, 2001; Aranguren et al., 2015), cloud forests and alpine environments with antarctic affinities (Morrone, 2001); they present high bio-diversity richness as well as high percentages of endemicity.

Between Andes and Coastal Cordillera, it is located the "transitional" (as it is usually reported in literature) Lara-Falcón system, or transitional zone of Lara and Falcón states. This region is mainly hilly, but has three main higher blocks (Serranía de Ziruma, Serranía de San Luis, and Sierra de Aroa, exceeding 1,500-2,000 m at their highest points). This is the youngest moun-tain range of Venezuela, reaching its present landscape configuartion just ca. 1 My, during Pleistocene. Biogeographic affinities are markedly Caribbean, and have being assigned to the Costa Venezolana biogeographic province by Morrone (2001). Main ecosystems include deciduous, semideciduous and cloud forests, as much as xerophilous and coastal vegetation.

The Llanos region (*Los Llanos*) was developed over a Tertiary geo-syncline, progressively filled by sediments from adjacent Guayana Shield, Coastal Cordillera, and the Andes until Pleistocene. Topography is very homogeneous and flat and it is extended to eastern Colombia, which is also included in the "Llanos Venezolanos" biogeographic province of Morrone (2001), with strong Caribbean affinities. Westernmost Western Llanos and northernmost Central Llanos were, until the twentieth century, densely cov-ered by different types of deciduous and semi-deciduous forests, interspersed with savanna strips, but are presently almost completely cleared, especially in areas previously covered by semi-deciduous associations. There thrive, however, some novel forests and forest plantations alongside introduced pastures. Southernmost Central and Western Llanos, as much as Eastern Lla-nos, are covered by a mosaic of alluvial wetlands, including riverine forests, as well as seasonal and hyperseasonal savannas.

Venezuelan Coastal Cordillera (*Cordillera de la Costa*) was uplifted dur-ing Paleogene as an island, separated from Guayana Shield by an inland sea corresponding to the present-day Llanos area. Because its antiquity and dif-ferent orientation (east-to-west) it can be easily separated from Venezuelan Andes, with which it is usually merged. Maximum heights are reached at the central and southeastern areas, where some massifs can exceed 2,000 m. It is usually divided in two parallel ranges, Serranía del Litoral, by the north, and Serranía del Interior, by the south. They are divided, in turn, in two main blocks, separated by the Unare depression. Eastern block includes Mar-garita island, Serranía de Paria and Serranía de Turimiquire ("Turumiquire" in earlier works; max. alt.: Cerro Tristeza, 2,660 m), which embraces the Turimiquire (primarily constituted by siliceous rocks and sandstone) and Car-ipe (primarily constituted by limestone, with massive caves and cliffs) massifs. Western block includes Western and Central Coastal Cordillera. Western Coastal Cordillera includes, in its northern range, Serranía de Santa María or Serranía de Nirgua (with Cerro La Chapa and Cerro Zapatero [1,400 m] as its major heights) and Serranía de Sorte (1,180 m), while in its southern range, it is only included the Cerro Azul massif (1,780 m). Central Coastal Cordillera includes Serranía del Ávila, a narrow mountain range between Caracas and Caribbean Sea, with the mountains of El Ávila, La Silla and Naiguatá (2,765 m) as its major heights; Serranía del Interior is quite discontinuous here, and only isolated massifs are found. They include the Cerros del Bachiller, Golfo Triste, and Cerro Platillón massifs, the latest reaching the highest point, with 1,930 m. Venezuelan Coastal Cordillera is considered as part of the Costa Venezolana biogeographic province, with clear Caribbean, but also Andean, Amazonian and even Guayanese affinities (Steyermark, 1974; Morrone, 2001; Meier and Berry, 2008). Venezuelan Coastal Cordillera includes mainly deciduous (especially in hills and lower mountains) and semideciduous (highly degraded/destroyed, especially at the bottom of main valleys) forests, but also evergreen lowland forests (in Barlovento and Guatopo areas), cloud forests and small areas of subalpine subpáramos (in central area, with signs of past glaciation in the Naiguatá mountain), as well as other highland shrublands (located in eastern area).

Guayana Shield (usually called Guiana Shield in literature in English, *Escudo Guayanés* in Spanish) is the most extensive physiographic unit of Venezuela. It is located south of the Orinoco river, and includes precambric rocks, locally buried by recent sediments. Spectacular tabletop mountains, known as tepuis (Figure 11.1a), are made up primarily by sandstone of Roraima formation, but are intruded by diabase, which has permitted their preservation from a former extensive table of more than 1,200,000 km^2, according to the classic theory of Gansser (1955); an alternative theory was proposed recently by Aubrecht et al. (2013), who argue for delayed erosion of lateritized materials as the main geomorphic process. Tepuis include impressive cave systems (Brewer-Carías and Audy, 2011), and can be grouped in ca. 50 more or less isolated mountain clusters that may be very high, as the Duida-Marahuaka, Parú, Jaua-Sarisariñama, Auyán, Chimantá, Aprada and Eastern Tepui Chain massifs (>2,000 m), and Sierra de la Neblina (surpassing 3,000 m). Lower tepuis, <2,000 m like the Guaiquinima and <1,000 m like Chako, however, are also common. Cristalyne massifs, made up of granite, include the Cuao-Sipapo and Maigualida ranges (>2,000 m), with batholites and lacolithes also of Precambrian lithology; lower granitic ranges include Serranía Pakaraima and Sierra Parima, below

1,500 m. Isolated granitic mountains, like Cerro Aratitiyope, or granitic *inselbergs*, especially along the course of the Orinoco, are typical geographical features of this region. Despite they have been considered "islands of time," with very ancient and singular biota, several recent radiations from groups that arose in the nearby lowlands, have been investigated recently (Salerno et al., 2012 for *Tepuihyla* frogs, and Smith, 1997 for the Bromeliaceous genus *Brocchinia*). Current biogeographic opinion includes Guayana Shield within the Amazonian subregion, including 5 biogeographic provinces; northernmost part of the central area, however, have been recently considered as a district of the Llanos province, in the Caribbean subregion (Boom, 1990; Huber, 1995; Morrone, 2001). Main vegetation types include savannas, deciduous, and semi-deciduous forests to the north and western Parima range, and evergreen lowland forests, cloud forests, several types of lowland, upland and highland shrublands and scrubs, alluvial wetlands, and peatlands in central and southern areas. Levels of endemism are very high; the frog *Oreophrynella huberi*, in fact, is perhaps the vertebrate species with the narrowest distributional range in the world, just 0.6 km^2! (Rojas-Suárez et al., 2009).

The Orinico river delta (*Delta del Orinoco*) is a megafan filled by sediment deposition from the Orinco river at its mouth, including the San Juan river estuary, nearby areas of Gulf of Paria, and flooded areas of eastern Monagas state. It is very recent in geologic time, dating back only from Pleistocene, and it is related, from a biogeographic point of view, with adjacent Trinidad island and nearby Guianas (González Boscán, 2003; Colonnello, 2004; Grande and Colonnello, 2016). Orinoco river delta physiographic province is included, however, within the Llanos Venezolanos biogeographic province in the system of Morrone (2001). Major vegetation types are alluvial and estuarine wetlands, but savannas may develop to the west.

Venezuela has more than 300 islands, islets, and keys. Despite being presently an island, Margarita can be considered a part of adjacent Coastal Cordillera, and was indeed connected to the continent until Pleistocene. Venezuelan Coastal Cordillera, in fact, extends to adjacent Trinidad island, where the Arima range is just an extension of Serranía de Paria, divided by a depression presently covered by the Caribbean Sea. Paraguaná peninsula was originally isolated from the continent, but it is presently connected to it by a narrow strip of stabilized sand. La Tortuga, La Orchila and La Blanquilla are the largest oceanic islands, while Los Roques, Los Monjes, Los Testigos, Los Hermanos, Los Frailes, and Islas Caracas are the main archipelagos. Keys may be seen at Los Roques, Isla Larga, and Morrocoy

National Park, in western half of the country. All Venezuelan islands are formally included within Costa Venezolana biogeographic province, typically Caribbean and with affinities with both the Lesser and Greater Antilles, as well as to adjacent Coastal Cordillera (Morrone, 2001). Marine and estuarine areas have been divided into 15 ecoregions; they are, according to Miloslavich et al. (2003): oceanic realm (pelagic and abyssal areas, part of Southern Caribbean Sea major ecoregion), oceanic islands (aquatic ecosystems surrounding them), Orinoco river delta, Lake Maracaibo, upwelling [eastern] region (including Margarita and adjacent islands), Cariaco Trench, Unare-Píritu, Costa Central, Golfo Triste, Tocuyo River, Golfete de Coro, Golfo de Venezuela, and Paraguaná.

11.1.2 Climate

Continental Venezuela is mainly influenced by Intertropical Convergence Zone (ITCZ) of the trade winds from the northeast and southeast, but marine regions are dependent also on northeast trade winds and several upwelling zones (MARNR-FMAM-PNUD, 2000; Miloslavich et al., 2003). Because its outstanding geographic diversity, main air annual temperatures are from 28.7°C in Las Piedras, near sea level, to <5°C at the highest peaks of the Andean mountains (Marrero, 1964). Mean annual precipitation is between 442 mm in Coro (21 m, one of the northernmost cities), to 3,336 mm in San Carlos de Río Negro (117 m, near southern tip of the country). Major cyclic climatic oscillations include El Niño Southern Oscillation (ENSO; relatively dry for Venezuela) and associated La Niña (relatively humid for Venezuela) in the short-term, and glaciation events in the long (geologic) term.

11.1.3 Previous Biodiversity Studies

In spite of the rich and complex ethnoecological and ethnotaxonomical knowledge treasured by aboriginal groups, biodiversity studies in Venezuela began, according to western scientific tradition, with the third voyage of Admiral Columbus, made in 1498, as it can be read in his diary (http:// www.cervantesvirtual.com/obra-visor/los-cuatroviajes-del-almirante-y-su-testamento-0/html/ff83a3be-82b1-11df-acc7-002185ce6064_5.html).

Biodiversity studies were scarce, but fruitful, during the era of Spanish domination, which spanned for about three centuries (from 1498 to 1811). They include some early encyclopedic works, as the *Historia General y Natural de las Indias* (1851–1855) by Fernández de Oviedo, and *Iter Hispanicum* (1758), the (postume) *magnum opus* of Pehr Löfling. Most important works from that time, however, were those of Alexander von Humboldt and Aimée Bonpland; the narrative of their *Voyages* (von Humboldt and Bonpland, 1807), made between 1799 and 1804, is a masterpiece of world literature, and inspired Charles Darwin to travel around the world as a naturalist (http://darwin-online.org.uk/EditorialIntroductions/Chancellor_Humboldt.html). Additional published studies from colonial times with relevant information on biodiversity include that of de Oviedo y Baños (1723), Gilij (1780–1784), Gumilla (1731), and Depons (1806), but several additional naturalists were also visiting Venezuela, and their studies were published, usually within larger works, elsewhere. First recorded attempts to study Venezuelan biodiversity, once emancipated, are those of José María Vargas, who was in contact with several of most renowned scientists of his time, including the botanist Augustin Pyramus de Candolle (Guerrero, 2003; Lindorf, 2008). After the trips of several European naturalists, travelers and agents of plant breeders, including Karl Moritz, the Schomburgk brothers, Jean Jules Linden, Hermann Karsten and Alfred Russell Wallace, in 1861 arrived from Germany Adolf Ernst (Lindorf, 2008). He, his pupils (especially Alfredo Jahn), and several foreign naturalists (e.g., Theophile Raymond and Pierre Henri George Bourgouine, also a pharmacist) made collections and studies while living in the country. After founding a Natural History Cathedra in the Caracas University (presently Central University of Venezuela) in 1874, Ernst published the first studies on national flora and fauna (Ernst, 1877a, b, c, 1891). Despite his efforts, those of Caracciolo Parra Olmedo in Mérida University (presently University of the Andes; Chalbaud, 1965), and animal collections prepared by Salomon Briceño Gabaldón and José Briceño Gabaldón (La Marca, 2003a), it was not until the establishment of Henri F. Pittier in the country, in 1920, when systematic collection of (botanical) specimens began all around the country (Lindorf, 2008).

Although some pioneer works, as those of Deflandre (1928) on freshwater green algae, and Chardon and Toro (1934) on fungi and lichens are worth of mention, modern biodiversity studies were markedly skewed to medical issues in earlier times. They include important contributions to parasitology, protistology and microbiology by José María Vargas (cited above), Rafael Rangel (a pupil of José Gregorio Hernández, pioneer of microbiological studies in the country), Enrique Tejera, José Francisco Torrealba, Arnoldo Gabaldón (controlling Malaria between 1936 and 1965) and José Vicente Scorza, among others. One of the pupils of Pittier, Tobías Lasser, founded formal studies in Biology at Central University of Venzuela in 1958. Since then, and especially after the so-called *renovación univer-*

sitaria of 1969, many Professors, mainly from Europe and South Cone of South America, went to the country. They include plant ecologist Volkmar Vareschi (Austria), plant anatomist Ingrid Roth (Germany), lichenologist Manuel López Figueiras (Spain), and plant ecologists Maximina Monsterio and Guillermo Sarmiento (Argentina), who spent much of their academic careers on national universities and research institutions. With the joint work of European, American (USA), Latin American, and local experts, prospecting of Venezuelan biodiversity have been intensifying since. They include eminent botanists as Julian A. Steyermark, Basset Maguire, John J. Wurdack, and Stephen Tillett, ornithologists as William H. Phelps, Jr., herpetologists as Juan Arturo Rivero and Stefan Gorzula, ichthyologists as Barry Chernoff and Antonio Machado-Allison, and phytogeographers as Otto Huber, among others, who have joined many expeditions.

Previous biodiversity summaries for Venezuela include the MARNR-FMAM-PNUD official report for the elaboration of the national plan on biodiversity (*Estrategia Nacional de Diversidad Biológica* or ENDIBIO; MARNR-FMAM-PNUD, 2000), followed by two publications on the state of knowledge of vascular plants (Huber et al., 1998) and fungi (Iturriaga et al., 2000), the encyclopedic *Biodiversidad en Venezuela*, in two volumes (Aguilera et al., 2003a), two works of Miloslavich et al. (2003, 2005), on marine areas, and an official website of the national Ministry of Environmental Issues (http://diversidadbiologica.minamb.gob.ve). Specific studies have been usually centered on taxonomy and distributional data of isolated taxonomic groups (mainly plants, birds, reptiles, fishes, and amphibians), but the works of FLASA (1953), Tello (1968), Brewer-Carías (1988), Huber (1992), Rosales and Huber (1996), Huber and Rosales (1997), Chernoff et al. (2003), Lasso et al. (2004a, 2009b), Lasso and Señaris (2008; only animals), Barreto et al. (2009), and Rial et al. (2010) include plenty of information from different taxonomic groups in an rather integrative manner. Genetic diversity is still poorly explored, and besides some information from several studies on different taxonomic groups (usually neither centered nor made in Venezuela) and a recent review on amphibians (vid. infra), there are no synoptical treatments for this biodiversity component. Under taxonomic diversity, phylogenetic diversity is usually implicit, but estimates for it are lacking for all Venezuelan ecosystems.

Relevant theoretical work on biodiversity has also been done in the country. They include the panbiogeographic method of the famous Italo-Venezuelan botanist and biogeographer León Croizat, a diversity index for biological communities (Bulla, 1994), a unit to measure morphological clinal variation in geographical contexts (Molinari, 2007), innovative criteria for inferring phylogenies and the delimitation of taxonomic groups (Grande, 2016), and a standard procedure to evaluate the conservation status of global ecosystems (Rodríguez et al., 2011). Major dissemination work have been made by the magazines *Natura* (now extinct) and *Río Verde*.

11.2 Diversity of Ecosystems

In the present revision, Venezuelan biomes are divided into three main categories: aquatic ecosystems, terrestrial ecosystems and wetlands. Terrestrial ecosystems and wetlands are ordered from the sea to inland areas, while aquatic ecosystems are listed in the opposite direction. Aquatic ecosystems are divided according to the chemical composition of water. That way, they are classified as: (1) marine, with a high concentration of salts, particularly NaCl, (which is 3–5%), (2) freshwater, associated with terrestrial environments, and with low to very low salt concentration (but including some salty inland lakes), and (3) estuarine, a combination of former two types. Terrestrial ecosystems and wetlands are classified, moreover, based on altitudinal range. Traditionally, they have been grouped according to physiognomy and floristic composition, but this could result inadequate, at least in some instances. Depending upon which criteria or combination of criteria are employed, they can be classified in different ways, but many of the resulting classifications will agree because of strong correlations between biotic and abiotic factors. Wetlands, here defined as temporally to permanently flooded terrains with rooted and emergent vascular plants, are transitional areas between typical terrestrial and typical aquatic ecosystems, variously defined regarding on the objectives and field of interest of the user (cf. Mitsch and Gosselink, 2015). Further subdivisions of major ecosystems here proposed are not offered, but they could be established based on similar criteria.

Additional biomes as yet not reviewed for Venezuela, and thus not considered here, include subterranean and aerial ecosystems. There exists, however, an important body of literature on geology, hydrology and zoology of local cave systems; biospeleothems (a especial type of stromatolite out of the reach of flowing water; Figure 11.3c) were recently reviewed for Guayana Shield by Aubrecht et al. (2008). Information on remaining cave biota is included within the headings "animals", "crustaceans", "insects" and "myriapods" (vid. infra). Deep-buried substrates have not been studied at any length, while aerial ecosystems were considered, but not described, as a 14th aerial zone in a schematic transect from Caribbean Sea to the Llanos in Beebe and Crane (1948). Present-day extension and conservation

Table 11.1 Current Status of Venezuelan Ecosystems and Biomes

	Ecosystems		Source	Surface	Conservation status
Aquatic ecosystems	Freshwater ecosystems	Lake Valencia	González Jiménez, 2003	912 km²	–
		High Andean lakes	González Jiménez, 2003	100 km²	–
		Glaciers (1)	Braun and Bezada, 2013	ca. 0.11 km²	CR
		Dams	González Jiménez, 2003	10,782 km²	–
	Estuaries	Orinoco delta	Miloslavich et al., 2003	30,000 km²	–
			González Jiménez, 2003	20,642 km²	–
		Lake Maracaibo	González Jiménez, 2003	12,870 km²	–
	Coastal ecosystems	Sandy beaches	Miloslavich et al., 2003	–	–
		Rocky shores	Miloslavich et al., 2003	–	–
		Coastal lagoons	Miloslavich et al., 2003	6,737 km²	–
		Coastal-marine	González Jiménez, 2003	25,061 km²	–
		Continental coasts	González Jiménez, 2003	2,015 km²	–
		Insular coasts	González Jiménez, 2003	2,439 km²	–
		Seagrass prairies (2)	Miloslavich et al., 2003	>3,73 km²	–
		Coral reefs (3)	Miloslavich et al., 2003	>1,800 km²	–
	Pelagic and abyssal ecosystems	Pelagic realms	Miloslavich et al., 2003	900,000 km²	–
Terrestrial ecosystems	Coastal vegetation (4)		Oliveira-Miranda et al., 2010	1,738 km²	VU (EN, CR)
	Xerophilous vegetation		Oliveira-Miranda et al., 2010	20,297 km²	EN (CR)

	Ecosystems	Source	Surface	Conservation status
Savannas	Seasonal savannas (5)	Oliveira-Miranda et al., 2010	99,412 km²	VU (LC, EN, CR, EL)
	Hyperseasonal savannas (6)	This work	18,130 km²	NT
Lowland and upland shrublands and scrubs	Guayana shrublands and scrubs (7)	Oliveira-Miranda et al., 2010	11,134 km²	LC
	Psammophilous (Guayanese) scrubs	Oliveira-Miranda et al., 2010	8,083 km²	LC
Deciduous forests	—	Oliveira-Miranda et al., 2010	28,858 km²	CR (VU, EN, EL)
Semideciduous forests	—	Oliveira-Miranda et al., 2010	56,691 km²	EN (LC, VU, CR)
Evergreen lowland forests	—	Oliveira-Miranda et al., 2010	311,496 km²	VU (NT, EN, CR)
Montane forests	Cloud forests	Oliveira-Miranda et al., 2010	7,078 km²	VU (EN, CR)
Tepui shrublands and scrubs (8)	—	Oliveira-Miranda et al., 2010	9,863 km²	arb. LC (VU); her. NT (VU)
Páramos	—	Oliveira-Miranda et al., 2010	4,117 km²	VU (CR)
Alpine deserts (9)	—	This work	ca. 30 km²	EN
Wetlands / Coastline wetlands	Mangroves	Miloslavich et al., 2003	3,000 km²	–
Estuarine wetlands	Herbaceous vegetation	Oliveira-Miranda et al., 2010	8,530 km²	LC (EN, CR)
	Palm and swamp forests	Oliveira-Miranda et al., 2010	5,960 km²	LC (NT, VU)
	Palm and swamp forests	Oliveira-Miranda et al., 2010	18,611 km²	LC (NT)
Alluvial wetlands	Palm and swamp forests (10)	Oliveira-Miranda et al., 2010	4,491 km²	LC (VU, EN, CR)
	Mauritia palm swamps	González Jiménez, 2003	1,200 km²	–

Table 11.1 (Continued)

	Ecosystems	Source	Surface	Conservation status
	Riverine forests	Oliveira-Miranda et al., 2010	66,707 km^2	VU (LC, NT, EN, CR)
	Riverine shrublands	Oliveira-Miranda et al., 2010	12,017 km^2	LC
	Banco, bajio and *esteros* "savannas"	González Jiménez, 2003	25,300 km^2	–
	Paspalum fasciculatum "savannas"	González Jiménez, 2003	25,380 km^2	–
	Sabanas húmedas (Barinas state)	González Jiménez, 2003	2,080 km^2	–
	Sabanas veraneras (Guayana)	González Jiménez, 2003	1,402.5 km^2	–
	Orinoco delta "savannas"	González Jiménez, 2003	227,62 km^2	–
Lacustrine wetlands	Rice fields	González Jiménez, 2003	1,638 km^2	–

Sources of reported information are included in cited literature.

(1) Estimated extension for 2011, including active glaciers and remaining ice/firn fields; conservation status: CR A2-3, C2(a), B1, D.

(2) Data available only for Morrocoy National Park.

(3) Data available only for Los Roques (1,500 km^2) and Las Aves (300 Km^2) archipelagos.

(4) Including 307 km^2 from Zulia state and 6 km^2 from Sucre state, reported by the authors as "sabanas abiertas."

(5) 87,767 Km^2 of shrubby and/or woodland savannas, plus 11,645 km^2 of "sabanas abiertas" from Anzoátegui, Monagas, and Bolívar states.

(6) Estimated as 1/3 of 54,320 km^2 of wet grassland reported by González Jiménez (2003); conservation status according to criteria A1 and C2(a).

(7) Including associated wetlands and half of ca. 2,000 km^2 of saxicolous vegetation reported by authors.

(7) Including associated wetlands and half of ca. 2,000 km^2 of saxicolous vegetation reported by authors.

(9) Conservation status: C2(c) and D.

(10) Not including estuarine wetlands from Zulia and Delta Amacuro states.

Figure 11.1 Examples of Venezuelan ecosystems: (a) Roraima tepui, showing from base to top, a mosaic of evergreen upland forests, shrublands and savannas, followed by cloud forests, tepui shrublands and scrubs, tepui peatlands (lower areas of the mountain summit), and highland lakes (photo: Charles Brewer-Carías); (b) Semi-deciduous forests and alluvial wetlands along the Tocuyo river (Caribbean basin), including one *cañabraval*, a dense clump of the reed *Gynerium sagittatum*, on the right bank of the river (photo: José Grande); (c) Caribbean coast at Morro de Choroní (Aragua state), terrestrial ecosystem is xerophilous scrub, while marine ecosystems include rocky beaches and pelagic realm (photo: José Grande); (d) Alpine desert and glaciers, in Sierra Nevada National Park, between Pico Bolívar (in the foreground) and Pico Humboldt (at the bottom; photo: Yeni Barrios, 16 XI 2016); (e) Chaguaramal (*Roystonea oleracea* palm forest) in a semi-deciduous forest life zone at the Tocuyo river basin, near Las Colonias, Falcón state (photo: Giuseppe Colonnello); (f) Páramo ecosystem in Sierra de Calderas, near Páramo de Guirigay, *Espeletia schultzii* is in the foreground (photo: José Grande); (g) Vegetation mosaic during dry season in northeastern Bolívar state, from bottom to foreground may be observed deciduous and semi-deciduous forests, *Trachypogon* savannas, riverine semi-deciduous forest, and alluvial wetlands with *Roystonea oleracea* and *Mauritia flexuosa* (photo: Giuseppe Colonnello); (h) Upland shrubland over sandstone at the base of Sororopán-tepui near Mantopai, southeastern Bolívar state (photo: José Grande).

status of Venezuelan ecosystems are available in Table 11.1. Examples of some of them are included in Figure 11.1.

11.2.1 Aquatic Ecosystems

Venezuelan aquatic ecosystems include several types of marine, freshwater and euryhaline water bodies. Major marine ecosystems are seagrass prairies, coral reefs, and coastal, pelagic and abyssal realms. Venezuelan marine waters are located in a transitional zone between two main biogeographic provinces, the Caribbean and the Brazilian, separated by the freshwater outflows of the Orinoco and Amazon rivers (Caballer Gutiérrez et al., 2015). Fairly atypical and uncommon are fumaroles and Andean lakes from xerophytic uplands, the last of them with relatively high salt concentration.

The outcome of a workshop on marine zoological diversity held in Guanare in 1999, in which more than 30 Venezuelan researchers participated, is offered in Table 11.4 of Miloslavich et al. (2003: 294); there, 2,697 species, 1,382 genera and 558 families are reported, but further information on them and updated numbers are available in Miloslavich et al. (2005). Species lists could be retrieved, although for selected groups only, from http://www.marinespecies.org/.

11.2.1.1 Freshwater Ecosystems

Included here are lotic (springs, rivulets, and rivers), and lentic (lakes and lagoons) water bodies, including sandy and muddy beaches surrounding them, within six main watersheds (Orinoco, Amazonas [through the Casiquiare branch], Lake Maracaibo, Caribbean, Cuyuní-Esequibo, and Lake Valencia). Glaciers, although moving downwards, are in solid phase, and are considered here, thus, separately. Additional information for Venezuelan freshwater ecosystems is available in González, E. et al. (2015).

Lotic ecosystems (Figure 11.1b) have been recently reviewed, with emphasis on conservation concerns, in the series *Ríos en riesgo de Venezuela* (Rodríguez-Olarte, 2017 onwards). They can be classified according to water chemical composition which, in turn, affect biological communities and ecosystem function. Black, clear and white waters may be distinguished according to sediment budget, conductivity, pH, and organic matter content. River bottoms and nearby beaches have been studied more or less intensively for water quality assessments; a recent study of the meiofauna of the Apure river, however, have revealed an interesting and diverse ensemble of organisms, including previously unrecorded phyla (vid. infra, under "Other animal groups"). Information on subterranean rivers is reviewed in Galán and Herrera (2017). Venezuela present no active volcanoes, but several hot springs are distributed over northern mountain ranges. Most of them only produce minor effects on adjacent vegetation, and a few of aquatic, usually widespread plants, are either floating, rooted or submerged. Hot springs from Sanare (locally called *Fumarolas de Sanare*) and small patches recently discovered between Lara and Portuguesa states, however, can produce major alterations to adjacent environments. Vegetation surrounding them is zonal, but some small patches are devoid of vegetation and can be humid throughout the rainy season. They are probably inhabited by an interesting microbiota, including thermophilic archaebacteria, as yet to be studied.

Lentic ecosystems include lakes, with a stable water level throughout annual cycle, and lagoons, with major water level fluctuations year-round. Lake Valencia is the largest endorheic water body of Venezuela; a rich benthic and planktonic fauna, and two endemic fishes (*Atherinella venezuelae*, endangered *fide* Campo and Ortaz, 2015; and *Lithogenes valencia*, reported as extinct in Provenzano, 2015) have been thriving there and in its tributaries, but pollution from adjacent cities, plantations and industries, as much as the introduction of commercial fishes from Africa since 1958 (especially *Oreochromis mossambicus*; Provenzano, 2015), have been seriously disturbing this unique ecosystem. Small lakes (locally known as *lagunas*), are common in Andean region (Gordon, 2016); they include highland glacial lakes (some of them still being formed, as glacier retreat continues) and several upland small units, from xerophilous areas in the Chama valley. The last of them have been experiencing a steady desiccation process since nineteenth century, but Laguna de Urao is now being rehabilitated by local authorities and citizens (http://infoviasvzla.com.ve/merida-los-meridenos-emprenden-recuperacion-de-la-laguna-de-urao/); some highland smaller lakes from Pantepui (Fig. 11.1 a) are also worth of mention, but information on them is very scarce to negligible. A major system of natural lagoons develops along the Cariaco-Casanay depression including Laguna Campoma and Laguna Buena Vista, both ca. 1 m depth. Their maximum amplitude is reached during rainy season, when water drains towards Gulf of Cariaco through a narrow natural channel; ornamental fish *Poecilia wingei* is endemic to the basin. Ephemeral lagoons present extreme water-level fluctuations, being completely dry to muddy during dry season. They include aquatic plants, fishes (some of them endemic) and, probably, a rich and as yet unexplored microbiota. Only one summary on vegetation from the granitic inselbergs of Guayana Shield, however, has been published (Lasso et al., 2014b). Several dams have been built in Venezuela for agriculture, hydroelectricity and/ or water supply since the beginnings of the twentieth century. Major units include Lake Guri and the Calabozo dam, with many square kilometers, but many others, as Taiguaiguay and Zuata dams, in the Aragua valleys, are smaller.

Surrounding the two highest peaks of Sierra Nevada de Mérida there still exist small to tiny remnants of the once extensive glaciers that covered, until the first years of the twentieth century, its five main massifs (Marrero, 1964; Braun & Bezada, 2013). The only remaining active glacier is at the base of Pico Humboldt (Figure 11.1d), but additional ice remnants and firn fields are at the base of Pico Bolívar (Braun and Bezada, 2013). To date, the only known inhabitants of Venzuelan glaciers are several classes of cryophilic bacteria (Ball et al., 2013; Rondón et al., 2016). Study of glacial ice is of great importance for paleoclimatological studies, but to our knowledge this has never been done for Venezuela. Using the same criteria as for terrestrial ecosystems, Venezuelan glaciers ranks as CR (Table 11.6), being the most endangered ecosystem in the country.

11.2.1.2   Estuaries

Major Venezuelan estuaries are located at the Maracaibo system (including Lake Maracaibo, Gran Eneal and Laguna de Sinamaica estuaries, among other subsystems; Rodríguez, 1973 and Medina and Barboza, 2006) and the Orinoco river delta (reviewed in Lasso et al., 2004a, and Miloslavich et al., 2011), the last of them with *ca.* 300 km of coast and some 100 minor subsystems (Cervigón, 2003). Economic activities include oil exploitation and fisheries, especially in Lake Maracaibo; there, *Callinectes sapidus* and *Litopenaeus schmitti* are valuable native species, but cultivation of exotic shrimps (*Litopenaeus vannamei*; also in adjacent Falcón state) is widespread. Additional disturbance is produced by the salt wedge that penetrates through the navigation channel built for the oil shipping, and by the introduction of fishes *Cichla orinocensis* and *C. temensis* from Orinoco river for the sport fishing; adjacent wetlands and upstream terrestrial ecosystems gives off sediments and nutrients, which, given recent deforestation, plantations and ranching, have produced duckweed flourishing and additional pollution. Fish communities of Venezuelan estuaries are differentiated according to water depth and coastal geomorphology, with characteristic species in each of the resulting environments (Cervigón, 2003).

11.2.1.3   Coastal Ecosystems

Venezuela possesses more than 4,000 km of coastlines. Rocky coasts (Figure 11.1c), are common in the Gulf of Santa Fé, in northern Araya-Paria peninsula, Margarita and Cubagua islands, Central Coastal Cordillera (locally called *Litoral Central*), Cerro Chichiriviche, northern Paraguaná peninsula and Los Monjes, Gran Roque, and La Blanquilla islands; remaining coastlines are occupied mainly by sandy beaches (Conde and Carmona-Suárez, 2003). Continental bioregion includes all coastal environments between 0 and −100 m; temperatures are high (28°C), and surrounding terrestrial ecosystems are mangrove forests (Miloslavich et al., 2003), as much as strand vegetation and xerophilous shrublands. Insular bioregion is made up of 314 islands and islets. Isla de Patos is the only federal dependency in the Atlantic Ocean, the remainder being located in the Caribbean Sea; major ecosystems are coral reefs, seagrass beds, sandy beaches, and rocky shores (Miloslavich et al., 2003), usually surrounded by xerophilous vegetation and with a very low input of freshwater. Southern Caribbean is very rich ecologically and taxonomically because of two upwelling zones, one in Colombia, the other in eastern Venezuela, which influence nearby pelagic ecosystems. Coastal lagoons are especially common in Unare-Píritu region, of which Laguna de Píritu is experiencing higher sedimentation rates (Conde and Carmona-Suárez, 2003). Especial ecosystems are developed over rocky platforms in central coasts (Ardito et al., 1995) and Paraguaná peninsula, and they harbor a very rich and highly characteristic macrophycobiota.

11.2.1.4   Seagrass Prairies

Developed in soft-bottom sublittoral environments, seagrass prairies are, despite their reduced size, one of the most productive ecosystems on Earth. Distribution and general information at the national level are available in Conde and Carmona-Suárez (2003), surface estimates in Miloslavich et al. (2003), conservation status in Lentino and Bruni (1994), and species composition of epiphytic macroalgae in Vera (1993; 57 spp. listed); high number of mollusks and microorganisms have been recorded (Vera, 1992; Conde and Carmona-Suárez, 2003). Seagrass beds serve as refuge, food supply and nest sites for a variety of animals, including commercial species of fishes and shrimps, as well as sea turtles. Native seagrass species include *Halodule wrightii, H. beaudettei, Halophila baillonis, H. engelmannii, H. decipiens, Ruppia maritima, Syringodium filiforme,* and *Thalasia testudinum* (which is the most abundant species) in at least 39 localities, usually near coral reefs

(Vera, 1992; Conde and Carmona-Suárez, 2003). Very regrettably, these communities have been destroyed at an increasing rate for the development of coastal resorts in the absence of regulatory legislation (Vera, 1992).

11.2.1.5 Coral Reefs

Coral reefs are constituted, mainly, by colonies of petreous hexacorals (class Anthozoa), fire corals (class Hydrozoa), and calcified red algae, differentially distributed along depth gradients that may vary according to geographical location. They are usually formed in transparent marine waters, *ca.* 20°C, from near sea surface to −50 − −70 m. Due to bulky freshwater input from continental lands, major coral reef formations in Venezuela are bordering oceanic islands. Most of the available information has been produced in studies conducted at Los Roques archipelago and Morrocoy National Park. Los Roques archipelago harbors the most important coral formation in southern Caribbean, with two large parallel bands of more than 20 km long and several minor reefs bordering each of the numerous keys, including 59–69 species of coral-forming cnidaria (Villamizar and Cervigón, 2017). Information on Venezuelan coral reefs was summarized in Conde and Carmona-Suárez (2003), while surface estimates were offered in Miloslavich et al. (2003). Massive bleaching has been reported locally, both by pollution and climate warming (Conde and Carmona-Suárez, 2003; Villamizar and Cervigón, 2017), as by the ENSO climatic events (del Mónaco et al., 2012). In Venezuela, cnidarian, fish and mollusk diversity is remarkable; regarding fishes, Cervigón (2003) quotes that in coral reefs occur the largest diversity in species, more than 300 only in Los Roques, with Gobiidae and Clinidae very abundant, and Chaetodontidae, Pomacentridae and Pomacanthidae very characteristic. Coral-forming species varies from 11 to 57, across 12 major localities in Venezuela (of a total of at least 15); they include, primarily, species of the genera *Millepora* (especially *M. alcicornis*), *Acropora, Agaricia, Diploria, Montastrea, Porites* and *Siderastrea*, but also of *Colpophyllia, Eunicea, Gorgonia, Meandrina, Plexaura, Plexaurella,* and *Pseudopterogorgia*; two coral reefs, both from Isla de Aves, are made up primarily by the reef-forming red algae *Lithotamnion* sp. (Conde and Carmona-Suárez, 2003). Coral reefs are threatened by uncontrolled tourist development, overfishing, oil refineries and petrochemical industry, deforestation of nearby wetlands, eutrophication, ship tank wash and dumping, overfishing, uncontrolled tourism, and occasional oil slick and shipwrecks (Weil, 2003).

11.2.1.6 Pelagic and Abyssal Ecosystems

Southern subarea of Caribbean Sea Large Marine Ecosystem (CSLME) encompasses most of continental and insular coasts of Venezuela, Trinidad & Tobago, Bonaire, Curaçao, Aruba and northeastern continental coast of Colombia. This is the only place of CSLME with important upwelling processes, which determine its very high productivity. Climate change, however, has weakened these phenomena during several years the last two decades, with severe consequences in fish production; the recent crisis of the sardine is perhaps the clearest example of it (Villamizar and Cervigón, 2017).

Twelve upwelling zones have been identified along the coast of Venezuela by means of Advanced Very High Resolution Radiometer images (AVHRR). There, water temperature may decrease between 5°C and 7°C from normal temperatures of 24–26°C. On the west coast, the upwelling zones are located east of the La Guajira peninsula, including Gulf of Venezuela and three localities along central Falcón state (viz., near Punto Fijo, Cape San Román, and Puerto Cumarebo). In the central coast, the upwelling at Cape Codera is unusual since it has been observed only occasionally, when there were no other upwelling fronts. Most important upwelling systems are located along the eastern coasts, between Puerto La Cruz and Cumaná, near Carúpano, and north of the Araya peninsula and Margarita island (Miloslavich et al., 2003). Upwelling occurs simultaneously at all these locations, driven by strong winds during first six months of the year, spreading over 200,000 km² and contributing to the most productive area in the Caribbean Sea. The productivity of the coastal waters is determined mainly by the upwelling dynamics, but also by the uptake of allochthonous material coming from large rivers and lakes. Venezuelan eastern coasts are the most productive, though seasonal, and its productivity is controlled both by the strength of upwelling and the permanent discharge from the Orinoco river; this area maintains seasonal concentrations of chlorophyll-*a* above 3 mg L⁻¹ and a total annual fish production of between 200,000 and 250,000 tons. Productivity of Gulf of Venezuela is mainly sustained by nutrients coming from adjacent Lake Maracaibo basin, but coastal areas have been affected there by oil extraction and the petrochemical industry for more than 60 years, with moderate to very high levels of pollution found in some areas such as the Tablazo and Amuay bays (Miloslavich et al., 2003). Biodiversity studies in pelagic domains have focused on fishes, but illuminating work on other taxonomic groups has been carried out in eastern Venezuela (Sánchez-Suárez and Díaz-Ramos, 2003; Zoppi de Roa, 2003; Zoppi de Roa and Pardo, 2003).

Benthopelagic (–50 – –1,000 m) and abyssal (>–500 m) Venezuelan ecosystems are virtually unexplored, with only few studies in the Cariaco Trench. This is an anoxic basin from 375 meters, to the bottom at *ca.* –1,400 m. Although supposed as very similar to other abyssal areas in tropical and even temperate zones, it could harbor an interesting and largely unexplored biota, including fishes (cf. Cervigón, 2003). Biodiversity of the Atlantic seabed has been reported as relatively poor, but very few studies have been done. There, local habitats develop predominantly on sand and mud.

11.2.2 *Terrestrial Ecosystems*

Traditionally, classification of terrestrial ecosystems is based upon vegetation, and makes use of a hierarchic system based on the concept of plant formation (e.g., Beard, 1944; UNESCO, 1973), in which physiognomy and floristics are the most important factors. Classification of Venezuelan terrestrial ecosystems, based on those criteria, have been worked out by Huber and Alarcón (1988), who listed 150 vegetation units and include distributional data, associated physiographic features and main floristic elements, Huber et al. (1998), with 650 vegetation types non-explicitly listed, but with gross estimates of the number of vascular plant species in each major unit, and Huber and Oliveira-Miranda (2010), with 12 major units plus 15 subunits, including aquatic and degraded vegetation as distinct types. These revisions, that of González Boscán (2013), only for ecosystems north of the Orinoco river, and revisions of Soriano and Ruiz (2003), Silva (2003), González Boscán (2003a, b), Hernández and Demartino (2003), Ataroff (2003), Riina and Huber (2003), and Azócar and Fariñas (2003) are, to date, the most updated and used references. Phytosociological studies are still scarce, but classifications are available for several savannas, páramos, alluvial wetlands, and anthropic vegetation. A summary for the Llanos region is available in Guevara et al., 2013, and for Andean region in Cuello and Cleef, 2009a, b and c; additional works include Vareschi (1968), for savannas and secondary grasslands, Velásquez (1965), for seasonal savannas, and Meier (2004) for cloud forests.

In the present revision, major terrestrial ecosystems are classified integrating bioclimate, physiognomy and floristics in the same essential way as in Huber and Oliveira-Miranda (2010), but wetlands are set apart form (typical) terrestrial ecosystems, while riverine forests are considered either as semi-deciduous or evergreen terrestrial forests when not under cyclic and prolonged flooding. Pioneer formations are considered as *facies* of major ecosystems, in which they are spatially included, while saxicolous vegetation, excluding dry vegetation refugia (as deciduous forests in granitic inselbergs of northern`Amazonas and northwestern Bolívar states), are included within adjacent units. Lowlands (0–500 m), uplands (500–1,500 m) and highlands (above 1,500 m) are defined as in Huber (1995), whereas montane and alpine belts are considered, respectively, as those areas below and above the tree line. Forests are defined, as it is usual, as a plant formation where trees (plants with a definite trunk, (3–)5(–8–10) m height) have their crowns interlocking (UNESCO, 1973), i.e. forming a canopy. Tepui meadows are here considered as peatlands. The physiognomic types "shrublands", "scrubs", "heaths" and "woodlands" tend to be interchangeably used. The term "heath" is primarily used for shrublands and scrubs, but also for (gnarled) montane forests, while "woodland" tend to be used for savannas and xerophilous vegetation, more or less densely covered by small trees. In the present revision, terrestrial ecosystems are differentiated based on most important growth form, with shrublands dominated by shrubs ≥ 1.2–1.5 m, above the breast height, and scrubs dominated by subshrubs, usually < 1 m height. Herbs may be present, sometimes even in higher proportion than ligneous or subligneous species (e.g., psammophilous scrubs dominated by Rapateaceae; vid. infra, under "lowland and upland sbrublands and scrubs"), but they are never dominated by grasses. Venezuelan alpine (microthermic) and subalpine (submicrothermic) belts are dominated by shrublands, scrubs, meadows and highland wetlands, and may be divided in páramo (Andes and Central Coastal Cordillera) and tepui formations (Guayana Shield); Eastern Coastal Cordillera may harbor an additional group, with both Caribbean and Guayanan affinities, as yet not recognized. Forests are here classified according to leaf phenology (viz., deciduous, evergreen and semideciduous); they were classified recently, also, according to angiosperm tree phylogeny (Slik et al., 2018). From a total of five types, four are humid and one is dry ("dry tropical," both deciduous and semi-deciduous), the last one through America, Africa, Madgascar, and India. Further research is needed to confirm whether this indicates the existence of a global dry forest region with a shared biogeographic origin, or selection for drought and fire resistance has favored the dominance of similar plant lineages in such units around the world. Of remaining four types, American humid forests, more closely associated with African forests, are present in Venezuela. Humid American forests, however, are more closely related to dry units than to remaining humid forest zones, viz. the Indo-Pacific (including eastern Africa and Madagascar) and (boreal) Subtropical, the last type extending to montane forests of northern Andes outside Venezuela (Slik et al., 2018).

The only Venezuelan desert is alpine desert, recognized by Vareschi (1970), but usually neglected in recent literature. Sand dunes in the so-called *sabanas de médanos* of southeastern Apure and southern Guárico states were truly *ergs*, similar to that of the Sahara, but are presently fossilized, while coastal dunes of northern Falcón state, near Coro, should be considered a special type of coastal ecosystem within the xeric life zone.

Air temperature decreases ca. 0.67°C each 100 m (Jahn 1934, Röhl 1951). Pittier (1935), based on resulting climatic belts, coined the terms *tierra caliente, tierra templada, tierra fría* and *tierra gélida*. They are treated here as macrothermic (> 24°C of mean anual temperature), mesothermic (12°C–24°C), microthermic (5°C–12°C), and hypermicrothermic (< 5°C, including cryothermic and nival belts of Cuatrecasas, 2013); precipitations occur as pluviosity in the first three, and usually as snow in the last one. Dry months are defined as those with mean precipitation < 50 mm, humid months as those between 50 and 100 mm, and perhumid months as those with > 100 mm. The anual cycle of pluviosity may be dry or xeric (≥ 6 dry months, xerophilous), seasonal (2–6 dry months, tropophilous), and humid (< 2 dry months, ombrophilous), following the Köppen's classification system (Köppen, 1936 and 1948).

11.2.2.1 Coastal Vegetation

Under this category coastal sand dunes and strand vegetation are here included. Strand vegetation could be herbaceous or shrubby. Shrubby subtype could be relatively tall near mangroves, with the arborescent species *Conocarpus erectus* and *Talipariti pernambucensis* ssp. *pernambucensis* usually common. Villareal et al. (2014) made a significant contribution to the study of coastal vegetation of the Zulia state, distinguishing sandy beaches, active sand dunes, primary and secondary (stabilized) sand dunes, coastal prairies in saturated soil, mangroves, and vegetation surrounding coastal lagoons (the last two, wetlands). Strand or, more properly, psammohalophilous coastal vegetation, may extend several kilometers inland, as in La Guajira peninsula, where they form extensive meadows dominated by *Sporobolus* spp. (Tamayo, 1964). The so-called "desert" of the Médanos de Coro Natural Monument, in northern Falcón state are, in fact, coastal sand dunes within a xeric shrubland area; the parasitic plant *Lennoa madreporoides* have been, to date, only found there. Main threatens for this ecosystem include aesthetic purposes, as usually beaches are cleared for tourism, as well as coconut plantations, but invasion by alien species, as *Calotropis* spp. (in sand dunes), and *Scaevola taccada* (in coastal scrubs; Grande and Nozawa,

2009) are increasingly important. Strand vegetation in sandy beaches is very endangered since human pressure is growing, and protected areas are scarce and frequently visited by tourists.

11.2.2.2 Xerophilous Vegetation

Macothermic (widespread) and submesothermic (restricted to Andean ranges) xerophilous plant formations, with pluviosity <600 mm, have been reported in northern Venezuela (Oliveira-Miranda et al., 2010). This vegetation type is usually called *espinares* or *cujizales* (dominated by leguminose shrubs to small trees as *Prosopis juliflora* and/or *Vachellia* spp., locally named *cujíes*) and *cardonales* (dominated by columnar cacti, locally named *cardones*), sometimes *espinares-cardonales* or *matorrales* (the last of them a fuzzy term also applied to anthropic vegetation by local scholars). Strictly speaking, they are xeromorph phytocenoses dominated by shrubs and subshrubs but, occasionally, also small trees, with a high percentage of spiny and evergreen ligneous species. They can be found near the marine coasts, but also in arid areas inland, including arid pockets, as much as leeward valleys and slopes. Reviews for this biome include Mateucci (1986), and Soriano and Ruiz (2003), which is centered on azonal Andean mid-elevation arid pockets, and include relevant data on zoogeography, interaction ecology and conservation status. Globural and nopal-like cacti, spinescent rosettes of the genera *Agave* and *Furcraea*, some grasses, reptiles (mainly lizards and lethal species of the viper genus *Crotalus*), rodents, bats, birds and arthropods are common inhabitants of these ecosystems. Based on personal observations, xerophilous vegetation seems to be very limited near the sea and associated marshes, probably because of the effects of higher concentrations of the sodium ion in saline spray, but also by the excessive accumulation of sand. There, over rocky places, in initial successional stages, and in disturbed areas, a xerophilous scrub may develop. Main threats to this biome include goats and invasive plant species, especially from families Apocynaceae subfamily Asclepiadoideae, Crassulaceae and Euphorbiaceae. Figure 11.1c shows a xerophilous scrub bordering rocky beaches in Central Coastal Cordillera.

11.2.2.3 Savannas

Savanna is one of the most debated ecosystems on Earth, with different definitions and disputed origins. Regarding local classifications, Tamayo (1964) includes all the macrothermic herbaceous formations dominated by grasses, even the halophytic meadows from coastal zones. Ramia (1967), focusing in the Llanos, proposed the *Trachypogon, Paspalum fasciculatum* and *banco, bajío* and *estero* savannas, which were associated to Tertiary sedimentary tables, riverbanks, and flooding plains, respectively; Ramia himself, however (Ramia, 1993, 1997), grouped them later into humid and dry, depending on whether or not they are subject to flooding, and according to their nutritional content (relatively high for flooded or humid savannas). Sarmiento (1990), instead, classified savannas according to seasonality and duration of flooding, dividing them into seasonal (with two marked seasons, one dry, the other humid), semiseasonal (with extensive flooding and modest dry season) and hyperseasonal (with two humid seasons separated by one dry and one perhumid season). Silva (2003), finally, divided them into seven landscape areas and two major biogeographic units (Guayana and Llanos), following the geographical distribution of three main types proposed by Sarmiento. Considering that semiseasonal savannas are better classified as wetlands, savannas *sensu stricto* (Figure 11.1g) are here defined as tropical ecosystems with a lower homogeneous layer of bunch grasses of C_4 metabolism, usually with an upper stratum of gnarled shrubs and/or small trees. They include, thence, from sclerophyllous woodland of *Curatella americana* (locally called *chaparrales*), with a very scarce but continuous grass layer, to open *Trachypogon* and *Axonopus* savannas, but also the *Mesosetum* spp. and *Anthaenanthia lanata* hyperseasonal units, under periodic flooding, and with *Caraipa llanorum* as a common tree species. Regarding eutrophic seasonal savannas, Grande (2007) offered a geobotanic study for two contrasting types in Central Coastal Cordillera; one of them, the *Agave* savannas, develops over slightly alcaline soils, and it is generated by anthropic burning from xeric shrublands over limestone.

Major threats to this biome are agriculture and ranching, with the substitution of native grasses and forbs by African forage species. Secondary grasslands, sometimes with arboreal and/or shrubby elements are somewhat common near true savannas, but soils and floristic composition are very different. The Aguaro-Guariquito National Park (north of the Orinoco river, including natural and seminatural ecosystems) and Canaima National Park (south of the Orinoco river, mainly anthropic) are the major protected areas in Venezuela. Recovery of tree layer after fire suppression have been studied from 1960 at the Calabozo biological station; the results of such investigations have been summarized by Fariñas (2013).

11.2.2.4   Lowland and Upland Shrublands and Scrubs

In contrast with true xerophytic vegetation, azonal shrublands and scrubs are the product of mechanical and/or nutritional soil constraints. Included here are most of Guayana Shield nonforested ecosystems developed over granite, sandstone (Figure 11.1h) and white sand. This vegetation type shows relations with eastern paleotropics, and extends into adjacent Colombia and Brazil, as well as Perú. Psammophilous scrubs (*herbazales arbustivos sobre arena* in Huber and Oliveira-Miranda, 2010) are located in western Amazonas state (southern Venezuela), primarily along mid and lower Ventuari, lower Casiquiare and Atabapo and Guainía rivers penneplains. Despite they occupy only *ca.* 8,083 km² (or 0.9% of Venezuela land surface), they are ecologically and floristically very important, with ancient taxa and high endemicity (Huber, 1995; Riina and Huber, 2003). Lowland and upland Guayana shrublands are closely related with tepui shrublands and scrubs, both in physiognomy and floristics.

Over karstic terrains, in northern Venezuela, some shrublands also develop. Most salient examples include those over calcite and limestone, transforming into secondary *Trachypogon-Agave* savannas once they are repeatedly burned (Grande, 2007; vid. supra), and *Euphorbia caracasana-Agave cocui* shrublands over arrecifal limestone on central Aragua and northern Guárico states. Major threat for Venezuelan shrublands and scrubs is mining. Calcicolous ecosystems of northern Venezuela in particular, are seriously threatened by mining for cement and lime, but also for anthropic burnings, only rarely associated with agriculture or cattle ranching.

11.2.2.5 Deciduous Forests

Deciduous forests are usually mixed with semi-deciduous forests, especially the tropophilous subtype, savannas (Figure 11.1g) and shrublands in a rather complex mosaic. Current tendency called them Seasonally Dry (Tropical) Forests (SDTFs; after Bullock et al., 1995), and could be differentiated into macrothermic and submesothermic, according to altitude and associated climate (Oliveira-Miranda et al., 2010). Despite traditional consideration of them as relictual, according to the Pleistocene Arch Theory (PAT, updated in Mogni et al., 2015), Mayle (2004) consider them as originally different, but similar by posterior dispersal events. In Neotropical SDTF, population migration and long-distance dispersal may have occurred across the Cerrado and the Chaco, via smaller peripheral areas of SDTF on calcareous soils, but there is no strong molecular evidence for this (Mogni et al., 2015). Most common floristic elements include families Leguminosae (especially trees of the genera *Senegalia* and *Pithecellobium*), Burseraceae, Capparaceae and Cactaceae (both trees and shrubs), but also Bromeliaceae, including terrestrial and epiphytic herbs.

Extensive *Handroanthus* forests in northern Anzoátegui and Monagas states are one of the most beautiful ecosystems of Venezuela, with amazing bing-bang flowering at the end of the dry season (Oliveira-Miranda et al., 2010). They are, very regrettably, not formally protected. Despite being very disturbed, and considered as the most endangered terrestrial ecosystem of Venezuela (Ibid.), several non-forested ecosystems, as alpine desert and strand vegetation could be even more endangered.

11.2.2.6   Semi-Deciduous Forests

This is, perhaps, the most overlooked forest type of Venezuela. Tropical (and subtropical) semi-deciduous forests, in fact, have been merged with deciduous forests in global (e.g., Slik et al., 2018), Neotropical (e.g., Gentry, 1995; DRYFLOR, 2016) and national (Aymard, 2011; Aymard et al., 2011) studies. In contrast to deciduous forests, usually under 800–1,200 mm of pluviosity, and >4 dry months, semi-deciduous forests are developed between 1,400–1,800 mm of pluviosity and 2–4 dry months. They are located within macrothermic (>24°C) and submesothermic (18–24°C) zones, from sea level to above 1,200 m.

Confusion between deciduous and semi-deciduous forests is due, perhaps, to traditional classification of tree-dominated ecosystems as either dry or humid (as in Bullock et al., 1995). The most conspicuous feature of semi-deciduous forests is that tree canopy loses 25%–75% of foliage during dry season. Despite deciduous and semi-deciduous (tropical) forests have many similarities, both in plant phylogeny and bioclimatic conditions, developing on relatively fertile soils, they are different enough to be considered as distinct ecosystems and even biomes. Semi-deciduous forests, in fact, are taller and have a more complex structure (generally 15–30 m height), with three strata, relatively rich in epiphytes (especially in montane-submesothermic types), and dominated by brevideciduous trees and evergreen understory herbs. Deciduous forest canopies show, in contrast, 5–15 m height and 2 strata, are poor in epiphytism, and are dominated by typically deciduous species, with a deciduous understory dominated by subshrubs. Montane semi-deciduous forests are floristically and spatially connected with lowland (macrothermic) units through valleys and associated alluvial ecosystems, but usually merged with adjacent cloud forests or considered as "transitional" (e.g., Steyermark and Huber, 1978; Huber, 1986). True transitional zones, however, are very reduced, and restricted to ecotones between these quite distinct ecosystems. Further discrimination between tropophilous and ombrophilous semi-deciduous forests are, based on available data, still problematic, and seem to be especially similar in Venezuelan Andes. Based on personal observation of presently small and usually unconnected remnants, near half of the area of the Lake Maracaibo basin was originally covered by semi-deciduous ombrophilous (towards south and southwest) and tropophilous (towards east) semi-deciduous forests, with evergreen lowland formations restricted to southwestern basin, at the base of Serranía de los Motilones (southernmost part of Cordillera de Perijá). Lake Maracaibo basin and Western Llanos units are perhaps the most endangered forests of Venezuela, with high risk of complete elimination in few years if serious efforts are not made. Typical floristic elements are, in northern Venezuela, *Erithryna poeppigiana*, *Heliocarpus americanus*, and *Myrcia fallax* for montane forests, and *Samanea saman*, *Enterolobium* spp., *Anacardium excelsum*, and *Tabebuia rosea* for lowland units.

Seasonally dry tropical forests of Latin America and the Caribbean, including deciduous and semi-deciduous types, are seriously threatened, with less than 10% of original extension still remaining in many countries (DRYFLOR, 2016). In Venezuela, in fact, 96% of deciduous and 83% of semi-deciduous forests are disturbed, very disturbed or completely destroyed (Oliveira-Miranda et al., 2010). At least for northern Venezuela, conservation status is, as in deciduous forests, critical. Perhaps the most studied case of forest destruction in Venezuela is that of the Western Llanos, where from *ca.* 3,000,000 ha remains, for 2017, less than 50,000 ha; of them, less than 20,000 ha are included under protected or management areas (Caparo Forestry Reserve, Caimital forest, and Río Viejo National Park), but even there they are quite isolated, reduced and/or threatened. Major remaining montane semi-deciduous forests are located in Mérida state, but continuous invasion for cattle and agriculture is seriously threatening this ecosystem, even in the Tapo-Caparo National Park. Examples of Venezuelan semi-deciduous macrothermic forests are shown in Figures 11.1b and g.

11.2.2.7 Evergreen Lowland Forests

This biome is located in macrothermic zones, with pluviosity usually above 1,800 mm (reaching 4,000 mm in southern Bolívar state), and less than two dry months. Evergreen lowland forests are characterized by a gradual substitution of the foliar mass throughout the year, with less than 25% of tree individuals losing their foliage during drier months (Oliveira-Miranda et al., 2010). They are found primarily south of the Orinoco river, but large extensions may also be observed in southwestern Lake Maracaibo basin; further locations in Venezuela include the Guatopo National Park, in southern

Miranda state, and adjacent hills and lower mountains, still very scarcely known (Aymard, 2011, Oliveira-Miranda et al., 2010). Only lowland evergreen forests south of the Orinoco were considered by Slik et al. (2018), who included them within the American realm, closer to the African lowland forests.

This is the most extensive terrestrial ecosystem of Venezuela, covering ca. 34% of the national territory (311,496 km²). Its main extension is located south of the Orinoco river, in Bolívar, Amazonas and Delta Amacuro states (90% of present cover, inlcuding both lowland and montane evergreen formations), with only ca. 2,000 km² in northern Venezuela (Oliveira-Miranda et al., 2010). Evergreen lowland forests from Lake Maracaibo basin are seriously threatened, but still is time to preserve them, if a national park, considering their original inhabitants (the Barí people), is included as part of that beautiful and interesting area. The evergreen lowland forests from Guatopo are formally protected under the Guatopo National Park, but adjacent forests of Barlovento are seriously threatened and unprotected. Exploration of this biome is still incomplete, and given its high richness in almost all taxonomic groups, it is expected that many new species will be discovered.

11.2.2.8 Montane Forests

Montane forests extend across all major Venezuelan mountain systems. Classical studies using transects, include Huber (1986), for Venezuelan Coastal Cordillera, and Ataroff and Sarmiento (2004), for Venezuelan Andes. In the Guayana highlands, above 1,500 m, they usually are restricted to protected areas, like canyons, but are homogeneously distributed in remaining environments; at least some of the so-called "*Bonnetia* forests", at the top of some tepuis, could be considered, in fact, a tall *facies* of local shrublands. Montane forests of Venezuela has not been evaluated in a phylogenetic context (cf. Slik et al., 2018), but other mountain forests in adjacent northern South America were classified as (boreal) Subtropical, probably with Laurasian affinities. Antarctic affinities, especially with the Valdivian forests, have been reported elsewhere.

Two distinct types of montane forests, dry and humid, may be differentiated. Montane and basimontane dry evergreen forests (5–15 m height) are located at the drier slopes of Cordillera de Mérida; because of restricted geographical distribution, as much as expanding agriculture and cattle grazing in nearby areas, they are highly threatened (Ataroff and Sarmiento, 2004). Evergreen montane forests or "cloud forests" have a canopy generally above 15 m height, and develop at the condensation zone of mountains. In wind-swept ridges and upper slopes of mountains, gnarled or "elfin" forests (Beard, 1944) and woodlands ≤ 5 m tall may be present. Altitudinal range, however, depends greatly on the *massenerhebung* effect. That way, cloud forest could appear from 500 m in isolated to 1,800 m in major ranges (Ataroff, 2003; Ataroff & García-Núñez, 2013). Isolated mountains, near the sea, were studied by Sugden (1983), while cloud forests from Coastal Cordillera, including floristic patterns were reviewed by Meier (2011). Cloud forests are prime water and biodiversity reservoirs, and despite being well represented in the national system of protected areas, as a biome, some of them, as those from Serranía de Nirgua, should be included in order to preserve ecosystem services and local endemic species. Azonal *Polylepis* forests and related *chirivitales* (dominated by *Escallonia tortuosa*) are interspersed within Páramo matrices from 3,000 to above 4,000 m.

11.2.2.9 Tepui Shrublands and Scrubs

Covering the summits of spectacular tabletop mountains of Guayana Shield (Figure 11.1a) develops, above 1,300–1,500 m, a very interesting and unique biogeographic unit, usually called Pantepui (Mayr and Phelps, 1967; Huber, 1987). First biological collections from that area were made by Robert Schomburgk in 1838, while first ascension to the top of a tepui (as these mountains are usually called) was made by Everard F. Im Thurn and Harry I. Perkins in 1884. Previous vegetation classifications considered Guayana highlands, above 1,500 m, as a tropical-alpine plant formation. It is questionable, however, if vegetation from Pantepui is really alpine, at least for its vast majority; only at the southernmost tip of Venezuela, in La Neblina massif, it is reached 3,000 m, the treeline for humid slopes in Venezuela (Sarmiento, 1971). At any rate, and above *ca.* 2,000 m, vegetation similar to certain tropical alpine systems develops. There, like in Páramos, there are caulirosules, but of family Bonnetiaceae besides Compositae. Distinctive floristic units include *Bonnetia maguireorum* shrublands, over shallow organic soils from La Neblina massif (Brewer-Carías, 2012), and *Chimantaea* spp. scrubs, on peatlands of the Chimantá massif (Riina and Huber, 2003).

Tepui shrublands and scrubs are usually mixed with peat bogs, which are dominated by herbs and forbs growing on deep water-saturated soils, and large extensions of bare rock. They are very variable in species composition, even within a single tepui. Soils are shallow, and may be very incipient to more or less developed, usually with high organic content. Grasses are scarce, although *Cortaderia roraimensis* can form large-sized colonies, and some bamboos, like several species of the endemic genus

Myriocladus, can grow in dense clumps that cover more or less extensive areas; a true alpine prairie or meadow is developed only in some peatlands of Sierra de Maigualida (Huber, 1995), but this association is more properly classified as a wetland (vid. infra, under "Peatlands"). Highland tepui vegetation and biota is very rich and still scarcely studied. As in other highland ecosystems of Venezuela, endemism is very high, and many new species are expected to be found as biodiversity prospecting continues. As for the páramos, tepui shrublands and scrubs (as well as associated peatlands) could be considered as authentic biogeographic archipelagos, with a very rich and diversified biota.

11.2.2.10   Páramos

Páramos are complex vegetation units. Cuatrecasas (1968) offered its classical definition, while Monasterio (1980) and Vareschi (1982) added further subdivisions. They key out as a distinct plant formation, apart from (alpine) tundra in the system of Beard (1944), but both are merged into a single biome in Wielgolaski (1997). The flora of the Páramos, including algae, fungi and plants is summarized in Vareschi (1970), angiosperms in Briceño and Morillo (2002, 2006), and monocotyledons in Morillo et al. (2010–2011); páramos's fauna is summarized in Díaz et al. (1997). Three main subdivisions were established by Cuatrecasas (1968): subpáramo (a shrubland), páramo *sensu stricto* (a complex unit including shrublands, grasslands and peatlands), and superpáramo (semi-desertic shrublands and scrubs). Caulirosules, mainly of family Compositae subtribe Espeletiinae, are common in páramo *sensu stricto* and superpáramo belts. *Escallonia* shrublands and woodlands (*chirivitales*) and *Polylepis* forests (*bosques de coloradito*) are scattered within Páramo matrices wherever loose accumulations of rocks are available.

Although mostly confined to Cordillera de Mérida, Sierra de Perijá and the Tamá massif, subpáramo vegetation, with diagnostic Espeletiinae (viz., *Libanothamnus neriifolius*) are present in Central Coastal Cordillera, in Aragua (Ortíz and Fernández-Badillo, 2017), Miranda, Distrito Capital and Vargas (Steyermark and Huber, 1978). Floristics of Venezuelan superpáramo, with incipient mineral soils subject to solifluction, was reviewed by Ricardi et al. (1997). Conservation status is considered, for all Venezuela, as VU, but may be locally CR (Oliveira-Miranda et al., 2010; Table 11.1). Despite subpáramo vegetation of Coastal Cordillera has been considered CR by Oliveira Miranda et al. (2010), it seems to be well preserved and protected. Eastern Coastal Cordillera subalpine shrublands and grasslands show strong

Caribbean and Guayanan affinities, rather than Antarctic, like true páramo, and are completely devoided of Espeletiinae. An example of a Venezuelan páramo is in Figure 11.1f.

11.2.2.11 Alpine Deserts

Bordering glaciers, and over some nearby areas of Sierra Nevada de Mérida, as much as in the highest peaks of Sierra del Norte or Sierra de La Culata, are located the only true deserts of Venezuela. Previous uses of the terms "desert" and "desertic" for local alpine ecosystems include the *desierto periglacial* and *páramo desértico* vegetation units of Monasterio (1980), who merged them with adjacent superpáramo and considered elevational ranges from 3,900–4,000 to 4,600–4,800 m. The first term has been applied only for alpine desert of Sierra Nevada, mainly on bare rocks, the second to similar ecosystems, on solid or more or less disaggregated rock, intermingled at its lower slopes with typical superpáramo vegetation. This plant formation is scarcely inhabited to completely devoid of vascular plants (vegetation cover of <5% to <1%), and harbors no Espeletiinae (according to Cuatrecasas [2013], *Coespeletia moritziana* may reach sometimes alpine deserts). Floristic composition, especially in small rocky refugia is, based on personal observations, very similar to most exposed areas of adjacent superpáramo (in agreement with the páramo *sensu lato* concept of Cuatrecasas, 1968), but at least one diagnostic species, *Draba chionophila*, is present both in Sierra Nevada and Sierra de La Culata. Previous studies cited lichenized fungi, bryophytes, insects and spiders, and considered upper limit of Espeletiinae (and thus, of páramo) at 4,600 m, upper limit of vegetal life (including fungi and bryophytes) at 4,800 m, glacial climatic limit at 4,850, and glacial orographic limit at 4,700 m, with only 20 species growing above 4,750 m (Vareschi, 1970). Despite accelerated glacier retreating since 1910 (Braun and Bezada, 2013), altitudinal belts or vegetation zones, including lower limit of glaciers, seems to be not changed so sharply. Areas until recently covered by glaciers, however, are currently alpine deserts.

This is the only hypermicrothermic Venezuelan ecosystem. Recent atmospheric warming, due to climate change, had probably caused reduction of alpine deserts, with vertical migration of species of grasses (e.g., *Festuca fragilis*) and composites (e.g., *Pentacalia* sp.) usually found in nearby superpáramo scrubs. Further studies should determine the validity of this biocenosis as either an independent plant formation or just as the most extreme facies of the páramo *sensu lato*. These studies are urgent, since this ecosystem could be disappearing and remaining glaciers are expected to completely vanish in 10–20 years (*glacierhub.org/2017/11/09/ venezuela-losing-last-glacier/*). Other names probably more appropriate, but larger, that could be applied to this ecosystem include cold tropical desert or cold (tropical, rocky) mountain desert. Actual extension of alpine deserts in Sierra Nevada is quite similar to that formerly covered by glaciers until 1910 (*ca.* 10 km² *fide* Braun and Bezada, 2013), plus uppermost exposed areas of Sierra de La Culata, about the double of this area (totalizing *ca.* 30 km²). According to the IUCN criteria (Rodríguez et al., 2011), alpine desert and associated glaciers are the most endangered ecosystems in the country (cf. Table 11.1).

11.2.3 Wetlands

A recent study for the Orinoco river basin, including Colombia and Venezuela, accept 49 different types of natural, and seven of artificial or semi-natural, wetlands (Lasso et al., 2014a). Although not so rich in taxonomic diversity and endemicity (except in highlands), Venezuelan wetlands are rich in phylogenetic diversity, with several phyla as yet not reported for adjacent aquatic and terrestrial ecosystems (vid. infra, under "other animal groups"), and very rich and complex from an ecological point of view. Some *facies* of several forests, shrublands, and scrubs, should be accepted only tentatively as wetlands if the aquatic phase is only temporal, and rooted plants are shared with adjacent terrestrial ecosystems. Wetlands are sometimes difficult to define, also, regarding salinity. Mangroves, in fact, can develop over a wide range of NaCl concentration, and may be either coastal (marine) or estuarine; the so-called *eneales*, dominated by the cattail *Typha domingensis* could thrive in both estuarine and freshwater ecosystems, and even in some marshes. Alluvial, madicolous, lacustrine and peatland wetlands could be very similar and intergrading, and could be merged into a single super unit, or freshwater wetlands.

11.2.3.1 Coastline Wetlands

They include mangroves (at least 34 localities in northern Venezuela *fide* Conde and Carmona-Suárez, 2003) and brackish coastal units, as Laguna de Tacarigua (central coasts) and Laguna de los Patos (eastern coasts), with mangroves and sedge-dominated *albuferas*, surrounded by typical terrestrial ecosystems as strand vegetation and xerophilous shrublands under marine influence. Venezuelan mangroves were reviewed by Conde and Alarcón

(1993), Barreto and González Boscán (1994), and Conde and Carmona-Suárez (2003), but additional references include Medina and Barboza (2003), on Lake Maracaibo basin, and Oliveira-Miranda et al. (2010), where biotic and abiotic patterns, as well as conservation concerns, are updated. The largest concentration of mangroves in the country is in the Orinoco river delta, but important communities develop also in the Lake Maracaibo system (constituted, from north to south, by the Gulf of Venezuela, Tablazo bay, Maracaibo strait, and Lake Maracaibo), Morrocoy National Park, and Margarita island. Dwarf mangroves develop in Laguna Aguas Blancas, in northern Turuépano National Park, where it is assumed there is a high concentration of Aluminium (Oliveira-Miranda et al., 2010), and in the keys of Morrocoy National Park. Sponges and crustaceans are very common and diversified in the roots of the species of the genus *Rhizophora*. Ecosystem services of coastal wetlands include coastline stabilization as much as refuge and food for fishes and crustaceans of commercial value, including their nurseries. An additional study on mangroves (Lazo, 2009, cited by Huber and Oliveira-Miranda, 2010), including geographical distribution and conservation status is still unpublished.

Venezuelan marshes are located along seacoast, and are technically termed *albuferas* or *lagunas costeras* in Spanish. The term *marisma* (marsh), could also apply, but it has been used to name sea passages near mangroves (e.g., Barreto et al., 2009), and could be confused with local *Spartina* spp. meadows, from estuarine environments. Consequently, the term "brackish coastline wetlands" is here proposed. Graminoid herbs, mainly of the Cyperaceae, but also of Poaceae and Typhaceae, are usually dominant. Few studies provide information on these ecosystems; they include Steyermark et al. (1994), Barreto et al. (2009) and Villareal et al. (2014).

11.2.3.2 Estuarine Wetlands

They are located, primarily, on major estuarine systems of Venezuela, Orinoco river delta coastal plain (ORDCP; reviewed in González Boscán, 2003b, 2011 and 2013, and Colonnello, 2004), and Lake Maracaibo system (reviewed in Medina and Barboza, 2006). Alluvial wetlands and mangroves may be closely associated, and may even intergrade. In the Orinoco river delta, in fact, *Euterpe*, *Manicaria*, *Mauritia*, and *Roystonea* palm forests, *Pterocarpus officinalis* swamps, and mangroves can coexist and intermingle within a single locality. Special types of estuarine wetlands are located in the C–3 (lower delta) subregion of the ORDCP. They include the *Spartina* (presently *Sporobolus* sect. *Spartina*) and *Crenea maritima* grassy marshes, plus the *Chrysobalanus icaco* shrublands and woodlands; *C. icaco* formations are reviewed in González Boscán (2011, 2013a). *Pterocarpus officinalis* forests and broad-leaved meadows of *Montrichardia* spp. (as defined by Huber, 1995), also known as *rabanales*, are perhaps the most extended estuarine wetlands of Venezuela; they are particularly well studied for the Orinoco delta (González Boscán, 2011).

11.2.3.3 Alluvial Wetlands

Included in this category are a wide variety of wetlands, all of them developed over flooding plains. Alluvial wetlands are usually associated with lotic ecosystems, but they may be more or less transitional with typical lacustrine units, as in the *esteros* (permanently flooded terrains within alluvial landscapes), and at the mouth of tributaries of larger rivers and lakes.

So-called humid or wet savannas ("semiseasonal savannas" of Sarmiento, 1990) are the largest alluvial wetlands in Venezuela. According to Oliveira-Miranda et al. (2010), they occupy 44,906 km^2 (54,390 km^2 *fide* González-Jiménez, 2003a), and are considered as EN (locally CR) according to the criteria of Rodríguez et al. (2011). They include, at least, two different types of grasslands: the *banco, bajío* and *estero* savannas, a grassland landscape developed on flood plains, including banks, shallows and ponds, which is flooded by rainwater and indirect overflow by damming of rivers during the rainy season, and the *Paspalum fasciculatum* savanna, flooded by direct river overflow (Ramia, 1967; González Jiménez, 2003a). The first type has received different names according to geographical location; thus, it has been called *sabanas veraneras* for eastern Venezuelan Guayana (Ramia, 1967), *sabanas húmedas* for southern Barinas and southern Cojedes states (Sarmiento, 1971 and Ramia, 1993, respectively), and *sabanas de pasto nativo* for Delta Amacuro state (four publications of Trujillo Arroyo in 1968, cited in González Jiménez, 2003a).

Seasonally flooded forests include *várzeas* (along white-water rivers) and *igapós* (along black-water rivers) of Brazilian literature, reviewed for Venezuela by Rosales (2003) and Marrero (2011), as well as seasonally flooded shrublands of western Amazonas state (reviewed by Oliveira-Miranda et al., 2010). Palm forest typically develops along rivers and floodplains, and several types can be differentiated according to dominant palm species. The so-called *morichales*, dominated by *Mauritia flexuosa*, are common in eastern and southern Venezuela. A complete review of this ecosystem in a supranational context was made by González Boscán (2016), while important contributions on ichthyology, hydrology, landscape ecology, conservation and biocultural concerns (for Venezuela) are included in Marrero and Rodríguez-Olarte (2014). Floristics, biogeography and conservation of *cha-guaramales* and remaining *Roystonea oleracea* communities (Figure 11.1e and g), functionally similar but biotically very different from *morichales*, were reviewed by Colonnello et al. (2016a and b) and Grande and Colonnello (2016). Synthetic works on *Copernicia tectorum* wetlands (which are completely depleted of superficial water during dry season) and remaining types of palm swamps in the country are still to be done. Palm brakes, palm marshes and palm swamps were differentiated as distinct plant formations by Beard (1944), but they seem to merge at least in some instances.

Also included in this category are several tall-graminoid communities, including the so-called *cañabravales* (Figure 11.1b, a small patch on the right-side border of the river), dense clumps of the reed *Gynerium sagittatum*, and the *guaduales* or *guafales* (Marrero, 2011), dominated by the bamboo *Guadua angustifolia*. Peculiar riverine communities, dominated by *Typha domingensis* and *Phragmites australis*, are located in dry Andean valleys along Chama river and Lake Caparú (Aranguren et al., 2015). Springs usually do not produce major changes in surrounding zonal vegetation, but aquatic plants, usually of widespread geographical distribution, may be found there (Marrero, 2011; Lasso et al., 2014a); *Pluchea amorifera*, a delicate herb of family Compositae, however, it is known only from the hot springs of Paria peninsula.

11.2.3.4 Lacustrine Wetlands

Wetlands surrounding Lake Valencia are inhabited by *Amaranthus australis*, *Gynerium sagittatum*, *Typha domingensis* and several additional grasses, shrubs and forbs, as well as *Limnobium laevigatum*, floating *Eichhornia crassipes*, *Pistia stratiotes* and several Lemnaceae (presently in Araceae, subfamily Lemnoideae); *Vallisneria americana*, locally extinct, and reduced populations of *Potamogeton pectinatus* have been characteristic submerged species (González-Jiménez, 2003). Regarding Andean lakes, illuminating studies have been conducted by Rico et al. (1996; upland arid pockets), Vareschi (1970; páramos), and Gordon (2016; both environments). Glacial lakes, like Laguna de Mucubají and Laguna Negra, above 3,500 m, are surrounded by herbaceous communities of *Callitriche heterophylla*, *Carex* sp., *Crassula venezuelensis*, *Elatine fassetiana*, *Epilobium denticulatum*, *Isoetes lechleri*, *Juncus* spp., *Limosella acaulis*, *Mimulus glabratus*, *Montia meridensis*, *Pilularia americana*, *Ranunculus flagelliformis* and *R. limoselloides*, and the subshrub *Hypericum brathys* (Velásquez, 1994). Lagoon system of Cariaco-Casanay depression includes herbaceous wetlands, dominated by sedges (*Cyperus articulatus, Eleocharis* spp. and *Fuirena umbellata*), *Amaranthus australis, Typha domingensis* (angiosperms), and *Acrostichum danaeifolium* (a fern). Ephemeral lagoons are also common in Venezuela, but they have been reviewed at any length only for the granitic inselbergs of Guayana Shield (Lasso et al., 2014b).

According to Marrero (2011), four different artificial lentic wetlands are present in Venezuela. They include *fosas* (pits), *lagunas de oxidación* (oxidation ponds), *abrevaderos* (water troughs), and *módulos* (polders). Polders are artificial systems of lagoons closed by earth dams, with drains controlled by gates through which the excess water from flooding and precipitation is regulated; in the 1970 s an extensive system of this type was built near the town of Mantecal and nearby areas of Apure state; there, the predominant vegetation consists of floating and rooted aquatic plants. Open shallow pits are usually built collaterally when excavations for road embankments are carried out and are very common in the Venezuelan Llanos; these systems can maintain water after the end of the rainy season, and are of great importance for fishes and wild terrestrial fauna. Additional artificial wetlands include dams and rice fields, the last of which cover 1,638 km^2 (González Jiménez, 2003a; Table 11.1).

11.2.3.5 Madicolous Ecosystems

This type of wetland is defined by the permanent flux of freshwater over rocks and/or rooted vegetation, forming a water-film and not properly a lotic system. So-called *escurrideros*, from areas of moderate to steep slopes where conditions of substrate allow to form a plug or blocking zone, are common in páramo landscapes; the plugs propitiate the accumulation of detritus, which over time retains a considerable amount of organic matter and sediments (Marrero, 2011). The draining area serves as a nucleus for the formation of organic soils, but these sites are frequently drained in order to recover a portion of dry soil for agriculture (Marrero, 2011). These alpine ecosystems are usually inhabited by *Carex* spp. and *Agrostis trichodes*, which may form true meadows (Azócar, 1974); *Acaulimalva purdiei* (Malvaceae) and several species of Apiaceae and Asteraceae are also common along the borders, and they are inhabited by many insects and chelicerates. Additional madicolous ecosystems are common at the margins of montane lotic ecosystems; there thrive some aquatic species, as *Lemna* spp., and properly madicolous species, as some liverworts and algae.

11.2.3.6 Peatlands

Peatlands develop within alpine (microthermic and submicrothermic) shrublands and meadows, including Páramo and Pantepui biomes. They are made up by shrubs, subshrubs and herbs over a layer of peat-forming mosses, especially of the genus *Sphagnum*. Tepui grasslands of Sierra de Maigualida, dominated by a species of *Axonopus* (a C_4 grass genus), could be considered true prairies, but they develop over peat beds ≥ 2 m depth, water-saturated over major part of the annual cycle (Huber et al., 1997), and are here considered, thus, as wetlands. Additional tepui peatlands include scrubs dominated by Rapateaceae (widespread), described as meadows by Huber (1995), and *Chimataea* spp. thickets in the Chimatá massif. Páramo peatlands are usually dominated by *Agrostis* spp. and *Aciachne pulvinata* (Azócar, 1974), the last species quite common in intensively pastorated localities. As it is usual in páramos (cf. Azócar and Fariñas, 2003), highland tepui peatlands are intermingled with shrublands, scrubs and forests, in a highly complex vegetation mosaic still scarcely understood.

11.3 Diversity of Taxonomic Groups

Taxonomic groups are presented throughout text as in traditional classifications, but they are ordered in Table 11.2 according to time of origin, following the system of Ruggiero et al. (2015). Valuable information on pathogenic viruses and bacteria, and eucaryotic parasites is available in Berger (2018). Six out of seven kingdoms, and at least 47 phyla, 423 orders, 2,408–2,440 families, 13,672–13,830 genera, and 50,583–54,826 species have been recorded for Venezuela, but more than 176,707–263,271 species are expected to occur (Table 11.2). Ordering of classes is still the matter of much of dispute, or information was not available in reviewed literature; they are, consequently, not included in Table 11.2.

Biological collections are powerful means for *ex situ* conservation of both genetic and taxonomic components of biodiversity, and provide information for basic research on morphology, anatomy, taxonomy, ecology, physiology and biotechnology. Venezuela has 25 herbariums (15 of them registered in the *Index Herbariorum*), which hold 626,000 specimens and 5,494 nomenclatural types, several zoological museums and collections, with a total of 2,080,000 vertebrate specimens and about 3,000,000 specimens of invertebrates (including many types), and 10 botanic gardens plus 18 zoos, with important collections of both native

Table 11.2   Number of Phyla, Orders, Families, Genera, Registered Species ("Species") and Expected Species ("Species est.") of Taxonomic Groups Reported from Venezuela

Kingdom	Taxonomic group	Phyla	Orders	Families	Genera	Species	Species est.
/	Viruses	/	–	–	–	>100	–
Bacteria	Bacteria (non-Cyanophyta)	–	–	–	–	>100	–
	Blue-green algae (Cyanophyta)	Cyanobacteria	4	20	≥103	>266	–
Protozoa	Diplomonadida	Metamonada	1	1	2	2	–
	Other Metamonada	Metamonada	–	–	2	16	–
	Mastigophorea	Euglenozoa	–	13	22	52	–
	Phytomonadida	Euglenozoa	–	1	1	1	–
	Euglenida	Euglenozoa	–	3	7	33	–
	Kinetoplastidia	Euglenozoa	1	1	1	1	–
	Other Euglenozoa	Euglenozoa	–	–	5	34	–
	Myxomycota	Amoebozoa	–	–	28	111	–
	Other Amoebozoa	Amoebozoa	1	1	1	10	–
	Choanoflagellida	Choanozoa	1	1	1	1	–
Chromista	Cryptomonadida	Cryptista	–	1	2	2	–
	Coccolithophorids	Haptophyta	2	10	27	38	–
	Ciliatea, Ciliates	Ciliophora	9	24	44	104	–
	Dinoflagellida, Dinoflagellates	Miozoa	6	22	38	173	–
	Eucoccidida	Miozoa	–	–	–	109	–
	Other Miozoa	Miozoa	–	–	18	86	–
	No previous name	Bigyra	1	1	1	7	–
	Phaeophyta, Brown algae	Ochrophyta	9	12	27	71	–

Kingdom	Taxonomic group	Phyla	Orders	Families	Genera	Species	Species est.
	Diatoms	Ochrophyta	32	51	123	389	–
	Silicoflagellata	Ochrophyta	1	1	1	1	–
	Oomycota	Pseudofungi	–	–	8	15	–
	Hyphochitridiomycota	Pseudofungi	–	–	4	8	–
	Plasmodiophoromycota	Cercozoa	–	–	2	4	–
	Rhizopoda	Retaria	–	–	–	>100	–
Plantae	Red algae	Rhodophyta	15	35	114	269	–
	Green algae	Chlorophyta	18	43	128	515	–
	Charophytes	Charophyta	1	1	2	37	–
	Liverworts	Marchantiophyta	–	34	146	731	–
	Mosses	Bryophyta	19	70	280	991	1,010–1,020
	Hornworts	Anthocerotophyta	2	2	5	5	–
	Lycophytes	Tracheophyta	–	3	5	146	–
	True ferns (Monilophyta)	Tracheophyta	–	28	116	1.009	–
	Gymnosperms	Tracheophyta	3	3	5	28	–
	Angiosperms	Tracheophyta	54	228	2,480	15,820	16,820–18,985
Fungi	Ascomycota	Ascomycota	–	>85	>809	5,209-8,009	>96,000
	Basidiomycota	Basidiomycota					
	Chytridiomycota	Chytridiomycota					
	Glomales	Glomeromycota					
	Zygomycota	Zygomycota					

Table 11.2 (Continued)

Kingdom	Taxonomic group	Phyla	Orders	Families	Genera	Species	Species est.
Animalia	Sponges	Porifera	≥13	≥41	≥64	163	ca. 250
	Cnidarians	Cnidaria	19	57	105	191	>561
	Comb jellies	Ctenophora	1	1	1	1	1
	Dicyemids	Rhombozoa	1	1	1	3	–
	Flat worms	Platyhelminthes	>7	>14	>25	> 50	>1,200
	Arrow worms	Chaetognatha	–	3	3	12	–
	Nematodes	Nematoda	–	≥89	≥172	596	>5,299
	Velvet worms	Onycophora	1	1	2	4	>5
	Tardigrades	Tardigrada	–	–	–	–	–
	Chelicerates	Arthropoda	11	75	445	>968	–
	Myriapods	Arthropoda	4	11	22	90	–
	Crustaceans	Arthropoda	–	>165	>591	>1,266	–
	Insects	Arthropoda	25	506-535	>5,136–5,285	>14,212–15,646	>41,803–126,192
	Gastrotrichs	Gastrotricha	–	–	–	–	–
	Rotifers	Rotifera	–	21	44	268	–
	Bryozoans	Bryozoa	–	–	–	–	–
	Nemertina	Nemertea	1	1	1	1	–
	Annelids	Annelida	–	>52	–	>535	–
	Sipunculids	Sipuncula	–	4	7	14	–
	Brachiopods	Brachiopoda	–	–	4	5	–
	Mollusks	Mollusca	>21	215	483	1,157	–

Kingdom	Taxonomic group	Phyla	Orders	Families	Genera	Species	Species est.
	Echinoderms	Echinodermata	18	39–42	76–85	139–148	—
	Cephalochordata	Chordata	—	2	2	2	—
	Urochordata	Chordata	—	11	25	44	—
	Marine fishes	Chordata	33	148	529	791	ca. 1,000
	Freshwater fishes	Chordata	10	58	380	1,000	>1,000
	Amphibians	Chordata	3	18	64	333	>400
	Reptiles	Chordata	3	30	122	370	>400
	Birds	Chordata	27	88	618	1,381	1,432
	Mammals	Chordata	14	47	184	390	436
		>47	**>423**	**>2,408–2,440**	**>13,672–13,830**	**>50,583–54,826**	**>176,707–263,271**

Sources are included under corresponding headings along text.

and exotic living specimens (Aguilera, 2000 cited in Miloslavich et al., 2003).

11.3.1 Microorganisms (Viruses and Bacteria, Excluding Cyanophyta)

Available information on Venezuelan microorganisms is rather scattered and usually not directly related with biodiversity, focusing on functional traits rather than in taxonomy. Regarding bacteria, there is an important body of literature, with both subkingdoms (viz., Negibacteria, including Cyanobacteria or Cyanophyta, and Posibacteria, including Actinomyce-tales) reported for the country. Information regarding viruses, however, has been largely limited to publications on human health (e.g., Silva et al., 2004; Antón et al., 2011) and phytopathology (e.g., Garrido and Brito, 2016). Both for viruses and for bacteria (not including Cyanophyta), and based upon information available on literature, it is estimated that studied species are above two orders of magnitude; potential numbers should be above three orders of magnitude, but this is, still, highly speculative. To date, no species of Archaebacteria has been recorded from Venezuela, but since they have been recently found as a widespread kind of microorgan-ism in oceans, soils, and even ruminant and human colons, it would be not surprising they are a rather common group of organisms in the country. Several recent references have stressed potentialities of Venezuelan micro-organisms in biotechnology (e.g., Rodríguez and López, 2009; Balcázar et al., 2015), public health (e.g., Silva et al., 2004), and bioremediation (Zamora et al., 2017). The Yanomami people, from southern Venezuela, is the human group with the richest bacteriome in the world (Clemente et al., 2015); recent investigations on them shed light on human antibiotic resistance (http://www.medicaldaily.com/yanomami-amazon-tribe-could-provide-answers-antibiotic-resistance-329670).

11.3.2 Algae (Including Cyanophyta)

Photosynthetic unicellular and thallose organisms have been considered tra-ditionally as algae, both prokaryotic (Cyanophyta) and eukaryotic (remain-ing groups). Ganesan (1989) and Bellorín (2003) offered the last updated lists of marine benthic macroalgae, while Yacubson (1972, 1974) offered the last available lists for Cyanophyta and Chlorophyta from continental waters (including freshwater, estuarine and coastal lagoon environments). Genera and species number of Charophyta are reported in Rodríguez et al.

(2017), including 19 species of *Chara* and 18 species of *Nitella* (both in family Characeae, order Charales). Scorza and Arcay (2003) cited several taxa of green algae and dinoflagellates, presumably from freshwater environments; they are added to those cited in Sánchez-Suárez and Díaz-Ramos (2003), exclusively marine, and those corresponding to green algae cited by Yacubson (1974). Of special relevance for Venezuelan phycology were the German-Venezuelan limnological expedition of 1952, which results were published in Gessner and Vareschi (1956). High-Andean diatoms are summarized in Rumrich et al. (2000), which could not be reviewed. Recent and interesting additions to Venezuelan algae come from the Guayana highlands, showing a previously unexpected level of endemism for microscopic organisms (Kaštovský et al., 2011, 2016). All phycological groups here considered are arranged above ordinal level according to the system of Ruggiero et al. (2015), while orders, families, genera and species are reported as cited in reviewed literature. Further information on the distribution and taxonomy of Venezuelan algae may be retrieved online from Gómez et al. (2015 onwards).

11.3.3 Fungi and Lichens

Modern mycological studies in Venezuela began with the works of Chardon and Toro (1934) and Denis (1970). Volmark Vareschi, Manuel López Figueiras and Vicente Marcano made relevant contributions to Venezuelan lichenology (summarized in Marcano, 2003), while Iturriaga et al., (2000) offered the first attempt to synthesize information on both non-lichenized and lichenized fungi, which is available online in Iturriaga and Minter (2006 onwards). Iturriaga and Ryvarden (2003) and Hernández et al. ([2014] 2016) include additional information on nonlichenized fungi. According to Iturriaga et al. (2000), of all the samples hitherto collected in the country, a large percentage (*ca.* 50%) has not yet been identified; the taxonomic study of the remaining material yielded a number of approximately 6,500 spp., which are reported in various publications, including numerous new species for science. It is estimated that since the first collections made by Humboldt and Bonpland, some 40,000 specimens of fungi have been collected in Venezuela; they are deposited, since 1975, mainly in national herbaria (Iturriaga et al., 2000).

Venezuela harbors a very rich mycobiota, from numerous microfungi in freshwater environments and soils, to *Macrocybe titans* (Basidiomycota: Agaricales), with fruiting bodies up to 40 cm high (Iturriaga and Ryvarden, 2003). Fungi from Venezuela include representatives of the five known phyla: Ascomycota, Basidiomycota, Chytridiomycota, Zygomycota, and Glomeromycota, the last of them previously treated as Glomales, and including arbuscular mycorhizal fungi. The zoosporic aquatic "fungi" comprise a group of ubiquitous organisms, belonging to the phyla Chytridiomycota (true fungi), Oomycota, Hyphochytridiomycota, Thraustochytridiomycota and Labyrinthulomycota (presently in kingdom Chromista, either as classes or orders), last two of them as yet not recorded from Venezuela. They are found as saprotrophs or parasites on plenty of organic substrates, such as plant remains and insect exudates, but also as parasites of aquatic macrophytes, algae, fishes, protozoans and invertebrates in terrestrial, freshwater and marine environments. Regarding Oomycota, 5 species of the plant pathogen *Phytophthora* (*P. palmivora*, *P. infestans*, *P. capsici*, *P. cinnamomi*, and *P. nicotianae* [including var. *parasitica*]), the animal and human pathogen *Pythium insidiosum*, the fish pathogen *Saprolegnia* sp. (Bermúdez, 1980), and at least 10 species in the genera *Achlya* (1 species), *Aphanomyces* (2), *Leptolegnia* (1), *Leptolegniella* (1), *Plectospira* (1), and *Pythium* (>4 additional species) have been reported (Karling, 1981a). It has been reported, also, the presence of at least 8 species in 4 different genera (*Anisolpidium*, *Hyphochytrium*, *Rhizidiomyces*, and *Rhizophidium*) of Hyphochitridiomycota (Karling, 1981a and b), 2 genera and 4 species of Plasmodiophoromycota (Karling, 1981a, all of them saprobes), and 28 genera plus 111 species of Myxomycota (Lado and Wrigley de Basanta, 2008). Lichens have been used as biological indicators of air quality (reviewed in Fernández et al., 2013). Examples of Venezuelan fungi are in Figure 11.2f-g. Species estimates, only including true fungi, are above 96,000 (Lucena, 2009).

11.3.4 Plants

True plants or embryophytes include phyla Marchantiophyta, Bryophyta, Anthocerotophyta, and Tracheophyta (see Table 11.2; ordered according to the systems of Christenhusz et al., 2011 and APG IV, 2016). Vascular plants reported by Ulloa et al. (2017; 15,116 species, 3,359 endemic) differs from those reported in the last available catalog of the vascular flora of Venezuela (Hokche et al., 2008; 17,003, 2,964 endemic, here accepted), partially reviewed in the series *Flora de Venezuela*, and regional studies as Flora of the Venezuelan Guayana (Steyermark et al., 1995–2005) and *Flora de Los Llanos* (Duno et al., 2007; with a recent supplement published by Aymard, 2017). According to Duno et al. (2009) there are 10,300 species (2,136 endemics) in Guayana, including Delta Amacuro state, ca. 7,000

Figure 11.2 Examples of Venezuelan vegetal taxa: (a) *Heliamphora nutans*, a carnivorous pitcher plant from the summit of Mount Roraima; a species of family Tipulidae (Hexapoda: Diptera) is on the ventral side of the reddish appendage (photo: Charles Brewer-Carías); (b) *Dalechampia dioscoreifolia* near Los Pijiguaos, northwestern Bolívar state (photo: José Grande); (c) *Brownea macrophylla*, from El Ávila National Park (photo: Bruno Manara); (d) *Monochaetum humboldtianum* from El Ávila the National Park, near Caracas (photo: Bruno Manara); (e) *Handroanthus guayacan*, in the garden of a house in Caracas (photo: José Grande); (h). *Habenaria pauciflora* from a *Trachypogon* savanna in El Ávila National Park (photo: Bruno Manara); (f) a species of the lichenized fungi *Cladonia s.l.* Ascomycota: Lecanoromycetes), from an upland shrubland over sandstone near Sororopán-tepui (photo: José Grande); (g) *Cerrena unicolor* (mossy maze polypore), a fungi of phylum Basidiomycota, class Agaricomycetes, from El Ávila National Park (photo: Bruno Manara).

(506) in the Andes, ca. 4,000 (322) in Coastal Cordillera, and 3,200 (30) in the Llanos. Expected (total) number of species varies from 21,070 (*fide* Davis et al., 1997) and 22,671 (assuming ¼ of the species to be described, following the estimates of Thomas, 1999), to 20,000–30,000 (Steyermark, 1977). Arborescent flora, including ferns, monocots and dicots was reviewed

by Aristeguieta (2003), while aquatic plants are included in the book of Velásquez (1994). Examples of Venezuelan plants are in Figures 11.2a-e, h and 3b. History of botanical exploration is lavishly treated in Texera (1991) and Lindorf (2008).

11.3.4.1 Bryophytes

Bryophytes *sensu lato* include three phyla (Marchantiophyta, Bryophyta, and Anthocerotophyta), all of them present in Venezuela. Liverworts (Marchantiophyta) have about 731 species, in 146 genera and 34 families, while hornworts (Anthocerotophyta), so far reported only by two genera in the families Anthocerotaceae and Dendrocerotaceae (León and Rico, 2003), are represented by at least one species in each of five genera (*Anthoceros, Dendroceros, Megaceros, Phaeoceros* and *Phaeomegaceros*), in two families and two orders according to the classification of Frey and Stech (2005; Ricardo Rico, pers. comm.). Regarding mosses (Bryophyta s.s.) León et al. (2015) include 19 orders, 70 families, 280 genera, and 991 species, increasing the number of families and genera, but decreasing the number of species, with respect to the previous national inventory (León and Rico, 2003; 53 families, 232 genera and 1,012 species); expected number of species, however, is 1,010–1,020 (Jesús Delgado, pers comm.). Information on Venezuelan mosses is permanently updated in the website *Musgos de Venezuela* (MdV, http://www.musgos.cecalc.ula.ve).

Regarding endemism, there is still no comprehensive studies. It is known, however, that 9 genera of liverworts are endemic to Venezuela (León and Rico, 2003), while 43 species of mosses and 23 species of liverworts are considered as endangered and are to be included in the second edition of the red book of the Venezuelan flora (http://briologiaenvenezuela.com.ve/?page_id=32). History of Venezuelan bryology and bryological exploration is reviewed in Moreno (1992) and León and Rico (2003).

11.3.4.2   Ferns and Lycophytes ("Pteridophytes")

Venezuelan pteridophytes are very diversified, including taxa from all extant supraordinal lineages. Lycophytes are represented by families Isoetaceae, Lycopodiaceae and Selaginellaceae, with 5 genera and 146 species; remaining species are included under Monilophyta (Tables 11.2 and 11.3). Vareschi (1969) made the last available revision for Venezuelan species, besides many collections and ecological observations (for Volkmar Vareschi's contribution to Venezuelan pteridology see Mostacero, 2013).

Table 11.3 Vascular Plants of Venezuela

			Orders	Families	Genera	Species	Endemicity
Euphyllophyta	Lycophyta		3	3	5	146	29 (19.86%)
	Monilophyta		11	33	116	1,009	126 (12.49%)
	Gymnosperms	Gnetopsida	1	1	1	7	1 (14.29%)
		Pinopsida	1	1	3	17	4 (23.53%)
		Cycadopsida	1	1	1	4	–
	Angiosperms	Nymphaeales clade	1	2	3	16	–
		Magnoliids	4	11	59	685	144 (21.02%)
		Chloranthales clade	1	1	1	11	1 (9.09%)
		Monocotyledons	9	43	601	4,131	823 (19.92%)
		Ceratophyllales clade	1	1	1	2	–
		Eudicotyledons	54	170	1,815	10,975	1,836 (16.73%)
			87	267	2,606	17,003	2,964 (17,43%)

Source: Hokche et al., (2008). Classification according to Christenhusz et al. (2011) and APG IV (2016). Endemicity refers to number of endemic species.

Most speciose families (>50 species) are, according to Hokche et al. (2008), Dryopteridaceae (225 species), Pteridaceae (117), Hymenophyllaceae (102), Selaginellaceae (88), Polypodiaceae (80), Grammitidaceae (79), Thelypteridaceae (77), Aspleniaceae (72), Cyatheaceae (63), Dennstaedtiaceae (57), and Lycopodiaceae (51), but species numbers are, following the system of Christenhusz et al. (2011) 175 for Dryopteridaceae, 119 for Pteridaceae, 159 for Polypodiaceae (including Grammitidaceae), 71 for Cyatheaceae (including Hymenophyllopsidaceae), and 24 for Dennstaedtiaceae. Family Hymenophyllopsidaceae (1 genus and 8 species, 5 of them endemic), once considered as restricted to Guayana Shield, have been found recently in sandstone mountains between Ecuador and Perú. Arborescent genera include *Alsophila* (3 species), *Cnemidaria* (8 species), *Cyathea* (52 species), and *Lophosoria* and *Marattia* (1 species each), all of them in the Monilophyta. Fossil arborescent Lycophyta are known from Paleozoic rocks of several Andean sites. Conservation status of Venezuelan pteridophytes is reported in Table 11.6.

11.3.4.3 Gymnosperms

Native gymnosperms include members of family Podocarpaceae, with valuable timber trees in genera *Retrophyllum* (1 sp.), *Podocarpus* (14 spp., sometimes shrubs in *P. buchholzii*), and *Prumnopytis* (2 spp.) thriving in mountain ranges around the country. Family Gnetaceae (7 species of lianas in the genus *Gnetum*) and Zamiaceae (5 species, all native and with high potential as ornamentals) are distributed mainly in the Guayana Shield. Several exotic Araucariaceae, Cupressaceae, Cycadaceae, Ephedraceae, Pinaceae, and Taxodiaceae, are cultivated as ornamentals or in forestry plantations.

11.3.4.4 Angiosperms

Although very likely total number of angiosperm species in Venezuela is widely surpassed by several groups of animals (especially insects, chelicerates and nematodes) they are, based on reported countings, the most numerous group of living beings inhabiting the country. According to current evidence, Angiosperms could not be divided, as in traditional classifications, into dicotyledons (reviewed for Venezuela by Morillo, 2003) and monocotyledons (reviewed for Venezuela by Ramia and Stauffer, 2003), but into eudicotyledons, monocotyledons and several minor monophyletic groups (APG IV, 2016). Eudicotyledons are the most numerous and, probably, more important group of Venezuelan plants. Expected number of species are here calculated from personal estimates based on herbarium work on nonorchid taxa during past 15 years (+ 600 species; 500 new to science, 100 new records), expected number of additional orchid species (+400 species, some new; Carnevali et al., 2005, 2008), species number reported in Hokche et al.(2008; 15,820), and percentage of species to be discovered in the Neotropics according to Ulloa et al. (2017; 16.67%).

Major angiosperm families in Venezuela (> 200 species) are, according to Hokche et al. (2008), Orchidaceae (1,506 species), Fabaceae (996) Compositae or Asteraceae (784), Rubiaceae (777), Gramineae or Poaceae (740), Melastomataceae (650), Cyperaceae (430), Bromeliaceae (374), Euphorbiaceae (356), Araceae (281), Piperaceae (272), Lauraceae (211), Solanaceae (210), and Myrtaceae (209). Although there is some controversy in the real identity of the involved species, the official binomial for the national tree is *Handroanthus chrysanthus* (formerly in genus *Tabebuia*). As presently understood, in fact, its delimitation is rather arbitrary, and matter of dispute, with possible hybridization and introgression (Gentry, 1982). The national flower is the locally named *flor de mayo* ("mayflower"), *Cattleya mossiae* (Orchidaceae), which specific delimitation is also controversial; the official flora of Venezuela treats this species as a variety of the widespread *Cattleya labiata* (Foldats, 1970). Important Venezuelan angiosperms include the cocoa (*Theobroma cacao* L., from Mesoamerica to the Amazon basin and tropical humid environments west of the Andes, now cultivated elsewhere; Franceschi, 2017), pineapple (*Ananas comosus* L., with very promising cultivars from Guayana Shield), cassava (*Manihot esculenta*), many as yet undervalued medicinal and edible species, and hundreds of potentially ornamental plants, with exotic flowers painted in astonishing colors (Figure 11.2b-e, and 11.3b).

11.3.5 Protozoans

This group, as traditionally considered, is polyphyletic. Monophyletic groups, previously treated as protozoans, fungi or algae, are included as distinct taxa in Table 11.2. Protozoans have been only roughly studied for Venezuela, with the vast majority of works centered on medical or public health issues. Díaz-Ungría (1981) made the first survey for Venezuela, followed by an update of Scorza and Arcay (2003). Additional monographic studies of selected groups include that of Tello (1981), in extant Rhizaria. Dinoflagellates have been summarized by Scorza and Arcay (2003) for freshwaters (as Dinoflagellida), and by Sánchez-Suárez and Díaz-Ramos (2003) for marine environments (as marine microalgae). The few studies on ecological protozoology in the country include those of Scorza (1997) and Scorza and Núñez-Montiel (1954), in which a protozoological biota with strong cosmopolitan affinities has been shown. Besides the Ciliophora cited by Scorza and Arcay (2003; as Ciliatea), Bermúdez (1980) added *Ichthyophthyrius multifiliis*, a fish pathogen.

11.3.6 Animals

Despite Venezuelan fauna has been studied since colonial times, first zoological checklists were published rather lately, by Adolf Ernst (1877 b, 1891). A work rarely cited, but encyclopedic and very useful, is the *Catálogo de la fauna venezolana,* of Jaime Tello, published in several volumes. Additional general studies include that of Gremone (1986), on vertebtrates, and those of Díaz-Ungría (1971, 1973), a valuable source of information on endoparasitic worms, including flatworms and nematodes. Cave fauna of Venezuela is summarized in Galán and Herrera (2006). For a history of Venezuelan zoology in the twentieth century the reader is referred to the work of Texera (2003).

Following paragraphs consider only most studied Venezuelan groups. For animals merely reported in literature or only scarcely studied, the reader is welcome to see the "other animal groups" heading, at the end of this section.

11.3.6.1 Sponges

Venezuelan sponges are relatively poorly known if compared with remaining Caribbean and South American countries. After the publication of Pauls (2003), who cites 11 orders, 34 families, 57 genera and 93 species, several new records have been published. Most important publications not included in that revision include the work of Volkmer-Ribeiro and Pauls (2000; cited there as "in press"), on freshwater sponges, and those of Sutherland (1980) and Díaz and Rützler (2009), on sponges from Caribbean mangroves. Although almost all Venezuelan sponges are of class Demospongiae, it has been reported at least one species of the genus *Sycon*, from class Calcarea (Díaz and Rützler, 2009). It is known, also, that some marine species from classes Hexactinelida have been observed or collected, but they are as yet not formally reported (Pauls, 2003). Recent work on class Calcarea from adjacent island of Curaçao (Cóndor-Luján et al., 2018) shows 16 new records, including 10 new species probably shared with adjacent Venezuela.

Freshwater sponges include 3 families, 9 genera and 17 species in the order Haplosclerida, 15 of them in the Orinoco river basin, and just one for each Lake Maracaibo and Unare river basins (Volkmer-Ribeiro and Pauls, 2000). To those numbers it should be added a species of the genus *Ephydatia* from Lake Valencia (Frost, 1981), and a recently described endemic species, thriving at the highest altitude in the world, and with affinities with species from the mountains of Mexico and southeastern Brazil (*Racekiela andina*; Quintero Hernández and Niño Barreat, 2017; Figure 11.3f). Total numbers of Venezuelan Porifera are summarized in Table 11.2, including 144 marine species (after Miloslavich et al., 2005). Expected number of species considers diversity in nearby areas of the Caribbean Sea and South American freshwater ecosystems, and assume that total number from the Caribbean Sea (*ca.* 250 spp. *fide* Pauls, 2003), very homogeneous and with low endemism, it is a good estimate for the country. Some Venezuelan sponges are promising in pharmacological research (Pauls, 2003).

11.3.6.2 Cnidarians

Losada and Pauls (2003) made the last comprehensive synopsis on this group for Venezuela, where only marine species were considered. These include classes Hydrozoa, Scyphozoa and Anthozoa (class Cubozoa as yet not reported for the country). The Zoantharia (47 spp.) and Ceriantipatharia (8 spp.) are among the most diverse of the Caribbean and American Atlantic, but more sampling should be done, especially below –20 m, in oceanic islands. Venezuelan coral reefs include 57 species of scleractinian corals (subclass Hexacorallia), 6 of them ahermatypic, which are present in greater proportion towards west of the country, especially in the archipelagos of Los Roques and Aves; its bathymetric distribution reaches –50 m. Subclass Octocorallia include 40 species of gorgonaceous corals, 2 species of telestaceous corals and 2 species from family Pennatulaceae, mostly above –20 m, although *Callogorgia verticillata* has been found below –50 m in trawl samples from Golfo Triste; as for subclass Hexacorallia, the greatest number of species and studies correspond to the western and central parts of the country.

11.3.6.3 Flatworms

Marine worms of class Turbellaria include *Syndisyrinx collongistyla*, recently studied by Delgado (2015) from seagrass prairies. Showy, rich-colored and larger worms are common on coral reefs, and showy to poorly-colored terrestrial species are usually found in evergreen ombrophilous forests but they have not been as yet formally reported for the country.

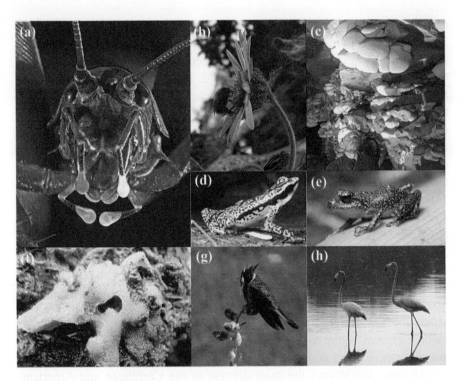

Figure 11.3 Examples of Venezuelan animal and microbial taxa (a) *Hydrolutos breweri*, a cave cricket from Chimantá massif, Venezuelan Guayana (photo: Charles Brewer-Carías). (b) *Bombus rubicundus*, pollinating *Senecio formosus* at the subpáramo of Sierra Nevada National Park, Mérida state (photo: Yeni Barrios); (c) Ball-shaped biospeleothelms from a cave in the Chimantá massif (photo: Charles Brewer-Carías).; (d) *Atelopus cruciger* (Harlequim toad), a critically endangered frog from Venezuelan Coastal Cordillera (photo: Celsa Señaris); (e) *Pristimantis muchimuk*, from Chimantá massif, Bolívar state (photo: Charles Brewer-Carías); (f) *Racekiela andina*, a freshwater sponge from glacial lakes of Cordillera de Mérida (photo: Víctor Quintero); (g) *Oxypogon lindenii* (white-bearded Helmetcrest), from superpáramo vegetation in Sierra de La Culata (photo: Oriana Manrique); (h) *Phoenicopterus ruber* (American flamingo), from Ciénaga El Mene, Lake Maracaibo basin (photo: Yeni Barrios).

Beside some general references (e.g., Díaz-Ungría, 1971, 1973; Guerrero, 1985; Berger, 2018), primarily on parasites of classes Trematoda and Cestoda, information on Platyhelminthes is rather scattered and scarce. One recent publication on fish parasites of class Monogenea, in fact, suggests the necessity of further studies (Figueredo Rodríguez et al., 2017). Given that Venezuelan Platyhelminthes are still poorly studied, and assuming a projected number of 1,000 marine fishes (Table 11.5), supposing at least one (different) species of class Monogenea for each host, and ca. 200 species from remaining classes, expected number of flat worms for Venezuela reaches 1,200 species as a conservative estimate (Table 11.2).

11.3.6.4   Nematodes

Despite this is, perhaps, the most speciose phylum of organisms, especially common in soils and soft-bottom seafloor, to date there are only four reviews in Venezuelan nematodes. Parasitic taxa have received greater attention, and were summarized by Corozzoli (2002, plus an addenda in 2015) for plants ($\geq$ 3 families, 35 genera and 133 species), and by Guerrero (2003) for animals, including humans (58 families and 270 [5,000 predicted] species). To date, just one publication on benthonic marine Nematoda in the Venezuelan Caribbean have been conducted... but 162 species are there reported! (Tiejten, 1984). Milosalvich et al. (2003) recorded 28 families, 79 genera and 166 species, apparently all of them free-living. Assuming that species overlap between the works of Miloslavich et al. (2003) and remaining revisions is negligible, a total of $\geq$89 families, >172 genera and 596 species is here reported (Table 11.2).

11.3.6.5   Annelids

Studies on Venezuelan annelids have been scarce, and skewed to class Polychaeta. Díaz and Liñero (2011) offered the latest review on this marine and estuarine group for the whole country, but more recently, Díaz et al. (2017) offered an updating of the state of knowledge, including information of this organisms as biological indicators of several parameters traditionally associated with human-generated pollution (as heavy metals and high hydrocarbon concentration), as well as normal parameters proper of non-disturbed ecosystems (like water clarity, salinity, depth, and community structure and function). These worms thrive particularly well on sandy beaches and seagrass prairies, and have been primarily studied on northeastern and (to a lesser extent) mid-western Venezuelan coasts. Knowledge on Venezuelan Polychaeta has grown steadily, from 1863 (when reported by the first time, with only two species) to 2017 (Díaz et al., 2017), with 49 families and 532 species, but the total number should be considerably higher (Ibid.).

Additional studies on the phylum include those of Díaz et al. (1997), who reported 1 earthworm species (*Rhinodrilus venezuelanus*, frequently reaching 50 cm long) for the Venezuelan páramos; other as yet unidentified giant earthworms are inhabitants of the Guayana highlands (Charles Brewer-Carías, pers. comm.). Several studies on earthworms and soil fertility have also been conducted, but taxonomic coverage is limited. Oligochaet annelids, with no further taxonomic information, are reported by Pinto and Montoya (2012) and Díaz and Najt (1990), while class Hirudinea has been reported from the Andes, where they have been used as indicators of water quality (Álvarez et al., 1995), and from the Orinoco delta; the leech *Blanchardiella fuhrmani* is abundant near creeks in the Colombian and Venezuelan páramos (Ringuelet, 1974). Aguilera et al. (2003b) reported 116 families, 269 genera and 344 species of phylum Annelida for Venezuela. Since that work was based exclusively on marine Polychaeta, these numbers are somewhat hazy and, at any rate, higher than the estimates for that year (cf. figure 11.3 of Díaz et al., 2017, where *ca.* 150 species are reported). Possible cause of this discrepancy is the sum of both Polychaeta and non-Polychaeta annelids so far known from Venezuela; this is, however, only hypothetical, and was not considered in the estimation of totals of Table 11.2.

11.3.6.6 Chelicerates

This group is relatively well documented, primarily by the works of the late Manuel A. González-Sponga (1929–2009). Many new species, however, are found after each field trip or revisionary study. Information on order Scorpionida (scorpions; 4 families, 19 genera, and 125 species) is summarized in González-Sponga (2003); in Venezuela, only *Ropalorus laticauda*, from dry areas, is not endemic. Order Opiliones (harvesters; 8, 96, 265) has been partially reviewed (González-Sponga, 2003 and references cited therein). Order Araneae (spiders; 49, 288, 423) is, based on current knowledge, the most diverse group in the country, inhabiting from coastal terrestrial ecosystems to alpine deserts (some Salticidae, according to Vareschi, 1970), and tend to be common as domiciliary species (mainly of family Pholcidae); best-studied are large spiders, including the greenbottle blue tarantula (*Chromatopelma cyaneopubescens*), endemic of Paraguaná peninsula, and the monkey spider, consumed by some Amerindian groups. Order Uropygi (whip scorpions; 1, 1, 1) has been reported only south of the Orinoco River. Order Solifugae (camel spiders; 1, 5, 5) has been rarely reported from natural environments, but is quite common in wholesale markets, such as Mercado de Quinta Crespo, in Caracas (González-Sponga, 2003). Díaz et al. (1997) reported >25 species of Acari from the Venezuelan páramos, while Jones et al. (1972) and Guerrero (1996) made reviews on the ticks (Ixodoidea) of Venezuela, including human, pet and cattle parasites. Important groups of Acari in Venezuelan agriculture were reviewed by Doreste (1968) and Aponte and McMurtry (1993), including spider mites (Tetranychidae) and phytoseiid mites (Phytoseiidae), with further additions in Moraes et al. (2004) for the last group; in total, more than 100 species have been recorded for the country. Subclass Pycnogonida (sea spiders) was reported by Miloslavich et al. (2003), including 4 families, 10 genera, and 10 species. Additional orders reported from the country include Pseudoscorpionida (false scorpions; 4, 20, 29). Amblypygi (tailless whipscorpions; 2, 3, 4), Schizomida (shorttailed whipscorpions; 1, 2, 5) and Ricinulei (hooded tickspiders; 1, 1, 1). Order Palpigradi (microwhip scorpions) has been as yet not reported for Venezuela.

11.3.6.7 Crustaceans

Last comprehensive treatment for Venezuelan crustaceans is that of Rodríguez and Suárez (2003). The number of taxa reported in Table 11.2 was obtained from the works of Miloslavich et al. (2003) for marine taxa (161 families, 498 genera, and 1,024 species), and Rodríguez and Suárez (2003) for freshwater and terrestrial species. Fourteen (14) out of 25 extant crustacean classes have been reported for the country. Class Notostraca includes only the cosmopolitan *Triops longicaudatus*, although a species of the genus *Lepidurus*, presently known from Argentina, could be widely distributed in the Andean zone of all South America, including Venezuela (Löfler, 1981). Class Conchostraca includes 3 genera, 1 species each; of them, *Leptestheria venezuelica* is endemic. Class Anostraca includes 3 genera and 6 species; one of them, *Artemia salina* is widely distributed in the world and is consumed by flamingos, due to which they have their characteristic coloration (Figure 11.3h). Class Branchiura have 1 family, 3 genera, and 6 species, all from freshwater environments; only one of them (*Argulus ernsti*) is endemic. Class Copepoda is particularly well diversified, both in fresh and marine waters, and presents 3 orders, 59 genera and 157 species in the country, 15 genera and 42 species mainly from lakes, wells and reservoirs, and 8 endemic. Species from class Cirripedia are included in two major groups: Thoracica, with families Lepadidae (probably 3 genera, 5 species) and Balanidae (4, 12), free-living and sessile, and Ascothoracica, parasites of echinoderms and coelenterate anthozoa and very poorly studied. Class Stomatopoda has at least 3 genera and 8 species. Class Ostracoda is scarcely known, but all reported species are endemics of the genera *Eucypris*, *Heterocypris*, *Physocypra* and *Potamocypra* plus *Cytheridella boldii*, from freshwater environments, and the cavernicolous *Strandesia venezolana,* also endemic; there are, however, no studies of the marine species, the largest group in the class. Class Mysidacea includes 9 genera; only 1 species (*Antromysis merista*), from freshwater environments, is endemic; remaining species are from marine-littoral or brackish waters. Class Amphipoda includes 57 genera and 65 species; at least 17 species are endemic, including *Hyalella cavernicola* and *Metaniphargus venezolanus*, both cave-dwelling, but only the first present no eyes and is depigmented. Class Isopoda includes 82 genera and 87 species; order Calabozoidea is endemic to Guárico and Aragua states, with only 1 species (*Calabozoa pellucida*), described in the 1980 s from wells and groundwater. Class Euphausiacea (best-known from the *krill* of polar pelagic zones) have 5 genera and 13 species. Class Decapoda include shrimps, lobsters and crabs; in Venezuela 75 families, 227 genera and 555 species (38 of them of freshwater crabs), have been reported.

11.3.6.8 Insects

Insects (class Hexapoda) is the most diversified group of living organisms on Earth, but available lists indicate that vascular plants barely outnumber them in Venezuela (cf. Tables 11.2 and 11.4); only for the order Coleoptera, expected number of species is >100,000! (Aguilera et al., 2003b). Most of the insect orders were reviewed in Aguilera et al. (2003), including Orthoptera (Cerdá, 2003a), Ephemeroptera (Segnini et al., 2003), Dictyoptera (Cerdá, 2003b), Isoptera (Rosales, 2003), Dermaptera (Savini and Joly, 2003), Hemiptera (Osuna, 2003), Homoptera (Gaiani, 2003), Megaloptera and Plannipennia (most commonly known as Neuroptera; De Marmels and Gaiani, 2003), Trichoptera (Cressa and Holzenthal, 2003), Lepidoptera (Clavijo, 2003), Coleoptera (Joly, 2003), Diptera (Delgado Puchi, 2003), Hymenoptera (García, 2003), and Plecoptera (Cressa and Stark, 2003); Collembola (springtails), treated as a separate class in Ruggiero et al. (2015), has been reported only from Páramos (Díaz and Najt, 1995). The number of families, genera and species of Venezuelan insects is summarized in Table 11.4. Additional information is available at http://insectoid.info/checklist/insecta/venezuela/ (last accessed 11/03/2018) and the website of the Museum of the Institute of Agricultural Zoology "Francisco Fernández Yépez" (http://www.miza-ucv.org.ve/), also considered for the species countings. Pioneer works on Venezuelan insects include that of Raymond (1994), on order Lepidoptera (butterflies and moths), which is illustrated with beautiful watercolors.

Insects are key elements of both natural and anthropized ecosystems; they act as pollinators and seed dispersers, but also as agricultural pests and vectors of human and animal parasites, and may cause damage to struc-

Table 11.4 Insects of Venezuela (Including Reported Species ("Species") and Expected Species ("Species est.") Numbers.

Orders	Families	Genera	Species	Species est.
Collembola	9	35	60	–
Zygentoma	1	1	1	–
Ephemeroptera	6	17	26	–
Odonata	13	124	527	–
Dictioptera	8	83	170	–
Isoptera	3	25	42	–
Dermaptera	6	33	99–103	–
Plecoptera	1	3	31	–
Orthoptera	14	178	303	–
Phasmida	–	–	–	–
Psocoptera	–	–	–	–
Thysanoptera	>1	>4	>5	–
Phthiraptera	–	–	–	–
Hemiptera	38	475	828	–
Homoptera	30	478	972	–
Megaloptera	2	3	16	–
Neuroptera	11	44	96	–
Coleoptera	112	1,461	3,763	>15,611–100,000
Strepsiptera	3	3	3	–
Mecoptera	9	41	160	–
Diptera	40–69	391–540	1,855–1,912	–
Trichoptera	12	47	311	–
Lepidoptera	85	1,108	2,500–3,873	–
Hymenoptera	69	578	1,470	16,749
	>505–534	**>5,135–5,284**	**>14,211–15,645**	**>41,802–126,191**

Sources are included in text.

tures such as houses and public buildings. Termites (Isoptera), ants (Hymenoptera: Formicidae), and several insect larvae are eaten by native peoples. Recent additions to Venezuelan entomofauna include *Bactrophora dominans* (González et al., 2015), and many new species (and even new genera) of Lepidoptera from the Guayana region (https://www.researchgate.net/profile/Mauro_Costa10). Interesting biogeographic relations, as the pre-Cretaceous connections between Guayana Shield and the highland plateaus of

Nigeria and Cameroon have been reported (DeMarmels et al., 1996, cited in DeMarmels, 2003). Venezuelan Odonata has been particularly well studied, with two published revisions (De Marmels, 1990 and 2003). There are some records of the orders Phasmida, Phtiraptera and Psocoptera, but information is scant and rather scattered, and national revisions are in order. Examples of Venezuelan insects are in Figures 11.2a and 11.3a-b.

11.3.6.9 Myriapods

To date, only class Chilopoda (centipedes) have been studied for Venezuela. Cazorla Perfetti (2012) listed the Venezuelan species, while Schileyko (2014) describes 11 species of Scolopendromorpha and included a key to the 46 species recognized by him. Centipede bites in humans are frequent and may potentially result in fatal consequences; (Cazorla Perfetti et al., 2012). One Venezuelan species, *Scolopendra gigantea* (giant scolopendra) is a common species in the lowlands, including Margarita island, as much as Nicaragua, Colombia, Dominican Republic and the islands of Trinidad, Jamaica, Aruba, and Curaçao; it is a carnivore, and feeds mainly on arthropods such as cockroaches, scorpions, crickets, grasshoppers and butterflies, but also on tarantulas and vertebrates, including lizards, mices, and even bats! (Molinari et al., 2005). Half of the species of Venezuelan Chilopoda were described between 1997 and 2006, by the late Venezuelan arachnologist and myriapodologist Manuel González-Sponga (Cazorla Perfetti, 2012). Class Diplopoda is also present in the country (pers. obs.).

11.3.6.10 Mollusks

Venezuelan mollusks include classes Polyplacophora, Scaphopoda, Gastropoda, Bivalvia, and Cephalopoda. The work of Tello (1975) is, to date, the only available revision including all reported groups, including extant and extinct terrestrial, freshwater, estuarine, and marine species. Concerning the study of mollusks, however, Venezuela is probably the poorest known country in the Caribbean (Caballer Gutiérrez et al., 2015).

Perhaps the best-studied group of mollusks in Venezuela are gastropods. They are summarized for Venezuela in Martínez (2003), where freshwater and terrestrial gastropods and freshwater bivalves from the Unare and Orinoco river basins (29 families, 64 genera and 147 species), and an addenda for marine gastropods, by Pablo E. Penchaszadeh and Claudio Paredes, (77 families, 239 genera, and 610 species) are included. Marine gatropods of continental shelf are rather different from those of oceanic islands. Common and ornamental marine gastropods include *Strombus gigas*, consumed by aborigins from central coast of Venezuela as a source of protein before the arrival of the Europeans, and *Melongena melongena* (Caribbean crown conch), especially common in the waters of the Gulf of Venezuelan. Terrestrial gastropods include the highly invasive African species *Achatina fulica*, which may be also a vector of several animal and human pathogens. Despite opisthobranchs are not a monophyletic group, they still tend to be used as a "natural" taxon in scientific literature; they are an important source of bioactive chemicals with potential medical applications, and some species inhabiting Venezuela have been cited in chemical studies. The best compilation of opisthobranch species known for Venezuela (Caballer Gutiérrez et al., 2015) cites 7 orders, 40 families, 79 genera, and 134 species, about 40% of the entire diversity known from the Caribbean. The state of knowledge of Venezuelan Cepahalopoda is still poor, but several works include valuable information on this class; Robaina (1986), for example, offered a guide for the identification of the Octopoda of Venezuela. Additional works on Venezuelan mollusks include Cummings and Mayer (1997), Lodeiros et al. (1999), Breure (2009), and Lasso et al. (2009a). All of them, plus Miloslavich et al. (2003), were taken into account for countings offered in Table 11.2.

11.3.6.11 Echindoderms

This is an exclusively marine group, with no known parasites and pentameral symmetry. In Venezuela, 5 classes, 18 orders, 39 families, 76 genera and 148 species have been reported (Penchaszadeh, 2003; Miloslavich et al., 2003). Taxonomic studies on this phylum have been scarce, especially for classes Crinoidea and Ophiuroidea (Penchaszadeh, 2003).

Class Asteroidea (sea stars) is perhaps the best studied and, includes 17 species; especially common are *Oreaster reticulatus*, an omnivorous star from seagrass prairies, several carnivorous species of the genera *Astropecten* and *Luidia*, and the large predator of other echinoderms *Tethyaster vestitus*. Classes Crinoidea (sea lilies, feather stars; 5 species), Ophiuroidea (brittle stars; 62), Echinoidea (sea urchins; 24) and Holothuroidea (sea cucumbers; 40) have been less studied, but are also important in coastal ecosystems, coral reefs, and seagrass prairies. Except for some sea urchins and holothuroids, this group is not consumed by local population. Some Asteroidea, however, are marketed for their ornamental value, but with no reports of extinction or overexploitation. In Venezuela, as well as in the rest of the Caribbean, the black sea urchin, *Diadema antillarum,* has almost completely disappeared, affected by a massive mortality in the 1980s (Greenstein, 1989).

11.3.6.12   Fishes (Marine and Freshwater)

Fishes is a paraphyletic group including bony (or Osteichthyes, containing amphibians, reptiles, birds and mammals, plus several ancestral groups) and cartilaginous (or Chondrichthyes) vertebrates from aquatic environments. They are considered under two different headings in Table 11.5, as it is usually done in other biodiversity studies and national reviews; some "freshwater" species, however, spend part of their life cycles in marine waters, as several Perciformes, Pleuronectiformes, Clupeiformes and Tetraodontiformes, while several "marine" species on freshwaters and/or estuaries.

Venezuelan freshwater ichthyofauna is one of the most important in South America (Machado-Allison, 2003; Lasso et al., 2004b). It is composed, primarily, by species of the orders Siluriformes and Characiformes (41% each, *fide* Machado-Allison, 2003). Marine ichthyofauna is quite representative of the Caribbean region (cf. Cervigón, 2003). A detailed and illustrated catalog of marine species is offered by Román (1980), and the monumental work of Cervigón, with two available editions, the last of which is from the 1990 s (Cervigón [1991–1996], updated in Cervigón and Rodríguez, 1997 and Cervigón, 2003); additional contributions are summarized in Cervigón (2003). Subsequent studies, especially in fishes from estuarine environments and inland waters (e.g., Lasso and Sanchez-Duarte, 2011, Ortega-Lara et al., 2012), suggest a greater number of species in Venezuela. According to FishBase (http://www.fishbase.org/, last updated June 27, 2017), there are 815 marine, and 929 freshwater species, but the total number is reported as 1,717, which suppose 27 species from both types of environment. There is only one endemic marine species (*Sparisoma griseorubra*, Cervigón, 2003), but many from freshwaters, including five main watersheds; a detailed list, however, is not available, and several species are waiting to be described. Major endemism occurs in the Casiquiare and Río Negro basins, but endemic species are usually shared with adjacent Colombia and/or Brazil; Orinoco watershed, by far the most extensive in the country, include the highest number of both total and endemic species.

Venezuelan fishes include many commercial species, used as food and aquarium pets. They include marine taxa as *Xiphias gladius* (swordfish), and species of genera *Sardinella* (sardine), *Caranx* (tuna), *Lutjanus*, *Mugil*, and many ornamental taxa from coral reefs. Species of genus *Serrasalmus* (locally known as *caribes*, elsewhere as *piranhas*), Candirú (a species of the genus *Vandellia*, a fish gall parasite), several ornamental microcharacids, the guppy (*Poecilia reticulata*, perhaps the most famous aquarium fish in the world), and species of genera *Pellona*, *Prochilodus*, *Curimata*, *Hydro-*

Table 11.5 Vertebrates of Venezuela

		Orders	Families	Genera	Species/group	Species/total	Species est.
Fishes	Marine	33	148	529	791 (1)	ca. 1,791 (–)	>2,000
	Freshwater	10	58	380	1,000		
Amphibians	Gymnophiona	/	2	7	10 (2)	333 (57.36%)	>400
	Caudata	/	1	1	5(4)		
	Anura	/	15	56	318 (185)		
Reptiles	Testudines	/	8	16	24 (2)	370 (32.70%)	>400
	Squamata	/	20	103	337 (119)		
	Crocodylia	/	2	3	5 (0)		
Birds	Passerines	/	29	335	783	1,381 (3.26%)	1,432
	non Passerines	26	59	283	598		
Mammals	Marsupials	2	2	11	35 (5)	390 (7.69%)	436
	Placentals	12	45	173	355 (25)		

Sources are included under headings "fishes," "amphibians," "reptiles," "birds," and "mammals".

lichus, Hoplias, Myleus, Mylossoma, Cichla, Brachyplatystoma (some individuals >100 kg), and *Pseudoplatystoma*, consumed by local inhabitants, are important freshwater fishes. Chondrichthyes include many marine species of chimaeras, stingrays and sharks (Cervigón, 2003), but only 6 species, in 2 genera of 1 family of order Myliobatiformes (stingrays), from freshwater environments (Machado-Allison, 2003).

11.3.6.13 Amphibians

Account of Molina et al. (2009) is the last comprehensive treatment for Venezuelan amphibians. Previous revisions include Ginés (1959), La Marca (1997a, b, 2003a, b) and Péfaur and Rivero (2000), while an updated checklist is offered in Barrio-Amarós (2009). Besides those works, it is also worth to mention a beautiful field guide recently edited for Canaima National Park, in southern part of the country (Señaris et al., 2014). Important Venezuelan groups include frogs from families Hylidae (80 species), Strabomantidae (57; Figure 11.3e), Leptodactylidae (24), Aromobatidae (50), Bufonidae (27; Figure 11.3d), Centrolenidae (26) and Hemiphractidae (21), with many endemic species and several endemic genera. Caecilians are represented by 2 families, 7 genera and 10 species, 2 of which are endemic, while Salamanders include 1 genus and 5 described species (Barrio-Amorós, 2009; Table 11.5). The revision of Lampo (2009) is the only comprehensive genetic study that has been published to date for a major taxonomic group of living beings inhabiting Venezuela.

Conservation status of Venezuelan amphibians is critical. The fungus *Batrachochytrium dendrobatidis*, which produces cutaneous chytridiomycosis, is responsible for the extinction of species of several frog genera, especially *Atelopus* (Bufonidae). Among the carrier agents is another frog, *Lithobates catesbeianus* (bullfrog), introduced into the Andean region for the meat rearing. *Atelopus vogli* is considered as extinct, since it has been not reported since the date of its original description, in 1933; some authors, however, doubt of its taxonomic validity, and until recently it was considered as a mere variation or subspecies of the more widely distributed striped toad, *A. cruciger* (the only Venezuelan species of *Atelopus* that still survives in Coastal Cordillera; Figure 11.3d). In relation to the Andean species of *Atelopus*, all would qualify in the CR category. The only species of another genus classified as CR in the country is the collared toad of Rancho Grande (*Mannophryne neblina*), from Henri Pittier National Park. Although its population status is presently unknown, its distribution would be very restricted, and could even be extinct, since it has not been seen since its original description in 1956. Another amphibian taxon that deserves urgent attention is the endemic *Aromobates nocturnus,* which has been classified as endangered; it is only known from cloud forests at 2,250 m, in the vicinity of Agua de Obispos (Trujillo state, Andean region), where it was common at the time of its description (Rojas-Suárez et al., 2009).

11.3.6.14 Reptiles

As presently understood, class Reptilia includes subclasses Crocodylomorpha, Rhyncocephalia, Squamata, Testudinata, and Aves (Ruggiero et al., 2015), but the last of them, commonly known as birds, have been usually set apart as a different class (as it is followed here; remaining subclasses, then, also to be considered separately). Previous national-level studies include La Marca (1997a, b, 2003a, b) and Péfaur and Rivero (2000), and one regional review, from the Andes region, edited by Enrique La Marca and Pascual Soriano (2004). According to Rivas et al. (2012), the last comprehensive treatment for all the country, there are 3 orders, 30 families, 122 genera (one of them exotic) and 370 described species (four of them exotic).

Venezuela is the richest country in the world regarding species of genus *Gonatodes* (17 described species, 9 of them endemic), and perhaps also of *Anadia, Cnemidophorus,* and *Phyllodactylus* (Rivas et al., 2012). Charismatic species include the largest predator of South America, *Crocodylus intermedius,* locally known as *caimán del Orinoco* (Barrio-Amorós, 2007), and four additional crocodylids, the iguana (*Iguana iguana*), and 24 turtles, including five marine species (*Caretta caretta, Chelonia mydas, Dermochelys coriacea, Eretmochelys imbricata,* and *Lepidochelys olivacea*). Beautiful and exotic Venezuelan snakes include the anaconda (*Eunectes murinus*) and *Corallus caninus* (Boidae), *Leptophis occidentalis* (Colubridae), and pure yellow individuals of *Bothriechis schegelii* (Viperidae). Poisonous species include members of families Elapidae (13 species of genera *Micrurus* and *Leptomicrurus*) and Vipereidae (12 species of genera *Bothriechis, Bothriopsis, Bothrops, Crotalus, Lachesis,* and *Porthidium*). Lizards are also very diversified; most speciose families are Dipsadidae (107 species), Gymnophthalmidae (52), Sphaerodactylidae (25), Polychrotidae (24), and Teiidae (17), with many endemic species and one endemic genus (*Ardecosaurus*; Rivas et al., 2012).

11.3.6.15 Birds

Venezuela is one of the most diverse countries in the world regarding birds, with 27 orders, 88 families, 618 genera, and 1,381 species (according to Clements, 2018; Ascanio et al., 2017, 2018 including 1,382 species), 18 of them hypothetical (i.e., recorded but with "no tangible evidence" according to the SACC [South American classification committee of the American Ornithological Society]), 45 endemic, and 1 regionally extinct (*Margarops fuscatus*, from Los Hermanos archipelago, but still inhabiting the nearby Antilles). According to Ascanio et al. (2017, 2018), 1,244 species are breeding in the country, 108 species occur regularly but only during nonbreeding season, 31 species are only vagrants, and 6 species have been introduced by humans (including the widespread *Columba livia*, plus *Lonchura malacca*, *Padda oryzivora*, *Passer domesticus*, *Ploceus cucullatus*, and *Psittacula krameri*). Species density tend to be also high. Thus, for example, in the Henri Pittier National Park it has been recorded 582 (5.4 spp.km^{-2}), while in the Tamá National Park, with 555 spp., species density reaches 3.9 spp. km^{-2}; these numbers, in fact, are only surpassed by a locality in Uganda, with 7.3 spp.km^{-2} (Lentino, 2003). Paleognathae includes only order Tinamiformes (tinamous), with 14 species in genera *Cryptellus* (9), *Nothocercus* (2), and *Tinamus* (3), all of them of family Tinamidae. Passerine birds, in contrast, are represented by 783 species in 335 genera and 29 families, 191 genera and 406 species in 9 families suboscine, the remaining oscine.

This is the best-known major taxonomic group in the country, with estimates reaching ca. 95% of reported species (Aguilera et al., 2003b), but taxonomic problems remains, and new taxa have still to be described (Ascanio et al., 2017, 2018). Lentino (1997, 2003) offered updated revisions for Venezuelan birds, but six field guides, with relevant taxonomic and distributional information, both at the supranational (Restall et al., 2007) and national (Meyer de Schauensee and Phelps, 1978 [first edition] and Hilty, 2003[second edition]; Phelps and Meyer de Schauenseee, 1979 [first edition] and 1994 [second edition]; Ascanio et al., 2017) levels, are commonly used as authorized references. Troupial (*Icterus icterus*) is the national bird, but charismatic species also include American flamingo (*Phoenicopterus ruber*; Figure 11.3h), four large (*Ara ararauna*, *Ara militaris*, *Ara chloropterus*, and *Ara macao macao*) and three smaller (*Ara severus*, *Orthopsittaca manilatus*, and *Diopsittaca nobilis*) macaws, red-fan parrot (*Deroptyus accipitrinus*), hoatzin (*Opisthocomus hoatzin*), oilbird (*Steatornis caripensis*), Harpy eagle (*Harpia harpija*), and 101 different hummingbirds, including sword-billed hummingbird (*Ensifera ensifera*), spangled and tufted coquettes (*Lophornis stictolophus* and *L. ornatus*), and *Oxypogon lindenii* (Figure 11.3g). Recent range expansions include large macaws (*Ara ararauna, A. chloropterus* and *A. macao*), which have been naturalizing in Caracas (http://www.bbc.com/mundo/noticias/2014/10/140930_venezuela_guacamayas_caracas_dp), and some species, previously recorded from Los Llanos, that are reaching the nearby Andes. Recent colonization of cities by *Coragyps atratus*, which feeds on waste places, is also worthy of mention.

11.3.6.16 Mammals

Mammals are the only living line of synapsids, the sister group of remaining living amniots. This group includes two monographs (Tello, 1979; Linares, 1998) and two recent exhaustive reviews (Soriano and Ochoa, 1997; Ochoa and Aguilera, 2003). The largest species density is in the Guayana region, followed by northern-coastal (including Coastal Cordillera and transitional zone of Lara-Falcón), Llanos, Andes, and Lake Maracaibo basin zoogeographical regions (Ochoa and Aguilera, 2003). Marine mammals represent only 5.7% of Venezuelan species, and several of them (especially mysticets) are migrants. Some species still present taxonomic problems, as those of the deer genus *Odocoileus*. In this genus, 1 (as *O. virginianus*; Moscarella et al., 2003) to 3 species (*O. cariacou* [widespread, mainly lowlands], *O. lasiotis* [páramos] and *O. margaritae* [Margarita Island]; Molinari, 2007), have been reported; the study of Molinari (2007), however, does not consider critical populations from other neotropical countries, and made use of phenetic rather than phylogenetic methods. Despite last comprehensive treatment for *Odocoileus virginianus sensu lato* (Hewitt, 2011) does not accept the endemic taxa from the páramos and Margarita, much more taxonomic work is to be done before definite conclusions can be drawn.

Recent description of several new species (e.g., Quiroga-Carmona and Molinari, 2012 and Quiroga-Carmona and Woodman, 2015), suggest that our knowledge is still incomplete. Representative mammal species present in Venezuela include tapir or *danta* (*Tapirus terrestris*), chigüire (*Hydrochaeris hydrichaeris*), jaguar (*Panthera onca*), three-toothed sloth (*Bradypus tridactylus*), cachicamo (*Dasypus novemcinctus*), and manatee (*Trichechus manatus*), 34 species of marsupials (5 of them endemic), 97 species of rodents (14 of them endemic), and 165 species of bats (4 of them endemic). Charismatic, ecologically relevant, and endangered is the only extant bear species of Latin

America, spectacled bear, *oso frontino* or *oso de anteojos* (*Tremarctos ornatus*), which inhabits Andean cloud forests and nearby páramos.

11.3.6.17   Other Animal Groups

Besides groups mentioned above, several other animal phyla have been reported for Venezuela. Since they are usually understudied, or reported number of species is low, they are grouped here under a separate heading.

Dicyemids (phylum Rhombozoa) are parasites of the kidney appendages of the cephalopods, with 3 species of the genus *Dicyema* reported from *Octopus vulgaris* populations of central coast (Gómez and Pechaszadeh, 2003). Phylum Onychophora (velvet worms) is represented in Venezuela by 4 species in 2 genera, both within family Peripatidae (de Sena Oliveira et al., 2012); at least 1 additional species and, probably, also 1 additional genus are present, however, in the surroundings of the city of Mérida, within Andean region (Luis Ángel Niño, pers. comm.). Although once considered restricted to particular environments, as matt-forming mosses, phylum Tardigrada (water bears) is presently considered almost ubiquitous; to date, nonetheless, they have been recorded only from sandy beaches of the Apure river, in the Llanos region (Pinto and Montoya, 2012). Venezuelan rotifers include 21 families, 44 genera and 268 species (Table 11.2) from freshwater, estuarine and marine environments; they have been reviewed, for Venezuela, by Zoppi de Roa and Pardo (2003). Benthic bryozoans (phylum Bryozoa) haven been dragged from Caño Macareo (Capelo et al., 2008), while some gastrotrichs, without additional taxonomic information, have been reported recently by Pinto and Montoya (2012). Extant members of phylum Brachiopoda includes 4 genera and 5 species (Cooper, 1977). Phyla Ctenophora, Sipuncula and Nemertea were reported by Miloslavich et al. (2003, 2011; numbers in Table 11.2). Chaetognathes, hemichordates and urochordates have been reviewed by Zoppi de Roa and Pardo (2003), but additional countings are offered in Miloslavich et al. (2003). To date, and according to the classification of Ruggiero et al. (2015), there are no records for 13 animal phyla (Placozoa, Orthonectida, Kinorhyncha, Loricifera, Nematomorpha, Priapula, Acanthocephala, Cycliophora, Entoprocta, Gnathostomulida, Micrognathozoa, Phoronidea, and Xenocoelomorpha), but most of them could be found as field trips and revisionary work continues.

11.4 Conservation Concerns

Since its inception (Myers, 1988), biodiversity hotspots have been used for biodiversity prospecting and conservation. Venezuela harbors northeastern tip of "Tropical Andes", an extension of former "Uplands of Western Amazonia" hotspot (Myers et al., 2000), which includes both the Andes and Coastal Cordillera. The "Caribbean Islands" hotspot (Mittermeier et al., 2004, 2011), includes all Venezuelan islands except those of Gulf of Venezuela, the Islas Caracas archipelago and Nueva Esparta state; Margarita island, however, should be included within the Tropical Andes hotspot, together with Coastal Cordillera, as explained above (see section 11.1.1). Venezuela harbors, also, two of the wilderness areas of the Americas (viz., "Amazonia" and "Llanos," the first including Guayana Shield; Mittermeier et al., 2003), and extensive areas of still pristine or only slightly modified lands (Sloan et al., 2014). As stressed by Marchese (2015), delineating hotspots includes quantitative criteria along with subjective considerations, with the risk of neglecting areas not so "hot", or with other kinds of values. Nowadays, in fact, it is widely acknowledged that biodiversity it is much more than the number of species present in a region, and a thorough conservation strategy cannot be based merely on it or their risk of elimination, but in ecosystem fragility, presence of endemic taxa, and particular genotypes of widespread species, as well as the status of biological interactions. The hotspot concept, in fact, is flawed in some aspects because of associated oversimplifications, but it is still a powerful tool for the conservation of global biodiversity. Detecting micro and nano-hotspots within larger areas, in fact, is a useful tool for setting conservation efforts, as has been shown recently by Cañadas et al. (2014), but this is still to be done in the country.

Venezuelan biodiversity, so rich and diverse, is seriously endangered (cf. Table 11.6), especially in its northernmost part. Conservation studies, however, have been focused in individual charismatic species, mainly homeothermic vertebrates (i.e., birds and mammals), as it had been the global tendency in last three decades (Di Marco et al., 2017). Conservation concerns about Venezuelan biodiversity have been reviewed by Romero (2011) and Madi et al., (2011), while conservation status, according to the criteria of the IUCN are summarized in Tables 11.1 and 11.6, from Rodríguez et al. (2010; terrestrial ecosystems), Llamozas et al. (2003; vascular plants; second edition in progress [http://www.lrfv.org/]), and Rodríguez et al. (2015; animals). According to Oliveira-Miranda et al. (2010), deciduous forests are the most threatened terrestrial ecosystem at the national level. This is true, in our opinion, only if human direct influence is considered, and non-forested

Table 11.6 Conservation Status of Selected Taxonomic Groups (According to the IUCN Red List Criteria (IUCN, 2012a and b))

Taxonomic group	NE	DD	NT	VU	EN	CR	RE	EW	EX
Lycophyta	1	–	–	–	1	1	–	–	–
Monilophyta	–	–	–	–	–	–	–	–	–
Gymnosperms	2	–	12	3	–	–	–	–	–
Magnoliids	4	8	41	9	3	4	–	–	–
Chloranthales	–	–	3	–	–	–	–	–	–
Monocotyledons	28	36	199	93	34	19	–	–	–
Eudicotyledons	132	66	622	86	29	27	–	–	1
Cnidaria	–	–	–	3	–	–	–	–	–
Chelicerates	–	12	–	–	2	–	–	–	–
Crustaceans	–	18	3	9	1	–	–	–	–
Insects	–	37	16	19	29	3	–	–	–
Mollusks	–	2	2	4	–	–	–	–	–
Fishes	–	68	17	30	25	4	–	–	1
Amphibians	–	76	76	13	12	18	–	–	1
Reptiles	–	27	2	11	12	2	–	–	–
Birds	–	57	115	21	19	4	1	–	–
Mammals	–	69	25	30	17	1	1	–	–
Total	**167**	**476**	**1,133**	**331**	**184**	**83**	**1**	**–**	**3**

Sources: Llamozas et al. (2003; plants) and Rodríguez et al. (2015; animals); DD, data deficient; NT: near threatened, NE: near endangered, VU: vulnerable, EN: endangered, CR: critically endangered, RE: regionally extinct, EW: extinct in the wild, EX: extinct.

ecosystems are excluded. Strand vegetation and alpine deserts seem to be, in fact, even more endangered. Deciduous forests are, at any rate, the most endangered major forest type in the country, perhaps only surpassed by dry montane forests. Extinction is already a problem in Venezuela. According to current knowledge, there are at least three extinct species, all of them endemic and from most populated areas in the country, Central an Western Coastal Cordillera: *Lithogenes valencia* (a freshwater fish presumably from Lake Valencia basin), *Atelopus vogli* (a frog, perhaps only a variation of *A. cruciger*, from Lake Valencia basin) and *Hunzikeria steyermarkii* (an angiosperm from the surroundings of Puerto Cabello and Morón, in northern Carabobo state). An additional species, the bird *Margarops fuscatus,* have been reported as regionally extinct, but it is still present in adjacent Lesser Antilles.

Environmental restoration has been practiced in Venezuela with some success, especially by the national oil industry (Rocha, 2003). Two recent publications summarize the restoration efforts and associated research in the country (Rocha, 2003; Herrera and Herrera, 2011).

11.4.1 National Laws and International Obligations

National Constitution (published in Gaceta Oficial Extraordinaria 5453, 24 III 2000) include, in Title III, Chapter IX, Articles 127–129, Venezuelan "environmental rights." They referred to a "safe, healthy and ecologically balanced" environment, protecting "biological diversity, genetic resources, ecological processes, national parks and natural monuments, as well as other areas of special ecological importance", establishing the obligation of the State to promote community participation. Specific laws on biodiversity include the Approbing Law of the Convention for the Protection of Flora, Fauna and Scenic Beauties of the Americas (Gaceta Oficial 20643, 13 XI 1941), Organic Law for the Organization of the Territory (Gaceta Oficial Extraordinaria 3238, 11 VIII 1983), Law of Coastal Zones (Gaceta Oficial de la República Bolivariana de Venezuela 37349, 19 XII 2001), Law on Fisheries and Aquaculture (Gaceta Oficial de la República Bolivariana de Venezuela 37727, 8 VII 2004, prohibiting trawling), Organic Law of the Environment (Gaceta Oficial de la República Bolivariana de Venezuela [extraordinary] 5833, 22 XII 2006), Law of Waters (Gaceta Oficial de la República Bolivariana de Venezuela 35595, 2 II 2007), Law of Management of Biological Diversity (Gaceta Oficial de la República Bolivariana de Venezuela 39070, 1 XII 2008), and Law of Forests (Gaceta Oficial de la República Bolivariana de Venezuela 40222, 6 VIII 2013). Additional details about some of these laws may be consulted in a review of Venzuelan legislation on biodiversity, published 15 years ago (González Jiménez, 2003b).

Venezuela is also signatory of, at least, 13 international conventions which are relevant for biodiversity: Convention for the Protection of Flora, Fauna and Scenic Beauties of the Americas (1941), Convention on International Trade in Endangered Species of Wild Fauna and Flora (CITES) (1977), Convention for the Protection and Development of the Marine Environment in the Wider Caribbean Region (WCR) or Cartagena Convention (1986), from which the Protocol on Oil Spills, Specially Protected Areas and Wildlife (SPAW) and the Ramsar Convention of 1988 derive, Convention on Biological Diversity (1994) from which decision 391 of the Board of the Cartagena Agreement on the Common Regime for Access to Genetic Resources and the Biosafety Protocol derive, International Tropical Timber Agreement (1994, 2006), Inter-American Convention for the Protection and Conservation of Sea Turtles (IAC) (1998), Kyoto Protocol to the United Nations (UN) Framework Convention on Climate Change (2004), UN Climate Change Conference, COP 21 or CMP 11, held in Paris (2015), Rotterdam Convention on the Prior Informed Consent Procedure for Certain Hazardous Chemicals and Pesticides in International Trade (2004), International Treaty on Plant Genetic Resources for Food and Agriculture (2004), and the Stockholm Convention on Persistent Organic Pollutants (2004). Venezuela actively participated in the 1992 Nairobi Convention for the establishment of Agenda 21 of UN and in the elaboration of the Agreement on Biological Diversity itself, which was signed at the Convention on Biological Diversity at the Earth Summit held in Rio de Janeiro in 1992. Despite this, national government does not develop actions aimed at the improvement of its environmental standards, or when applied, they have not been successful. Venezuela, in fact, is the Latin American country that most pollutes and contributes to climate change, with almost double the average of energy consumption (2,271 kg oil eq. *per capita)* and almost 3 times the amount of CO_2 *per capita* (6.4 metric tons) in the region, above the global average, and have a rate of deforestation that remains almost the regional average, with forest losses of 2,749.3 km² per year and one of the lowest forest productivity in the Americas (WBG, 2016: 223). Weak monitoring of biological and cultural diversity, deficiencies in processing petitions for research permission, decreasing support from government to national research institutions, and delayed participation in international conventions, through which Venezuela could integrate into a sustainable economic dynamic, have propitiated recent cases of biopiracy, as the use of alternative sources of Paclitaxel from illegal gatherings of fungi from Guayana Shield, and the appropriation of traditional knowledge from native peoples (http://www.globalissues.org/news/2011/02/09/8470).

11.4.2 Biodiversity Threats

Major biodiversity loss and environmental disturbance come from deforestation, indiscriminate hunting and fishing, the plant and pet trade, logging, ranching, and mining, both for local consumption and for exportation.

Causes and extension of deforestation in Venezuela have been reviewed recently (Pacheco et al., 2011). According to these authors, the causal factor with greater influence is the agricultural expansion, the more important underlying factor is demographic growth (each one with 41.8% influence), and the interaction pattern of the most frequent causal factors was derived from regional, national and international migrations, especially in southern Lake Maracaibo basin and Western Llanos. The flawed figure of Forestry Reserves (steadily becoming "Cattle Reserves"), have generated many and complex threats to national ecosystems, including previously pristine and remote areas, as Imataca (northeastern Guayana Shield) and Caparo (Western Llanos). Expanding frontiers of agriculture, especially by traditional slash and burn subsistence units or "conucos" made by creole population is especially harmful in Coastal Cordillera. Forestry plantations, however, seems to be not so harmful; pine plantations in southeastern Venezuelan Llanos, in fact, are highly reversible and, when properly managed and maintaining adjacent pristine areas untouched, may even preserve local biota (Suárez et al., 2000).

Venezuela's rich deposits of oil, gas, gold, diamond and several strategic minerals, like coltan (the so-called "blue gold") are an opportunity for national development, but also an endless source of potential destruction of natural ecosystems. Recent efforts of the national government to promote exploitation of super-heavy oil (with enormous quantities of coke production) and the megaproject *Arco Minero del Orinoco* ("Orinoco Mining Arc"), ratified in a recent executive order (decree 2,248, Gaceta Oficial de la República Bolivariana de Venezuela 40855, 24 II 2016) with the aim to exploit strategic minerals (primarily gold) for the reactivation of suddenly impoverished economy, have caused international anger and alarm (https://periodismocc.wordpress.com/2017/04/28/green-light-for-destruction-mega-mining-in-venezuela/; Naveda, 2017).

Regarding wetlands and aquatic ecosystems, human impacts include the Caño Mánamo closing for agricultural activities in the decade of the

1970 s, with important landscape and vegetation deterioration, soil hyper-acidification and social changes (http://www.abrebrecha.com/articulos. php?id=24853), the Mantecal polders for pasture management, and destruction of mangroves, including RAMSAR areas (like Morrocoy National Park), especially in central and mid-western coasts. Climate change is one the main threats for glaciers, alpine deserts and highland tepui ecosystems, as well as for coastal wetlands (Rodríguez et al., 2013; Marrero and Rodríguez-Olarte, 2017), including towns and cities along coastline. Effluents on beaches and sublittoral ecosystems, like coral reefs and seagrass prairies, shrimp farms and breeding of aggressive fishes, especially in the estuarine and marine-coastal environments of Falcón and Zulia states and Lake Valencia, are also worth of mention.

Invasive species of Venezuela and their harmful effects on natural ecosystems were summarized by Ojasti et al. (2001) and Pérez et al. (2007). According to a recent report, there are at least 94 invasive species in the country, including several of the most dangerous worldwide, as the gastropod *Achatina fulica* and the frog *Lithobates catesbeianus* (http://www. el-nacional.com/noticias/ciencia-tecnologia/especies-invasoras-amenazan-biodiversidad-del-pais_81471). Because accelerated and nonplannified development of Venezuelan aquaculture from mid-twentieth century, inlcuding the uncontrolled introduction of exotic species, like trouts in mountain rivers and lakes, and the above-mentioned examples from Lake Maracaibo and Lake Valencia, some critical situations, such as in Laguna de Los Patos, a coastal lagoon near Cumaná, have been produced; there, for example, introduced *Oreochromis mossambicus* caused the local extinction of 13 of the 23 original fish species in just 12 years (Nirchio and Pérez, 2002). Most dangerous invasive fish species, however, is red lionfish (*Pterois volitans*), while dove (*Columba livia*), a widespread introduced bird species, may serve as a vector of human diseases. Several agricultural weeds and pests, including eight of the top-ten world's worst weeds (*sensu* Holm et al., 1977) and African grasses like *Melinis minutiflora*, *Hyparrhenia rufa*, *Megathyrsus maximus*, *Cenchrus clandestinus* and *C. purpureus*, which invade degraded forests, delay natural succesion and contribute to continuous burning, have also been reported. *Cenchrus ciliaris* and several succulent plants, including *Kaloanchoe x houghtonii* (previously reported as *K. daigremontiana*), *Stapelia gigantea* (Herrera and Nassar, 2009), and *Euphorbia* sp., are invading xerophilous areas. *Calotropis* spp. are especially harmful in degraded agricultural lands, deciduous forest zones, and coastal vegetation. The exotic seaweed *Kappaphycus alvarezii*, introduced for cultivation, is affecting the corals of Cubagua, while soft coral *Xenia* sp., from Indonesia, has reduced marine diversity by displacing native corals in the Mochima National Park. The exotic tree *Mimosa bracaatinga* is an invasive species in Coastal Cordillera and the Andes, especially in southernmost Cordillera de Mérida, between Mérida and Táchira states (pers. ob.). Bad reforestation practices, with exotic species like *Eucalyptus* spp. and *Pinus* spp., have produced also environmental deterioration and biotic impoverishment.

11.4.3 Protected Areas

Venezuela has a total of 386 ABRAEs (*Áreas Bajo Régimen de Protección Especial* or "Areas under Special Administration Regime"), 95 of which are in coastal and marine areas (Oliveira-Miranda et al., 2010; Villamizar and Cervigón, 2017). They include 44 national parks, 17 natural monuments, 6 wild fauna refuges, and 2 biosphere reserves (protected areas with priority on conservation purposes, or PAs). Caura National Park is the second largest terrestrial PA of the world (7,533,952 ha), created in 2017 by Gaceta Oficial de la República Bolivariana de Venezuela 41118 in one of the last pristine river basins of Earth; this park, and the Alto Orinoco-Casiquiare Biosphere Reserve, protect large areas of Guayana Shield. Remaining ABRAEs were conceived as proteced lands for the purpose of current or potential use, including forestry, and wild fauna and hydraulic reserves, among others. Perhaps most relevant Venezuelan marine PA is the Archipiélago de Los Roques National Park, where most diverse marine communities are present, including coral reefs, seagrass prairies and mangroves (Villamizar and Cervigón, 2017). This PA, created in 1972, is the largest marine national park in Venezuela and the Caribbean Sea, with 221,220 ha. Politically, this was regarded as a commitment by Venezuela to conserve marine ecosystems, honoring the Second World Conference on National Parks held in Yellowstone, in 1972.

Despite Venezuelan ABRAEs are among most diverse and extensive conservation areas of Latin America, covering about 75,400,000 ha (ca. 42% of the country), they have been the subject of severe criticism (e.g., Yerena and Naveda, 2014; Naveda, 2017), because of inappropriate management. Marine and coastal areas, moreover, are underrepresented. Marine ABRAEs, in fact, cover only 0.6% (5,470 km²) of Venezuela's territorial sea, while both coastal and marine ABRAEs include 3,777,787.47 ha, less than 8.70% of protected areas of the country. Coastal PAs incorporate 18.2% (682 km) of the 3,964 km of Venezuelan continental coastline, with almost all of the major coral reef ecosystems included and protected, but strictly marine areas comprise only 3.84% of the total PAs, indicating a strong bias towards terrestrial and, to a lesser extent, coastal ecosystems. All Venezuelan coastal and marine PAs are in the central Caribbean marine ecoregion, which is ranked as being one of the highest priority sites for conservation among coastal biogeographical provinces of Latin America and the Wider Caribbean (Miloslavich et al., 2003; García et al., 2011; Vilamizar and Cervigón, 2017). Little or no coordination between central and local authorities makes difficult to solve or at least diminish current threats to ABRAEs, while institutional strength and professional capability are the most urgent needs to be afforded by their management agencies (Miloslavich et al., 2003; Villamizar and Cervigón, 2017; Naveda, 2017).

Current international conservation projects include the so-called Triple A initiative, or Andes-Amazon-Atlantic Ecological Corridor, the world's largest area planned ever to preserve biodiversity and mitigate climate change (http://www.gaiafoundation.org/gaia-Amazonas-inspires-climate-change-action-and-the-worlds-largest-ecological-corridor/, http://desarrollosustentable.com.ve/wataniba-corredor-ecologicocultural-Amazonas-andes-atlanticoaaa/). As originally planned, it will span 135 million ha (521,240 sq. mi) of rainforest, 4% of which are located in Venezuela. Other international, very promising project, is that of the Jaguar Corridor Initiative, which is working to preserve the genetic integrity and future of the species by connecting and protecting core populations from Mexico to Argentina, including Venezuelan Guayana and Western Llanos (http://www.panthera.org/initiative/jaguar-corridor-initiative).

Future conservation planning should consider several types of reserves, including micro and nano-reserves and not only mid-sized or huge mega-reserves, like it is usual in developing countries. Total area to be preserved, however, depends strongly on ecosystem type, as has been stressed in previous studies, usually between 10% to 30% for tropical ecosystems (cf. Françoso et al., 2015). An interesting approach to conservation was proposed recently by Sanz (2007), based on the distribution of endemic, threatened, spatially restricted and/or habitat-specialist vertebrate taxa.

11.5 Investigation Priorities

Venezuela has a rich and potentially valuable biodiversity. To know, to preserve and to rationally use its potentialities must be the three major investigation priorities.

As has been stressed by many authors, basic research on taxonomy, biogeography and ecology should be conducted in order to complete the still very preliminary list of species and ecosystem mapping, including the collection, curation and revision of biological specimens and imagery. Particularly poorly studied in Venezuela are microorganisms (viruses and bacteria), followed by nematodes, fungi, flatworms, non-polychaetan annelids, arthropods (especially myriapods, achari and insects), and mollusks.

Conservation biology, including the widening of criteria to establish strategic protected areas, integrated with the national culture and local economy, is perhaps the most important scientific endeavor to accomplish; the proper interaction of humans with the environment perhaps the most crucial issue on the topic. Conservation efforts are urgently needed for deciduous, semi-deciduous and evergreen lowland forests of northern Venezuela, but culture is also endangered, especially those of reduced indigenous peoples. Further research, thus, is guaranteed on biocultural diversity, both for indigenous and nonindigenous citizens. Ecosystem services, especially water, are currently severely endangered.

Biotechnology, mainly with bacteria (e.g., Balcázar et al., 2015), fungi (both lichenized and nonlichenized; Marcano, 2003; Iturriaga and Ryvarden, 2003), and plants (González Rosquel, 2003), including pharmacological prospecting and bioremediation protocols based on local experimental research (e.g., Zamora et al., 2017) have been worked out in national institutions, and are very promising for the future. Perhaps the most important biotechnological achievement of Venezuelan science, in fact, is related to biodiversity, and it is due to the late Jacinto Convit, who inoculated *Mycobacterium leprae* in *Dasypus* sp. (Mammalia: Xenarthra), and obtained a serum which mixed with the BCG (the tuberculosis vaccine) result in leprosy immunization. This method, further improved by World Health Organization (WHO) is still in use and is very effective in its control. Agrobiodiversity, including local cultivars of cocoa, pineapple, cassava, potato, sweet potato, cashnut, corn, chilies, and pijiguao palm, developed for millenia by ancestral aboriginal peoples, are also of great relevance. Wild birds (e.g., *Penelope* spp.), reptiles (e.g., *Caiman crocodylus*) and mammals (e.g., *Hydrochaeris hydrochaeris*) are an important source of protein and leather for local people, but could be reared also for massive consumption (cf. González Jiménez and Szeplaki, 1998). Local races of domestic mammals include the Mucuchíes dog (*Canis lupus familiaris*; https://es.wikipedia.org/wiki/Mucuch%C3%ADes_(raza_de_perro)) and Carora cows (*Bos primigenius taurus*; the only tropical race of cattle for milk production; https://www.razacarora.com.ve/), but also wild herds of Spanish horses (*Equus ferus caballus*), which are still grazing in Western Llanos. Additional valuable resources include aquatic fauna and local fisheries; aquaculture, including shrimp cultivation in particular, have been a very productive practice in recent history, and could be extended to native spescies, as *Litopenaeus schmitti* and *Callinectes sapidus*, as much as to the *cachama* or *tambaqui* (*Colossoma macropomum*), currently reared for local consumption, to meet the national demand for protein.

Acknowledgments

Otto Huber, Charles Brewer-Carías, Stephen Tillett, Mauricio Ramia, Alfredo Carabot, Winfried Meier, Bruno Manara, Gilberto Morillo, Valois González Boscán, María Beatriz Barreto, Carlos Reyes, Lilianyel Lucena, Edmundo Guerrero, Giuseppe Colonnello, Miguel Pietrangeli, Celsa Señaris, Egleé L. Zent, Gustavo Fermín, Ricardo Rico, Jesús Delgado, Shingo Nozawa, Ángel Villarreal, Yeni Barrios, Víctor Quintero Hernández, Oriana Manrique and Luis Ángel Niño Barreat provided advice, literature, pictures and/or relevant information on ecosystems and taxonomic groups. The Instituto Experimental Jardín Botánico de Caracas (formerly Fundación Instituto Botánico de Venezuela), Instituto de Zoología y Ecología Tropical (formerly Instituto de Zoología Tropical), Fundación La Salle de Ciencias Naturales, Instituto Jardín Botánico de Mérida, MERF Herbarium, Sistema Teleférico Mukumbarí, and Charles Brewer-Carías provided logistical support in the field trips. Thanks are extended to Thammineni Pullaiah for encouragement and editorial advise.

Keywords

- climate
- conservation concerns
- ecosystems
- taxonomic groups

References

Aguilera, M., Azócar, A., & González, J. E., (2003a). *Biodiversidad en Venezuela*, 2 voll. Fundación Polar, Ministerio de Ciencia y Tecnología, Fondo Nacional de Ciencia, Tecnología e Innovación (FONACIT), Caracas, Venezuela.

Aguilera, M., Azócar, A., & González, J. E., (2003b). Venezuela: Un país megadiverso. In: Aguilera M. M., Azócar, A., & González Jiménez, E., (eds.), *Biodiversidad en Venezuela,* vol. 2. Fundación Polar, Ministerio de Ciencia y Tecnología, Fondo Nacional de Ciencia, Tecnología e Innovación (FONACIT), Caracas, Venezuela. pp. 1056–1072.

Aguilera, M., (2000). *Especiación* y *Biodiversidad*. Trabajo de Ascenso, Universidad Simón Bolívar, Caracas, Venezuela.

Aguilera, M., Anderssen, M. R., et al., (2007). *Medio Físico y Recursos Ambientales*. Geo Venezuela, vol. 2. Fundación Empresas Polar, Caracas, Venezuela.

Álvarez, M., et al., (1995). Hirudíneos dulceacuícolas como indicadores de calidad del agua del río Albarregas, *Actas del III Congreso Latinoamericano de Ecología*, Mérida, Venezuela.

Antón, K., Guzmán, L. M., et al., (2011). Bacterias patógenas aisladas en la nasofaringe de niños indígenas Warao, estado Sucre, Venezuela. *Rev. Soc. Ven. Microbiol.*, *31*(2), 112–117.

APG [Angiosperm Phylogeny Group], (2016). An update of the Angiosperm Phylogeny Group classification for the orders and families of flowering plants: APG IV. *Bot. J. Linn. Soc.*, *181*(1), 1–20.

Aponte, O., & McMurtry, J. A., (1993). Phytoseiid mites of Venezuela (Acari: Phyoseiidae). *Int. J. Acarol.*, *19*(2), 149–157.

Aranguren, A., Costa, M., Guevara, J., & Carrero, O., (2015). Phytobiogeography of the species associated with dry intermountain valleys in the Chama river middle basin, Mérida, Venezuela. *Acta Bot. Venez.*, *38*(1), 63–85.

Ardito, S., Gómez, S., & Vera, B., (1995). Estudio sistemático de las macrilagas marinas bentónicas en la localidad de Taguao, Distrito Federal, Litoral Central, Venezuela. *Acta Bot. Venez.*, *18*(1–2), 53–56.

Aristeguieta, L., (2003). *Estudio Dendrológico de la Flora de Venezuela*. Monografías de la Academia de Ciencias Físicas, Matemáticas y Naturales, *38*.

Ascanio, D., Miranda, J., et al., (2018). *Species Lists of Birds for South American Countries and Territories: Venezuela*, Available at: http://www.museum.lsu.edu/~Remsen/SACCCountryLists.htm. (version 29 VI 2015, last accessed 04 III 2018).

Ascanio, D., Ramírez, G., & Restall, R., (2017). *Birds of Venezuela*, Helm Field Guides, Yale University Press, USA.

Ataroff, M., & García-Núñez, C., (2013). Selvas y bosques nublados de Venezuela. In: Medina, E., Huber, O., Nassar, J. M., & Navarro, P., (eds.). *Recorriendo el Paisaje Vegetal de Venezuela*. Ediciones IVIC, Instituto Venezolano de Investigaciones Cientificas (IVIC), Caracas, Venezuela. pp. 125–155.

Ataroff, M., (2003). Selvas y bosques de montaña. In: Aguilera M. M., Azócar, A., & González, J. E. (eds.). *Biodiversidad en Venezuela*, vol. 2. Fundación Polar, Ministerio de Ciencia y Tecnología, Fondo Nacional de Ciencia, Tecnología e Innovación (FONACIT), Caracas, Venezuela. pp. 762–810.

Ataroff, M., & Sarmiento, L., (2004). Las unidades ecológicas de los Andes de Venezuela. In: La Marca, E., & Soriano, P., (eds.). *Reptiles de Los Andes de Venezuela*, Fundación Polar, Conservación Internacional, CODEPRE-ULA, Fundacite Mérida, BIOGEOS, Mérida, Venezuela. pp. 10–26.

Aubrecht, R., Brewer-Carías, C., Šmída, B., Audy, M., & Kováčik., L'., (2008). Anatomy of biologically mediated opal speleothems in the world's largest sandstone cave: Cueva Charles Brewer, Chimantá Plateau, Venezuela. *Sediment. Geol.*, *203*, 181–195.

Aubrecht, R., Lánczos, T., Schlögl, J., Vlček, L., & Šmída, B., (2013). Arenitic caves in Venezuelan tepuis: what do they say about tepuis themselves?. Pp.: 221–226. *In: Karst and Caves in Other Rocks*, Pseudokarst –oral, ICS Proceedings, Brno, Czech Republic.

Aymard, G. A., (2011). Bosques húmedos macrotérmicos de Venezuela. In: Aymard C. G. A., (ed.). *Bosques de Venezuela: Un Homenaje a Jean Pierre Veillon*. BioLlania, ed. especial. pp. 33–46.

Aymard, G. A., (2017). Adiciones a la flora vascular de los Llanos de Venezuela: Nuevos registros yestados taxonómicos. In: Aymard, C. G. A., (ed.). *La Orinoquia Colombo-Venezolana*. BioLlania, ed. especial, *15*, pp. 1–296.

Aymard, G. A., Farreras, P. J. A., & Schargel, R., (2011). Bosques secos macrotérmicos de Venezuela. In: Aymard, C. G. A., (ed.). *Bosques de Venezuela: Un Homenaje a Jean Pierre Veillon*. BioLlania, ed. especial. pp. 155–177.

Azócar, A., & Fariñas, M., (2003). Páramos. In: Aguilera, M. M., Azócar, A., & González, J. E., (eds.), *Biodiversidad en Venezuela*, vol. 2. Fundación Polar, Ministerio de Ciencia y Tecnología, Fondo Nacional de Ciencia, Tecnología e Innovación (FONACIT), Caracas, Venezuela. pp. 716–733.

Azócar, A., (1974). *Análisis de las Características de Diferentes Hábitats en la Formación Páramo*, Facultad de Ciencias, Universidad de Los Andes, Mérida, Venezuela.

Balcázar, W., Rondón, J., Rengifo, M., Ball, M. M., Melfo, A., Gómez, W., et al., (2015). Bioprospecting glacial ice for plant growth promoting bacteria. *Microbiol. Res.*, *177*, 1–7.

Ball, M. M., Gómez, W., Magallanes, X., Rosales, R., Melfo, A., & Yarzábal, L. A., (2013). Bacteria recovered from a high-altitude, tropical glacier in Venezuelan Andes. *World J. Microbiol. Biotech.*, doi: 10.1007/s11274–013–1511–1.

Barreto, M. B., & González Boscán, V., (1994). Distribución, desarrollo estructural y ambientes geomorfológicos de los manglares de Venezuela. *Terra* 10(19): 83–111.

Barreto, M. B., Barreto, E., et al., (2009). Estudio integral del sistema lagunar Bajo Alacatraz-Mata Redonda-La Salineta de la península de Paria, estado Sucre, Venezuela: Geomorfología, hodrología, calidad del agua, vegetación y vertebrados. *Acta Biol. Venez.*, *29*(1–2), 1–59.

Barrio-Amorós, C. L., (2007). *Crocodylus intermedius*, el mayor predador de Suramérica. *Reptilia, 62, 32–39.*

Barrio-Amorós, C. L., (2009). Riqueza y endemismo. pp.: 25–39. In: Molina, C., Señaris, J.C., Lampo, M., & Rial, A. (eds.); *Anfibios de Venezuela: estado del conocimiento y recomendaciones para su conservación*. Ediciones Grupo TEI, Caracas, Venezuela.

Beard, J. S., (1944). Climax vegetation in tropical America. *Ecology, 25*(2), 127–158.

Beebe, W., & Crane, J., (1948). Ecología de Rancho Grande, una selva nublada subtropical en el norte de Venezuela. *Bol. Soc. Venez. Ci. Nat. 11*(73), 217–258.

Bellorín, A. M., (2003). Algas marinas bentónicas. pp. 94–103. In: Aguilera M. M., Azócar, A., & González Jiménez, E. (eds.); *Biodiversidad en Venezuela*, vol. 1. Fundación Polar, Ministerio de Ciencia y Tecnología, Fondo Nacional de Ciencia, Tecnología e Innovación (FONACIT), Caracas, Venezuela.

Berger, S., (2018). *Infectous Diseases of Venezuela,* Gideon E-Book series, Los Ángeles, California, USA, available at: www.gideononline.com/ebooks/.

Bermúdez, D., (1980). Preliminary experiences in the control of fish diseases in warm water aquaculture operations in Venezuela. *J. Fish Diseases*, *3*, 355–357.

Boom, B. M., (1990). Flora and vegetation of the Guayana-Llanos ecotone in Estado Bolívar, Venezuela. *Mem. N. Y. Bot. Gard.*, *64*, 254–278.

Braun, C., & Bezada, M., (2013). The history and disappearance of glaciers in Venezuela. *J. Latin Amer. Geogr.*, *12*(2), 85–124.

Breure, A. S. H., (2009). Radiation in land snails on Venezuelan tepui islands. In: Cohen, A., et al., (eds.). *Evolutionary Islands: 150 Years After Darwin, Abstract 38*, Leiden, The Netherlands.

Brewer-Carías, C. (1988). *Cerro de la Neblina, resultados de la expedición*, Caracas, Venezuela.

Brewer-Carías, C., & Audy, M., (2011). *Entrañas del Mundo Perdido*, Editorial Altholito, Caracas, Venezuela.

Brewer-Carías, C., (2012). La Neblina: El tepuy más alto y remoto. *Río Verde, 6*, 61–74.

Briceño, B., & Morillo, G., (2002). Catálogo abreviado de las plantas con flores de los páramos de Venezuela, parte I: Dicotiledóneas (Magnoliopsida). *Acta Bot. Venez., 25*(1), 1–46.

Briceño, B., & Morillo, G., (2006). Catálogo de las plantas con flores de los Páramos de Venezuela, parte II: monocotiledóneas (Liliopsida). *Acta Bot. Venez., 29*(1), 89–134.

Bulla, L., (1994). An index of eveness and its associated diversity measure. *Oikos, 70*, 167–171.

Bullock, S. H., Mooney, H. A., & Medina, E., (1995). *Seasonally Dry Tropical Forest*, Cambridge University Press, Cambridge, UK.

Caballer, G. M., Ortea, J., Rivero, N., Carías, T. G., Malaquias, M. A. E., & Narciso, S., (2015). The opisthobranch gastropods (Mollusca: Heterobranchia) from Venezuela: An annotated and illustrated inventory of species. *Zootaxa, 4034*(2), 201–256.

Campo, Z. M. A., & Ortaz, M., (2015). Tinícalo del lago de Valencia, *Atherinella venezuelae*. In: Rodríguez, J. P., García-Rawlins, A., & Rojas-Suárez, F., (eds.). *Libro Rojo de la Fauna Venezolana, 4th edn*. Provita y Fundación Empresas Polar, Caracas, Venezuela.

Cañadas, E. M., Fenu, G., et al., (2014). Hotspots within hotspots: Endemic plant richness, environmental drivers, and implications for conservation. *Biol. Conserv., 170*, 282–291.

Capelo, J. C., & Buitrago, J., (1998). Distribución geográfica de los moluscos marinos en el oriente de Venezuela. *Mem. Fund. La Salle Cien. Nat., 63*(150), 109–160.

Capelo, J. C., Gutiérrez, J., et al., (2008). Bentos. In: Lasso, C. A., & Señaris, J. C., (eds.). *Biodiversidad Animal del Caño Macareo, Punta Pescador y Áreas Adyacentes, Delta del Orinoco*, Fundación La Salle de Ciencias Naturales, StatoilHydro. pp. 31–52.

Carnevali, G., Gerlach, G., & Romero, G., (2008). Orchidaceae. In: Hokche, O., Berry, P. E., & Huber, O., (eds.). *Nuevo Catálogo de la Flora Vascular de Venezuela*, Fundación Instituto Botánico de Venezuela "Dr. Tobías Lasser," Caracas, Venezuela. pp. 753–789.

Cazorla, P. D. J., (2012). Listado de species de ciempiés (Myriapoda, Chilopoda) conocidas en Venezuela. *Bol. Malariol. Salud Ambient., 52*(2), 295–300.

Cazorla, P. D. J., Loyo, S. J. E., et al., (2012). Aspectos clínicos, epidemiológicos y de tratamiento de 11 casos de envenenamiento por ciempiés en Adícora, Península de Paraguaná, estado Falcón, Venezuela. *Acta Toxicol. Argent., 20*(1), 25–33.

Cerdá, F. J., (2003a). Dictiópteros. In: Aguilera, M. M., Azócar, A., & González, J. E., (eds.). *Biodiversidad en Venezuela*, vol. 1. Fundación Polar, Ministerio de Ciencia y Tecnología, Fondo Nacional de Ciencia, Tecnología e Innovación (FONACIT), Caracas, Venezuela. pp. 340–349.

Cerdá, F. J., (2003b). Ortópteros. In: Aguilera, M. M., Azócar, A., & González, J. E., (eds.). *Biodiversidad en Venezuela*, vol. 1. Fundación Polar, Ministerio de Ciencia y Tecnología, Fondo Nacional de Ciencia, Tecnología e Innovación (FONACIT), Caracas, Venezuela. pp. 350–361.

Cervigón, F., & Rodríguez, B., (1997). Lista actualizada de los peces marinos de Venezuela. In: La Marca, E., (ed.). *Vertebrados actuales y fósiles de Venezuela, Serie Catálogo Zoológico de Venezuela*, Museo de Ciencia y Tecnología, Mérida, Venezuela. pp. 17–52.

Cervigón, F., (1991–1996). *Los peces marinos de Venezuela, 2nd edn.* Fundación Científica Los Roques, Caracas, Venezuela (voll. 1–2 [1991–1993]), and Editorial Ex-Libris, Caracas, Venezuela (voll. 3–4 [1994–1996]).

Cervigón, F., (2003). Peces marinos. In: Aguilera, M. M., Azócar, A., & González, J. E., (eds.). *Biodiversidad en Venezuela*, vol. 2. Fundación Polar, Ministerio de Ciencia y Tecnología, Fondo Nacional de Ciencia, Tecnología e Innovación (FONACIT), Caracas, Venezuela. pp. 548–561.

Chalbaud, E., (1965). El rector heroico, *Universidad de los Andes, Publicaciones del Rectorado, Colección Ilusters Universitarios, 1,* Mérida, Venezuela.

Chardon, C., & Toro, R., (1934). Mycological explorations of Venezuela. *Monogr. Univ. Puerto Rico, B, 2*, 1–134.

Chernoff, B., Machado-Allison, A., Riseng, K., & Montambault, J. R., (2003). A biological assessment of the aquatic ecosystems of the Caura River basin, Bolívar State, Venezuela. *RAP Bull. Biol. Asses., 28.*

Christenhusz, M. J. M., Zhang, X.-C., & Schneider, H., (2011). A linear sequence of extant families and.genera of lycophytes and ferns. *Phytotaxa, 19*, 7–54.

CIEN (Comité Interfilial de Estratigrafía y Nomenclatura), Código Estratigráfico de las Cuencas Petroleras de Venezuela, *Petróleos de Venezuela, S. A. (PDVSA),* Available at: www.pdv.com/lexico/lexicoh.htm, s/d.

Clavijo, J. A., (2003). Lepidópteros. In: Aguilera, M. M., Azócar, A., & González, J. E., (eds.). *Biodiversidad en Venezuela*, vol. 1 Fundación Polar, Ministerio de Ciencia y Tecnología, Fondo Nacional de Ciencia, Tecnología e Innovación (FONACIT), Caracas, Venezuela. pp. 426–436.

Clemente, J. C., Pehrsson, E. C., Blaser, M. J. et al., (2015). The microbiome of uncontacted Amerindians. *Sci. Adv., 1,* e1500183

Clements, J. F., Schulenberg, T. S., Ilif, M. J., Roberson, D., Fredericks, T. A., Sullivan, B. L., et al., (2017). *The Bird/Clements Checklist of Birds of the World.* Available at: http://www.birds.cornell.edu/clementschecklist/download/.

Colonnello, G., (2004). Las planicies deltaicas del río Orinoco y golfo de Paria: aspectos físicos y vegetación. In: Lasso, C. A., Alonso, L. E., Flores, A. L., & Love, G., (eds.). Rapid Assessment of the Biodiversity and Social Aspects of the Aquatic Ecosystems of the Orinoco Delta and the Gulf of Paria, Venezuela. *RAP Bull. Biol. Asses., 37*, pp. 37–54.

Colonnello, G., Grande, J., & Márquez, I. (2016a). *Roystonea oleracea* (Arecaceae) communities in Venezuela. *Bot. J. Linn. Soc.* 82 (2): 439-450.

Colonnello, G., Zambrano-Martinez, S., & Grande, J., (2016b.) Conservacion de las comunidades de *Roystonea oleracea* (palma "chaguaramo" o "mapora") en Venezuela. In: Lasso, C. A., Colonnello, G., & Moraes, R. M., (eds.). *Morichales, Canaguchales y Otros Palmares Inundables de Suramérica, Parte II.* Colombia, Venezuela, Brasil, Perú, Bolivia, Paraguay, Uruguay y Argentina, Serie Recursos Hidrobiológicos y Pesqueros Continentales de Colombia, XIV. Instituto de Investigación de Recursos Biológicos Alexder von Humboldt, Bogotá, Colombia. pp. 546–569.

Conde, J. E., & Carmona-Sánchez, C., (2003). Ecosistemas marino-costeros. In: Aguilera, M. M., Azócar, A., & González, J. E., (eds.). *Biodiversidad en Venezuela*, vol. 2. Fundación

Polar, Ministerio de Ciencia y Tecnología, Fondo Nacional de Ciencia, Tecnología e Innovación (FONACIT), Caracas, Venezuela. pp. 862–883.

Cóndor-Luján, B., Louzada, T., Hajdu, E., & Klautau, M., (2018). Morphological and molecular taxonomy of calcareous sponges (Porifera: Calcarea) from Curaçao, Caribbean Sea. Zool. J. Linn. Soc., zlx082, https://doi.org/10.1093/zoolinnean/zlx082.

Cooper, A. G., (1977). Brachiopods from the Caribbean Sea and adjacent waters; *Studies in Tropical Oceanography 14*, University of Miami Press, Coral Gables, USA.

Corozzoli, R., (2002). Especies de nemátodos fitoparasíticos en Venezuela. *Interciencia, 27*(7), 354–364.

Cressa, C., & Holzenthal, R. W., (2003). Tricópteros. In: Aguilera, M. M., Azócar, A., & González, J. E., (eds.). *Biodiversidad en Venezuela,* vol. 1. Fundación Polar, Ministerio de Ciencia y Tecnología, Fondo Nacional de Ciencia, Tecnología e Innovación (FONACIT), Caracas, Venezuela. pp. 412–425.

Cressa, C., & Stark, B. P., (2003). Plecópteros. In: Aguilera, M. M., Azócar, A., & González, J. E. (eds.). *Biodiversidad en Venezuela,* vol. 1. Fundación Polar, Ministerio de Ciencia y Tecnología, Fondo Nacional de Ciencia, Tecnología e Innovación (FONACIT), Caracas, Venezuela. pp. 478–487.

Cuatrecasas, J., (1968). Páramo vegetation and its life forms. In: Troll, C., (ed.). *Geoecology of the Mountainous Regions of the Tropical Andes*, UNESCO. pp. 163–186.

Cuatrecasas, J., (2013). A systematic study of the subtribe Espeletiinae. The New York Botanical Garden Press, New York, 689 pp.

Cuello, N. & Cleef, A.M. 2009a. The forests of Ramal de Guaramacal in the Venezuelan Andes. *Phytocoenologia 39*(1): 109-156.

Cuello, N. & Cleef, A.M. 2009b. The páramo vegetation of Ramal de Guaramacal, Trujillo state, Venezuela. I. Zonal communities. *Phytocoenologia 39*(3): 295-329.

Cuello, N. & Cleef, A.M. 2009c. The páramo vegetation of Ramal de Guaramacal, Trujillo state, Venezuela. II. Azonal vegetation. *Phytocoenologia 39*(4): 389-409.

Cummings, K. S., & Mayer, C. A., (1997). *The Freshwater Mussels (Bivalvia: Etherioidea) of Venezuela,* Available at: http://www.inhs.uiuc.edu/animals_plants/molul sk/SA/Ven. html.Illinois Natural History Survey.

Davis, S. D., Heywood, V. H., Herrera-MacBride, O., Villalobos, J., & Hamilton, A. C., (1997). *Centres of Plant Diversity,* vol. 3. The Americas, Information Press, Oxford, UK.

De Moraes, G. J., McMurtry, J. A., Denmark, H. A., & Campos, C. B., (2004). A revised catalog of the mite family Phytoseiidae. *Zootaxa, 434,* 1–494.

De Sena Oliveira, I., Morley, St. J., Read, V., & Mayer, G., (2012). A world checklist of Onycophora (velvet worms), with notes on nomenclature and status of names. *Zookeys, 211,* 1–70.

Deflandre, G., (1928). Algues d'éau douce du Venezuela (Flagellés et Chlorophycées). *Rev. Algol., 3,* 211–241.

Del Mónaco, C., Haiek, G., Narciso, S., & Galindo, M., (2012). Massive bleaching of coral reefs induced by the, (2010). ENSO, Puerto Cabello, Venezuela. *Rev. Biol. Trop., 60*(2), 527–538.

Delgado, H. N. M., (2015). *Syndisyrinx collongistyla* (Turbellaria: Umagillidae) y su relación con el hábitat del erizo *Lytechinus variegatus* en la ensenada Los Juanes, Parque Nacional Morrocoy, *Trabajo Especial de Grado,* Universidad Central de Venezuela, Facultad de Ciencias, Escuela de Biología, Caracas, Venezuela.

Delgado, P. N., (2003). Dípteros. In: Aguilera, M. M., Azócar, A., & González, J. E., (eds.). *Biodiversidad en Venezuela,* vol. 1. Fundación Polar, Ministerio de Ciencia y Tecnología, Fondo Nacional de Ciencia, Tecnología e Innovación (FONACIT), Caracas, Venezuela. pp. 448–461.

DeMarmels, J., & Gaiani, M., (2003). Megalópteros y planiplenios. In: Aguilera, M. M., Azócar, A., & González, J. E., (eds.). *Biodiversidad en Venezuela,* vol. 1. Fundación Polar, Ministerio de Ciencia y Tecnología, Fondo Nacional de Ciencia, Tecnología e Innovación (FONACIT), Caracas, Venezuela. pp. 396–410.

DeMarmels, J., (1990). An updated checklist of the Odonata of Venezuela. *Odonatologica, 19*(4), 333–345.

DeMarmels, J., (2003). Odonatos. In: Aguilera, M. M., Azócar, A., & González, J. E., (eds.). *Biodiversidad en Venezuela,* vol. 1. Fundación Polar, Ministerio de Ciencia y Tecnología, Fondo Nacional de Ciencia, Tecnología e Innovación (FONACIT), Caracas, Venezuela. pp. 312–325.

DeMarmels, J., Clavijo, J. A., & Chacín, E., (1996). A new subspecies of *Xylophanes tersa* (Sphingidae) from Venezuela. *J. Lepidopterologists' Soc., 50*(4), 303–308.

Denis, R. W. G., (1970). Fungus flora of Venezuela and adjacent countries. *Kew Bull. Add. Ser., 3,* 1–530.

Depons, F., (1806). Voyage a la partie oriental de la Terre Ferme, dans l'Amérique Méridionale, fait pendant les années 1801, 1802, 1803, et 1804, contenant la description de la capitaniere générale de Caracas composée des provinces de Venézuéla, Maracaibo, Varinas, la Guiane espagnole, Cumaná et de l'ile de la Marguerite, 2 voll., *París, Impr. de Fain et Cie., Paris, France.*

Di Marco, M., Chapman, S., et al., (2017). Changing trends and persisting biases in three decades of conservation science. *Glob. Ecol. Conserv., 10,* 32–42.

Díaz, A., & Najt, J., (1990). Etude des peuplements des microarthropodes da. ns deux paramos de l'état de Mérida (Venezuela), l: Abondance et phénologie. *Rev. Ecol. Biol. Sol., 27,* 159–184.

Díaz, A., & Najt, J., (1995). Collemboles (Insecta) des Andes vénézuéliennes. *Bull. Mus. Natl. Hist. Nat., 4e sér.,* Sect. *A, (2–4),* 417–435.

Díaz, A., Péfaur, J. E., & Durant, P., (1997). Ecology of South American Páramos, with emphasis on the Fauna of the Venezuelan Páramos. In: Wielgolaski, F. E., (ed.). *Polar and Alpine Tundra, Ecosystems of the World, Part 3,* Elsevier, Amsterdam, The Netherlands. pp. 263–310.

Díaz, D. O., & Liñero, A. I., (2011). *Poliquetos de Venezuela, Universidad de Oriente, Editorial Universitaria,* Cumaná, Venezuela.

Díaz, D. O., Bone, D., & Rodríguez, C. T., (2017). Estado del conocimiento de los poliquetos en Venezuela. In: Díaz, D. O., (ed.). *Poliquetos de Sudamérica, Publicación Especial del Boletín del Instituto Oceanográfico de Venezuela,* Universidad de Oriente (UDO), Cumaná, Venezuela. pp. 127–143.

Díaz, M. C., & Rützler, K., (2009). Biodiversity and abundance of sponges in Caribbean mangrove: Indicators of environmental quality. *Smithson. Contrib. Mar. Sci., 38,* 151–172.

Díaz-Ungría, C., (1971). *Parasitología de los Animales Domésticos de Venezuela,* vol. 2, Consejo de Desarrollo Científico y Humanístico, Universidad del Zulia, Maracaibo.

Díaz-Ungría, C., (1973). Helmintos endoparásitos de Venezuela. *Cien. Vet. Maracaibo, 3,* 37–243.

Díaz-Ungría, C., (1981). Protozoos de Venezuela. *Kasmera, 9,* 147–215.

Doreste, E., (1968). Primera lista de acaros de importancia agricola en Venezuela. *Agron. Trop., 18*(4), 449–460.

DRYFLOR [Banda, K., 62 additional authors], (2016). Plant diversity patterns in neotropical dry forests and their conservation implications *Science, 353*(6306), 1383–1387.

Duno de Stefano, R., Aymard, C. G. A., & Huber, O., (2007). *Catálogo anotado e ilustrado de la flora vascular de los Llanos de Venezuela,* FUDENA, Fundación Empresas Polar. Caracas, Venezuela.

Duno de Stefano, R., Stauffer, F., Riina, R., Huber, O., Aymard, A., Hokche, O., et al., (2009). Assessment of vascular plant diversity and endemism in Venezuela. *Candollea, 64*(2), 203–212.

Ernst, A., (1877a). Estudios sobre la flora y fauna de Venezuela. In: *Primer Anuario Estadístico de Venezuela.* Imprenta Nacional, Caracas, Venezuela. pp. 211–316.

Ernst, A., (1877b). Filices venezuelanae o sea enumeración sistemática de los helechos de la flora de Venezuela. In: *Primer Anuario Estadístico de Venezuela.* Imprenta Nacional, Caracas, Venezuela. pp. 236–248.

Ernst, A., (1877c). Idea general de la fauna de Venezuela. In: *Primer Anuario Estadístico de Venezuela.* Imprenta Nacional, Caracas, Venezuela. pp. 274–292.

Ernst, A., (1891). Idea general de la fauna de Venezuela. *Boletín del Ministerio de Obras Públicas, 81–100,* 22–74.

Fariñas, M., (2013). Cambios en la sabana de la Estación Biológica de los Llanos "Francisco Tamayo," Sociedad Venezolana de Ciencias Naturales, Calabozo, Venezuela. In: Medina, E., Huber, O., Nassar, J. M., & Navarro, P., (eds.). *Recorriendo el Paisaje Vegetal de Venezuela.* Ediciones IVIC, Instituto Venezolano de Investigaciones Cientificas (IVIC), Caracas, Venezuela. pp. 231–247.

Fernández de Oviedo, V. G., (1851–1855). Historia general y natural de las Indias, islas y tierra-firme del mar océano. In: Amador de los Rios, J., (ed.). *Real Academia de la Historia, Madrid, España.*

Fernandez, R., Galarraga, F., Hernandez, J., Gonzalez, R., & Benzo, Z., (2013). Los líquenes en el estudio de la contaminación atmosférica en la ciudad de Caracas, Venezuela. In: Medina, E., Huber, O., Nassar, J. M., & Navarro, P., (eds.). *Recorriendo el paisaje vegetal de Venezuela.* Ediciones IVIC, Instituto Venezolano de Investigaciones Cientificas (IVIC), Caracas, Venezuela. pp. 293–304.

Figueredo Rodríguez, A., Rodríguez, E., Espinoza, H., & Fuentes, J. L., (2017). Algunas adiciones a la fauna monogenética (Platyhelminthes) de Venezuela. *Saber, 29,* 461–471.

Foldats, E., (1970). Orchidaceae. In: Lasser, T., (ed.). *Flora de Venezuela, 15*(3), 1–522.

Franceschi, B. J. V., (2017). El cacao en la vida venezolana y en el mundo, *Fondo Editorial Tropykos.*

Françoso, R. D., Brandão, R., Nogueira, C. C., Salmona, Y. B., Bomfim, M. R., & Colli, G. R., (2015). Habitat loss and the effectiveness of protected areas in the Cerrado Biodiversity Hotspot. *Natureza & Conservação, 13,* 35–40.

Frey, W., & Stech, M., (2005). A morpho-molecular classification of the Anthocerotophyta (hornworts). *Nova Hedwigia, 80*(3–4), 541–545.

Frost, T. M., (1981). Analysis of ingested particles within a freshwater sponge. *Trans. Am. Microsc. Soc., 100*, 271–277.

Gaiani, P. M. A., (2003). Homópteros. In: Aguilera, M. M., Azócar, A., & González, J. E., (eds.). *Biodiversidad en Venezuela*, vol. 1. Fundación Polar, Ministerio de Ciencia y Tecnología, Fondo Nacional de Ciencia, Tecnología e Innovación (FONACIT), Caracas, Venezuela. pp. 388–394.

Galán, L., & Herrera, F.F., (2006). Fauna cavernícola de Venezuela: una revisión. Bol. Soc. Venez. Espel. 40: 39–57.

Galán, L., & Herrera, F. F., (2017). Ríos subterráneos y acuíferos kársticos de Venezuela: Inventario, situación y conservación. In: Rodíguez-Olarte, D., (ed.), *Ríos en riesgo de Venezuela*, vol. I. Colección Recursos Hidrobiológicos de Venezuela, Universidad Centroccidental Lisandro Alvarado, Barquisimeto, Venezuela. pp. 153–171.

Ganesan, E. K., (1989). *A catalog of benthic marine algae and seagrasses of Venezuela*, Fondo Editorial CONICIT, Caracas, Venezuela.

Gansser, A., (1955). The Guiana Shield (S. America): geological observations. *Eclogae Geol. Helv. 47*(1), 77–112.

García, R. J. L., (2003). Himenópteros. In: Aguilera, M. M., Azócar, A., & González, J. E., (eds.). *Biodiversidad en Venezuela*, vol. 1. Fundación Polar, Ministerio de Ciencia y Tecnología, Fondo Nacional de Ciencia, Tecnología e Innovación (FONACIT), Caracas, Venezuela. pp. 462–477.

García, S., Araujo, J., Reid, J., Castillo, A., Duarte, J., & Méndez, P., (2011). Marine and coastal protected areas System in Venezuela. *J. Coast. Res., Spec. Issue, 64*, 1975–1978.

Garrido, M. J., & Brito, M., (2016). Transmisión de virus de plantas por coleópteros. *Rev. Fac. Agron. (UCV), 42*(2), 61–74.

Gentry, A. H., (1982). Bignoniaceae. In: Luces de Febres, Z., & Steyermark, J. A., (eds.). *Flora de Venezuela, 8*(4), 1–433.

Gentry, A. H., (1995). Diversity and floristic composition of neotropical dry forest. In: Bullock, S. H., Mooney, H. A., & Medina, E., (eds.). *Seasonally Dry Tropical Forest*. Cambridge University Press, Cambridge, UK.

Gessner, F., & Vareschi, V., (1956). *Ergebnisse der Deutschen limnologischen Venezuela-Expedition*. Deutscher Verlag der Wissenschaften, *1*.

Gilij, F. S., (1780–1784). *Sagio di storia americana, o sia, storia naturale, civile e sacra de regni, e delle provincie spagnuole di Terra-Ferma nell'America Meridionale descritto dall'abate* F. S. Gilij, 4 voll., Perigio, Rome, Italy.

Ginés, Fr., (1959). Familias y géneros de anfibios (Amphibia) de Venezuela. Mem. Soc. Cien. Nat. La Salle, *19*(53), 85–146.

Gómez, J. M., & Penchaszadeh, P., (2003). Diciémidos. In: Aguilera, M. M., Azócar, A., & González, J. E., (eds.). *Biodiversidad en Venezuela*, vol. 1. Fundación Polar, Ministerio de Ciencia y Tecnología, Fondo Nacional de Ciencia, Tecnología e Innovación (FONACIT), Caracas, Venezuela. pp. 220–226.

Gómez, S., Carballo, B. Y., García, M., & Gil, N., (2015). *Web Ficoflora Venezuela: Catálogo Digital de la Ficoflora de Venezuela*. Universidad Central de Venezuela, Caracas, Venezuela, electronic publication continuously updated. Available at http://www.ciens.ucv.ve/ficofloravenezuela, onwards.

González Boscán, V., (2003a). Bosques secos. In: Aguilera, M. M., Azócar, A., & González, J. E., (eds.). *Biodiversidad en Venezuela*, vol. 2. Fundación Polar, Ministerio de Ciencia y Tecnología, Fondo Nacional de Ciencia, Tecnología e Innovación (FONACIT), Caracas, Venezuela. pp. 734–744.

González Boscán, V., (2003b). Delta del Orinoco. In: Aguilera, M. M., Azócar, A., & González, J. E., (eds.). *Biodiversidad en Venezuela*, vol. 2. Fundación Polar, Ministerio de Ciencia y Tecnología, Fondo Nacional de Ciencia, Tecnología e Innovación (FONACIT), Caracas, Venezuela. pp. 900–917.

González Boscán, V., (2011). Los bosques del delta del Orinoco. In: Aymard, C. G. A., (ed.). *Bosques de Venezuela: Un homenaje a Jean Pierre Veillon*. BioLlania, ed. especial, *10*. pp. 197–240.

González Boscán, V., (2013a). *La vegetación de Venezuela al Norte del río Orinoco*, Fundación Instituto Botánico Dr. Tobías Lasser, Caracas, Venezuela.

González Boscán, V., (2013b). Los matorrales de pantano de *Chrysobalanus icaco* del delta del Orinoco. In: Medina, E., Huber, O., Nassar, J. M., & Navarro, P., (eds.). *Recorriendo el Paisaje Vegetal de Venezuela*. Ediciones IVIC, Instituto Venezolano de Investigaciones Cientificas (IVIC), Caracas, Venezuela. pp. 79–95.

González Boscán, V., (2016). Los palmares de pantano de *Mauritia flexuosa* en Suramérica: una revisión. In: Lasso, C. A., Colonnello, G., & Moraes, R. M., (eds.). *Morichales, Canaguchales y Otros Palmares Inundables de Suramérica, Parte II*. Colombia, Venezuela, Brasil, Perú, Bolivia, Paraguay, Uruguay y Argentina, Serie Recursos Hidrobiológicos y Pesqueros Continentales de Colombia, XIV. Instituto de Investigación de Recursos Biológicos Alexder von Humboldt, Bogotá, Colombia. pp. 45–83.

González, E., Malaver, N., & Naveda, J., (2015). Los ecosistemas acuáticos de Venezuela y su conservación. In: Gabaldón, A., Rosales, A., Buroz, E., Córdoba, J., Uzcátegui, G., & Iskandar, L., (eds.). *El Agua en Venezuela: Una riqueza Escasa, 2 voll*. Fundación Empresas Polar, Editorial Exlibris, Caracas, Venezuela. pp. 187–251.

González, J. C., Arozena, I., & Picard, X., (1980). *Geología de Venezuela y de sus Cuencas Petrolíferas*, Ediciones Foninves, Caracas, Venezuela.

González Jiménez, E., & Szeplaki, O. E., (1998). *Manejo y Producción de Chigüires a Nivel de Fincas*. Papeles de Fundacite Aragua, 1–13.

González, J. E., (2003a). Humedales continentales. In: Aguilera, M. M., Azócar, A., & González, J. E. (eds.). *Biodiversidad en Venezuela*, vol. 2 Fundación Polar, Ministerio de Ciencia y Tecnología, Fondo Nacional de Ciencia, Tecnología e Innovación (FONACIT), Caracas, Venezuela. pp. 884–898.

González, J. E., (2003b). Marco jurídico e institucional. In: Aguilera, M. M., Azócar, A., & González, Jiménez, E., (eds.). *Biodiversidad en Venezuela*, vol. 2. Fundación Polar, Ministerio de Ciencia y Tecnología, Fondo Nacional de Ciencia, Tecnología e Innovación (FONACIT), Caracas, Venezuela. pp. 982–993.

González, J. M., Shea, J., & Brewer-Carías, C., (2015). First reports of *Bactrophora dominans* Westwood, (1842). (Orthoptera: Romaleidae) from Venezuela and French Guiana (South America), with comments on biology, ecology and distribution of the species. *Check List*, *11*(2)-1614, 1–6.

González, R. V., (2003). Biotecnología: Clave para el aprovechamiento de la biodiversidad. In: Aguilera M. M., Azócar, A., & González, J. E., (eds.), *Biodiversidad en Venezuela*, *vol. 2*. Fundación Polar, Ministerio de Ciencia y Tecnología, Fondo Nacional de Ciencia, Tecnología e Innovación (FONACIT), Caracas, Venezuela. pp. 998–1011.

González-Sponga, M. A., (2003). Quelicerados. In: Aguilera, M. M., Azócar, A., & González, J. E. (eds.). *Biodiversidad en Venezuela*, vol. 1. Fundación Polar, Ministerio de Ciencia y

Tecnología, Fondo Nacional de Ciencia, Tecnología e Innovación (FONACIT), Caracas, Venezuela. pp. 274–286.

Gordon, C. E., (2016). Riqueza y composición de la vegetación acuática de algunas lagunas en los Andes Venezolanos. *Ecotrópicos, 29*(1–2), 1–27.

Grande, J. & Colonnello, G., (2016). Comunidades de *Rosytonea oleracea* en Venezuela: Aspectos fitogeográficos y principales componentes florísticos. In: Lasso, C. A., Colonnello, G., & Moraes R. M. (eds.). *Morichales, Canaguchales y Otros Palmares Inundables de Suramérica, Parte II.* Colombia, Venezuela, Brasil, Perú, Bolivia, Paraguay, Uruguay y Argentina, Serie Recursos Hidrobiológicos y Pesqueros Continentales de Colombia, XIV, Instituto de Investigación de Recursos Biológicos Alexder von Humboldt, Bogotá, Colombia. pp. 517–545.

Grande, J., & Nozawa, S., (2009). Notas sobre la naturalización de *Scaevola taccada* (Gaertn.) Roxb. (Goodeniaceae) en las costas de Venezuela. *Acta Bot. Venez., 33*(1), 33–40.

Grande, J., (2007). *Estudio ecológico de las sabanas del cerro "El Picacho" (municipio Santiago Mariño, estado Aragua, Venezuela), Trabajo Especial de Grado,* Universidad Central de Venezuela. Facultad de Ciencias, Escuela de Biología, Caracas, Venezuela.

Grande, J., (2016). Novitates agrostologicae, V: Generic mergers in the tribe Olyreae. *Bol. Centro Invest. Biol. Univ. Zulia, 50*(1), 19–43.

Greenstein, B. J., (1989). Mass mortality of the West-Indian echinoid *Diadema antillarum* (Echinodermata: Echinoidea): a natural experiment in taphonomy. *Palaios, 4,* 487–492.

Gremone, C., (1986). *Fauna de Venezuela: Vertebrados,* Editorial Biosfera, Caracas, Venezuela.

Guerrero, R., (1985). Parasitología. In: *Fondo Editorial Acta Científica Venezolana, ASOVEM,* El estudio de los mamíferos en Venezuela, evaluación y perspectivas, Caracas, Venezuela. pp. 35–91.

Guerrero, R., (1996). Las garrapatas de Venezuela (Acarina: Ixodoidea). Listado de species y claves para su identificación. *Bol. Malariol. Salud Ambient., 36*(1–2), 1–24.

Guerrero, R., (2003). Nemátodos (Zooparasíticos). In: Aguilera, M. M., Azócar, A., & González, J. E. (eds.). *Biodiversidad en Venezuela,* vol. 1. Fundación Polar, Ministerio de Ciencia y Tecnología, Fondo Nacional de Ciencia, Tecnología e Innovación (FONACIT), Caracas, Venezuela. pp. 254–263.

Gumilla, J., (1731). *El Orinoco Ilustrado y Defendido: Historia Natural, civil y geográfica de este gran río y de sus caudalosas vertientes,* Madrid, Spain.

Hackley, P. C., Urbani, F., Karlsen, A. W., & Garrity, C. P., (2006). *Geologic Shaded Relief Map of Venezuela,* 1:750,000, 2 sheets, available at: https://pubs.usgs.gov/of/2006/1109/.

Hernández, L., & Demartino, A., (2003). Bosques y selvas (ombrófilos) de tierras bajas. In: Aguilera, M. M., Azócar, A., & González, J. E., (eds.). *Biodiversidad en Venezuela,* vol. 2. Fundación Polar, Ministerio de Ciencia y Tecnología, Fondo Nacional de Ciencia, Tecnología e Innovación (FONACIT), Caracas, Venezuela. pp. 746–761.

Hernández, M. J. E., Fernández, R., & Lucena, L., (2016). Análisis de la colección de hongos del Herbario Nacional de Venezuela (VEN) y sus muestras tipo. *Anartia [2014], 26,* 136–146.

Herrera, F., & Herrera, I., (2011). La *restauración Ecológica en Venezuela: Fundamentos y Experiencias,* Ediciones IVIC, Instituto Venezolano de Investigaciones Científicas (IVIC), Caracas, Venezuela.

Herrera, I., & Nassar, J. M., (2009). Dos plantas exóticas suculentas invaden zonas áridas en el P. N. Cerro Saroche (Venezuela). *Bol. Soc. Latin. Carib. Cact. Suc., 6*(2), 8–10.

Hewitt, D. G., (2011). *Biology and Management of White-Tailed Deer*, CRC Press, Boca Raton, USA.

Hilty, S. L., (2003). *Birds of Venezuela, 2nd edn.* Princeton University Press, Princeton, USA. (also published by Helm Field Guides in 2002).

Hokche, O., Berry, P. E., & Huber, O., (2008). Nuevo Catálogo de la Flora Vascular de Venezuela, *Fundación Instituto Botánico de Venezuela "Dr. Tobías Lasser,"* Caracas, Venezuela.

Holm, L. R. G., Plucknett, D. L., Pancho, J. V., & Herberger, J. P., (1977). *The World's Worst Weeds, Distribution and Biology*, University Press of Hawaii, Honolulu, USA.

Huber, O., & Alarcón, C., (1988). *Mapa de Vegetación de Venezuela, Ministerio del Ambiente y los Recursos Naturales Renovables (MARNR),* The Nature Conservancy, Fundación Bioma, Caracas, Venezuela.

Huber, O., (1986). Las selvas nubladas de Rancho Grande: Observaciones sobre su fisionomía, estructura y fenología. In: Huber, O., (ed.). *La selva nublada de Rancho Grande, Parque Nacional "Henri Pittier," el ambiente físico, ecología vegetal y anatomía vegetal.* Fondo Editorial Acta Científica Venezolana, Seguros Anauco, C. A., Caracas, Venezuela. pp. 131–170.

Huber, O., (1987). Consideraciones sobre el concepto de Pantepui. *Pantepui, 2,* 2–10.

Huber, O., (1992). *El Macizo del Chimantá, Escudo de Guayana, Venezuela: Un ensayo ecológico tepuyano.* Oscar Todtmann, Caracas, Venezuela.

Huber, O., (1995). Vegetation. In: Berry, P. E., Holst, B. K., & Yatskievych, K., (eds.). *Flora of the Venezuelan Guayana*, vol. 1-*Introduction*, Timber Press, Portland, Oregon, USA, and Missouri Botanical Garden, St. Louis, USA. pp. 97–160.

Huber, O., Duno de Stefano, R. R., Riina, R., Stauffer, F., Pappaterra, L., Jiménez, A., et al., (1998). *Estado actual del Conocimiento de la Flora en Venezuela.* Documentos Técnicos para la Estrategia Nacional de Diversidad Biológica 1, Ministerio del Ambiente y de los Recursos Naturales Renovables (MARNR), Fundación Instituto Botánico de Venezuela (FIBV), Caracas, Venezuela.

Huber, O., & Rosales, J., (1997). *Ecología de la Cuenca del Río Caura, Venezuela. II.* Estudios especiales. *Scientia Guaianæ, 7,* 441–468.

Huber, O., Rosales, J., & Berry, P. E., (1997). Estudios botánicos en las montañas altas de la cuenca del río Caura (Estado Bolívar, Venezuela). In: Huber, O., & Rosales, J., (eds.). *Ecología de la Cuenca del Río Caura, Venezuela. II.* Estudios especiales. *Scientia Guaianæ, 7,* 441–468.

INE [Instituto Nacional de Estadística], (2018). Available at: www.ine.gob.ve. (Last accessed 14/ I/.2018).

Iturriaga, T., & Minter, D. W., (2006). *Hongos de Venezuela.* Available at: www.cybertruffle.org.uk/venefung.

Iturriaga, T., & Ryvarden, L., (2003). Hongos. In: Aguilera, M. M., Azócar, A., & González, J. E. (eds.). *Biodiversidad en Venezuela*, vol. 1. Fundación Polar, Ministerio de Ciencia y Tecnología, Fondo Nacional de Ciencia, Tecnología e Innovación (FONACIT), Caracas, Venezuela. pp. 62–87.

Iturriaga, T., Páez, I., Sanabria, N., Holmquist, O., Bracamonte, L., & Urbina, H., (2000). *Estado Actual del Conocimiento de la Micobiota en Venezuela.* Documentos Técnicos para la Estrategia Nacional de Diversidad Biológica 2, Ministerio del Ambiente y de los

Recursos Naturales Renovables (MARNR), Fundación Instituto Botánico de Venezuela (FIBV), Caracas.

IUCN [International Union for the Conservation of Nature], (2012a). *IUCN Red List Categories and Criteria, version 3. 1, 2nd edn.*, IUCN, Gland, Switzerland and Cambridge, UK.

IUCN International Union for the Conservation of Nature, (2012b). *Guidelines for Application of IUCN Red List Criteria at Regional Levels, version 4. 0*, IUCN, Gland, Switzerland and Cambridge, UK.

Jahn, A., (1934). Las temperaturas medias y extremas de las zonas altitudinales de Venezuela. *Bol. Soc. Venez. Ci. Nat., 2*(14), 135–172.

Joly, L. J., (2003). Coleópteros. In: Aguilera, M. M., Azócar, A., & González, J. E., (eds.). *Biodiversidad en Venezuela*, vol. 1. Fundación Polar, Ministerio de Ciencia y Tecnología, Fondo Nacional de Ciencia, Tecnología e Innovación (FONACIT), Caracas, Venezuela. pp. 438–447.

Jones, E., Clifford, C., Keirans, J., & Kohls, G., (1972). The Ticks of Venezuela (Acarina: Ixodoidea) With a Key to the Species of Amblyomma in the Western Hemisphere. *Brigham Young Univ. Sci. Bull., Biol. Ser. 17*, 4.

Karling, J. S., (1981a). Some zoosporic fungi of Venezuela. *Nova Hedwigia, 34*, 645–668.

Karling, J. S., (1981b). *Rhizidiomyces bullatus* (Sparrow) *comb. nov.*, and other anisochytrids from Venezuela. *Nova Hedwigia, 34*, 669–678.

Kaštovský, J., Fučíková, K., Hauer, T., & Bohunická, M., (2011). Microvegetation on the top of Mt. Roraima, Venezuela. *Fottea, 11*(1), 171–186.

Kaštovský, J., Veselá, J., Bohunická, M., Fučíková, K., Štenclová, L., & Brewer-Carías, C., (2016). New and unusual species of cyanobacteria, diatoms and green algae, with a description of a new genus *Ekerewekia gen. nov.* (Chlorophyta) from the table mountain Churí-tepui, Chimantá Massif (Venezuela). *Phytotaxa, 247*(3), 153–180.

Köppen, W., (1936). Das geographische System der Klimate, vol. 1, part C. Pp.: 1–44. In: Köppen, W.; and Geiger, R. (eds.), *Handbuch der Klimatologie;* Bornträger, Berlin, Germany.

Köppen, W., (1948). *Climatología, con un estudio de los climas de la tierra;* Fondo de Cultura Económica, Ciudad de México y Buenos Aires.

La Marca, E., & Soriano, P., (2004). *Reptiles de los Andes de Venezuela.* Fundación Polar, Conservación Internacional, CODEPRE-ULA, Fundacite Mérida, BIOGEOS, Mérida, Venezuela.

La Marca, E., (1997a). Lista actualizada de los anfibios de Venezuela. In: La Marca, E., (ed.). *Vertebrados Actuales y Fósiles de Venezuela*, vol. 1. Serie catálogo zoológico de venezuela, museo de ciencia y tecnología de mérida, Mérida, Venezuela. pp. 103–120.

La Marca, E., (1997b). Lista actualizada de los reptiles de Venezuela. In: La Marca, E., (ed.). *Vertebrados Actuales y Fósiles de Venezuela*, vol. 1. Serie catálogo zoológico de venezuela, Museo de Ciencia y Tecnología de Mérida, Mérida, Venezuela. pp. 123–142.

La Marca, E., (2003a). Anfibios. In: Aguilera, M. M., Azócar, A., & González, J. E., (eds.). *Biodiversidad en Venezuela*, vol. 2. Fundación Polar, Ministerio de Ciencia y Tecnología, Fondo Nacional de Ciencia, Tecnología e Innovación (FONACIT), Caracas, Venezuela. pp. 582–595.

La Marca, E., (2003b). Reptiles. In: Aguilera, M. M., Azócar, A., & González, J. E., (eds.). *Biodiversidad en Venezuela*, vol. 2. Fundación Polar, Ministerio de Ciencia y Tecnología, Fondo Nacional de Ciencia, Tecnología e Innovación (FONACIT), Caracas, Venezuela. pp. 596–608.

Lado, C., & Wrigley, B. D., (2008). A review of neotropical myxomycetes (1828–2008). *Anales Jard. Bot. Madrid 65*(2), 211–254.

Lampo, M., Chacón, A. E., & Nava, F., (2009). Acervo genético. In: Molina, C., Señaris, J. C., Lampo, M., & Rial, A., (eds.). *Anfibios de Venezuela: Estado del Conocimiento* y *Recomendaciones Para su Conservación*, Ediciones Grupo TEI, Caracas, Venezuela. pp. 57–63.

Lasso, C. A., & Sánchez- Duarte, P., (2011). *Los Peces del Delta del Orinoco: Diversidad, Bioecología, Uso* y *Conservación*, Fundación La Salle de Ciencias Naturales, Chevron C. A. Venezuela, Caracas, Venezuela.

Lasso, C. A., Alonso, L. E., Flores, A. L., & Love, G., (2004a). Evaluación rápida de la biodiversidad y aspectos sociales de los ecosistemas acuáticos del delta del río Orinoco y golfo de Paria, Venezuela. *RAP Bull. Biol. Asses., 37.*

Lasso, C. A., Lew, D., Taphorn, D., Do Nascimiento, C., Lasso Alcalá, O. M., Provenzano, F., et al., (2004b). Biodiversidad Ictiológica Continental de Venezuela, Parte I, lista de e species y distribución por cuencas. *Mem. Fund. La Salle Cien. Nat., 159–160*, 105–195.

Lasso, C. A., Martínez, E. R., Capelo, J. C., Morales, B. M., & Sánchez-Maya, A., (2009a). Lista de los moluscos (Gastropodos: Bivalvia) dulceacuícolas y estuarinos de la cuenca del Orinoco (Venezuela). *Biota Col., 10*(1–2), 63–74.

Lasso, C. A., Rial, A., Colonnello, G., Machado-Allison, A., & Trujillo, F., (2014a). *Humedales de la Orinoquia (Colombia- Venezuela), Serie Editorial Recursos Hidrobiológicos* y *Pesqueros Continentales de Colombia*, vol. XI, Instituto de Investigación de Recursos Biológicos Alexander von Humboldt (IAvH). Bogotá DC, Colombia.

Lasso, C. A.; Morales-Betancourt, M. A.; Colonnello, G.; Fernández, Á.; and Grande, J., (2014b). Charcos temporales en afloramientos rocosos del Escudo Guayanés. In: Lasso, C. A., Rial, A.; Colonnello, G.; Machado-Allison, A.; and Trujillo, F. (eds.), *Humedales de la Orinoquia (Colombia- Venezuela), Serie Editorial Recursos Hidrobiológicos* y *Pesqueros Continentales de Colombia*, vol. XI, Instituto de Investigación de Recursos Biológicos Alexander von Humboldt (IAvH). Bogotá DC, Colombia. pp.: 151–153

Lasso, C. A., & Señaris, J. C., (2008). *Biodiversidad animal del caño Macareo, Punta Pescador* y *áreas adyacentes, delta del Orinoco.* Fundación La Salle de Ciencias Naturales, StatoilHydro, Caracas, Venezuela.

Lasso, C. A., Señaris, J. C., Rial, A., & Flores, A. L., (2009b). Evaluación rápida de la biodiversidad de los ecosistemas acuáticos de la cuenca alta del río Cuyuní, Guayana Venezolana. *RAP Bull. Biol. Asses., 55.*

Lazo, R., (2009). *Diagnóstico* y *Evaluación de Áreas de Manglar de la República Bolivariana de Venezuela*, Instituto Venezolano de Investigaciones Científicas (IVIC).

Lentino, R. M., & Bruni, A. R., (1994). *Humedales Costeros de Venezuela: Situación Ambiental*, Sociedad Conservacionista Audubon de Venezuela, Caracas, Venezuela.

Lentino, R. M., (2003). Aves. In: Aguilera, M. M., Azócar, A., & González, J. E., (eds.). *Biodiversidad en Venezuela*, vol. 2. Fundación Polar, Ministerio de Ciencia y Tecnología, Fondo Nacional de Ciencia, Tecnología e Innovación (FONACIT), Caracas, Venezuela. pp. 610–648.

Lentino, R., M., (1997). Catálogo de las aves de Venezuela. In: La Marca, E., (ed.). *Vertebrados Actuales y Fósiles de Venezuela.* Serie Catálogo Zoológico de Venezuela, vol. 1, Museo de Ciencia y Tecnología de Mérida, Mérida, Venezuela. pp. 143–202.

León, V. Y., Ussher, M. M. S., Rojas, R. C. A., & Delgado, R. J. F., (2015). Musgos de venezuela (MDV): Una base de datos para conocer y conservar la briodiversidad. *Acta Bot. Venez. 38*(2), 169–179.

León, V.Y., & Rico, G. R. R., (2003). Briofitos. In: Aguilera, M. M., Azócar, A., & González, J. E., (eds.). *Biodiversidad en Venezuela,* vol. 1. Fundación Polar, Ministerio de Ciencia y Tecnología, Fondo Nacional de Ciencia, Tecnología e Innovación (FONACIT), Caracas, Venezuela. pp. 122–135.

Lindorf, H., (2008). Historia de las exploraciones botánicas en Venezuela. In: Hokche, O., Berry, P. E., & Huber, O., (eds.). *Nuevo Catálogo de la Flora Vascular de Venezuela.* Fundación Instituto Botánico de Venezuela "Dr. Tobías Lasser," Caracas, Venezuela. pp. 17–40.

Llamozas, S., Duno de Stefano, R., Meier, W., Riina, R., Stauffer, F., Aymard, G., et al., (2003). *Libro Rojo de la Flora Venezolana.* Fundación Polar, Fundación Instituto Botánico de Venezuela "Dr. Tobías Lasser," Provita, Caracas, Venezuela.

Lodeiros, S., Cesar, J. M., Marín, E., Bauma, R., & Prieto, A. A., (1999). *Catálogo de Moluscos Marinos de las Costas de Venezuela: Clase Bivalvia,* Universidad de Oriente, Cumaná, Venezuela.

Löfler, H., (1981). Notostraca y Conchostraca. In: Hurlbert, S. H., Rodríguez, G., & Dias dos Santos, N. (eds.). *Aquatic Biota of Tropical South America,* San Diego State University, USA. pp. 3–4.

Losada, F. J., & Marques, P. S., (2003). Cnidarios. In: Aguilera, M. M., Azócar, A., & González, J. E., (eds.). *Biodiversidad en Venezuela,* vol. 1. Fundación Polar, Ministerio de Ciencia y Tecnología, Fondo Nacional de Ciencia, Tecnología e Innovación (FONACIT), Caracas, Venezuela. pp. 228–241.

Lucena, L., (2009). *Contribución al conocimiento de los hongos de la familia Xylariaceae asociados a madera en descomposición en un bosque tropical,* Trabajo Especial de Grado, Universidad Central de Venezuela, Facultad de Ciencias, Escuela de Biología, Caracas, Venezuela.

Machado-Allison, A., (2003). Peces de agua dulce. In: Aguilera, M. M., Azócar, A., & González, J. E., (eds.). *Biodiversidad en Venezuela,* vol. 1. Fundación Polar, Ministerio de Ciencia y Tecnología, Fondo Nacional de Ciencia, Tecnología e Innovación (FONACIT), Caracas, Venezuela. pp. 562–581.

Madi, Y., Vázquez, J., León, A., & Rodrígues, J., (2011). Estado de conservación de los bosques y otras formaciones vegetales en Venezuela. In: Aymard, C. G. A., (ed.), *Bosques de Venezuela: Un Homenaje a Jean Pierre Veillon.* BioLlania, ed. especial, *10,* pp. 303–324.

Marcano, V., (2003). Líquenes. In: Aguilera, M. M., Azócar, A., & González, J. E., (eds.). *Biodiversidad en Venezuela,* vol. 1. Fundación Polar, Ministerio de Ciencia y Tecnología, Fondo Nacional de Ciencia, Tecnología e Innovación (FONACIT), Caracas, Venezuela. pp. 104–120.

Marchese, C., (2015). Biodiversity hotspots: A shortcut for a more complicated concept. *Glob. Ecol. Conserv., 3,* 297–309.

MARNR-FMAM-PNUD, (2000). *Primer Informe de Venezuela Sobre Diversidad Biológica,* Ministerio del Ambiente y de los Recursos Naturales, Oficina Nacional de Diversidad Biológica, Caracas, Venezuela.

Marrero, C., & Rodríguez-Olarte, D., (2014). Ríos de morichal de la Orinoquia Venezolana: Modeladores del paisaje, soportes de biodiversidad, flujo geohídrico e identidad cultural, *Editorial Académica Española,* Saarbrücken, Germany.

Marrero, C., & Rodríguez-Olarte, D., (2017). Los humedales costeros venezolanos en los escenarios de cambios climáticos: Vulnerabilidad, perspectivas y tendencias. In: Botello, A. V., Villanueva, S., Gutiérrez, J., & Rojas, G. J. L., (eds.). *Vulnerabilidad de las zonas costeras de Latinoamérica al cambio climático,* UJAT, UNAM, UAC. pp. 461–476.

Marrero, L., (1964). Venezuela y sus recursos: Una geografía visualizada física, humana, económica y regional, *Cultural Venezolana, S. A,* Caracas, Venezuela.

Martínez, R., (2003). Moluscos. In: Aguilera, M. M., Azócar, A., & González, J. E., (eds.). *Biodiversidad en Venezuela,* vol. 1 Fundación Polar, Ministerio de Ciencia y Tecnología, Fondo Nacional de Ciencia, Tecnología e Innovación (FONACIT), Caracas, Venezuela. pp. 488–513.

Mateucci, S.D., (1986). Las zonas áridas y semiáridas de Venezuela. *Zonas Áridas, 4,* 39–48.

Mayle, F. F., (2004). Assessment of the neotropical dry forest refugia hypothesis in the light of paleoecological data and vegeatation model simulations. *J. Quaternary Sci., 19*(7), 713–720.

Mayr, E., & Phelps, W. H., (1967). The origin of the bird fauna of the south Venezuelan highlands. *Bull. Am. Mus. Nat. Hist., 136,* 269–328.

Medina, E., & Barboza, F. (2003). Manglares del Sistema del Lago de Maracaibo: Caracterización Fisiografica y Ecológica. *Ecotropicos* 16(2): 75-82.

Medina, E., & Barboza, F., (2006). Lagunas Costeras del lago de Maracaibo: Distribución, estatus y perspectivas de conservación. *Ecotrópicos* 19(2), 128–139.

Meier, W. 2004. *Flora y vegetación del Parque Nacional El Ávila (Venezuela, Cordillera de la Costa), con especial énfasis en los bosques nublados. Freiburg, Germany. Spanish version of Meier, W. 1998."* Flora und Vegetation des Ávila - Nationalparks (Venezuela/ Küstenkordillere) unter besonderer Berücksichtigung der Nebelwaldstufe". Diss. Bot. 296. 485. J. Cramer in der Gebr. Borntraeger Verlagsbuch-handlung – Berlin – Stuttgart.

Meier, W., & Berry, P. E., (2008). *Ampelozizyphus guaquirensis* (Rhamnaceae), a new tree species endemic to the Venezuelan Coastal Cordillera. *Bittonia, 60*(2), 131–135.

Meier, W., (2011). Los bosques nublados de la Cordillera de la Costa en Venezuela. In: Aymard, C. G. A., (ed.). *Bosques de Venezuela: Un Homenaje a Jean Pierre Veillon.* BioLlania, ed. especial, *10,* pp. 106–121.

Meyer de Shauensee, R., & Phelps., W. H., (1978). *A Guide to the Birds of Venezuela,* Princeton University Press, Princeton, USA.

Miloslavich, P., Díaz, J. M., Klein, E., Alvarado, J., Díaz, C., Gobin, J., Escobar-Briones, E., Cruz-Motta, J. J., Weil, E., Cortés, J., Bastidas, A. C., Robertson, R., Zapata, F., Martín, A., Castillo, J., Kazandjian, A., & Ortiz, M., (2015). Marine biodiversity in the Caribbean: regional estimates and distribution patterns. *Plos One* 5(8), e11916.

Miloslavich, P., Klein, E., Martin, A., Bastidas, C., Marín, B., & Spiniello, P., (2005). Venezuela. In: Miloslavich, P., & Klein, E., (eds.). *Caribbean Marine Ecosystems: The Known and the Unknown, Part I: Marine Biodiversity Reviews,* Destesh Publicactions, Lancaster, USA. pp. 109–136.

Miloslavich, P., Klein, E., Yerena, E., & Martin, A., (2003). Marine biodiversity in Venezuela: Status and perspectives. *Gayana, 67*(2), 275–301.

Miloslavich, P., Martín, A., Klein, E., Díaz, Y., & Lasso, A. O. M., (2011). Biodiversity and conservation of the estuarine and marine ecosystems of the Venezuelan Orinoco Delta. In: Grillo, O., & Venora, G., (eds.). *Ecosystems Biodiversity*, Intech. pp. 67–90.

Mitsch, W. J. J., & Gosselink, G., (2015). *Wetlands, 5th edn.*, John Willey & Sons Inc., Hoboken, USA.

Mittermeier, R. A., Mittermeier, C. G., Brooks, T. M., Pilgrim, J. D., Konstant, W. R., Fonseca, G. A. B., et al., (2003). Wilderness and biodiversity conservation. *Proc. Nat. Acad. Sci., 100*(18), 10309–10313.

Mittermeier, R. A., Robles-Gil, P., Hoffmann, M., Pilgrim, J., Brooks, T., Mittermeier, C. G., et al., (2004). *Hotspots Revisited: Earth's Biologically Richest and Most Endangered Terrestrial Ecoregions,* CEMEX, Ciudad de Mexico, Mexico.

Mittermeier, R. A., Turner, W. R., Larsen, F. W., Brooks, T. M., & Gascon, C., (2011). Global biodiversity conservation: The critical role of hotspots. In: Zachos, F. E., & Habel, J. C., (eds.). *Biodiversity Hotspots,* Springer Publishers, London, UK. pp. 3–22.

Mogni, V. Y., Oakley, L. J., & Prado, D. E., (2015). The distribution of woody legumes in Neotropical dry forests: The Pleistocene arc theory 20 years on. *Edin. J. Bot., 72*(1), 35–60.

Molina, C., Señaris, J. C., Lampo, M., & Rial, A., (2009). *Anfibios de Venezuela: Estado del Conocimiento y Recomendaciones Para su Conservación,* Ediciones Grupo TEI, Caracas.

Molinari, J., (2007). Variación geográfica en los venados de cola blanca (Cervidae, *Odocoileus*) de Venezuela, con énfasis en *O. margaritae,* la especie enana de la isla de Margarita. *Mem. Fund. La Salle Cien. Nat., 167,* 29–72.

Molinari, J., Gutiérrez, E. E., Ascenção, A. A., Nassar, J. M., Arends, A., & Márquez, R. J., (2005). Predation by giant centipedes, *Scolopendra gigantea,* on three species of bats in a Venezuelan cave. *Caribbean J. Sci., 41*(2), 340–346.

Monasterio, M., (1980). Las formaciones vegetales de los páramos de Venezuela. In: Monasterio, M., (ed.). *Estudios Ecológicos en los Páramos Andinos*, Universidad de los Andes, Mérida, Venezuela. pp. 93–158.

Moreno, E. J., (1992). Revisión histórica de la briología en Venezuela. *Tropical Bryol., 6,* 139–145.

Moreno, E., & Hernández, J., (2013). Volkmar Vareschi y su extraordinaria contribución a la liquenología en Venezuela. In: Medina, E., Huber, O., Nassar, J. M., & Navarro, P., (eds.). *Recorriendo el Paisaje Vegetal de Venezuela*, Ediciones IVIC, Instituto Venezolano de Investigaciones Cientificas (IVIC), Caracas, Venezuela. pp. 265–278.

Morillo, G., (2003). Dicotiledóneas. In: Aguilera, M. M., Azócar, A., & González, J. E., (eds.). *Biodiversidad en Venezuela,* vol. 1. Fundación Polar, Ministerio de Ciencia y Tecnología, Fondo Nacional de Ciencia, Tecnología e Innovación (FONACIT), Caracas, Venezuela. pp. 164–193.

Morillo, G., Briceño, B., & Silva, J. F., (2010–2011). *Botánica y Ecología de las Monocotiledóneas de los Páramos en Venezuela,* voll. 2. ICAE/ULA-Facultad de Ciencias, Mérida, Venezuela.

Morrone, J. J., (2001). *Biogeografía de América Latina y El Caribe;* Manuales y Tesis SEA 3, CYTED, ORCYT-UNESCO, Sociedad Entomológica Aragonesa, Zaragoza, España.

Moscarella, R. A., Aguilera, M., & Escalante, A. A., (2003). Phytogeography, population structure, and implications for conservation of white-tailed deer (*Odocoileus virginianus*) in Venezuela. *J. Mammalogy, 84*, 1300–1313.

Mostacero, J., (2013). Los helechos de Volkmar Vareschi, pensamiento y legado. In: Medina, E., Huber, O., Nassar, J. M., & Navarro, P., (eds.). *Recorriendo el Paisaje Vegetal de Venezuela*. Ediciones IVIC, Instituto Venezolano de Investigaciones Cientificas (IVIC), Caracas, Venezuela. pp. 249–263.

Myers, N., (1988). Threatened biotas: hotspots in tropical forests. *Environmentalist, 8*, 178–208.

Myers, N., Mittermeier, R. A., Mittermeier, C. G., Fonseca, G. A. B., & Kent, J., (2000). Biodiversity hotspots for conservation priorities. *Nature, 403*, 853–858.

Naveda, S. J. A., (2017). *La Paradoja del Parque Nacional Caura, 7(533). 952 contradicciones y retos,* Availbale at: http://desarrollosustentable.com.ve/la-paradoja-del-parque-nacional-caura-7-533-952-contradicciones-y-retos-jorge-naveda/?platform=hootsuite.

Nirchio, M., & Pérez, J. E., (2002). Riesgos del cultivo de tilapias en Venezuela. *Interciencia, 27*(1), 39–44.

Ochoa, G. J., & Aguilera, M. M., (2003). Mamíferos. In: Aguilera, M. M., Azócar, A., & González, J. E., (eds.). *Biodiversidad en Venezuela,* vol. 2. Fundación Polar, Ministerio de Ciencia y Tecnología, Fondo Nacional de Ciencia, Tecnología e Innovación (FONACIT), Caracas, Venezuela. pp. 650–672.

Ojasti, J., González, J. E., Szeplaki, O. E., & García, R. L. B., (2001). *Informe Sobre las species Exóticas en Venezuela,* Ministerio del Ambiente y de los Recursos Naturales, Caracas, Venezuela.

Oliveira-Miranda, M. A., Huber, O., Rodríguez, J. P., Rojas-Suárez, F., De Oliveira-Miranda, R., Hernández-Montilla, M., et al., (2010). Riesgo de eliminación de los ecosistemas terrestres de Venezuela. In: Rodríguez, J. P., Rojas-Suárez, F., & Giraldo, H. D., (eds.). *Libro Rojo de los Ecosistemas Terrestres de Venezuela*, Provita, Shell Venezuela, Lenovo (Venezuela), Caracas, Venezuela. pp. 109–235.

Ortega-Lara, A., Lasso Alcalá, O. M., Lasso, C. A., Andrade de Pasquier, G., & Bogotá-Gregory, J. D., (2012). Peces de la cuenca del río Catatumbo, cuenca del Lago de Maracaibo, Colombia y Venezuela. *Biota Col., 13*(1), 71–98.

Ortiz L., I., & Fernández-Badillo, A., (2017). Flora y fauna del subpáramo costanero del Parque Nacional Henri Pittier, estado Aragua, Venezuela. *Mem. Fund. La Salle Cien. Nat., 75*(183), 19–33.

Osuna, E., (2003). Hemípteros. In: Aguilera, M. M., Azócar, A., & González, J. E., (eds.). *Biodiversidad en Venezuela,* vol. 1. Fundación Polar, Ministerio de Ciencia y Tecnología, Fondo Nacional de Ciencia, Tecnología e Innovación (FONACIT), Caracas, Venezuela. pp. 378–386.

Oviedo, B. J., (1723). *Historia de la Provincia de Venezuela*, Madrid, Spain.

Pacheco, C., Aguado, I., & Mollicone, D., (2011). Las causas de la deforestación en Venezuela: un estudio retrospectivo. In: Aymard, C. G. A., (ed.). *Bosques de Venezuela: Un Homenaje a Jean Pierre Veillon*. BioLlania, ed. especial, *10*, pp., 281–292.

Pauls, S., (2003). Esponjas. In: Aguilera, M. M., Azócar, A., & González, J. E., (eds.). *Biodiversidad en Venezuela,* vol. 1. Fundación Polar, Ministerio de Ciencia y Tecnología, Fondo Nacional de Ciencia, Tecnología e Innovación (FONACIT), Caracas, Venezuela. pp. 210–219.

Péfaur, J. E., & Rivero, J., (2000). Distribution, species-richness, endemism, and conservation of Venezuelan amphibians and reptiles. *Amph. Rept. Conserv., 2(2), 42–70.*

Penchaszadeh, P. E., (2003). Equinodermos. In: Aguilera, M. M., Azócar, A., & González, J. E., (eds.). *Biodiversidad en Venezuela,* vol. 1. Fundación Polar, Ministerio de Ciencia y Tecnología, Fondo Nacional de Ciencia, Tecnología e Innovación (FONACIT), Caracas, Venezuela. pp. 514–521.

Pérez, J. E., Alfonsi, C., Salazar, S. K., Macsotay, O., Barrios, J., & Martínez, E. R., (2007). species marinas exóticas y criptogénicas en las costas de Venezuela. *Bol. Inst. Oceanogr. Venez., 46(1),* 79–96.

Pérez, O. J., Hoyer, M., Hernández, J. N., Rodríguez, C., Márques, V., Sué, N., et al., (2005). Alturas del Pico Bolívar y otras cimas andinas venezolanas a partir de observaciones GPS. *Interciencia, 30(4),* 213–216.

Phelps, W. H., & Meyer, S. R., (1979). *Una guía de las Aves de Venezuela,* Gráficas Armitano, Caracas, Venezuela.

Phelps, W. H., & Meyer, S. R., (1994). *Una guía de las Aves de Venezuela, 2nd edn.,* Editorial ExLibris, Caracas, Venezuela.

Pinto, M., & Montoya, J., (2012). Composición de la comunidad de meiofauna de las playas arenosas del sector Isla Apurito (río Apure, Venezuela), Primer Congreso Venezolano de Ciencia, Tecnología e Innovación en el marco de la LOCTI y del PEII, *Libro de Resúmenes,* vol. 1, Caracas, Venezuela, 118.

Pittier, H., (1935). Apuntaciones sobre geobotánica de Venezuela. *Bol. Soc. Venez. Ci. Nat., 3(23),* 93–114.

Provenzano, F., (2015). Corroncho desnudo del lago de Valencia, *Lithogenes valencia.* In: Rodríguez, J. P., García-Rawlins, A., & Rojas-Suárez, F., (eds.). *Libro Rojo de la Fauna Venezolana, 4th edn.,* Provita, Fundación Empresas Polar, Caracas, Venezuela.

Quintero, H. V. M., & Niño, B. J. G., (2017). *Racekiela andina sp. nov.* (Spongillida: Spongillidae): First report of a freshwater sponge from the Venezuelan Andes. *Zootaxa, 4341(2),* 275–278.

Quiroga-Carmona, M., & Molinari, J., (2012). Description of a new shrew of the genus *Cryptotis* (Mammalia: Soricomorpha: Soricidae) from the Sierra de Aroa, an isolated mountain range in northwestern Venezuela, with remarks on biogeography and conservation. *Zootaxa, 3441,* 1–20.

Quiroga-Carmona, M., & Woodman, N., (2015). A new species of *Cryptotis* (Mammalia, Eulipotyphla, Soricidae) from the Sierra de Perijá, Venezuelan-Colombian Andes. *J. Mammalogy, 96(4),* 800–809.

Ramia, M., & Stauffer, F., (2003). Monocotiledóneas. In: Aguilera, M. M., Azócar, A., & González, J. E., (eds.). *Biodiversidad en Venezuela,* vol. 1. Fundación Polar, Ministerio de Ciencia y Tecnología, Fondo Nacional de Ciencia, Tecnología e Innovación (FONACIT), Caracas, Venezuela. pp. 152–162.

Ramia, M., (1967). Tipos de sabanas en Los Llanos de Venezuela. *Bol. Soc. Venez. Ci. Nat., 27(112),* 264–288.

Ramia, M., (1993). *Ecología de las Sabanas del Estado Cojedes: Relaciones Vegetación-Suelo en Sabanas Secas.* Colección cuadernos FLASA. Serie Ciencia y Tecnología, N° 4.

Ramia, M., (1997). *Ecología de las Sabanas del Estado Cojedes: Relaciones Vegetación-Suelo en Sabanas Húmedas*. Colección cuadernos FLASA. Serie Ciencia y Tecnología, N° 9.

Raymond, T., (1994). *Venezuelan Butterflies*, Armitano Editor, Caracas, Venezuela. (Translation of 1982 Venezuelan edition, in Spanish).

Restall, R., Rodner, C., & Lentino, M., (2007). *Birds of Northern South America, an Identification Guide*, voll.2. Helm Field Guides, Yale University Press, USA.

Rial, B. A., Señaris, J. C., Lasso, C. A., & Flores, A. L., (2010). Evaluación rápida de la biodiversidad y aspectos socioecosistémicos del Ramal Calderas, Andes de Venezuela. *RAP Bull. Biol. Asses., 56.*

Ricardi, S. M. H., Gaviria, J., & Estrada, J., (1997). La flora del superpáramo venezolano y sus relaciones fitogeográficas a lo largo de los Andes. *PlantULA, 1*(3), 171–187.

Rico, R., Rodríguez, L. E., Pérez, R., & Valero, A., (1996). Mapa y análisis de la vegetación xerófila de las lagunas de Caparú, cuenca media del río Chama, estado Mérida. *PlantULA, 1*(1), 83–94.

Riina, R., & Huber, O., (2003). Ecosistemas exclusivos de la Guayana. In: Aguilera, M. M., Azócar, A., & González, J. E., (eds.). *Biodiversidad en Venezuela*, vol. 2. Fundación Polar, Ministerio de Ciencia y Tecnología, Fondo Nacional de Ciencia, Tecnología e Innovación (FONACIT), Caracas, Venezuela. pp. 828–861.

Ringuelet, R. A., (1974). Los hirudíneos terrestres del género *Blanchardiella* Weber del páramo norandino de Colombia. *Physis, 33*, 63–69.

Rivas, G. A., Molina, C. R., Ugueto, G., N., Barrios, T. R., Barrio-Amoros, C. L., & Kok, P. J. R., (2012). Reptiles of Venezuela: an updated and commented checklist. *Zootaxa, 3211*, 1–64.

Robaina, G., (1986). Guía práctica para el conocimiento de los pulpos de las costas de Venezuela. *Contribuciones Científicas, 10*, 2–27.

Rocha, C., (2003). Restauración ambiental. In: Aguilera, M. M., Azócar, A., & González, J. E., (eds.). *Biodiversidad en Venezuela*, vol. 2. Fundación Polar, Ministerio de Ciencia y Tecnología, Fondo Nacional de Ciencia, Tecnología e Innovación (FONACIT), Caracas, Venezuela. pp. 1012–1027.

Rodríguez, B., & López, M., (2009). Evaluación de la fertilización biológica del frijol con cepas nativas de *Rhizobium* aisaladas de un ultisol de la altiplanicie del estado Guárico. *Agron. Trop., 59*(4), 381–386.

Rodríguez, G., & Suárez, H., (2003). Crustáceos. In: Aguilera, M. M., Azócar, A., & González, J. E. (eds.). *Biodiversidad en Venezuela*, vol. 1. Fundación Polar, Ministerio de Ciencia y Tecnología, Fondo Nacional de Ciencia, Tecnología e Innovación (FONACIT), Caracas, Venezuela. pp. 288–311.

Rodríguez, G., (1973). *El sistema de Maracaibo*, Instituto Venezolano de Investigaciones Científicas (IVIC), Caracas, Venezuela.

Rodríguez, J. P., García-Rawlins, A., & Rojas-Suárez, F., (2015). *Libro Rojo de la Fauna Venezolana, 4th edn.*, Provita, Fundación Empresas Polar, Caracas, Venezuela.

Rodríguez, J. P., Rodríguez-Clark, K. M., Baillie, J. E. M., Ash, N., Benson, J., Boucher, T., et al., (2011). Establishing IUCN Red List criteria for threatened ecosystems. *Conserv. Biol., 25*(1), 21–29.

Rodríguez, J. P., Rojas-Suárez, F., & Giraldo, H. D., (2010). *Libro rojo de los ecosistemas terrestres de Venezuela, Provita, Shell Venezuela, Lenovo (Venezuela)*, Caracas, Venezuela.

Rodríguez, R. J. C., Guilarte, B. A. J., Marcano, A., & Velásquez-Boadas, A. J., (2017). Contribución al conocimiento de los carófitos recientes (Characeae, Charophyta) registrados en la República Bolivariana de Venezuela., *Libro de Resúmenes del XII Congreso Venezolano de Botánica,* Universidad Bolivariana de Venezuela, sede Monagas, 11 al 15 de julio de 2017, p. 77.

Rodríguez-Olarte, D., (2017). *Ríos en riesgo de Venezuela,* Colección Recursos Hidrobiológicos de Venezuela, Universidad Centroccidental Lisandro Alvarado, Barquisimeto, Venezuela.

Röhl, E., (1951). Sobre el gradiente térmico vertical de Venezuela. *Bol. Soc. Cien. Fís. Mat. Nat., 14*(44), 3–60.

Rojas-Suárez, F., Señaris, J. C., Molina, C., & Barrio-Amorós, C., (2009). Estado de conservación. In: Molina, C., Señaris, J. C., Lampo, M., & Rial, A., (eds.). *Anfibios de Venezuela: Estado del Conocimiento y Recomendaciones Para su Conservación,* Ediciones Grupo TEI, Caracas, Venezuela. pp. 75–81.

Román, B., (1980). *Peces Marinos de Venezuela,* Fundación La Salle de Ciencias Naturales, Caracas, Venezuela.

Rondón, J., Gómez, W., et al., (2016). Diversity of culturable bacteria recovered from Pico Bolívar's glacial and subglacial environments, at 4, 950 m, in Venezuelan Tropical Andes. *Canadian J. Microbiol., 62*(11), 904–917.

Rosales, C. J., (2003). Isópteros. In: Aguilera, M. M., Azócar, A., & González, J. E., (eds.). *Biodiversidad en Venezuela,* vol. 1. Fundación Polar, Ministerio de Ciencia y Tecnología, Fondo Nacional de Ciencia, Tecnología e Innovación (FONACIT), Caracas, Venezuela. pp. 362–369.

Rosales, J., (2003). Bosques y selvas de galería. In: Aguilera, M. M., Azócar, A., & González, J. E., (eds.), *Biodiversidad en Venezuela,* vol. 2. Fundación Polar, Ministerio de Ciencia y Tecnología, Fondo Nacional de Ciencia, Tecnología e Innovación (FONACIT), Caracas, Venezuela. pp. 812–826.

Rosales, J., & Huber, O. (1996). *Ecología de la Cuenca del Río Caura, Venezuela. I. Caracterización general. Scientia Guaianæ, 6,* 1–131.

Ruggiero, M. A., Gordon, D. P., Orrell, T. M., Bailly, N., Bourgoin, T., Brusca, R. C., et al., (2015). A Higher Level Classification of All Living Organisms. *PLoSONE, 10*(4), e0119248.

Rumrich, F. E., Lange-Bertalot, H., & Rumrich, M., (2000). Diatomeen der Anden von Venezuela bis Patagonien/Feurland. In: Lange-Bertalot, H., (ed.). *Iconographia Diatomologica, Annotated Diatom Micrographs,* vol. 9, Koeltz Scientific Books, Köenigstein, Germany. pp. 1–673.

Salerno, P. E., Ron, S. R., Señaris, J. C., Rojas-Runjaic, F. J., Noonan, B. P., & Cannatella, D. C., (2012). Ancient tepui summits harbor young rather than old lineages of endemic frogs. *Evolution, 66*(10), 3000–3013.

Sánchez-Suárez, I. G., & Díaz-Ramos, R., (2003). Microalgas marinas. In: Aguilera, M. M., Azócar, A., & González, J. E., (eds.). *Biodiversidad en Venezuela,* vol. 1. Fundación Polar, Ministerio de Ciencia y Tecnología, Fondo Nacional de Ciencia, Tecnología e Innovación (FONACIT), Caracas, Venezuela. pp. 88–93.

Sanz, V., (2007). Son las áreas protegidas de la isla de Margarita suficientes para mantener su biodiversidad? Análisis espacial del estado de conservación de sus vertebrados amenazados. *Mem. Fund. La Salle Cien. Nat., 167,* 111–130.

Sarmiento, G., (1971). Investigaciones ecológicas en los Llanos venezolanos. In: Dirección de Cultura, (ed.). *La Ciencia en Venezuela*. Ediciones Universidad de Carabobo, Valencia, Venezuela. pp. 320–333.

Sarmiento, G., (1990). Ecología comparada de ecosistemas de sabanas en América del Sur. In: Sarmiento, G., (ed.). *Las Sabanas Americanas, Aspectos de su Biogeografía, Ecología y Utilización*, Fundación Fondo Editorial Acta Científica Venezolana, Caracas, Venezuela. pp. 15–56.

Savini, V., & Joly, L. J., (2003). Dermápteros. In: Aguilera, M. M., Azócar, A., & González, J. E., (eds.). *Biodiversidad en Venezuela*, vol. 1. Fundación Polar, Ministerio de Ciencia y Tecnología, Fondo Nacional de Ciencia, Tecnología e Innovación (FONACIT), Caracas, Venezuela. pp. 370–377.

Schargel, R., (2011). Una reseña de la geografía física de Venezuela, con énfasis en los suelos. In: Aymard, C. G. A., (ed.). *Bosques de Venezuela: Un Homenaje a Jean Pierre Veillon*. BioLlania, ed. especial, *10*, pp. 178–188.

Schileyko, A. A., (2014). A contribution to the centipede fauna of Venezuela (Chilopoda: Scolopendromorpha). *Zootaxa*, *3821*(1), 151–192.

Scorza, J. V., & Arcay, L., (2003). Protozoos. In: Aguilera, M. M., Azócar, A., & González, J. E., (eds.). *Biodiversidad en Venezuela*, vol. 1. Fundación Polar, Ministerio de Ciencia y Tecnología, Fondo Nacional de Ciencia, Tecnología e Innovación (FONACIT), Caracas, Venezuela. pp. 194–208.

Scorza, J. V., & Núñez-Montiel, O., (1954). Estudio experimental sobre la sucesión de protozoarios que se desarrolla en las infusiones de musgo y de las variaciones de pH que la acompañan. *Acta Biol. Venez.*, *1*, 213–230.

Scorza, J. V., (1997). Introducción al conocimiento microscópico de ambientes astáticos, 1. Protozoos- Humedales de Venezuela. *Talleres*, *5*, 13–62.

Segnini, F. S. E., Chacón, M. M., & Domínguez, E., (2003). Efemerópteros. In: Aguilera, M. M., Azócar, A., & González, J. E., (eds.). *Biodiversidad en Venezuela*, vol. 1. Fundación Polar, Ministerio de Ciencia y Tecnología, Fondo Nacional de Ciencia, Tecnología e Innovación (FONACIT), Caracas, Venezuela. pp. 326–339.

Señaris, J. C., Lampo, M., Rojas-Runjaic, F. J. M., & Barrio-Amorós, C. L., (2014). Guía ilustrada de los anfibios del Parque nacional Canaima, Venezuela, *Ediciones IVIC, Instituto Venezolano de Investigaciones Científicas (IVIC)*, Caracas, Venezuela.

Silva, J. F., (2003). Sabanas. In: Aguilera, M. M., Azócar, A., & González, J. E., (eds.). *Biodiversidad en Venezuela*, vol. 2. Fundación Polar, Ministerio de Ciencia y Tecnología, Fondo Nacional de Ciencia, Tecnología e Innovación (FONACIT), Caracas, Venezuela. pp. 678–695.

Silva, J., Ramírez, L., Alfieri, A., Rivas, G., & Sánchez, M., (2004). Determinación de microorganismos indicadores de calidad sanitaria. Coliformes totales, coliformes fecales y aerobios mesófilos en agua potable envasada y distribuida en San Diego, estado Carabobo, Venezuela. *Rev. Soc. Ven. Microbiol.*, *24*(1–2), 46–49.

Slik, J. W. F., et al., (2018). Phylogenetic classification of the world's tropical forests. *Proc. Natl. Acad. Sci. U. S. A.*, *115*(8), 1837–1842.

Sloan, S., Jenkins, C. N., Joppa, L. N., Gaveau, D. L. A., & Laurance, W. F., (2014). Remaining natural vegetation in the global biodiversity hotspots. *Biol. Conserv.*, *177*, 12–24.

Smith, J. F., (1997). Molecular evolution and adaptive radiation in *Brocchinia* (Bromeliaceae: Pitcairnioideae) atop tepuis of the Guayana Shield. In: Givnish, T. J., & Sytsma, K.

J., (eds.). *Molecular Evolution and Adaptive Radiation.* Cambridge University Press, Cambridge, UK.

Soriano, P. J., & Ruiz, A., (2003). Arbustales xerófilos. In: Aguilera, M. M., Azócar, A., & González, J. E., (eds.). *Biodiversidad en Venezuela,* vol. 2. Fundación Polar, Ministerio de Ciencia y Tecnología, Fondo Nacional de Ciencia, Tecnología e Innovación (FONACIT), Caracas, Venezuela. pp. 696–715.

Soriano, P., & Ochoa, G., J., (1997). Lista actualizada de los mamíferos de Venezuela. In: La Marca, E., (ed.). *Vertebrados Actuales y Fósiles de Venezuela.* Serie Catálogo Zoológico de Venezuela, vol. I, Museo de Ciencia y Tecnología de Mérida, Mérida, Venezuela. pp. 205–227.

Steyermark, J. A., & Huber, O., (1978). *Flora del Ávila.* Sociedad Venezolana de Ciencias Naturales, Ministerio del Ambiente y los Recursos Naturales Renovables (MARNR), Caracas, Venezuela.

Steyermark, J. A., (1974). Relación florística entre la Cordillera de la Costa y la zona de Guayana y Amazonas. *Acta Bot. Venez., 9,* 245–252.

Steyermark, J. A., (1977). Future outlook for threatened and endangered species in Venezuela. In: Prance, G. T., & Elias, T., (eds.). *Extinction is Forever.* The New York Botanical garden, New York. pp. 128–135.

Steyermark, J. A., (1994). Collaborators, In: Bruno Manara, (ed.). *Flora del Parque Nacional Morrocoy.* Agencia Española de Cooperación Internacional, Fundación Instituto Botánico de Venezuela, Caracas, Venezuela.

Steyermark, J. A., Berry, P. A., & Holst, B. K., (1995–2005). *Flora of the Venezuelan Guayana,* voll. 9, Missouri Botanical Garden Press and Timber Press, St. Louis and Portland, USA.

Suárez, L., Molina, C. R., Bulla, L. A., & Francisco, V., (2000). Efecto de plantaciones de *Pinus caribaea* sobre la herpetocenosis en Uverito, Venezuela. *Ecotrópicos, 13*(2), 67–74.

Sugden, A. M., (1983). Determinants of species composition in some isolated neotropical cloud forests. In: Sutton, S. L., Whitmore, T. C., & Chadwick, A. C., (eds.). *Tropical Rain Forest: Ecology and Management.* Blackwell Scientific Publications, Oxford, UK. pp. 43–56.

Sutherland, J. P., (1980). Dynamics of the epibenthic community on roots of the mangrove *Rhizophora mangle* at Bahía de Buche, Venezuela. *Mar. Biol., 58,* 75–84.

Tello, J., (1968). *Historia Natural de Caracas,* Consejo Municipal del Distrito Federal, Caracas, Venezuela.

Tello, J., (1975). *Catálogo de la fauna venezolana, VIII: Mollusca,* Publicaciones de la comision organizadora de la III Conferencia de las Naciones Unidas sobre elderecho del mar, Caracas, Venezuela.

Tello, J., (1979). *Mamíferos de Venezuela,* Fundación La Salle de Ciencias Naturales, Artes Gráficas Toledo, Caracas, Venezuela.

Tello, J., (1981). *Catálogo de la Fauna Venezolana, II: Sarcodina.* 1-Rhizopoda, *b*-Foraminiferida, Editorial Sucre, Caracas, Venezuela.

Texera, A. Y., (1991). *La exploración Botánica en Venezuela (1754–1950).* Fondo Editorial Acta Científica Venezolana, Caracas, Venezuela.

Texera, A. Y., (2003). *La Zoologia en Venezuela 1936–1970 una Historia Social,* Universidad Central de Venezuela, Fundación Polar, Caracas, Venezuela.

Thomas, W. W., (1999). Conservation and monographic research on the flora of Tropical America. *Biod. Conserv., 8*, 1007–1015.

Tietjen, J. H., (1984). Distribution and species diversity of deep-sea nematodes in the Venezuela basin. *Deep Sea Res., 31*, 119–132.

Tomayo, F., (1964). *Ensayo de Clasificación de Sabanas de Venezuela*, Escuela de Geografía, Facultad de Humanidades y Educación, Universidad Central de Venezuela, Caracas, Venezuela.

Ulloa, C., Acevedo-Rodríguez, P., et al., (2017). An integrated assessment of the vascular plant species of the Americas. *Science, 358*, 1614–1617.

UNESCO [United Nations Educational, Scientific, and Cultural Organization], (1973). *International Classification and Mapping of Vegetation, Series 6*, Ecology and Conservation, Paris, France.

Urbani, F., (2011). Un esbozo de la geología de Venezuela. In: Aymard, C. G. A., (ed.). *Bosques de Venezuela: Un Homenaje a Jean Pierre Veillon*. BioLlania, ed. especial, *10*, pp. 27–32.

Vareschi, V., (1968). Sabanas del valle de Caracas. In: *Estudio de Caracas*, vol. 1: Ecologia vegetal-Fauna. Universidad Central de Venezuela, Caracas, Venezuela. Pp. 17-122.

Vareschi, V., (1969). Helechos. In: Lasser, T., (ed.). *Flora de Venezuela, 1*(1–2), 1–1033.

Vareschi, V., (1970). *Flora de los páramos de Venezuela*, Universidad de Los Andes, Ediciones del Rectorado, Mérida, Venezuela.

Vareschi, V., (1982). *Ecología tropical*, Sociedad Venezolana de Ciencias Naturales. Caracas, Venezuela.

Velásquez, J., (1965). Estudio fitosociológico de los pastizales de las sabanas de Calabozo, estado Guárico. *Bol. Soc. Ven. Cienc. Nat.* 109: 59-101.

Velásquez, J., (1994). *Plantas Acuáticas Vasculares de Venezuela,* Consejo de Desarrollo Científico y Humanístico, Universidad Central de Venezuela, Caracas, Venezuela.

Vera, B., (1992). Seagrasses of venezuelan coast: Distribution and components. In: *Coastal Plant Communities of Latin America,* Academic Press Inc. pp. 135–140.

Vera, B., (1993). Contribución al conocimiento de las macroalgas asociadas a las praderas de *Thalassia testudinum* König. *Acta Bot. Venez., 16*(2), 19–28.

Villamizar, E., & Cervigón, F., (2017). Variability and sustainability of the southern subarea of the Caribbean sea large marine ecosystem. *Environmental Development, 22*, 30–41.

Villarreal, Á., Gil, B., Bastidas, D., Wingfield, R., & Rodríguez, J., (2014). Composición florística, ambientes y vegetación de la isla Zapara, estado Zulia, Venezuela. *Bol. Centro Invest. Biol. Univ. Zulia, 48*(1), 1–20.

Volkmer-Ribeiro, C., & Pauls, S. M., (2000). Esponjas de agua dulce (Porifera: Demospongiae) de Venezuela. *Acta Biol. Venez., 20*(1), 1–28.

Von Humboldt, A., & Bonpland, A., (1807). *Le Voyage aux Régions Equinoxiales du Nouveau Continent [1799–1804]*, voll.30, París, France.

WBG [World Bank Group], (2016). *The Little Green Data Book 2016*, Publishing for The World Bank, Washington, DC.

Weil, E., (2003). The corals and coral reefs of Venezuela. In: Cortés, J., (ed.). *Latin American Coral Reefs,* Elsevier Science, B. V. pp. 303–330.

Wielgolaski, F. E., (1997). Polar and alpine tundra, *Ecosystems of the World, Part 3*. Elsevier, Amsterdam, The Netherlands.

Yacubson, S., (1972). Catálogo e iconografía de las Cyanophyta de Venezuela. *Bol. Centro Invest. Biol. Univ. Zulia, 5*, 1–78.

Yacubson, S., (1974). Catálogo e inconografía de las Chlorophyta de Venezuela. *Bol. Centro Invest. Biol. Univ. Zulia, 11*, 1–143.

Yerena, E., & Naveda, J., (2014). *The Real Extension of Venezuelan Protected Areas: Lessons to Assess Aichi Target 11*. IUCN World Parks Congress 2014, Sydney, Australia.

Zamora, F. A. C., Ramos, O. J. R., Arias, M., & Hernández, V. I., (2017). Respuesta de la comunidad microbiana al biotratamiento de un suelo contaminado con un crudo mediano. *Revista Internacional de Contaminación Ambiental, 33*(4), 629–639.

Zoppi de Roa, E., & Pardo, M. J., (2003). Rotíferos. In: Aguilera, M. M., Azócar, A., & González, Jiménez, E., (eds.). *Biodiversidad en Venezuela*, vol. 1. Fundación Polar, Ministerio de Ciencia y Tecnología, Fondo Nacional de Ciencia, Tecnología e Innovación (FONACIT), Caracas, Venezuela. pp. 242–253.

Zoppi de, R. E., (2003). Hemicordados, cordados y quetognatos. In: Aguilera, M. M., Azócar, A., & González, J. E., (eds.). *Biodiversidad en Venezuela*, vol. 1. Fundación Polar, Ministerio de Ciencia y Tecnología, Fondo Nacional de Ciencia, Tecnología e Innovación (FONACIT), Caracas, Venezuela. pp. 522–532.

CHAPTER 12

Biodiversity in Australia: An Overview

C. R. DICKMAN

Professor in Ecology, School of Life and Environmental Sciences, The University of Sydney, NSW 2006, Australia, E-mail: chris.dickman@sydney.edu.au

12.1 Introduction

12.1.1 Geography, Geology, Climate, and History

Australia has the unique distinction of being both a single country and an island continent. In addition to its continental land area of 7,595,342 km^2, at least 5,442 islands 0.01–5,786 km^2 in size (total area 32,969 km^2) lie offshore (SEWPAC, 2012; http://www.ga.gov.au/scientific-topics/national-location-information/dimensions/area-of-australia-states-and-territories). The island state of Tasmania, located off the southern coast of eastern Australia, covers an additional 64,519 km^2, excluding its surrounding islands. Australia's jurisdiction covers seven external territories, including almost 5.9 million km^2, or 42% of Antarctica, and also extends 200 nautical miles (370 kms) seaward from all land areas except in the north due to the close proximity of New Guinea. The vast area of ocean encompassed within the 200 nautical miles (~14 million km^2: Watford et al., 2005) is the Exclusive Economic Zone. Here, Australia has sovereign rights that include the conservation and management of natural resources in the waters superjacent to the seabed and of the seabed and its subsoil (http://www.ga.gov.au/scientific-topics/marine/jurisdiction/maritime-boundary-definitions#eez). Encompassing a very broad range of land systems and different climatic zones, and experiencing a prolonged period of isolation in its long geological history, Australia is home to an extraordinarily rich and largely endemic biota. For these reasons, it has been recognized by Conservation International as one of the world's 17 megadiverse countries (Mittermeier et al., 1997). To understand where Australia's biota came from and the forces that maintain its biodiversity now, we need to consider briefly the physical geography and history of this great southern continent.

Lying between latitudes 10° 40' and 43° 39' S and longitudes 113° 09' and 153° 39' E, the Australian land mass is bounded to the west by the Indian

Ocean and to the east by the Pacific Ocean. The tropical Timor Sea and Arafura Sea lap the northern coast, while the great Southern Ocean lashes the continent's southern shores. Renowned as the world's driest inhabited continent, Australia has few large freshwater lakes and relatively few major rivers. The largest body of water, sometimes referred to as the continental 'sump' owing to its elevation 15 m below sea level, is Lake Eyre. Although it covers up to 10,000 km^2 when full, this saline lake is ephemeral and fills only after prolonged rains throughout the catchment area.

As might be expected for a continental landmass, Australia is geologically complex and comprises rocks and minerals of hugely varied age and origin. Crystals of zircon discovered recently in Western Australia have been dated to 4.374 billion years. Forming just 125 million years or so after the creation of the Earth-moon system, these ancient crystals represent the oldest known pieces of the Earth's crust (Valley et al., 2014). Other materials, from the Yilgarn craton in the southwest of Western Australia and from the Pilbara craton in the state's northwest, are at least 3.5 billion years old (Griffin et al., 2004; Hickman and Van Kranendonk, 2008). The Pilbara region holds the record for the oldest exposed surface strata, with cherts embedded in volcanic basalts that date to 3.5 billion years. This region also hosts some of the world's earliest known sedimentary rocks. Aged 2.5–3.5 billion years, these strata confirm the presence of rivers and lakes or seas in which the eroded sediments were deposited (Johnson, 2004).

Around 1.3–1.4 billion years ago these ancient cratons—or 'proto-continents'—collided in the east with the North Australian and South Australian cratons, creating a landmass that now underlies much of the modern continent. These cataclysmic events resulted in the upthrust of mountain ranges, the eroded remnants of which can still be seen in northern South Australia, Western Australia and the Northern Territory (Johnson, 2004). More recently, just 80–100 million years ago, the Great Dividing Range began to arise near the eastern edge of the Australian landmass. This event accompanied the opening of the Tasman Sea and the rifting of New Zealand from Australia's east coast, resulting in uplift of the continent's eastern margin. The geology of the Great Dividing Range is exceedingly complex, with sedimentary strata and granites dating from more than 300 million years and more recent episodes of volcanism that have resulted in basaltic outcrops in many places (Johnson, 2004). Erosion of the Great Dividing Range has softened and lowered what would have been once a towering cordillera. The highest point now—Mount Kosciuszko—is 2,228 m a.s.l., with most of the range just 300–1600 m a.s.l. Nonetheless, running 150–400 km inland from the east coast and extending more than 3500 km from north Queensland to western Victoria, the Great Dividing Range is an outstanding and significant landform—the third longest mountain chain in the world after the Andes and the Rocky Mountains.

The great antiquity and stability of the Australian continent have exposed the land to prolonged weathering. This has resulted in the generally subdued topography of the country's landscapes, and soils that are leached and impoverished over vast areas. Two of the major processes that renew soils in other parts of the world have operated much less conspicuously in Australia. Firstly, orographic activity has been limited, with few episodes of mountain-building occurring since the uplift of the Great Dividing Range. Secondly, glaciation has been relatively inconsequential. Although Australia has experienced intense cycles of arid glacial and humid interglacial conditions, especially over the last 4–5 million years, ice accumulated primarily only in Tasmania and in the southeastern highlands of the Great Dividing Range (Johnson, 2004). Immense glaciers in other regions, especially in the Northern Hemisphere, scraped and gouged the landscapes as they advanced and left vast areas of 'new' soils as they retreated. Australia's modest glaciers were much more localized in their effects. The immense period of weathering and limited opportunities for soil renewal have left Australia with a legacy of shallow, skeletal soils that are depleted in many of the nutrients that are needed for plant growth. The nutrient-poverty of the soils, in turn, has had profound consequences for the evolution of Australia's biota (Augee and Fox, 2000; Orians and Milewski, 2007).

In addition to the physical environment, climate plays a key role in dictating the distribution and abundance of biota. Australia's continental climate has varied greatly over time, oscillating between tropical 'greenhouse' conditions that were associated with extensive forests, rivers and freshwater lakes, to cool, arid 'icehouse' conditions when freshwater was limited and landscapes were dominated by grasslands rather than by forest. The early Miocene, beginning 23 million years ago, marks the start of one of the lushest greenhouse phases. This lasted for 8 million years before giving way to an icehouse event that heralded a prolonged period of continental drying. This latter period has continued, with several oscillations, to the present, with characteristically arid landforms such as sand dunes developing over the last million years (Byrne et al., 2008). The intensification and spread of arid conditions have affected the evolutionary history of many groups of plants and animals, and contribute to the powerful imagery associated with Australia's expansive and iconic 'outback' landscapes.

Australia's current climate has many influences. At the largest geographical scale are massive oceanic and atmospheric circulation phenomena that affect flows of water and air, in particular, moist air that leads to rainfall. In the west, the Indian Ocean Dipole (IOD)—which is defined by relative differences in sea surface temperatures between two areas in the Indian Ocean—is associated with more rainfall (if the IOD is negative) or less rainfall (if the IOD is positive) across the Top End and southern Australia, as well as with differences in degree of cloud cover in the northwest (Bureau of Meteorology, 2012). In the east, the El Niño – Southern Oscillation (ENSO)—which is driven by relative differences in sea surface temperatures between the central and eastern tropical Pacific Ocean—can bring flooding rains to central and eastern regions in the La Niña (positive) phase and droughts in the El Niño (negative) phase. IOD and ENSO phenomena dominate regional climates every few years; the coincidence of both cycles can result in extreme and widespread rainfall or drought events. Near-shore currents affect weather more locally. Warm waters flowing down the east and west coast of Australia bring moisture that condenses on the cold land surface in winter, resulting in frequently heavy rains in this season (Bureau of Meteorology, 2012).

At the continental scale, rainfall is lowest in the arid regions of southern and central Australia, with annual averages as low as 150–200 mm. These regions are associated with high-pressure belts and descending dry air that reduces cloud cover; in consequence, levels of both solar radiation and evaporative water loss from the soil are high. Air temperatures on hot days can exceed 50°C, and surface temperatures of 60–70°C are not uncommon (Williams and Calaby, 1985). Rainfall increases with distance from the arid heartland, especially in coastal and subcoastal areas as proximity to both moisture sources and reliable rain-producing weather systems increases (http://www.bom.gov.au/climate/averages/maps.shtml). The exceptions to this pattern are the arid Pilbara region in the northwest and along parts of the equally dry Great Australian Bight in the south. Average annual rainfall is greatest at higher elevations; for example, mountainous parts of northeastern Queensland and western Tasmania receive average annual rainfall of 3,000 mm or more. In Tasmania and higher parts of the Great Dividing Range, precipitation in winter occurs as snow. In general, rainfall in northern Australia occurs in summer and is associated with tropical monsoons, whereas rainfall at higher latitudes occurs mostly in winter. Eastern Australia, especially east of the Great Dividing Range, is the only region that reliably receives both summer and winter rainfall, although heavy rainfall can occur unpredictably during any season in the central arid zone (http://www.bom.gov.au/climate/ averages/maps.shtml).

Both climate and physical geography shape patterns in the occurrence of present-day biota, but to understand where a biota comes from we must consider its historical biogeography. This is particularly important for a now-isolated island continent. If we step back to the beginning of the Triassic Period, some 250 million years ago, the world's landmasses were coalesced into a single supercontinent, termed Pangaea. Pangaea began to break up during the Triassic, with the vast northern landmass, Laurasia, separating from its southern counterpart, Gondwana (Johnson, 2004). These huge continents themselves also began to fracture, with Africa breaking free from the eastern Gondwanan landmass about 170 million years ago, and India, Madagascar and New Zealand separating in turn over the next 100 million years. Australia and eastern Antarctica separated some 38–45 million years ago (Long et al., 2002). South America and western Antarctica were the last components of the ancient Gondwanan landmass to dissociate, sundering perhaps 30 million years ago. Laurasia, comprising modern-day Asia, Europe and North America, remained united as the break-up of Gondwana took place. These titanic deep-time events occurred because the continental land masses sit on plates in the Earth's crust and are propelled by convection currents that lie deep beneath the surface. These currents were responsible for the collision of the cratons that form the bedrock of much of modern Australia and, since the sundering of the Australian plate with Antarctica, they continue to push the Australian continent northward at the rate of about six centimeters per year (White, 2006). Some 30 million years ago, the northward-drifting Australian plate collided with the Pacific plate, resulting in buckling and massive uplifting of land along the active edges. The uplifted land now forms part of present-day New Guinea. This rugged landmass was connected by land to Australia at the time of uplift 30 million years ago, but lower-lying areas have been submerged on several occasions since (White, 2006). The most recent land connection formed during the last ice age when sea levels fell by as much as 160 meters. New Guinea has been an island since rising sea levels breached the land bridge to form Torres Strait 15,000 years ago.

For Australia, the reshaping of the Earth's surface over the last 200 million years has had profound consequences. The sundering of Antarctica from Australia and South America allowed the formation of the circumpolar current, and with it, increased productivity of the southern oceans. This, in turn, appears to have fostered the evolution of large filter-feeding whales (Mysticetes) that now abound in the oceans around modern Australia (Fordyce and

Barnes, 1994; Fordyce, 2006). On land, elements of the biota such as many flowering plants and invertebrates have Pangaean origins and share deep affinities with relatives on other continents (Augee and Fox, 2000). Other elements, such as the marsupials, are descended from Gondwanan stock that entered Australia some 60–65 million years ago when the continent was connected by the Antarctic land bridge to South America (Archer and Kirsch, 2006). By contrast, the first Australian rodents invaded from the north, arriving via Melanesia some 5 million years ago by crossing the newly formed Torresian land bridge (Aplin and Ford, 2014). Still other biota arrived by rafting. For example, the ancestors of modern nasute termites arrived in Australia on 2–3 occasions between 10 and 20 million years ago, completing oceanic migrations from perhaps as far away as South America (Arab et al., 2017). The ancestors of a Kangaroo Island trapdoor spider, *Moggridgea rainbowi*, appear also to have made a long crossing over the Indian Ocean, arriving 2.27–16.02 million years ago (Harrison et al., 2017). Phylogenetic analyzes are being used increasingly to uncover the origins of Australia's biota, with the composition of many groups turning out to be a complex mix of Gondwanan endemism and more recent immigration (e.g., Sniderman and Jordan, 2011; Heads, 2014; Jarvis et al., 2014; Oliver and Hugall, 2017).

In essence, Australia's biota has diverse origins that comprise resident lineages of varied antiquity and immigrants that have arrived and settled at different times since at least the end of the Cretaceous 65 million years ago. Once in Australia, of course, organisms were then subject to forces that caused further divergence and contributed to the pattern of diversity that we see today (Heads, 2014). For example, many closely related species of plants and animals occur in southeastern and southwestern Australia, and represent taxa descended from ancestral populations that were split by marine inundation or the onset of hostile climatic regimes (Crisp and Cook, 2007). Seasonally arid conditions began to arise in northwestern Australia in the late Palaeocene and moved into inland areas towards the mid-Miocene by about 15 million years ago (Martin, 2006; Morton et al., 2011). Aridity intensified throughout the Pliocene, culminating in the last very arid glacial maximum 18,000–20,000 years ago (Hesse et al., 2005). Phylogenetic evidence indicates that many lineages of arid-adapted plants and animals began to diverge from mesic-adapted ancestors from mid-Miocene times as the climate began to dry, with speciation taking place both *in situ* and in association with severe climatic oscillations that repeatedly fractured and then reassembled species populations across the inland environment (Byrne et al., 2008, 2011; Martin, 2017). Many species persisted in evolutionary refugia in permanent freshwater habitats, in the central ranges, and in high elevation sites in eastern and western Australia, their isolation leading to high levels of disjunction and endemism (Davis et al., 2013). In addition to these large-scale forces, local climatic conditions, fire regimes, soils and interactions between species dictate the patterns of biodiversity at local and regional scales (Dickman et al., 2014; Gibson et al., 2017). I consider next the biomes that are so formed.

12.2   Major Ecosystems

Australia's great size, diverse landscapes and multiple climatic zones mean that it supports a bewilderingly large range of ecosystem types. Many formal classification systems have been proposed to organize and distinguish them. One of the earliest and simplest recognized four broad biogeographic regions: the Eyrean (encompassing the central deserts), Torresian (across the monsoonal north), the Bassian in the southeast and the Westralian in the southwest (Spencer and Gillen, 1912). More recent schemes recognize many more regions. For example, Olson et al. (2001) recognized 40 ecoregions within Australia and its island territories, out of a total 867 ecoregions worldwide. In Australia, however, perhaps the most widely used scheme is the Interim Biogeographic Regionalization for Australia (IBRA), which has now progressed to version 7.0 (https://environment.gov.au/land/nrs/science/ibra/australias-bioregions-maps). Earlier versions of IBRA helped to inform the ecoregions of Olson et al. (2001), which often represent clusters of two or more similar IBRA regions. IBRA uses a landscape-based approach to classify Australia's terrestrial environment and those of its external territories. In total, IBRA delineates 89 biogeographic regions and 419 subregions, each reflecting a unifying set of major environmental influences that shape the occurrence of flora and fauna and their interaction with the physical environment (Department of the Environment and Energy, 2017). Each IBRA subregion itself usually contains at least two regional ecosystems, with the result that thousands of this finest-scale ecosystem type can be recognized. For brevity here, I describe a higher-level grouping of the 89 biogeographic regions into biomes. Developed initially by the World Wildlife Fund, 14 terrestrial biomes are recognized worldwide, with eight of these in Australia (Olson et al., 2001; Figure 12.1). I also describe equivalent classification schemes that provide recognition of Australia's marine (Spalding et al., 2007) and freshwater (Abell et al., 2008) systems.

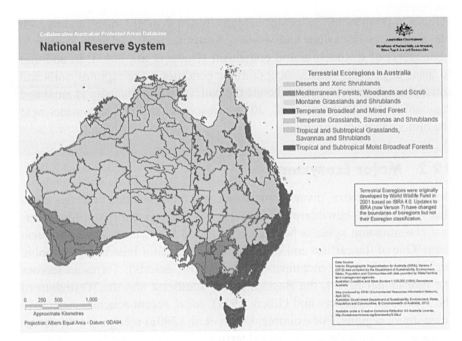

Figure 12.1   Terrestrial biomes in Australia. Biomes are represented by different shading, and Interim Biogeographic Regionalization for Australia (IBRA) regions by lines. Source: Australian Government Department of the Environment and Energy. Note that biomes are termed ecoregions in this figure, whereas this chapter uses the term 'biome' to accord with original usage (Olson et al., 2001).

12.2.1   Deserts and Xeric Shrublands

Covering 3,468,165 km², this is the largest of Australia's biomes. It spans the continent's arid center and encompasses 21 IBRA regions. This biome contains the continent's five major deserts: the Great Sandy Desert, Gibson Desert, Great Victoria Desert, Simpson Desert and Sturt's Stony Desert, which together occupy some 20% of Australia's landmass (Williams and Calaby, 1985). Not surprisingly, rainfall is generally low (< 375 mm) but also highly variable and greatly exceeded by potential evaporation. Spinifex or porcupine grass (*Triodia* spp.) dominates much of the central and northern parts of this biome, especially on nutrient-impoverished sand plains and sand dunes; the grassland in places has no shrubs or trees, but in most areas the cover of Spinifex is punctuated by patches of low woodland or scattered trees. Woodlands dominated by *Acacia* spp. occur in western and some eastern parts of this biome, and chenopod shrublands and mallee eucalypts in southerly areas (Williams and Calaby, 1985). The eastern-most part of the biome encompasses the Channel Country. Covering over 304,000 km², landscapes here are some of the most open in the entire country. It is possible to stand in parts of the Channel Country and see the horizon unobstructed by any large trees or shrubs in an arc of 360°. Rains that fall in catchments to the north of the Channel Country surge through this region in south to southwestern directions in river channels that give this region its name (Dickman, 2010).

Areas of rugged upland topography occur in western parts of this biome, as well as in the central MacDonnell Ranges and the Flinders Range in the southeast. Despite the harsh and ostensibly hostile character of the deserts and xeric shrublands biome, this great region sustains the world's richest desert reptile and insectivorous mammal faunas. Diverse and endemic flowering plants occur in all upland areas that have been studied, and endemic algae, aquatic vascular plants, fish, snails, crustacea and many insects are restricted to desert springs (Fensham, 2010). Among the most bizarre and highly derived fauna are the richly biodiverse stygofauna that occur in subterranean aquifers in the western part of this biome (Humphreys, 2006). Here, a projected 4,140 species of invertebrates—some with Gondwanan or Pangean ancestry—have been uncovered, making this the most diverse stygofauna that has yet been described (Guzik et al., 2010).

12.2.2 Tropical and Subtropical Grasslands, Savannas, and Shrublands

This immense biome encompasses almost all of northern Australia, from the western Kimberley to Cape York, with a southern extension into northern New South Wales. It encompasses 23 IBRA regions and occupies 2,407,400 km². This region receives higher rainfall than the deserts and xeric shrublands, in the order of 400 mm a year where the two biomes abut, but in the range of 900–1,600 mm a year in some coastal and subcoastal areas. Perennial grassland and sparse, low (4–8 m) species of *Eucalyptus* or *Acacia* dominate the lower rainfall areas, with the structural complexity and diversity of these communities increasing as rainfall increases. Among the most expansive of the grasslands is the Mitchell Grass Downs IBRA region (335,000 km²), which extends from the interior of the Northern Territory deep into central Queensland. Dominated by Mitchell grasses (*Astrebla* spp.), the heavy clay soils of the downlands crack and flex depending on climatic conditions. Trees do not establish readily due to the instability of the soils, but many vascular plants, invertebrates and some vertebrates (e.g., Julia Creek dunart, *Sminthopsis douglasi*) are endemic to Mitchell grass areas. On the eastern flank of the downlands are vast tracts of *Acacia*-dominated communities, especially brigalow *A. harpophylla*, which has its own endemic flora and fauna (Keith, 2004). Although much of this biome is characterized by undulating plains, areas of rugged and broken topography occur throughout the Kimberley and Top End, as well as around Mount Isa and in the desert uplands of Queensland.

As in the deserts and xeric shrublands, upland areas in the tropical and subtropical grasslands, savannas and shrublands biome harbor diverse and endemic flowering plants and invertebrates, including many insects and crustaceans in stygofauna of the Kimberley (Guzik et al., 2010). Patches of dry rainforest and vine thicket occur in many parts of this biome, especially in coastal locations. Along with large islands such as Groote Eylandt and the Tiwi Islands, these habitats contain many endemic plants, snails, earthworms and other invertebrates (McKenzie et al., 1991). Among the vertebrates, endemic taxa occur in each of the five classes. Many have links with New Guinea and Southeast Asia due to exchanges across Torres Strait. Large numbers of birds make regular crossings of the strait (Pratt and Beehler, 2015); mammals, such as tree-kangaroos (*Dendrolagus* spp.) and cuscuses (*Phalanger mimicus* and *Spilocuscus maculatus*), are the descendants of colonists that evolved in New Guinea (Heinsohn and Hope, 2006).

12.2.3 *Temperate Grasslands, Savannas, and Shrublands*

The temperate grasslands, savannas and shrublands are restricted to inland areas of southeastern Australia, abutting the deserts and xeric shrublands biome to the west and more heavily wooded biomes elsewhere. This biome differs from the tropical grasslands due to the generally cooler temperatures and wider annual temperature range, as well as the dominant vegetation formations. Rainfall averages 250 mm a year in the west, but can be as much as 650 mm a year at sites along the eastern margin; rainfall occurrence is unpredictable, but shifts from summer dominance in the northern parts of this biome to winter dominance in the south.

Encompassing four IBRA regions, almost half of the 529,779 km² extent of this biome is occupied by the Mulga Lands. Soils of the Mulga Lands are poor and dominated by mulga (*Acacia aneura*), but with significant patches of eucalyptus woodland with varied species of box and ironbark. Although generally flat with few hills, the Mulga Lands contain several important wetlands that flood on a seasonal basis and support large populations of wetland birds. To the south of the Mulga Lands lie the remaining IBRA regions that comprise this biome: the Cobar Peneplain, Darling Riverine Plains and Riverina. The Cobar Peneplain is characterized by rolling downs and flat plains, with stony ridges and some locally significant peaks, such as those in the Gunderbooka Range. Mulga infiltrates in northerly areas, with open eucalypt- and cypress-dominated woodlands elsewhere (Keith, 2004). The remaining IBRA regions in this biome are drained by major rivers, in particular, the Darling, Murray and Murrumbidgee Rivers. These form alluvial plains with productive soils that support extensive and well developed gallery forests dominated by river red gums (*Eucalyptus camaldulensis*), a wide variety of box-eucalypts and species of *Acacia*. Soils are generally more sandy and less productive away from the riparian areas, and support shrublands dominated by saltbush (*Atriplex* spp.), bluebush (*Maireana* spp.) and other chenopod species, as well as native grasslands (Keith, 2004). It is likely that this biome retains some endemic flora and fauna, especially in the less disturbed range country, but this has been little studied. Unfortunately, much of the native vegetation of this biome has been obliterated by cropping for wheat and cotton, and especially by the pastoral industry; sheep and goat grazing dominate in most areas. Even now, vast areas of mulga continue to be cleared annually in Queensland to run ever-greater flocks of domestic grazing animals (Cogger et al., 2017).

12.2.4   *Mediterranean Forests, Woodlands, and Scrub*

Restricted to the southwestern corner and parts of the southeastern mainland, this biome encompasses 14 IBRA regions and occupies 782,952 km². It is bounded inland mostly by the deserts and xeric shrublands biome; annual rainfall rises from a minimum of ~250 mm along the boundary to over 1,200 mm in some coastal and subcoastal sites. In general, the Mediterranean-type ecosystems are characterized by hot, dry summers and cool, moist winters; frosts occur infrequently. Only five regions in the world experience similar climatic conditions, ensuring that the habitats they contain are globally rare. Despite this, the Mediterranean forests, woodlands and scrub biome is characterized by extraordinarily high levels of diversity of uniquely adapted animal and plant species. Collectively, the five globally distributed regions of this biome harbor over 10 percent of the Earth's plant species (http://environment.gov.au/land/nrs/science/ibra/australias-ecoregions). Most plants are fire-adapted, and dependent on this disturbance for their persistence.

The Western Australian part of the Mediterranean forests, woodlands and scrub biome supports extensive areas of *Eucalyptus*-dominated woodlands and forests, with different mallee species, jarrah (*E. marginata*) and giant karri (*E. diversicolor*) being well represented. Diverse heath communities line the southwestern coastline and dominate the region's offshore islands. In areas near the coast, such as the Stirling Range, and inland in the Coolgardie IBRA region, some of the most diverse and magnificent floras can be found (Cunningham, 2005). Paralleled only by the South African Fynbos in numbers of species and species density, the shrublands of southwestern Australia have floras that are significantly more diverse than the other Mediterranean biomes (Gibson et al., 2017). More than 5,500 species of plants have adapted to the forests and scrubs of this region, with nearly 70% being endemic (Cunningham, 2005). Many fungi, invertebrates and small vertebrates are restricted also to these rich scrublands. In its northernmost reaches, the southwestern part of this biome harbors living representatives of some of the Earth's oldest known forms of life – stromatolites. Although fossil stromatolites from further north in Western Australia have been dated to 3.45 billion years old, they differ little in appearance from the living forms. Stromatolites and related microbialites are comprised mainly of cyanobacteria and are sometimes termed 'microbial mats' due to their rounded, rug-like appearance (Grey and Planavsky, 2009).

Although the biota on the eastern and western sides of the Mediterranean forests, woodlands and scrub biome share many affinities owing to their connection before the southward expansion of the deserts, the eastern biota is less diverse than its western counterpart. Nonetheless, well-developed shrubland and forest formations occur on the Eyre and Yorke Peninsulas and extend onto Kangaroo Island, Australia's third-largest island (4,405 km²). The eastern extent of this biome incorporates the expansive Murray-Darling Depression, an immense IBRA region (199,583 km²) that encompasses the confluence and lower floodplains of the Murray and Darling Rivers. River red gum forests dominate many areas along the rivers and associated wetlands (Keith, 2004). These areas provide habitats for rich communities of aquatic insects and other invertebrates, fish, frogs and water birds. South of the riparian areas, extending into Victoria, diverse shrublands that incorporate Spinifex grasses and a patchy overstory of mallee-form eucalypts occur in the Sunset, Big Desert and Little Desert areas. These sustain rich assemblages of small vertebrates, especially reptiles (Cogger, 2014).

12.2.5　Temperate Broadleaf and Mixed Forests

This is the most complex biome in Australia. Extending continuously from southeastern Queensland to the eastern edge of South Australia, and encompassing the island state of Tasmania, it contains 23 IBRA regions. Although not the largest biome (456,807 km²), the southeastern forests that define it experience highly varied temperature and rainfall regimes and occur at elevations from sea level to 1585 m in the Barrington Tops National Park. Tall forests dominated by *Eucalyptus* spp. cloak the slopes of the Great Dividing Range, especially in areas receiving higher rainfall; understories in these forests can be structurally complex and floristically diverse, with many species of *Acacia, Casuarina, Cassinia* and other shrubs, as well as numerous ferns, mosses and lichens. Higher elevations support tangled snow gum (*Eucalyptus pauciflora*) woodlands and, in sheltered locations on the mainland and in Tasmania, cool temperate rainforests dominated by evergreen Antarctic beech (*Lophozonia* spp.). Formerly placed within the genus *Nothofagus* (Heenan and Smissen, 2013), members of the beech group occur widely in southwestern Argentina, Chile, New Zealand, New Caledonia and New Guinea, indicating deep Gondwanan heritage. Warm temperate rainforests occur at lower elevations, often on poorer quality soils, and comprise usually 3–5 tree species in any locality. Subtropical rainforests occur in more northerly parts of the biome and are structurally more complex and species-rich than their temperate counterparts (Keith, 2004). In places where rainfall is less than 1,100 mm, 'dry' rainforests and littoral rainforests may occur, as well as extensive tracts of open forest and woodland dominated by different species of *Eucalyptus*. Other vegetation communities occur in this biome, including grasslands, heathlands and wet-lands (Keith, 2004).

Because of its rugged topography and structurally complex physical environment, the temperate broadleaf and mixed forests biome has provided evolutionary refugia for many groups of plants and animals during past epi-sodes of continental cooling and drying (Markgraf et al., 1995; Kershaw et al., 2017). In addition to the very wide range of contemporary habitats the biome provides, the refugia have ensured the persistence of a remarkably diverse spectrum of organisms that show high levels of local and regional endemism (Mackey et al., 2012). These organisms include many plants, fungi, invertebrates, and representatives among the frogs, reptiles, birds and mammals (e.g., Milner et al., 2012; Rix and Harvey, 2012; Baker and Dick-man, 2018). Further divergence and endemism has arisen due to the separa-tion of many taxa by river systems that flow eastward to the coast from the upper reaches of the Great Dividing Range. Among the best studied of these systems are the Hawkesbury, Hunter and Clarence Rivers, although many other rivers act as biogeographic barriers—and as habitats for allopatric spe-ciation of freshwater invertebrates, fish, turtles and other taxa—throughout the biome (Heatwole, 1987).

12.2.6   *Tropical and Subtropical Moist Broadleaf Forests*

This biome (34,591 km²) is restricted to central and northern coastal areas of Queensland north of the Tropic of Capricorn. The closed forests of this biome occur along the Great Dividing Range and represent a northerly extension of the temperate forests further south. The biome encompasses three IBRA regions—the Wet Tropics, Central Mackay Coast and Pacific Subtropical Islands (Lord Howe Island and Norfolk Island)—as well as many small off-shore islands such as those in the Whitsunday group. Although the climate usually is relatively stable with low variability in annual temperature and high levels of rainfall (usually more than 2,000 mm per year), the biome is subject to disturbance from cyclones over the summer months. Elevations range from sea level to 1,611 m (Mount Bartle Frere). Rainfall at the higher elevations is pronounced; annual rainfall at Mount Bellenden Ker (1,593 m) is over 8,300 mm, and up to 300–400 mm may fall per day during cyclones (http://www.bom.gov.au/climate/averages/maps.shtml). Despite poor gra-nitic soils on the peaks and lower ranges, the tropical and subtropical forests of this biome are extensive, complex and diverse, although most have well-developed climbing vines, creepers, epiphytes, palms and ferns. Depending on the classification scheme that is used, 12–16 different types of rainforest can be distinguished on the mainland, with further types present on Lord Howe Island and Norfolk Island (Pickard, 1983; Mills, 2007). Mangrove forests occur in coastal zones, and small patches of heath can be found at high elevation sites where soils are too thin to support forest cover.

Much like the temperate broadleaf and mixed forests biome, the tropical and subtropical moist broadleaf forests harbor many species and community types that hark back to earlier times when Australia was cooler and drier. Repeated contractions and expansions of biota to and from the refugia of this biome have provided fertile conditions for speciation, and it is a national biodiversity 'hotspot'. Many species claim Gondwanan heritage, such as the rare, large (40 m) stockwellia trees (*Stockwellia quadrifida*) which have affin-ities with *Eucalyptus*, while others share relatives with taxa in New Guinea that crossed or reverse-crossed Torres Strait during prior rafting or overland dispersal events (Ladiges et al., 2003). The Wet Tropics IBRA region alone boasts some 85 endemic plant species and 68 species of endemic terres-trial vertebrates (http://whc.unesco.org/en/list/486), including such iconic species as Victoria's riflebird (*Ptiloris victoriae*), tree-kangaroos (*Dendro-lagus* spp.), several species of rainforest possums, and the bizarre musky rat-kangaroo (*Hypsiprymnodon moschatus*). The rich biodiversity values of this biome have stimulated much research and ensured that attention is focused on conservation. Much of the remaining intact forest is protected, with some 8,940 km² listed as a World Heritage site (http://whc.unesco.org/en/list/486). Both Lord Howe Island and Norfolk Island also support many endemic species. At least 113 species of vascular plants and ~950 of 1,600 invertebrates are thought to be endemic to Lord Howe, compared with 50 plant species and ~100 of 700 invertebrates on Norfolk Island (Priddel and Wheeler, 2014). Lord Howe Island received World Heritage status in 1982, as did a historic precinct on Norfolk Island in 2010.

12.2.7 Montane Grasslands and Shrublands

Comprising only a single IBRA region, this biome is the smallest in Australia and covers just 12,330 km². It is distinguished primarily by elevation as it includes montane and alpine areas above 1,300 m. Although many peaks in Tasmania exceed this height, this biome is restricted formally to southeastern New South Wales (including parts of the Australian Capital Territory) and adjacent parts of Victoria. This is the highest region in the Great Dividing Range, and the only area where deep snow falls each year. Low woodland formations occur at elevations up to 1,600 m, or even 1,800 m in more sheltered sites. These are dominated by one or two species of *Eucalyptus*, usually the snow gums *E. pauciflora*, *E. debeuzevillei* or *E. niphophila*, as well as black sally *E. stellulata*. A moderately diverse understory of sclerophyllous shrubs is usually present under the low (5–15 m), open woodland canopy; ground cover is provided by tussock grasses and herbs (Keith, 2004).

Above the treeline, woody plants are represented only by short (0.2–1 m), slow-growing shrubs that are able to withstand the freezing winds and seasonal burial under snow. Coral heath (*Epacris microphylla*), yellow kunzea (*Kunzea muelleri*) and mountain plum pine (*Podocarpus lawrencei*) are among the species that are commonly encountered. Interspersed among these low shrubs are patches of tussock grass, alpine meadows dominated by daisies and other herbs, and bogs and fens that sustain sedges, rushes and mosses such as *Sphagnum cristatum*, especially in sites where drainage is poor. Fjaeldmark habitats occur in small patches in the most exposed alpine areas, and are characterized by small, ground-hugging shrubs and herbs that huddle in protected sites amid field of boulders and small rocks (Keith, 2004). At least 70 of the 200 or so species of vascular plants in Australia's alpine regions are locally endemic, but endemism at the generic level is lacking. All genera occur outside Australia, in the southern or northern hemisphere (Costin et al., 1979), reflecting the relatively recent development of the alpine environment since the Miocene (Kershaw et al., 2017). Endemic fauna of this biome include the mountain pygmy-possum (*Burramys parvus*), southern corroboree frog (*Pseudophryne corroboree*) and numerous species of invertebrates, including color-changing grasshoppers in the genus *Kosciuscola*.

12.2.8 Tundra

This last terrestrial biome is not represented in Australia, but rather in the vast area of Antarctica claimed as Australian Antarctic Territory and on sub-Antarctic islands such as Macquarie Island (125 km²) and the more remote Heard Island (368 km²) and the McDonald Island group (2.5 km²). The high latitudes of the tundra environments mean that temperatures are low year round, although the moderating effect of the ocean elevates temperatures above freezing for much of the year on the sub-Antarctic islands. Rainfall on Macquarie Island averages 986 mm annually and falls evenly throughout the year, although snow is common in winter (http://www.bom.gov.au/climate/averages/maps.shtml). Tundra vegetation is limited. Lichens and mosses occur in sheltered sites in Antarctica, but grasses, sedges, herbs, heaths and dwarf shrubs also occur patchily or form carpets on the sub-Antarctic islands. Trees are absent. Heard Island has 34-recorded species of lichen, 44 moss species, 12 liverworts and 12 species of vascular plant (http://heardisland.antarctica.gov.au/). By contrast, Macquarie Island supports at least 150 species of lichen, 81 species of moss, over 50 liverworts and 46 species of vascular plant (with three endemic); 127 species of algae are known, as are more than 200 species of fungus (Parks and Wildlife Service, 2006). About 10% of the ~300 invertebrate species on Macquarie Island are thought to be endemic and, apart from two now-extinct birds, there are no endemic terrestrial vertebrates. Nonetheless, the Antarctic tundra biome provides major breeding grounds for several species of seals and seabirds. For example, penguins are the most numerous of the 29 or so bird species that breed on Macquarie Island, with the endemic royal penguin (*Eudyptes schlegeli*) achieving a population of 850,000 breeding pairs. The outstanding natural values of Macquarie Island, Heard Island and the McDonald Islands led to their inscription on the World Heritage list in 1997 (Parks and Wildlife Service, 2006).

12.2.9 Freshwater Systems

Based on the distribution and composition of freshwater fish species, the Australian landmass has been split into 10 different regions, or ecoregions, out of 426 ecoregions identified worldwide (Abell et al., 2008; Figure 12.2). These ecoregions show some parallels with the country's terrestrial biomes, with the two largest—plus the Pilbara ecoregion—overlapping with the deserts and xeric shrublands and the Murray-Darling Basin coinciding in large part with the eastern section of the Mediterranean forests, woodlands and shrubs biome and the temperate grasslands, savannas and shrublands biome (cf. Figures 12.1 and 12.2). In broad terms, the ecoregional mapping (Abell et al., 2008) shows that Australia's western deserts sustain the smallest numbers of freshwater fish and fewest endemics, whereas the Kimberley, Top End and east coast support both the greatest numbers of fish species (67–101) and most endemics (20–27 species). Intriguingly, however, despite their modest fish diversity, most species (51–71%) in southwestern Western Australia and the immense Lake Eyre

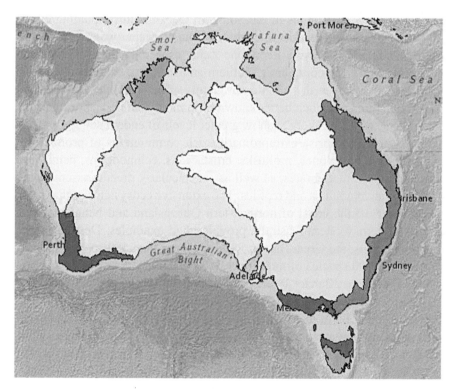

Figure 12.2 Freshwater ecoregions in Australia. Ecoregions are represented by different shading. Source: Freshwater Ecosystems of the World (http://www.feow.org./) and Abell et al. (2008).

Basin are endemic to those regions (Abell et al., 2008). Such high endemism reflects the isolation of freshwater bodies in both regions by the arid lands that surround them (Unmack, 2001). It is not clear how the distributions and compositions of other freshwater organisms align with the ecoregions defined by Abell et al. (2008) but, for species that are as confined for their entire life-cycle to freshwater as fish, substantial levels of concordance might be expected.

12.2.10 Marine Systems

A useful categorization of marine systems was proposed by Spalding et al. (2007), based primarily on the taxonomic configuration of benthic and shelf biota to depths of 200 m. Using this scheme, 10 ecoregions can be distinguished within five provinces in the waters surrounding northern Australia and its island territories in the Indian and Pacific Oceans. Another 10 ecoregions can be discerned in four provinces around southern continental waters, and three further ecoregions in the Southern Ocean realm (Spalding et al., 2007). Use of the 200 m isobath excludes biota of deeper waters that occur within Australia's Exclusive Economic Zone, especially those in the deep ocean troughs off the continent's southern and southeastern coast (Watford et al., 2005). However, this should not greatly affect estimates of biodiversity: marine biota are much better known in shallow than in deep waters, are more diverse and also likely to show greater levels of endemism. Australia's marine systems comprise extraordinarily rich communities of protozoans, algae, sponges, anemones, mollusks, crustaceans, echinoderms, hemichordates and other invertebrates, as well as spectacularly diverse assemblages of fish, marine mammals and reptiles. The extensive ecosystems of the Great Barrier Reef off the coast of northeastern Queensland and Ningaloo Reef in northwestern Western Australia provide good examples. Despite recent bleaching events, the Great Barrier Reef has more than 450 species of hard coral, over 3,000 species of mollusc, 1,300 crustaceans and ~1,625 species of bony fish (Great Barrier Reef Marine Park Authority, 2013). Deepwater (18–144 m) surveys at Ningaloo have recorded 155 species of sponge, 227 species of echinoderm and 236 molluscan species, and many more species in these groups in subsurface waters (http://www.ningaloo.org.au/). Other well-known and iconic species that ply the nearshore waters around Australia's coast include whale sharks (*Rhincodon typus*), manta rays (*Manta* spp.), dugong (*Dugong dugon*) and 58 species of dolphins, porpoises, baleen and toothed whales (Burbidge et al., 2014).

12.3 Organisms and Taxonomic Groups

As one of the world's 17 megadiverse counties (Mittermeier et al., 1997), Australia is highly biodiverse with large numbers of endemic species (to qualify formally as megadiverse, a country must have at least 5,000 endemic plant species and have marine ecosystems within its borders). Yet, estimates of the numbers of species vary widely. This is not surprising. On the one hand, many parts of Australia have been poorly surveyed, and even the better-known sites have usually been surveyed with particular target groups in mind. Relatively little focus has been placed on bacteria, which globally comprise the majority of living species (Larsen et al., 2017), or on protists, archaeans or fungi. Although recent advances in the collection and analysis of environmental DNA (eDNA) should speed the identification of new taxa via their genetic signatures (e.g., Barnes and Turner, 2016), this has been little used to date in Australia, and support for taxonomy generally is low and declining (Weaver, 2017). On the other hand, even in groups that are reasonably well known, recent research has begun to uncover high and previously unsuspected levels of cryptic biodiversity. For example, 1,050 species of frogs and reptiles were known in 2000, but by 2012 this number had increased to 1,218 (Cogger, 2014); many of the new species had been 'hidden' within taxa that had been thought to comprise just a single species. Similarly, the number of species within *Antechinus*—one of the most intensively studied genera of marsupials—increased from 10 to 15 between 2012 and 2016, again from the exposure of cryptic species (Baker and Dickman, 2018).

The following account draws heavily on data compiled by Chapman (2009), who estimated the number of species in Australia to be about 570,000, with 147,579 of these being formally described. Although updates on numbers of species are given where available, the caveats noted above suggest that estimates for some taxa will be based on best guesses rather than on completely reliable information. Underestimates are likely to be most acute for bacteria. Thus, whereas Chapman (2009) accepted a figure for the global number of all species to be about 11 million, Larsen et al. (2017) argued that the true figure is more likely to be 1–6 billion owing to the dominance of bacteria.

In the sections that follow, comments are made on each of the major taxonomic groups. Tables also are provided that summarize the numbers of species in each taxonomic group that are described and estimated to occur in Australia, as well as—where known—estimates of the percentage endemics in Australia and the percentage contribution that Australian species make to the global biota in each group. I acknowledge a great debt in presenting this information to the hundreds of experts whose work is so briefly summarized here, as well as to Chapman (2009) whose collation of a vast amount of scattered information has made the present overview possible. Viruses are the only group not included in the tabular summaries. Not only is there an ongoing debate about their status as living organisms, virus 'species' are notoriously difficult to specify. Chapman (2009) estimated that Australia might contain 200–400 virus species, with up to half of these being endemic, but cautioned that these figures should not be relied upon.

12.3.1   *Prokaryota, Protoctista, Chromista, and Others*

Species in this group are defined primarily by what they are not: plants, fungi or animals. The group contains single-celled autotrophs and heterotrophs that live in terrestrial, freshwater and marine environments, either as free-living organisms or as organisms that are associated with plants, fungi, animals or other single-celled organisms as endo- or exobionts. The group also contains much larger organisms such as kelps and other seaweeds that have been variously regarded at times as algae or as members of other groups; they are here regarded as chromistans, outside the Kingdom Plantae, following the arrangement of Chapman (2009). Both the species-level taxonomy and higher-level classification of these organisms remain fluid, and the number of species they represent is unclear. Nonetheless, based on extrapolations of known and described species, it is likely that organisms in this 'rag-bag' represent a sizeable component of the alpha-level diversity in many Australian systems (Table 12.1). Protoctista (protozoans, foraminiferans, flagellates, some slime molds and other eukaryotes) probably number in the tens of thousands of species in Australia; at least 40,000 Prokaryota (bacteria) also have been estimated (Saunders et al., 1996), but more recent estimates based on extrapolations of the numbers of unique bacterial species in the guts of insects and other multicellular host species (Larsen et al., 2017) suggest that this figure is a gross underestimate. Further research will probably uncover millions more prokaryote species, with this one group likely to contribute more species than all other groups to both Australian and global biodiversity (Larsen et al., 2017). Levels of endemism in Prokaryota are unknown (Table 12.1), but if bacteria are uniquely associated with host groups that are themselves confined largely to Australia, then high levels of endemism are likely.

Table 12.1   Prokaryota, Protoctista, Chromista and Other Species (Non-Plants, Fungi or Animals) Known and Estimated From Australia in Different Taxonomic Groups, the Percentage Contribution to their Respective Groups in the Global Biota, and the Percentage of Australia's Species that is Endemic

Taxon	Described from Australia	Estimated from Australia	% of global species	% Australian endemic*
Prokaryota (bacteria)	~40	40,000	0.5	>50
Cyanophyta	270	~500	10	?
Chromista	2,130	>15,000	8.5	?
Protoctista	>1,346	~65,000	4.7	?
Total	**>3,786**	**>120,500**	**<10**	**?**

Tabular data extracted from Chapman (2009).

* Extrapolated from known species.

12.3.2 Plants

Australian flora is relatively well known, at least in terms of alpha taxonomy, with ~24,716 described and accepted, and another 2,129 or so species estimated (Table 12.2). Among the major groups, Bryophyta is the least speciose, with mosses and liverworts comprising the great majority of the known and estimated flora. About a quarter of bryophytes are endemic, with many more species having ancient Gondwanan links or more recent links with Southeast Asia and the Pacific region. Some species, such as the moss *Thuidium cymbifolium*, are widespread throughout the Indo-Pacific; others, such as the leafy liverwort *Ptilidium ciliare*, occur from New Zealand to southern South America, or even in the ice-free parts of Antarctica (Schofield, 1992; Ramsay, 2006). In general, members of the Bryophyta achieve greater prominence in mesic areas where moisture is reliably available year round, especially in eastern, southwestern and northern Australia where they may grow on varied soil substrates, on logs and on the trunks of trees (e.g., Ramsay and Cairns, 2004). However, a few species, such as the moss (*Pottia latzii*) occur in sheltered sites in the deserts and xeric shrublands biome, or have distributions that encompass most or all biomes throughout the continent and on many of Australia's offshore islands (https://www.anbg.gov.au/bryophyte/bryogeography-a-intro.html).

Confined largely to freshwater and marine systems, but occurring also in moist soil, hypersaline environments and even in deeply frozen snow banks, algae form the second most speciose of Australia's major plant groups (Table

Table 12.2 Plant Species Known and Estimated From Australia in Different Taxonomic Groups, the Percentage Contribution to their Respective Groups in the Global Flora, and the Percentage of Australia's Species that is Endemic

Taxon	Described from Australia	Estimated from Australia	% of global species	% Australian endemic*
Algae	~3,236	~3,545	~29	?
Charophyta	(1040)	(1099)	(51.7)	?
Chlorophyta	(654)	(904)	(22.3)	?
Glaucophyta	(1)	(1)	(20.0)	?
Rhodophyta	(1541)	(1541)	(25.3)	?
Bryophyta	1,847	~2,200	11.4	25
Mosses	(976)	(>976)	(8.9)	(22.7)
Liverworts	(841)	(>841)	(16.8)	(23–28)
Hornworts	(30)	(>30)	(12.7)	(23–28)
Vascular plants	19,324	~21,645	6.9	91.8
Ferns and allies	(498)	(~525)	(4.2)	(33.8)
Gymnosperms	(120)	(~120)	(11.7)	(96)
Magnoliophyta	(18,706)	(~21,000)	(7.0)	(93.3)
Total	~24,716	26,845	7.9	~86

Tabular data extracted from Chapman (2009).

* Extrapolated from known species.

Bracketed figures indicate estimates for major groups within Algae, Bryophyta and vascular plants.

12.2). Recent surveys and taxonomic research suggest that many more species of algae may occur than the 3,500 or so that have been estimated (Table 12.2), especially in remote areas that have been previously under-sampled such as around Australia's northwestern coast (Huisman, 2015). Australian algae are estimated to comprise almost 30% of algae globally; this figure can be expected to rise when new species and genera are formally described. Among the currently described forms, Australia's largest group of algae is the Rhodophyta, or red algae. These are mostly marine and comprise both familiar seaweeds and coralline forms that play key roles in coral reef formation. Charophyta—freshwater green algae—comprise another large group within the Australian flora. These are widespread in natural freshwaters of the inland such as wetlands and riverine systems, but can be found also in such ubiquitous and disturbed sites as farm dams (Casanova and Brock, 1999), brackish, saline and even hyper-saline aquatic habitats (Casanova et al., 2011). These algae form important habitats for many aquatic invertebrates and food for invertebrates, fish, birds and other semiaquatic vertebrates. Species within Chlorophyta range from one-celled organisms to complex multi-celled plants, and are mostly marine. There are, however, green algae within Chlorophyta that occur in saline and hyper-saline situations on land, as well as representatives that live on trees and rocky substrates. One genus, *Halimeda*, is well known as a prominent component of reefs and lagoons in tropical and subtropical waters, where they are major contributors to reef formation (Cremen et al., 2016). Only one glaucophyte has been described from Australia, but more of these rare, unicellular, freshwater algae are likely to be found when further survey work has been undertaken (Guiry and Guiry, 2017).

Vascular plants dominate the Australian landscape, both in terms of species richness and their ubiquity throughout the terrestrial environment (Table 12.2). Magnoliophyta—flowering plants—are particularly prominent. The magnoliophyte family Myrtaceae, for example, contains some 1,850 species in about 70 genera in Australia, and is represented over vast areas of the continent by iconic species of *Eucalyptus*, *Corymbia*, *Melaleuca*, *Angophora* and *Kunzea*. Equally widely distributed are species within Mimosaceae, especially the acacias, or wattles. About 1,000 species of *Acacia* are known and described. They occur in rainforest, desert, alpine and forest habitats, sometimes forming the dominant tree over immense areas (e.g., mulga, *Acacia aneura*; brigalow, *A. harpophylla*) or as minor to major components of eucalypt-dominated woodlands and forests (e.g., silver wattle, *A. dealbata*). Other speciose families in Australia include the peas (Fabaceae), with about 1,500 species in 136 genera, orchids (Orchidaceae) with over 1,700 species, grasses (Poaceae, 1,300 species), and daisies, everlastings and relatives (Asteraceae, ~1,250 species) (https://environment.gov.au/science/abrs). Less speciose families, such as Proteaceae, also are often very well known because they contain emblematic taxa or species with showy flowers such as banksias, waratahs, needlewoods and grevilleas. In the southwest of Western Australia, members of all these families, and many more, co-occur north of Perth, in the Stirling Range and inland, forming a global biodiversity 'hotspot' (Myers et al., 2000). Other regions with rich communities of flowering plants include the Sydney Basin and many sites within the tropical and subtropical moist broadleaf forests biome (Keith, 2004).

Less speciose, but sharing similarly high levels of endemism as the magnoliophytes, are the gymnosperms (Table 12.2). Including cypress-pines, kauri pines, hoop pines, cycads and the ancient Wollemi pine (*Wollemia nobilis*), gymnosperms are most prominent on the slopes and in the montane regions of the Great Dividing Range, and also in Tasmania. Several species do, nonetheless, occur in the deserts and xeric shrublands biome, across the tropical north, or in the southwest of Western Australia. Gymnosperms usually occur in mixed communities with magnoliophytes, sometimes as scattered individuals but sometimes also as the dominant members of the community. In north-central New South Wales, for example, white cypress pine (*Callitris columellaris*) covers much of the 3,000 km² area of the Pilliga forest region, and is a common tree throughout the inland (Moore, 2005). Members of the genus *Araucaria* share a broadly southern distribution, from South America to New Caledonia and New Guinea, and may exceed 70 m in height (Silba, 1986). One species, the Norfolk Island pine (*A. heterophylla*), is endemic to Norfolk Island, but now also a popular cultivated tree in many parts of the world owing to its distinctive appearance and growth form. The remaining group of vascular plants in Australia, the ferns and fern allies, comprise over 500 species, about a third of which are endemic (Table 12.2). The fern allies are represented by at least 50 native species of lycophytes, psilophytes and sphenophytes, with the remainder being 'true ferns'. True ferns are found in most parts of the continent, but are most numerous in temperate, tropical and subtropical areas where rainfall is high and reliable. Among the best-known species are the giant tree ferns (*Cyathea* and *Dicksonia* spp.) and, in many inland areas, the 'water-clovers' (*Marsilea* spp.); sporocarps of one species—nardoo (*M. drummondii*)—have long been used as a food source by Aboriginal people (Allen, 2011).

12.3.3 Fungi

Research on Australia's fungi remains very much a work in progress, with the result that many species remain undescribed and the level of endemism is uncertain (Table 12.3). Nonetheless, Australia's described species account for about 12% of the world's estimated total number of fungi, and members of all the major fungal groups are represented (Table 12.3). Fungi are widespread and diverse in all environments and biomes throughout the continent (https://www.fungimap.org.au/), occurring in soil, on trees as well as in other organisms, although species richness is greatest in areas where moisture is reliable. Some species occur also in freshwater and marine systems (Hyde et al., 1998). Most of the species described so far from Australia are in the most diverse phyla, Ascomycota and Basidiomycota. The totals shown in Table 12.3 for these groups include 3,488 ascomycete species that are lichenized (i.e., in mutualistic relationships with algae or other organisms)

Table 12.3 Fungi Species Known and Estimated From Australia in Different Taxonomic Groups, the Percentage Contribution to their Respective Groups in the Global Fungal Biota, and the Percentage of Australia's Species that is Endemic

Taxon	Described from Australia	Estimated from Australia	% of global species	% Australian endemic*
Ascomycota	7,187		11.2	?
Basidiomycota	3,730	10,000	11.8	?
Blastocladiomycota	9		5.0	?
Chytridiomycota	15		2.1	?
Glomeromycota	28		16.6	?
Microsporidia	130			?
Neocallimastigomycota	2		10.0	?
Zygomycota	119		11.2	?
Unplaced to phylum	626			
Total	**11,846**	**50,000**	**11.9**	

Tabular data extracted from Chapman (2009).

* Estimates are not available for the phyla separately, but Williams (2001) estimated that about 90% of the species of fungi overall are endemic.

and seven basidiomycotes. P. McCarthy, cited in Chapman (2009), suggested that ~4,500 lichen-forming fungi may occur in Australia, with 34% of these being endemic. The Basidiomycota includes some of the most well-known fungi, including mushrooms, stinkhorns, jelly fungi, rusts and smuts. The Ascomycota includes penicillin-fungi (*Penicillium* spp.), some yeasts and species that are parasitic or pathogenic in plants and animals (e.g., *Candida* spp.), as well as species that produce large subterranean fruiting bodies, or truffles. In southwestern Australia, several species of potoroid marsupials specialize in eating truffles, and have dental and digestive adaptations suggesting that they have coevolved over long periods with rich communities of these fungi (Dickman, 2015). Truffles, other Ascomycota, some Glomeromycota and some Basidiomycota enter into associations with the roots of plants. These associations—termed mycorrhizal—are of critical importance for maintaining vascular plants because the hyphal filaments of these fungi greatly enhance the ability of plants to gather water and scarce mineral nutrients (Wang and Qiu, 2006). In the Australian region, where water and soil nutrients are often in short supply, the importance of mycorrhizal fungi cannot be overestimated. Members of the remaining fungal groups are mostly parasitic, such as the chytrid fungi in the phylum Blastocladiomycota; these have decimated frog communities in Australia and in other parts of the world.

12.3.4 Invertebrates

Nearly 100,000 invertebrates have been described and named from Australia, almost a third of the overall numbers that are estimated to occur (Table 12.4). As in most other parts of the world, one group is overwhelmingly dominant, the Class Insecta. Some 63 – 64% of Australia's known and estimated invertebrates are insects (Table 12.4) and, of these, almost half are Coleoptera (Yeates et al., 2003). Hymenoptera, Diptera and Lepidoptera are each expected also to number in the tens of thousands, with other insects from 23 further Orders making up the remainder (Austin et al., 2004; Raven and Yeates, 2007). The only Orders not known to occur in Australia are Grylloblattodea, a small group of extremophile insects that is restricted to icy montane environments, and Mantophasmatodea, a small group of carnivorous insects endemic to Africa (Cranston, 2010). A single species of zorapteran is known, and that is localized on Christmas Island (Chapman, 2009). Collectively, insects occur in all biomes, terrestrial and aquatic, throughout Australia, with up to 70% estimated to be endemic (Ridsdill-Smith, 2004). Some groups tend to predominate more in certain habitats than others, such as ants in arid habitats and caddis flies, dragonflies and other groups with an aquatic life history stage near freshwater. Surveys in forested habitats suggest that insect diversity may be especially high in the canopy (e.g., Kitching et al., 1993; Houston et al., 1999). Despite such broad patterns in occurrence, there is no doubt that insects dominate all other animal groups in almost all situations in Australia in terms of numbers of individuals and numbers of species, nor that their roles as predators, prey, decomposers, pollinators, seed and spore vectors are of fundamental importance for the functioning of ecological systems and human enterprises (Cranston, 2010).

Other invertebrates in the Australian region can be grouped into 16 further broad categories (Table 12.4). For convenience here, and following Chapman (2009), these include a mix of Classes, sub-Phyla and Phyla, as well as a 'rag-tag' group of Phyla—labeled 'others' in Table 12.4—with often uncertain taxonomic status but generally small numbers of species (e.g., arrow worms, or Chaetognatha, Mesozoa, Priapulida; Zhang, 2011a, b). Arachnida, Crustacea and Mollusca stand out as having the most numbers of described species in Australia, but large numbers are estimated in the taxonomically vexed Nematoda and Platyhelminthes (Table 12.4). Levels of endemism, where known, are varied; perhaps not surprisingly, the groups

Table 12.4 Invertebrate Species Known and Estimated From Australia in Different Taxonomic Groups, the Percentage Contribution to their Respective Groups in the Global Fauna, and the Percentage of Australia's Species that is Endemic

Taxon	Described from Australia	Estimated from Australia	% of global species	% Australian endemic*
Hemichordata	17	22	15.7	~25
Echinodermata	1,475	~2,000	21.1	31
Insecta	~62,000	~205,000	6.2	~70
Arachnida	6,615	31,338	6.5	?
Pycnogonida	215	?	16.0	~50
Myriapoda	553	~3,100	3.4	86
Crustacea	7,266	~9,500	15.5	?
Onychophora	71	~80	43.0	100
Hexapoda	338	~2,070	3.7	~17.6
Mollusca	~8,700	~12,250	10.2	38
Annelida	2,192	~4,230	13.1	67
Nematoda	~2,060	~30,000	8.2	?
Acanthocephala	56	~160	4.9	?
Platyhelminthes	1,593	~10,000	8.0	?
Cnidaria	1,705	~2,200	17.4	?
Porifera	1,476	~3,500	24.6	56
Others	~2,371	~5,015	18.7	?
Total	**~98,703**	**~320,465**	**7.3**	**?**

Tabular data extracted from Chapman (2009).

* Extrapolated from known species.

that contain marine representatives with powers of oceanic dispersal (e.g., Hemichordata, Echinodermata, Pycnogonida, Mollusca, Porifera) tend to have relatively low levels of endemism, whereas strictly terrestrial groups (e.g., Onychophora, Myriapoda) have more endemic representatives. Australian onychophorans, or velvet worms, appear to be wholly endemic and also represent over 40% of the world's known species (Table 12.4). Even without insects, the extraordinary diversity and sheer numbers of individuals and species represented by all the other invertebrates ensure that members of one or more groups will be found in all Australian aquatic and terrestrial environments. Surveys of Sydney Harbor alone, for example, uncovered 2,437 species within just the Echinodermata, Crustacea, Mollusca and Polychaeta (Hutchings et al., 2013). The authors emphasized that many more invertebrates in other taxa remained to be listed, and that many species even in the better-known groups awaited formal description and study.

12.3.5 Chordates

Chordates are relatively less speciose than the other major groups considered so far, but they are generally better known and contain some of the most familiar species that are associated with Australia, notably the marsupials. Almost 90% of the species that are estimated to occur across the eight major chordate groups in Australia have been described, and about 42% of these are endemic (Table 12.5).

Fishes comprise the largest chordate group in Australia, with 5,000 described species and another 750 or so species that are suspected but not yet described. Estimates provided by Hoese et al. (2006) suggest that most fish species in Australian waters are marine (85.4%) or estuarine (9.2%); the remaining species (5.4%) are confined to river systems, lakes, and even to small and highly localized artesian springs within desert regions (Kerezsy, 2011). Fish diversity varies locally. It is, not surprisingly, lowest in isolated inland freshwater bodies where just 1–2 species may occur, but can be as high as 70 or more species, as in the drainages of the Kimberley region (Morgan et al., 2014). Species such as the Murray cod (*Maccullochella peelii*) and yellowbelly (*Macquaria ambigua*) are well known for their contributions to the local inland economy. In marine systems, local and regional fish diversity can be very high. In Sydney Harbor, for example, almost 600 species of fish have been documented (Hutchings et al., 2013), and in the Great Barrier Reef system, some 1,625 species are known (Great Barrier Reef Marine Park Authority, 2013). Of the many iconic species of fish, the large game species, small but showy reef species, sharks and gigantic whale sharks are among the most well known, although the smaller species tend to be endemic. Other marine chordates include small numbers of marine Agnatha (hagfish and lampreys), poorly known Cephalochordata (lancelets) and some 850 species of Tunicata (sea squirts and relatives). At least half of the species in these latter three groups are thought to be Australian endemics (Table 12.5).

The remaining chordate groups have escaped complete dependency on water, but may nonetheless require freshwater for part of their life cycle (Amphibia) or fresh or marine waters for foraging (species in all groups). Amphibia in Australia are represented by five families of frogs (and the introduced cane toad, *Rhinella marinus*, in the toad family Bufonidae), the

Table 12.5 Chordate Species Known and Estimated from Australia in Different Taxonomic Groups, the Percentage Contribution to Their Respective Groups in the Global Fauna, and the Percentage of Australia's Species that is Endemic

Taxon	Described from Australia	Estimated from Australia	% of global species	% Australian endemic*
Mammals	397	~405	7.2	88
Birds	828	~900	8.3	45
Reptiles	985	~1050	11.3	94
Amphibia	237	~250	3.6	94
Fishes	~5,000	~5,750	16.0	24
Agnatha	5	~10	4.3	60
Cephalochordata	8	~8	24.2	50
Tunicata	757	~850	27.4	50
Total	**~8,217**	**~9,223**	**12.8**	**42**

Tabular data extracted mostly from Chapman (2009), with updates from Burbidge et al. (2014) and Cogger (2014).

* Extrapolated from known species.

great majority of which are endemic (Table 12.5). Frogs achieve their greatest diversity in higher rainfall areas in the Great Dividing Range, the southwest of Western Australia and across the Top End and Kimberley. Some species, such as the rocket frog (*Litoria nasuta*), have distributions that stretch for thousands of kilometers in coastal and subcoastal areas, whereas others, such as the 18 species of nursery frog (*Cophixalus* spp.), are confined to usually tiny ranges in montane rainforests in northeastern Queensland (Cogger, 2014). Other species, such as the green tree frog (*Litoria caerulea*), have ranges that extend from the eastern seaboard well into the deserts and xeric shrublands biome. It does this by using small and sometimes ephemeral bodies of water, even artificial sources such as water troughs, toilets and showers, dispersing into the broader landscape during periods of rain. Yet other frogs have the major parts of their ranges in arid regions. Species such as the water-holding frog (*Cyclorana platycephala*) and desert spadefoot frog (*Notaden nichollsi*) may spend years underground, emerging to feed and reproduce in large numbers when heavy rains break their 'dormancy' (Hillman et al., 2009).

Australia is sometimes known as the 'land of lizards' (Morton and Emmott, 2014), a title that reflects the many species that are estimated to occur nation-wide and the large numbers of individuals and species that often co-occur. But reptilian diversity encompasses not just seven families of lizards, but also six families of snakes, four families of turtles and one family of crocodiles; in total, over 1,000 species are estimated to occur (Table 12.5). Most species are endemic. Local reptilian diversity can attain levels of 50–60 species per km^2 in some desert areas (Dickman et al., 2014), with species ranging from tiny dwarf skinks (*Menetia greyii*) that weigh no more than 1 g, to perenties (*Varanus giganteus*) that are more than three orders of magnitude more massive. The distributions of individual species vary greatly. Large sea-turtles in the Family Cheloniidae occur in the waters of northern Australia, extending south of the Tropic of Capricorn on both sides of the continent. By contrast, several species of freshwater turtles are confined to a single river catchment or, in the case of the western swamp turtle (*Pseudemydura umbrina*), to one or two small swamps (Cogger, 2014). On land, contrasting species distributions can be found in many families. For example, while several species of skinks have almost pan-continental distributions (e.g., *M. greyii*; leopard skink *Ctenotus pantherinus*), the Pedra Branca cool-skink occupies a single, wave-swept rock off the southern coast of Tasmania, and the eastern Torres rainbow-skink (*Carlia quinquecarinata*) two small islands in Torres Strait (Cogger, 2014). Reptiles occupy all aquatic and terrestrial environments in Australia and, with species such as the saltwater crocodile (*Crocodylus porosus*), desert taipans (*Oxyuranus* spp.) and thorny devil (*Moloch horridus*), provide some of the most fearsome and bizarre creatures that stalk lands anywhere.

Almost as speciose as the reptiles, Australia's birds are arranged in some 90 families that represent over 8% of the world's avifauna (Table 12.5). Because of their ability to traverse bodies of water, proportionally fewer birds are endemic than the other, primarily terrestrial, chordate groups, but the 45% level of endemism for Australian birds is nonetheless high by comparison with most countries. It is likely that most or all endemic species have been discovered and described; estimates of further species (Table 12.5) include vagrants that wash up occasionally on Australia's mainland and island territories (Christidis and Boles, 2008; http://www.birdlife.org.au/conservation/science/taxonomy). Birds are exceedingly numerous and ubiquitous, living and breeding in Australia's Antarctic Territory and on its sub-Antarctic islands, on the tiniest rocks to the largest offshore islands, and in all terrestrial and freshwater biomes on the continent (Barrett et al., 2003). Species richness is highest in structurally complex forest habitats where sufficient food, foraging site, shelter and breeding resources are available to be partitioned, but large fluxes of mobile species occur also in inland areas in response to shifts in climatic conditions (Runge and Tulloch, 2017). Many species occupy, or have the potential to occupy, large geographical ranges, but some taxa are much more restricted. For example, 13 species are endemic to the Wet Tropics rainforests of northeastern Queensland (Crome and Nix, 1991), the noisy scrub-bird (*Atrichornis clamosus*) to a small area of coastal thicket in southwestern Western Australia, and several species of grasswren (*Amytornis* spp.) to small areas in central and northern Australia (Barrett et al., 2003). The Lord Howe woodhen (*Gallirallus sylvestris*) is confined to that small island. Birds are arguably the most conspicuous and frequently seen chordates in Australia, with such iconic species as emus (*Dromaius novaehollandiae*) observable in all rural areas and an extraordinary diversity of parrots, cockatoos, honeyeaters, kingfishers, finches, magpies, owls and more that occur in urban and nonurban habitats everywhere.

Australia's marsupial mammals are perhaps the best-known component of the country's fauna, but the oft-depicted kangaroos and koala (*Phascolarctos cinereus*) represent only a fraction of the overall marsupial fauna and of the mammals in their entirety. Some 397-mammal species are known, with a few more that are expected to be uncovered in ongoing genetic research (Table 12.5). The mammals provide an excellent example of the diverse origins of the Australian biota more broadly. Thus, ancestors of the country's two living egg-laying mammals—the monotremes (short-beaked echidna, *Tachyglossus aculeatus*; platypus, *Ornithorhynchus anatinus*)—were present on the Australian landmass during the early Cretaceous at least 110 million years ago (Long et al., 2002). Fifty million years later, ancestral marsupials arrived via the Gondwanan migration route through Antarctica from South America, with early bats entering the continent either via the land bridge or across stepping stone islands in the oceans between Australia and Asia (Long et al., 2002; Hand, 2006). The first whales appeared in Australian waters about 28 million years ago after the continent had separated from Antarctica (Fordyce, 2006). The first wave of rodents invaded Australia by 'island-hopping' from Southeast Asia perhaps 5 million years ago, and another, containing ancestral *Rattus*, half a million to a million years ago (Aplin, 2006). All other mammals have been brought in more recently by human activity.

Marsupials now dominate the Australian mammal fauna in terms of number of species (~170) and in their occupation of all terrestrial environments. Carnivorous and insectivorous dasyurid marsupials dominate in the deserts and xeric shrublands biome, whereas possums, gliders, pygmy-possums, kangaroos and rat-kangaroos tend to be more highly represented in all biomes with woodland or forest cover (Dickman, 2015). Bats (~80 species) also are more highly represented in forested areas, with a tendency for more species to occur in tropical regions across northern Australia than at cooler southern latitudes (Hand, 2006). Rodents (~60 species) occur ubiquitously, but with more species co-occurring in central arid and northern tropical regions than in temperate forest biomes (Aplin, 2006). The dugong, dolphins, porpoise, baleen and toothed whales (59 species) occur in all marine waters within Australia's jurisdiction, with the larger species (e.g., humpback whale, *Megaptera novaeangliae*; blue whale, *Balaenoptera musculus*) making extensive migrations between Antarctica and the nearshore and deeper waters of the Australian continent (Fordyce, 2006). Apart from one species of shrew (Christmas Island shrew, *Crocidura attenuata trichura*), which may have become recently extinct, Australia's other mammals—domestic animals, pet cats and dogs, livestock, game animals, commensal rodents and others—have been introduced by people. One of these mammals, the dingo (*Canis dingo*), was introduced at least 5,000 years ago and is regarded by many as an endemic, native species (Crowther et al., 2014).

12.4 Genetic Diversity

With representatives from all the major branches and most of the smaller branches of the tree of life, genetic diversity within Australia's biota is exceedingly high. Because of the country's generally high representation of endemic taxa (Tables 12.1–12.5), much diversity at the genetic level also is found only in Australia. This diversity provides unparalleled insights into the evolution of life on Earth (e.g., stromatolites) and into the origins of major taxa such as songbirds (Low, 2016), as well as glimpses of unique, ancient lineages that occur nowhere else (e.g., Wollemi pine, some stygofauna, monotremes). As might be expected, the biota of Australia's remote external territories is represented by many taxa that occur nowhere else, and on the continental landmass three regions are listed as global biodiversity hotspots because of their rich and highly endemic biotas: the southwest of Western Australia, the Wet Tropics, and the border ranges of northeastern New South Wales and southeastern Queensland (Cresswell and Murphy, 2017). More generally, using measures of phylogenetic diversity that incorporate species' evolutionary history and geographical rarity (Laity et al., 2015), eastern Australia stands out as having the highest levels of phylogenetic diversity for vascular plants and terrestrial vertebrates. The southwest of Western Australia, large areas near Darwin in the Top End and smaller areas in the Pilbara and central desert ranges are also very important for this key aspect of genetic diversity (Laity et al., 2015; Cresswell and Murphy, 2017).

Since European colonization, Australia's genetic diversity has been augmented by the introduction of large numbers of plants, animals, fungi and microbes from overseas. Some of these introductions have detrimentally affected the diversity and status of native species (see section 12.5, below), but others have been made with the deliberate intention of improving pastoral, agricultural and horticultural productivity. Introductions of new genetic stock and manipulations of existing genomes are made for many purposes, including to improve crop yield, speed crop growth and maturation, enhance the taste or flavor of the foods that are grown, increase resistance to pathogens, diseases and other pests, reduce the need for fertilizers, or expand the range of soil and climatic conditions under which the crops will productively grow (e.g., Heal et al., 2004; Cowling, 2007; Tester and Langridge, 2010; Tabashnik et al., 2013). Much research in cereal and fruit growing regions is now focusing on improving genetic resistance to abiotic stressors such as salinized soils and extreme climatic events (Mickelbart et al., 2015), as well as on conserving existing genetic resources and new resources (Mal et al., 2011). Similar efforts are being made to augment and improve genetic stock in aquaculture, viticulture, livestock and other industries that depend on natural resources.

12.5 Threats to Biodiversity

As in other parts of the world, biodiversity in Australia faces many threats. To a large extent, these threats have two common and related causes: human overpopulation, and the excessive demand for resources that people make on the natural world. Although overpopulation is not always considered to be as acute in Australia as in other parts of the world owing to the continent's relatively sparse population (3.3 people per km^2 compared with the world average of > 53 people per km^2; Dickman, 2014), three factors amplify the human impact. Firstly, *per capita* usage of resources is relatively high in Australia compared with most other countries, and resource consumption is exacerbated by the export of much food, gas, oil, coal, minerals and other products that sustain large human populations beyond Australian shores (O'Connor, 2014). Secondly, because the continent's soils are generally thin and nutrient-impoverished, attempts to farm them intensively in the style that was familiar to early European colonists have resulted in immense damage to landscapes over vast areas. Species, habitats and ecological communities have been degraded or destroyed as a consequence (Dickman, 2015). Thirdly, Australia has been more adversely affected by the impacts of invasive predators, competitors, weeds and other organisms than most other parts of the world, except for isolated island ecosystems (e.g., Salo et al., 2007). On average, over the last half-century, for every million people added to the Australian population, almost one (0.95) native vertebrate species has become extinct (Dickman, 2014). Since the arrival of the first European colonists in 1788, 30 species of native mammals alone have become extinct, the highest rate of extinction of mammals for any part of the world over the same period (Woinarski et al., 2014).

Threats affecting Australia's biodiversity have been listed recently (Cresswell and Murphy, 2017) under the major categories of 'pollution,' 'consumption and extraction of natural resources,' 'clearing and fragmentation of native ecosystems,' 'urban development,' 'pressures from livestock production,' pest species and pathogens,' altered fire regimes,' 'changed hydrology,' 'pressures facing aquatic ecosystems,' interactions among pressures,' and 'global climate change and climate variability'. Indifference about biodiversity, poor governance and continually declining funding for biodiversity conservation can be added to this list (Ritchie et al., 2013; Waldron et al., 2017). Of course, there are marked spatial and temporal differences in the intensity with which all these threats operate, making prioritization and conservation planning important components of biodiversity conservation strategies (Stow et al., 2014; Cresswell and Murphy, 2017). In 2016, 1,808 species were listed as threatened under Australian national conservation legislation (Cresswell and Murphy, 2017; and see below); recent advances in our ability to map and forecast changes to ecosystems indicate that many habitats, communities and ecosystems also will be added to lists of threatened entities (Keith et al., 2013).

12.6 National Laws and International Obligations

Every local, state and territory government has passed legislation or regulations that seek to conserve biodiversity, either through the identification and protection of threatened species, populations, habitats, or ecological communities, or the mitigation of key threats. At the state level, protected areas (e.g., National Parks, Nature Reserves, Conservation Reserves) can also be established. At the same time, much tension exists within state and local jurisdictions between conservation agencies and agencies whose mandate it is to develop and exploit natural resources or find new places for the expanding human population to occupy. At the national level, the key legislation relating to biodiversity is the *Environment Protection and Biodiversity Conservation (EPBC) Act 1999*. This allows for the listing of threatened species, threatened ecological communities and the key processes that afflict them. Although funding from the Australian government is limited, the listing of entities in the schedules of the *EPBC Act* requires proponents of any development or exploitative actions to consider their proposed actions and to mitigate them to reduce further impacts on biodiversity.

In addition to allowing for the listing of threatened species and communities, the *EPBC Act* provides protection for biodiversity under several other categories of national environmental significance. These are the identification, listing and management for conservation of world heritage properties, national heritage places, wetlands of international importance (as listed under the Ramsar Convention), migratory species protected under international agreements, Commonwealth marine areas, and the Great Barrier Reef Marine Park. The *EPBC Act* also seeks to regulate 'nuclear actions' (including, for example, any impacts of mining for ore, storage and safe disposal of nuclear wastes on biodiversity) and the use of water resources in relation to coal seam gas and large coal mining developments (http://www.environment.gov.au/epbc/what-is-protected). The provisions of the *EPBC Act* are variably successful. For example, generally strong protection is provided for Australia's 20 World Heritage-listed sites and 65 wetland sites of international importance, but government-supported proposals to allow fracking for coal seam gas and—especially—a huge coal mine in Queensland have downplayed the impacts of these developments on biodiversity and have been highly controversial (http://www.abc.net.au/news/2017-10-07/thousands-protest-adani-mine-across-australia/9026336). Other national legislation may bear upon biodiversity and its conservation. The *Water Act 2007*, for example, allowed the establishment of the Murray-Darling Basin Authority, a statutory body set up to sustainably manage the water resources and wetlands of the Murray and Darling Rivers and their catchment area.

At the global level, Australia is a signatory to several international agreements. Among these are the *Convention on Biological Diversity*, the *Strategic Plan for Biodiversity 2011–2020*, the *Paris Agreement on Climate Change*, and the international treaties noted above. As with many other nations that sign up to international environmental agreements, Australia's record in meeting its stated objectives under the agreements is patchy (Ulloa et al., 2018) and its allocation of appropriate levels of funding and resources inadequate (Waldron et al., 2017). Acceptance of the scientific evidence for climate change, and perhaps even of the importance of biodiversity, varies with political ideology in Australia (Fielding et al., 2012). This suggests that support for wise environmental stewardship will continue to wax and wane in line with the politics of the day.

12.7 State of the Environment

In 1996 the first independent national assessment of the state of Australia's environment was made, and every five years since an updated and official State of the Environment report has been made. The first report concluded that "some aspects of Australia's environment are in good condition by international standards. Unfortunately, the report also shows that Australia has some serious environmental problems" (State of the Environment Advisory Council, 1996, p 7). The State of the Environment report for 2016 (Cresswell and Murphy, 2017) indicates that little has changed. Among the key findings of the 2016 report are that the major pressures on biodiversity identified in previous reports have not decreased; that the number of threatened species and threatened ecological communities has increased since 2011; that reducing the impacts of feral predators and feral herbivores is an essential action for effective conservation; and that continuing population growth in urban and peri-urban areas is having increasing direct and indirect effects on surrounding natural ecosystems. Cresswell and Murphy (2017) also noted that it was not possible to assess the overall long-term effectiveness of management actions taken to limit the impacts of invasive species, and that the lack of data and information from long-term monitoring of biodiversity was universally acknowledged as a major impediment to biodiversity conservation. Unfortunately, in an act of breath-taking folly, the nation's Long Term Ecological Research Network—the only facility in the country tracking long-term changes in at least some components of biodiversity—was a victim of government funding cuts in 2017; it has been defunded, decommissioned and was set for final closure in June 2018.

On the brighter side, the State of the Environment report for 2016 highlights some improvements that herald a more optimistic outlook. Thus, the potential impacts of climate change on biodiversity are becoming better known; methods of 'future proofing' populations of native species by using *ex situ* management, translocation and other means are becoming used more widely; biodiversity monitoring projects involving 'citizen scientists' are contributing to our understanding of trends for selected species; and technological improvements are helping to uncover and understand Australia's species and genetic diversity (Cresswell and Murphy, 2017). Importantly too, an increasing area of Australia is coming under Indigenous management. This is enabling traditional practices to form the basis of new forms of contemporary, collaborative environmental and resource management, and helping to drive an increase in the overall area of the country within Australia's National Reserve System (NRS). Currently, some 10,500 areas are included within the NRS, covering over 1.5 million km² or 19.63% of the country (http://www.environment.gov.au/land/nrs). An intriguing trend over the last two decades has been the rise in privately owned land that is set aside for biodiversity conservation. Between them, for example, Bush Heritage Australia and the Australian Wildlife Conservancy have purchased some 55,500 km² of land that is now protected and managed to conserve native species and ecological processes. In addition, national nongovernment organizations such as WWF-Australia and the Australian Conservation Foundation, as well as local and regional alliances, are harnessing the efforts of citizens concerned about conserving the natural world. As stewards of a large country blessed with a spectacularly rich and highly endemic biota, Australians will need to continue to step up to ensure that this living heritage is sustained for generations to come.

Keywords

- genetic diversity
- threats

References

Abell, R., Thieme, M. L., Revenga, C., Bryer, M., Kottelat, M., Bogutskaya, N., et al., (2008). Freshwater ecoregions of the world: A new map of biogeographic units for freshwater biodiversity conservation. *BioScience, 58*, 403–414.

Allen, H., (2011). The space between: Aboriginal people, the Victorian exploring expedition and the relief parties. In: Joyce, E. B., & McCann, D. A., (eds.). *Burke & Wills: The Scientific Legacy of the Victorian Exploring Expedition*, CSIRO Publishing: Melbourne, Victoria, pp. 245–274.

Aplin, K. P., (2006). Ten million years of rodent evolution in Australasia: Phylogenetic evidence and a speculative historical biogeography. In: Merrick, J. R., Archer, M., Hickey, G. M., & Lee, M. S. Y., (eds.). *Evolution and Biogeography of Australasian Vertebrates*, Auscipub: Oatlands, New South Wales, pp. 707–744.

Aplin, K., & Ford, F., (2014). Murine rodents: Late but highly successful invaders. In: Prins, H. H. T., & Gordon, I. J., (eds.). *Invasion Biology and Ecological Theory: Insights From a Continent in Transformation*, Cambridge University Press: Cambridge, pp 196–240.

Arab, D. A., Namyatova, A., Evans, T. A., Cameron, S. L., Yeates, D. K., Ho, S. Y. W., & Lo, N., (2017). Parallel evolution of mound-building and grass-feeding in Australian nasute termites. *Biol. Lett.*, *13*, 20160665.

Archer, M., & Kirsch, J., (2006). The evolution and classification of marsupials. In: Armati, P. J., Dickman, C. R., & Hume, I. D., (eds.). *Marsupials*, Cambridge University Press: Cambridge, pp. 1–21.

Augee, M., & Fox, M., (2000). *Biology of Australia and New Zealand*, Benjamin Cummings, Sydney, New South Wales.

Austin, A. D., Yeates, D. K., Cassis, G., Fletcher, M. J., La Salle, J., Lawrence, J. F., et al., (2004). Insects 'Down Under' – Diversity, endemism and evolution of the Australian insect fauna: Examples from select orders. *Aust. J. Entomol.*, *43*, 216–234.

Baker, A., & Dickman, C., (2018). *The Secret Lives of Carnivorous Marsupials*, CSIRO Publishing, Melbourne, Victoria.

Barnes, M. A., & Turner, C. R., (2016). The ecology of environmental DNA and implications for conservation genetics. *Cons. Genetics*, *17*, 1–17.

Barrett, G., Silcocks, A., Barry, S., Cunningham, R., & Poulter, R., (2003). *The New Atlas of Australian Birds*, Royal Australasian Ornithologists Union, Melbourne, Victoria.

Burbidge, A. A., Eldridge, M. D. B., Groves, C., Harrison, P. L., Jackson, S. M., Reardon, T. B., et al., (2014). A list of native Australian mammal species and subspecies, *The Action Plan for Australian Mammals 2012*, CSIRO Publishing: Melbourne, Victoria, pp. 15–32.

Bureau of Meteorology, (2012). *Record-Breaking La Niña Events*, Australian Government Bureau of Meteorology, Canberra, Australian Capital Territory.

Byrne, M., Steane, D. A., Joseph, L., Yeates, D. K., Jordan, G. J., Crayn, D., et al., (2011). Decline of a biome: evolution, contraction, fragmentation, extinction and invasion of the Australian mesic zone biota. *J. Biogeog.*, *38*, 1635–1656.

Byrne, M., Yeates, D. K., Joseph, L., Kearney, M., Bowler, J., Williams, M. A. J., et al., (2008). Birth of a biome: Insights into the assembly and maintenance of the Australian arid zone biota. *Mol. Ecol.*, *17*, 4398–4417.

Casanova, M. T., & Brock, M. A., (1999). Charophyte occurrence, seed banks and establishment in farm dams in New South Wales. *Aust. J. Bot.*, *47*, 437–444.

Casanova, M. T., Bradbury, S., & Ough, K., (2011). Morphological variation in an Australian species of *Lamprothamnium* (Characeae, Charophyceae) in response to different salinities. *Charophytes*, *2*, 87–92.

Chapman, A. D., (2009). *Numbers of Living Species in Australia and the World, 2nd edn.* Department of the Environment, Water, Heritage and the Arts, Canberra, Australian Capital Territory.

Christidis, L., & Boles, W. E., (2008). *Systematics and Taxonomy of Australian Birds*, CSIRO Publishing, Melbourne, Victoria.

Cogger, H. G., (2014). *Reptiles and Amphibians of Australia, 7th edn.* CSIRO Publishing: Melbourne, Victoria.

Cogger, H. G., Dickman, C., Ford, H., Johnson, C., & Taylor, M. F. J., (2017). *Australian Animals Lost to Bulldozers in Queensland 2013–15*, WWF-Australia Technical Report, WWF-Australia, Sydney, New South Wales.

Costin, A. B., Gray, M., Totterdell, C. J., & Wimbush, D. J., (1979). *Kosciuscko Alpine Flora*, CSIRO Publishing, Melbourne, Victoria.

Cowling, W. A., (2004). Genetic diversity in Australian canola and implications for crop breeding for changing future environments. *Field Crops Res., 104*, 103–111.

Cranston, P. S., (2010). Insect biodiversity and conservation in Australasia. *Annu. Rev. Entomol., 55*, 55–75.

Cremen, M. C. M., Huisman, J. M., Marcelino, V. R., & Verbruggen, H., (2016). Taxonomic revision of *Halimeda* (Bryopsidales, Chlorophyta) in southwestern Australia. *Aust. Syst. Bot., 29*, 41–54.

Cresswell, I. D., & Murphy, H. T., (2017). *Australia State of the Environment 2016: Biodiversity*, Australian Government Department of the Environment and Energy, Canberra, Australian Capital Territory.

Crisp, M. D., & Cook, L. G., (2007). A congruent molecular signature of vicariance across multiple plant lineages. *Mol. Phylogen. Evol., 43*, 1106–1117.

Crome, F., & Nix, H., (1991). Birds. *Kowari, 1*, 55–68.

Crowther, M. S., Fillios, M., Colman, N., & Letnic, M., (2014). An updated description of the Australian dingo (*Canis dingo* Meyer, 1793). *J. Zool., 293*, 192–203.

Cunningham, I., (2005). *The Land of Flowers: An Australian Environment on the Brink*, Otford Press, Sydney, New South Wales.

Davis, J., Pavlova, A., Thompson, R., & Sunnucks, P., (2013). Evolutionary refugia and ecological refuges: Key concepts for conserving Australian arid zone freshwater biodiversity under climate change. *Glob. Change Biol., 19*, 1970–1984.

Department of the Environment and Energy, (2017). *Australia's Bioregions (IBRA)*, Department of the Environment and Energy, Canberra, Australian Capital Territory.

Dickman, C. R., (2010). Arriving in the Desert Channels country. In: Robin, L., Dickman, C., & Martin, M., (eds.). *Desert Channels: The Impulse to Conserve*, CSIRO Publishing, Melbourne, Victoria, pp. 1–23.

Dickman, C. R., (2014). Whither wildlife in an overpopulated world? In: Goldie, J., & Betts, K., (eds.). *Sustainable Futures: Linking Population, Resources and the Environment*, CSIRO Publishing, Melbourne, Victoria, pp. 13–23.

Dickman, C. R., (2015). *A Fragile Balance: The Extraordinary Story of Australian Marsupials, 2nd edn.* Australian Geographic, Sydney, New South Wales.

Dickman, C. R., Wardle, G. M., Foulkes, J., & de Preu, N., (2014). Desert complex environments. In: Lindenmayer, D., Burns, E., Thurgate, N., & Lowe, A., (eds.). *Biodiversity and Environmental Change: Monitoring, Challenges and Direction*, CSIRO Publishing, Melbourne, Victoria, pp. 379–438.

Fensham, R., (2010). Artesian springs. In: Robin, L., Dickman, C., & Martin, M., (eds.). *Desert Channels: The Impulse to Conserve*, CSIRO Publishing, Melbourne, Victoria, pp. 145–159.

Fielding, K. S., Head, B. W., Laffan, W., Western, M., & Hoegh-Guldberg, O., (2012). Australian politicians' beliefs about climate change: Political partisanship and political ideology. *Env. Politics, 21*, 712–733.

Fordyce, R. E., & Barnes, L. G., (1994). The evolutionary history of whales and dolphins. *Annu. Rev. Earth Planet. Sci., 22*, 419–455.

Fordyce, R. E., (2006). A southern perspective on Cetacean evolution and zoogeography, In: Merrick, J. R., Archer, M., Hickey, G. M., & Lee, M. S. Y., (eds.). *Evolution and Biogeography of Australasian Vertebrates*, Auscipub: Oatlands, New South Wales, pp. 755–778.

Global Biodiversity, Volume 4

Gibson, N., Prober, S., Meissner, R., & Van Leeuwen, S., (2017). Implications of high species turnover on the southwestern Australian sandplains. *PLoS ONE, 12*, e172977.

Great Barrier Reef Marine Park Authority, (2013). *Great Barrier Reef Biodiversity Conservation Strategy 2013*, Great Barrier Reef Marine Park Authority, Townsville, Queensland.

Grey, K., & Planavsky, N. J., (2009). *Microbialites of Lake Thetis, Cervantes, Western Australia – a Field Guide*, Geological Survey of Western Australia, Record 2009/11: Perth, Western Australia.

Griffin, W. L., Belousova, E. A., Shee, S. R., Pearson, N. J., & O'Reilly, S. Y., (2004). Archean crustal evolution in the northern Yilgarn Craton: *U–Pb* and *Hf*-isotope evidence from detrital zircons. *Precambrian Res., 131*, 231–282.

Guiry, M. D., & Guiry, G. M., (2017). *AlgaeBase: World-wide Electronic Publication.* http://www.algaebase.org.

Guzik, M. T., Austin, A. D., Cooper, S. J. B., Harvey, M. S., Humphreys, W. F., Bradford, T., et al., (2010). Is the Australian subterranean fauna uniquely diverse? *Invert. Syst., 24*, 407–418.

Hand, S. J., (2006). Bat beginnings and biogeography: The Australasian record. In: Merrick, J. R., Archer, M., Hickey, G. M., & Lee, M. S. Y., (eds.). *Evolution and Biogeography of Australasian Vertebrates*, Auscipub: Oatlands, New South Wales, pp. 673–705.

Harrison, S. E., Harvey, M. S., Cooper, S. J. B., Austin, A. D., & Rix, M. G., (2017). Across the Indian Ocean: A remarkable example of transoceanic dispersal in an austral mygalomorph spider. *PLoS ONE, 12*, e180139.

Heads, M., (2014). *Biogeography of Australasia: A Molecular Analysis*, Cambridge University Press, Cambridge.

Heal, G., Walker, B., Levin, S., Arrow, K., Dasgupta, P., Daily, G., et al., (2004). Genetic diversity and interdependent crop choices in agriculture. *Res. Energy. Econ., 26*, 175–184.

Heatwole, H., (1987). Major components and distributions of the terrestrial fauna. In: Dyne, G. R., & Walton, D. W., (eds.). *Fauna of Australia*, vol. 1A. *General Articles*, Australian Government Publishing Service, Canberra, Australian Capital Territory, pp. 101–135.

Heenan, P. B., & Smissen, R. D., (2013). Revised circumscription of *Nothofagus* and recognition of the segregate genera *Fuscospora, Lophozonia*, and *Trisyngyne* (Nothofagaceae). *Phytotaxa, 146*, 1–31.

Heinsohn, T., & Hope, G., (2006). The Torresian connections: Zoogeography of New Guinea. In: Merrick, J. R., Archer, M., Hickey, G. M., & Lee, M. S. Y., (eds.). *Evolution and Biogeography of Australasian Vertebrates*, Auscipub: Oatlands, New South Wales, pp. 71–93.

Hesse, P. P., Luly, J. G., & Magee, J. W., (2005). The beating heart: Environmental history of Australia's deserts. In: Smith, M., & Hesse, P., (eds.). *23°S: Archaeology and Environmental History of the Southern Deserts*, National Museum of Australia Press, Canberra, Australian Capital Territory, pp. 56–72.

Hickman, A. H., & Van Kranendonk, M. J., (2008). *Archean Crustal Evolution and Mineralization of the Northern Pilbara Craton – a Field Guide*, Geological Survey of Western Australia, Record, 2008/13: Perth, Western Australia.

Hillman, S. C., Withers, P. C., Drewes, R. C., & Hillyard, S. D., (2009). *Ecological and Environmental Physiology of Amphibians*, Oxford University Press, Oxford.

Hoese, D. F., Bray, D. J., Allen, G. R., Paxton, J. R., Wells, A., & Beesley, P. L., (2006). *Zoological Catalogue of Australia*, vol. *35*, *Parts 1–3*, CSIRO Publishing and Australian Biological Resources Study, Melbourne, Victoria, and Canberra, Australian Capital Territory.

Houston, W., Rayner, D., Melzer, A., Doyle, P., Coates, M., & Newby, B., (1999). Preliminary assessment of the biodiversity of arthropods of a central Queensland dry rainforest. In: Ponder, W., & Lunney, D., (eds.). *The Other 99%: The Conservation and Biodiversity of Invertebrates*, Royal Zoological Society of New South Wales: Sydney, New South Wales, pp. 107–110.

Huisman, J. M., (2015). *Algae of Australia: Marine Benthic Algae of Northwestern Australia, 1. Green and Brown Algae*, CSIRO Publishing, Melbourne, Victoria.

Humphreys, W. F., (2006). Aquifers: The ultimate groundwater-dependent ecosystems. *Aust. J. Bot.*, *54*, 115–132.

Hutchings, P. A., Ahyong, S. T., Ashcroft, M. B., McGrouther, M. A., & Reid, A. L., (2013). Sydney Harbour: Its diverse biodiversity. *Aust. Zool.*, *36*, 255–320.

Hyde, K. D., Gareth-Jones, E. B., Leaño, E., Ponting, A. D., & Vrijmoed, L. L. P., (1998). Role of fungi in marine ecosystems. *Biodiv. Cons.*, *7*, 1147–1161.

Jarvis, E. D., Mirarab, S., Aberer, A. J., Li, B., Houde, P., Li, C., et al., (2014). Whole-genome analyzes resolve early branches in the tree of life of modern birds. *Science*, *346*, 1320–1331.

Johnson, D., (2004). *The Geology of Australia*, Cambridge University Press, Cambridge.

Keith, D., (2004). *Ocean Shores to Desert Dunes: The Native Vegetation of New South Wales and the ACT*, Department of Environment and Conservation, Sydney, New South Wales.

Keith, D. A., Rodríguez, J. P., Rodríguez-Clark, K. M., Nicholson, E., Aapala, K., Alonso, A., et al., (2013). Scientific foundations for an IUCN Red List of ecosystems. *PLoS ONE*, *8*, e62111.

Kerezsy, A., (2011). *Desert Fishing Lessons: Adventures in Australia's Rivers*, University of Western Australia Press, Crawley, Western Australia.

Kershaw, A. P., Martin, H. A., & McEwen, M. J. R. C., (2017). The Neogene: A period of transition. In: Hill, R. S., (ed.). *History of the Australian Vegetation: Cretaceous to Recent*, University of Adelaide Press, Adelaide, South Australia, pp. 299–327.

Kitching, R. L., Bergelson, J. M., Lowman, M. D., McIntire, S., & Carruthers, G., (1993). The biodiversity of arthropods from Australian rainforest canopies: general introduction, methods, sites and ordinal results. *Aust. J. Ecol.*, *18*, 181–191.

Ladiges, P. Y., Udovicic, F., & Nelson, G., (2003). Australian biogeographical connections and the phylogeny of large genera in the family Myrtaceae. *J. Biogeog.*, *30*, 989–998.

Laity, T., Laffan, S. W., González-Orozco, C. E., Faith, D. P., Rosauer, D. F., Byrne, M., et al., (2015). Phylodiversity to inform conservation policy: An Australian example. *Sci. Tot. Env.*, *534*, 131–143.

Larsen, B. B., Miller, E. C., Rhodes, M. K., & Wiens, J. J., (2017). Inordinate fondness multiplied and redistributed: The number of species on Earth and the new pie of life. *Q. Rev. Biol.*, *92*, 229–265.

Long, J., Archer, M., Flannery, T., & Hand, S., (2002). *Prehistoric Mammals of Australia and New Guinea: 100 Million Years of Evolution*, University of New South Wales Press, Sydney, New South Wales.

Low, T., (2016). *Where Song Began*, Yale University Press, New Haven, Connecticut.

Mackey, B., Berry, S., Hugh, S., Ferrier, S., Harwood, T. D., & Williams, K. J., (2012). Eco-system greenspots: Identifying potential drought, fire, and climate-change microrefuges. *Ecol. Appl.*, *22*, 1852–1864.

Mal, B., Rao, R., Sthapit, B., & Sajise, P., (2011). Conservation and sustainable use of tropical fruit species diversity: Bioversity's efforts in Asia, the Pacific and Oceania. *Ind. J. Plant Genet. Res.*, *24*, 1–22.

Markgraf, V., McGlone, M., & Hope, G., (1995). Neogene paleoenvironmental and paleocli-matic change in southern temperate ecosystems – a southern perspective. *Trends Ecol. Evol.*, *10*, 143–147.

Martin, H. A., (2006). Cenozoic climate change and the development of the arid vegetation in Australia. *J. Arid Env.*, *66*, 533–563.

Martin, H. A., (2017). Australian Tertiary phytogeography: Evidence from palynology. In: Hill, R. S., (ed.). *History of the Australian Vegetation: Cretaceous to Recent*, University of Adelaide Press, Adelaide, South Australia, pp. 104–142.

McKenzie, N. L., Johnston, R. B., & Kendrick, P. G., (1991). *Kimberley Rainforests of Aus-tralia*, Surrey Beatty & Sons, Sydney, New South Wales.

Mickelbart, M. V., Hasegawa, P. M., & Bailey-Serres, J., (2015). Genetic mechanisms of abiotic stress tolerance that translate to crop yield stability. *Nature Rev. Genetics*, *16*, 237–251.

Mills, K., (2007). *The Flora of Norfolk Island*, The Author: Jamberoo, New South Wales.

Milner, M. L., Rossetto, M., Crisp, M. D., & Weston, P. H., (2012). The impact of multiple biogeographic barriers and hybridization on species-level differentiation. *Am. J. Bot.*, *99*, 2045–2057.

Mittermeier, R. A., Gil, P. R., & Mittermeier, C. G., (1997). *Megadiversity: Earth's Biologi-cally Wealthiest Nations*, Cemex, Mexico City.

Moore, P., (2005). *A Guide to Plants of Inland Australia*, Reed New Holland, Sydney, New South Wales.

Morgan, D. L., Unmack, P. J., Beatty, S. J., Ebner, B. C., Allen, M. G., Keleher, J. J., et al., (2014). An overview of the 'freshwater fishes' of Western Australia. *J. Roy. Soc. West. Aust.*, *97*, 263–278.

Morton, S. R., & Emmott, A. J., (2014). Lizards of the Australian deserts: Uncovering an extraordinary ecological story. *Hist. Rec. Aust. Sci.*, *25*, 217–226.

Morton, S. R., Stafford Smith, D. M., Dickman, C. R., Dunkerley, D. L., Friedel, M. H., McAllister, R. R. J., et al., (2011). A fresh framework for the ecology of arid Australia. *J. Arid Env.*, *75*, 313–329.

Myers, N., Mittermeier, R. A., Mittermeier, C. G., Fonseca, G. A. B., & Kent, J., (2000). Biodiversity hotspots for conservation priorities. *Nature*, *403*, 853–858.

O'Connor, M., (2014). What population growth will do to Australia's society and economy. In: Goldie, J., & Betts, K., (eds.). *Sustainable Futures: Linking Population, Resources and the Environment*, CSIRO Publishing, Melbourne, Victoria, pp. 37–45.

Oliver, P. M., & Hugall, A. F., (2017). Phylogenetic evidence for mid-Cenozoic turnover of a diverse continental biota. *Nature Ecol. Evol.*, *1*, 1896–1902.

Olson, D. M., Dinerstein, E., Wikramanayake, E. D., Burgess, N. D., Powell, G. V. N., Under-wood, E. C., et al., (2001). Terrestrial ecoregions of the world: a new map of life on Earth. *BioScience*, *51*, 933–938.

Orians, G. H., & Milewski, A. V., (2007). Ecology of Australia: The effects of nutrient-poor soils and intense fires. *Biol. Rev., 82*, 393–423.

Parks and Wildlife Service, (2006). *Macquarie Island Nature Reserve and World Heritage Area Management Plan 2006,* Parks and Wildlife Service, Tasmanian Government Department of Tourism, Arts and the Environment, Hobart, Tasmania.

Pickard, J., (1983). Vegetation of Lord Howe Island. *Cunninghamia,* 1, 133–266.

Pratt, T. K., & Beehler, B. M., (2015). *Birds of New Guinea, 2nd edn.*, Princeton University Press, Princeton, New Jersey.

Priddel, D., & Wheeler, R., (2014). New South Wales: State of the islands. *Australia's State of the Islands Report,* Australian Government, Canberra, Australian Capital Territory, pp. 24–40.

Ramsay, H. P., & Cairns, A., (2004). Habitat, distribution and the phytogeographical affinities of mosses in the Wet Tropics bioregion, northeast Queensland, Australia. *Cunninghamia, 8*, 371–408.

Ramsay, H. P., (2006). *Flora of Australia,* vol. 51, *Part 1 (Mosses),* CSIRO Publishing and Australian Biological Resources Study, Melbourne, Victoria, and Canberra, Australian Capital Territory.

Raven, P. H., & Yeates, D. K., (2007). Australian biodiversity: Threats for the present, opportunities for the future. *Aust. J. Entomol., 46*, 177–187.

Ridsdill-Smith, J., (2004). Entomology and the Australian Entomological Society. *Aust. J. Entomol., 43*, 211–215.

Ritchie, E. G., Bradshaw, C. J. A., Dickman, C. R., Hobbs, R., Johnson, C. N., Johnston, E. L., et al., (2013). Continental-scale governance and the hastening of loss of Australia's biodiversity. *Cons. Biol., 27*, 1133–1135.

Rix, M. G., & Harvey, M. S., (2012). Phylogeny and historical biogeography of ancient assassin spiders (Araneae: Archaeidae) in the Australian mesic zone: Evidence for Miocene speciation within Tertiary refugia. *Mol. Phylogen. Evol., 62*, 375–396.

Runge, C., & Tulloch, A. I. T., (2017). Solving problems of conservation inadequacy for nomadic birds. *Aust Zool., 39*, 279–294.

Salo, P., Korpimäki, E., Banks, P. B., Nordström, M., & Dickman, C. R., (2007). Alien predators are more dangerous than native predators to prey populations. *Proc. Roy. Soc. B., 274*, 1237–1243.

Saunders, D., Beattie, A., Eliott, S., Fox, M., Hill, B., Pressey, B., Veal, D., Venning, J., Maliel, M., & Zammit, C., (1996). *Biodiversity, Australia: State of the Environment 1996,* Australian Government: Canberra, Australian Capital Territory.

Schofield, W. B., (1992). Bryophyte distribution patterns. In: Bates, J. W., & Farmer, A. M., (eds.). *Bryophytes and Lichens in a Changing Environment,* Clarendon Press, Oxford, pp. 103–130.

SEWPAC., (2012). *Master Islands Dataset – Unpublished,* Department of Sustainability, Environment, Water, Population and Communities: Canberra, Australian Capital Territory.

Silba, J., (1986). An international census of the Coniferae. *Phytologia Memoirs, 8.*

Sniderman, J. M. K., & Jordan, G. J., (2011). Extent and timing of floristic exchange between Australian and Asian rainforests. *J. Biogeog., 38*, 1445–1455.

Spalding, M. D., Fox, H. E., Allen, G. R., Davidson, N., Ferdaña, Z. A., Finlayson, M., et al., (2007). Marine ecoregions of the world: A bioregionalization of coastal and shelf areas. *BioScience*, *57*, 573–583.

Spencer, W. B., & Gillen, F. J., (1912). *Across Australia*, vol. 1 and 2, Macmillan, London.

State of the Environment Advisory Council, (1996). *Australia State of the Environment 1996*, CSIRO Publishing, Melbourne, Victoria.

Stow, A., Maclean, N., & Holwell, G. I., (2014). *Austral Ark: The State of Wildlife in Australia and New Zealand*, Cambridge University Press: Cambridge.

Tabashnik, B. E., Brévauly, T., & Carrière, Y., (2013). Insect resistance to Bt crops: Lessons from the first billion acres. *Nature Biotech.*, *31*, 510–521.

Tester, M., & Langridge, P., (2010). Breeding technologies to increase crop production in a changing world. *Science*, *327*, 818–822.

Ulloa, A. M., Jax, K., & Karlsson-Vinkhuyzen, S. I., (2018). Enhancing implementation of the Convention on Biological Diversity: A novel peer-review mechanism aims to promote accountability and mutual learning. *Biol. Cons.*, *217*, 371–376.

Unmack, P. J., (2001). Biogeography of Australian freshwater fishes. *J. Biogeog.*, *28*, 1053–1089.

Valley, J. W., Cavosie, A. J., Ushikubo, T., Reinhard, D. A., Lawrence, D. F., Larson, D. J., et al., (2014). Hadean age for a post-magma-ocean zircon confirmed by atom-probe tomography. *Nature Geosci.*, *7*, 219–223.

Waldron, A., Miller, D. C., Redding, D., Mooers, A., Kuhn, T. S., Nibbelink, N., et al., (2017). Reductions in global biodiversity loss predicted from conservation spending. *Nature*, *551*, 364–367.

Wang, B., & Qiu, Y.-L., (2006). Phylogenetic distribution and evolution of mycorrhizas in land plants. *Mycorrhiza*, *16*, 299–363.

Watford, F. A., Beard, D. J., Hay, R. J., & Ambrose, K., (2005). Innovative visualization for Australia's marine zone. In: *Spatial Intelligence, Innovation and Praxis, Proceedings of SSC 2005*, The National Biennial Conference of the Spatial Sciences Institute, Melbourne, Victoria.

Weaver, H. J., (2017). State of the Science of Taxonomy in Australia: *Results of the 2016. Survey of Taxonomic Capacity*, Department of the Environment and Energy: Canberra, Australian Capital Territory.

White, M. E., (2006). Environments of the geological past. In: Merrick, J. R., Archer, M., Hickey, G. M., & Lee, M. S. Y., (eds.). *Evolution and Biogeography of Australasian Vertebrates*, Auscipub, Oatlands, New South Wales, pp. 17–50.

Williams, O. B., & Calaby, J. H., (1985). The hot deserts of Australia. In: Evenari, M., Noy-Meir, I., & Goodall, D. W., (eds.). *Hot Deserts and Arid Shrublands, A*, Elsevier: Amsterdam, pp. 269–312.

Woinarski, J. C. Z., Burbidge, A. A., & Harrison, P. L., (2014). *The Action Plan for Australian Mammals 2012*, CSIRO Publishing: Melbourne, Victoria.

Yeates, D. K., Harvey, M. S., & Austin, A. D., (2003). New estimates for terrestrial arthropod species-richness in Australia. *Rec. South Aust. Mus. Monograph Ser.*, *7*, 231–242.

Zhang, Z.-Q., (2011a). Animal biodiversity: An introduction to higher-level classification and taxonomic richness. *Zootaxa*, *3148*, 7–12.

Zhang, Z.-Q., (2011b). Phylum Arthropoda von Siebold, 1848. *Zootaxa*, *3148*, 99–103.

Index